Parasitic Protozoa

Volume III

Gregarines, Haemogregarines, Coccidia, Plasmodia, and Haemoproteids

CONTRIBUTORS

Masamichi Aikawa
Stephen C. Ayala
Richard Carter
William E. Collins
Sherwin S. Desser
Carter L. Diggs
J. P. Dubey
John V. Ernst
A. Murray Fallis
Reginald D. Manwell
W. M. Reid
Karl H. Rieckmann
M. D. Ruff
Thomas M. Seed
Paul H. Silverman
Kenneth S. Todd, Jr.

Parasitic Protozoa

Volume III

Gregarines, Haemogregarines, Coccidia, Plasmodia, and Haemoproteids

Edited by

Julius P. Kreier

Department of Microbiology
College of Biological Sciences
The Ohio State University
Columbus, Ohio

ACADEMIC PRESS New York San Francisco London 1977
A Subsidiary of Harcourt Brace Jovanovich, Publishers

ACADEMIC PRESS, INC.
111 Fifth Avenue, New York, New York 10003

United Kingdom Edition published by
ACADEMIC PRESS, INC. (LONDON) LTD.
24/28 Oval Road, London NW1

Library of Congress Cataloging in Publication Data

Main entry under title:

Parasitic protozoa.

Includes bibliographies and indexes.
CONTENTS:
–3. Gregarines, Haemogregarines, codeidia, plasmedia, and haemoproteids.
1. Protozoa. Pathogenic. 2. Parasites.
I. Kreier, Julius P. [DNLM: 1. Protozoa.
2. Protozoan infections. QX50 P223]
QR251.P24 593′.1′04524 76-13041
ISBN 0–12–426003–9

PRINTED IN THE UNITED STATES OF AMERICA

Contents

List of Contributors

Numbers in parentheses indicate the pages on which the authors' contributions begin.

Masamichi Aikawa (467), Institute of Pathology, Case Western Reserve University, Cleveland, Ohio

Stephen C. Ayala (267), Universidad del Valle Medical School, Tulane University, International Center for Medical Research, Cali, Colombia

Richard Carter (359), Laboratory of Parasitic Diseases, National Institute of Allergy and Infectious Diseases, National Institutes of Health, Bethesda, Maryland

William E. Collins (467), Vector Biology and Control Division, Bureau of Tropical Diseases, Center for Disease Control, Public Health Service, Department of Health, Education, and Welfare, Atlanta, Georgia

Sherwin S. Desser (239), Department of Microbiology and Parasitology, University of Toronto, Toronto, Ontario, Canada

Carter L. Diggs (359), Department of Immunology, Walter Reed Army Institute of Research, Washington, D.C.

J. P. Dubey (101), Department of Veterinary Pathobiology, College of Veterinary Medicine, The Ohio State University, Columbus, Ohio

John V. Ernst (71), Regional Parasite Research Laboratory, Agricultural Research Service, U.S. Department of Agriculture, Auburn, Alabama

A. Murray Fallis (239), Department of Microbiology and Parasitology, Faculty of Medicine, University of Toronto, Toronto, Canada

Reginald D. Manwell (1, 311), Department of Zoology, Syracuse University, Syracuse, New York

W. M. Reid (33), Department of Poultry Science, University of Georgia Athens, Georgia

Karl H. Rieckmann (493), Department of Biology, University of New Mexico, Albuquerque, New Mexico

M. D. Ruff (33),* Department of Poultry Science, University of Georgia, Athens, Georgia

Thomas M. Seed (311), Division of Biological and Medical Research, Argonne National Laboratory, Argonne, Illinois

Paul H. Silverman (493), Research and Graduate Affairs, University of New Mexico, Albuquerque, New Mexico

Kenneth S. Todd, Jr. (71), College of Veterinary Medicine, University of Illinois, Urbana, Illinois

* Present address: Animal Parasitology Institute, Agricultural Research Center, Beltsville, Maryland.

Preface

The parasitic protozoa are a large and diverse group. Many are of interest to physicians and veterinarians because they produce disease in man and his livestock. Others, which seldom produce disease, should be familiar to the practitioner of medicine and to the research scientist because they are present in the animal body and thus must be recognized to avoid a misdiagnosis, while still others, such as the intestinal and rumen protozoa, perform a useful function in the animal's economy, and their presence is an indication of health rather than disease.

I have included in these volumes protozoa parasitic in animals, such as fish and insects, which are not usually included in books on pathogenic protozoa. I did this because I believe veterinary medicine should concern itself with all species of animals, excepting man, whose care falls to the physician. From a more practical standpoint, I feel the inclusion of parasites of diverse species is appropriate in a book on protozoa of veterinary and medical interest because no matter how we set ourselves off from nature we remain a part of it, and thus we inevitably share parasites with the other species with which we live.

Because of the wide range of parasites and the volume of material available, no single author could hope to be qualified to write on all of them, thus I have chosen to have each chapter written by someone qualified in that area. This course of action, while it avoids the problems of the limitations of a single author, has problems of its own, the most serious being the variability in the authors' styles and attitudes which produces uneveness in the treatment of the contributions. For this I accept responsibility as editor. For all that is good and useful in these volumes I thank the authors of the chapters and the staff of Academic Press who have aided in the production of these volumes. I also wish to thank the Army Malaria Project, whose support of my research has made it possible for me to continue my interest in protozoology.

Julius P. Kreier

Contents of Other Volumes

VOLUME I

1

Gregarines and Haemogregarines

Reginald D. Manwell

I. Introduction

Many of the protozoans encompassed in the scope of this chapter have little or no practical importance; few are pathogens. Yet, they are all of interest to the parasitologist and to those biologists whose view of their specialty is not too parochial. It is also likely that some of the parasites considered may have significance in the maintenance of those biological balances which are of such importance in nature. This is especially true of the parasites of insects, which, collectively, are probably man's chief competitors.

II. Gregarines

A. Characteristics

Gregarines (order Gregarinida) make up a large group. At least several hundred species are known, divided among perhaps 150 genera, and it is quite certain that others exist. Many have been described within the last decade. The figures above may be compared with those given by Labbé in 1899. He listed "55 certain genera and 13 uncertain (containing) 105 certain species and 90 uncertain," Perhaps, however, the most remarkable thing is that our knowledge of these interesting animals had even then progressed so far.

According to Watson (1916), gregarines were probably first seen by Redi in 1708, though the earliest recorded description is that of Dufour in 1828. Kölliker (1848) published an "elaborate memoir" on the group, and was perhaps the first to see the organisms in syzygy (though he thought it longitudinal fission) (Fig. 2B). It is this characteristic which caused Dufour to coin the term "gregarines" for them 20 years earlier. Syzygy may, in fact, be a sexual response.

Hosts are invariably invertebrates and the parasites are best known from arthropods, which often show extremely heavy infections. The only invertebrate phyla not known to harbor gregarines are the Rotifera (Trochelminthes), Porifera, and Protozoa (Watson, 1916). It is hard to explain such a peculiar host distribution. The complexity of both gregarine morphology and life history makes it all but certain that they are a very ancient group, having had ample time to adapt to life in the higher vertebrates. Somehow an opportunity was missed!

Grassé (1953) has summarized their physical characteristics as follows: Sporozoa having "a single axis of symmetry, apicobasal polarity, a tendency to cephalization, particularly marked in the more highly evolved species. . . ." He might have added that the majority do not undergo asexual reproduction in their hosts (with the result that few are pathogens), and that they exhibit a very peculiar and still little understood form of motility often called "gregariniform locomotion." It is worth noting that the biological relationships of gregarines are very obscure. They do not resemble any other protozoa, free-living or parasitic.

The habit of syzygy already referred to involves more than the association in pairs which Dufour observed; they often form groups of threes, fours, and more.

Many gregarines are relatively large. There are species which attain a length of 10 mm and can be easily seen with the naked eye. Others, however, are among the smallest protozoa, scarcely larger than a mammalian erythrocyte, for example, *Lipotropha*, a parasite of blowfly larvae.

Although gregarines when free (known then as sporadins) are often very active creatures, they entirely lack locomotor organelles. However the cytoplasm is usually well provided with myonemes. It has recently become possible to examine them with the electron microscope and, at least in some cases, they appear to consist of bundles of microtubules.

Mitochondria, a Golgi apparatus, ribosomes, and other minute structures essential to the metabolic functioning of the cell are also present. It is of considerable interest that the nucleus (always in the posterior portion) was seen by von Siebold in 1839, at a time when microscopes were very primitive instruments and cytological technique even more so.

The body in the majority of gregarines is subdivided into several quite distinct regions by transverse constrictions, the most anterior of which is the epimerite, a specialized part of the protomerite which follows, and the deutomerite (Fig. 1). The former is often very complex in structure,

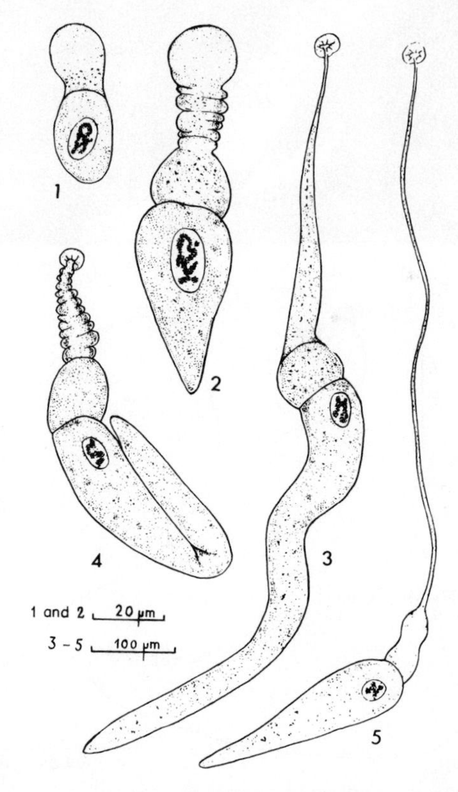

Fig. 1. *Acutispora pulchra* Ormières 1966, a cephaline gregarine (Eugregarina) with a remarkably long neck. These figures show the characteristic structure of cephaline gregarines: an anterior portion (protomerite) bearing the epimerite, or attaching organelle, and a posterior portion called the deutomerite. In the latter is the nucleus, clearly shown. This gregarine parasitizes the myriapod, *Lithobius.* (1) Young individual; (2) more advanced stage; (3) normal adult (sporadin); (4) with epimerite contracted; (5) the same, with the epimerite fully extended. (After Ormières, 1966.) Used by permission of *Protistologica.*

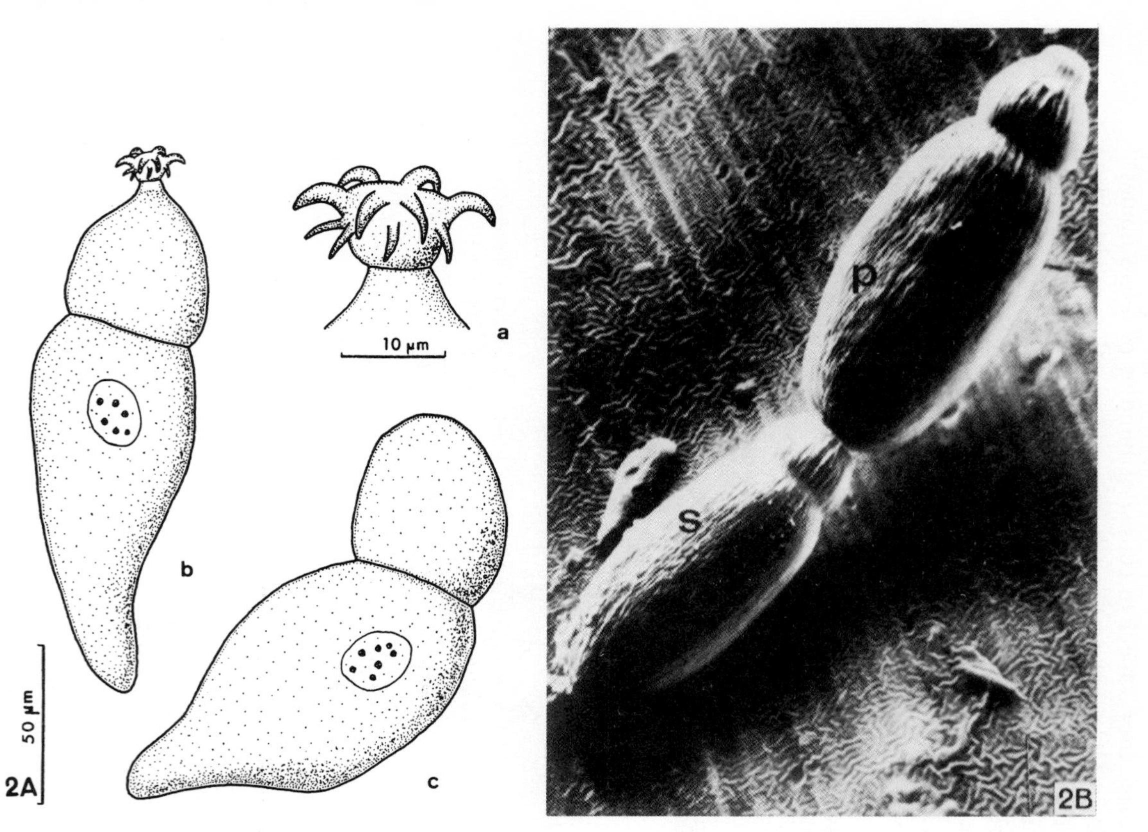

Fig. 2A. *Ancyrophora uncinata* Léger, 1892, a coleopteran cephaline gregarine (Eugregarinina). (a) Epimerite bearing twelve rather sharp hooklets, with which the parasite attaches itself to the gut wall of its host; (b and c) trophozoites, showing the characteristic division of the body into three parts. [From Baudoin, *J. Protozool.* (1971); used by permission.]

Fig. 2B. Syzygy of young gamonts of *Gregarina* sp. from the cricket *Udeopsylla nigra.* Note especially the fine (epicytic) folds of the cuticular surface. P, primite; S, satellite. ×1560. From Heller and Weiser, 1973; reproduced with permission of the *Jour. Protozool.*

serving the functions of attachment to the host cell and the absorption of nutriment (Fig. 2).

However, in the more primitive schizogregarines (often called archigregarines) the body seems quite simple. There are no subdivisions, although there may be a suckerlike structure at the anterior end, as in *Selenidium fallax* (Fig. 3). Another genus, *Ditrypanocystis,* which has an undulating membrane of unknown function (since it seems not to aid locomotion) is shown in Fig. 4.

B. Taxonomy

Taxonomists differ greatly in their classification of the group. The arrangement of higher taxa usually followed is that of Léger (1907), used with slight changes below. Grassé (1953) has suggested an alternate, based on the assumed evolutionary significance of the presence or absence of schizogony in the life cycles, and Weiser (1955) has also proposed a new scheme. However, Léger's scheme enjoys wider acceptance, and has the advantage of greater simplicity. As given below, it is largely quoted from Manwell (1968).

1. Léger's Scheme

Order: Gregarinida: Lankester, 1885. Sporozoa which (except in the schizogregarines) do not undergo schizogony. Adult trophozoites are motile, but lack locomotor organelles. Often associate in syzygy (Fig 2B). The presexual stages join and form a gametocyst, in which a number of gametes are ultimately produced.

"*Suborder 1:* Eugregarinina Doflein, 1910. Gregarines in which reproduction is limited to the sexual phase, without schizogony (merogony). The spores always contain 8 merozoites. Mostly parasitic in annelids and arthropods." Many species.

"*Tribe 1:* Acephalina Kölliker, 1848. Body not differentiated into clearly defined regions. Mainly occurring in the coelom. Early stages intracellular; later ones may be extracellular or not. About 10 families."

Tribe 2: "Cephalina Delage, 1896. Body differentiated into 2 main regions: the protomerite and deutomerite (anterior and posterior, respectively), with a septum between. Deutomerite may be further subdivided in some species. About 12 families."

"*Suborder 2:* Schizogregarinina Léger, 1907. Gregarines with reproduction in both asexual and sexual phases of the life cycle. Often intracellular in the former phase. A relatively small group. Two families, more or less."

Weiser's scheme would divide the two families of schizogregarines into six subfamilies, based on comparative morphology.

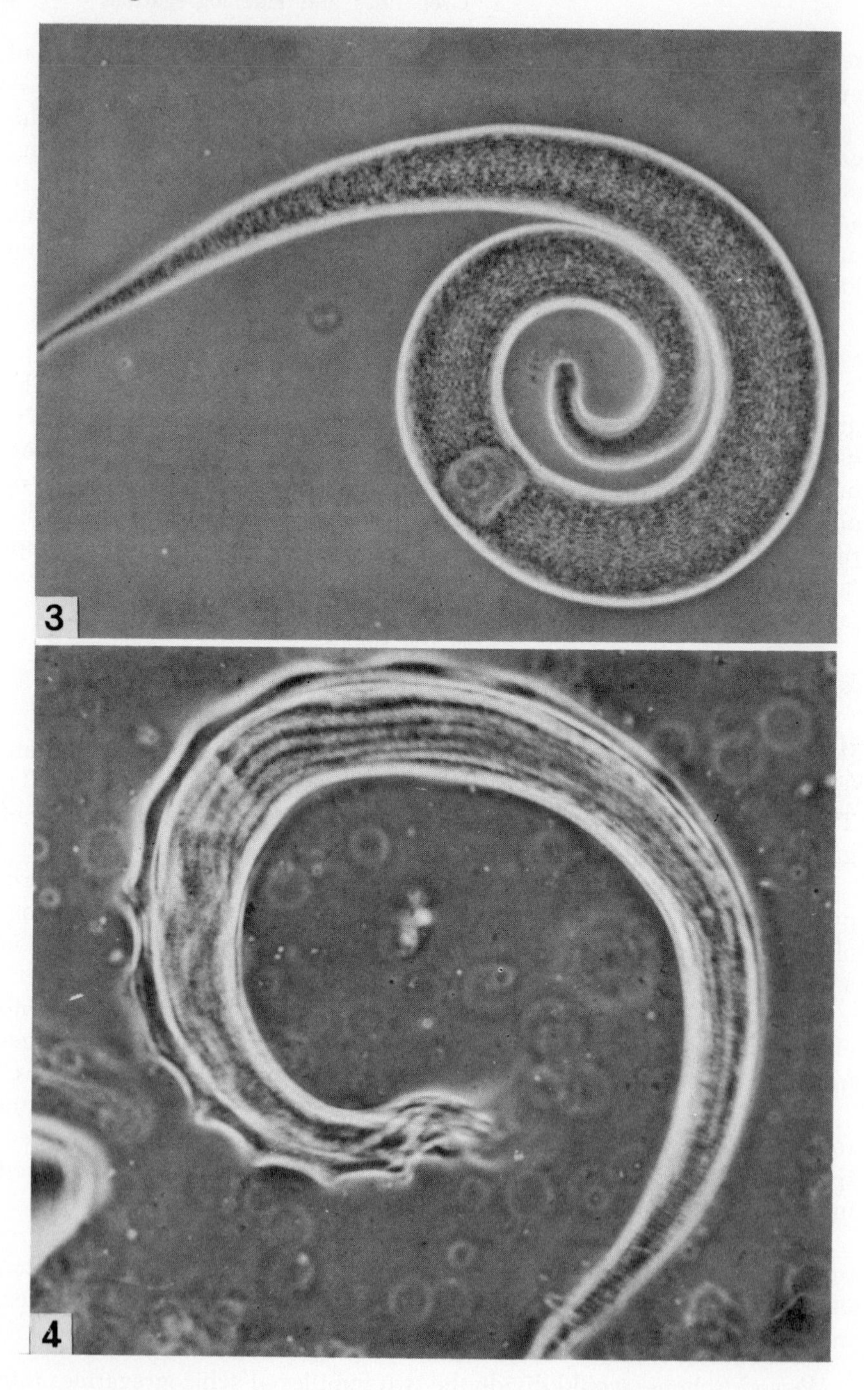
3
4

2. Grassé's Scheme

Grassé prefers to group all gregarines into a single class, and believes they should be split into three orders, the Archigregarina, Eugregarina, and the Neogregarina. There is certainly some justification for the former since gregarines differ greatly from other Sporozoa in almost every respect, but the latter change is more debatable.

He believes that in many species of schizogregarines schizogony is a primitive trait and he therefore proposes that they be called archigregarines. As evidence, he points to the fact that their hosts are still marine in habitat. Nonmarine hosts exhibit schizogony, which he thinks was secondarily acquired, and these he prefers to call neogregarines.

An example of the first order is *Selenidium* (a parasite of an annelid worm), of the second order is *Monocystis* (commonly occurring in earthworms), and of the third order is *Ophryocystis elektroschirrha* (of the monarch butterfly). All three are considered in greater detail below.

C. Eugregarines

Morphology varies so much that it is difficult to select a form which can be considered typical.

Acephalines have no very clearly defined body regions, except that at the anterior end there is usually an organelle of attachment. *Monocystis*, already mentioned, is an excellent example. There are many species of this parasite, and they are so common as to be easily obtainable for study.

Cephalines are also very common. *Lankesteria culicis*, a parasite of *Aedes aegypti* larvae, is an excellent and much studied example. It is almost as common in the warmer parts of the world as its host; Walsh and Callaway (1969) found it in 95% of the larvae collected in the area of Houston, Texas. Unfortunately, there is little reason to think it is a pathogen.

Cephaline gregarines have, in the adult stage, a rather complex structure, consisting of a "rostrum" at the anterior end, a "protomerite" just behind it, and finally a "deutomerite," in which is located the nucleus (Figs. 1 and 2A. There is usually a transverse septum behind the epimerite, which is an organelle serving to anchor the parasite to its host cell and un-

---

Fig. 3. *Selenidium fallax* Macgregor and Thomasson 1965. An archigregarine the body of which is not divided into regions and which lacks any organ of attachment. Nucleus is clearly visible.

Fig. 4. *Ditrypanocystis cirratuli* Burt, Denny, and Thomasson, 1963. An archigregarine remarkable for the possession of an undulating membrane. Since the membrane does not function in locomotion its real function remains unknown. Both of these species are parasites of a polychaete worm. (Figures 3 and 4 used by permission of the *J. Protozool.* From Macgregor and Thomasson, 1965.)

doubtedly also is important in the absorption of nutritive substances. Its form differs strikingly in different species.

Electron microscopy has shown that the surface of some gregarines displays minute longitudinal folds instead of having the smooth texture seen under the ordinary light microscope. There may also be structures such as the "permanent cytostomelike openings" detected by Walsh and Callaway (1969) on the surface of *Lankesteria culicis.*

1. Internal Structure

Gregarines show considerable internal differentiation. The outer layer of ectoplasm is usually divided into three layers: the "epicyte" (external), "sarcocyte" (intermediate), and "myocyte" (inner). There are also contractile fibrils or myonemes in the myocyte and, in certain species, fibrils believed to have a skeletal function.

Below the ectoplasm is the endoplasm, which has a complex structure of its own. In addition to the nucleus, there may be a nucleolus in older stages. In some species this is said to pulsate. There are also likely to be numerous granules of a special form of glycogen, known as "paraglycogen," a reserve foodstuff characteristic of gregarines. It may give the adult parasites a grayish, opaque appearance; iodine stains it an ocher brown.

In recent years, electron microscope studies have been made of an increasing number of gregarine species. These have revealed complex ultrastructure and, often, unexpectedly complex external morphology. A figure (2B) from a recent study by Heller and Weise (1973) of a gregarine of undetermined species from a cricket (*Udeopsylla nigra*) is reproduced here. Two parasites are shown in syzygy, a habit typical of gregarines. The scanning electron microscope is especially effective for studies of this type. The pellicular folds which it reveals are possibly related to motility, as indicated below, and also to physiological functions, such as respiration and nutrition. (Vivier, 1968).

2. Movement

Most gregarines exhibit a uniform gliding movement when free of their host cells, often leaving a trail of mucus behind. So far there have been many theories but no really satisfactory explanation of how this is achieved. Mackinnon and Hawes (1961) comment in this connection "It must be admitted that it is easier to find fault with any of these theories than to improve on them." The ejection of mucus may itself be a factor. Vavra and Small (1969) were able to demonstrate in the epicyte a series of minute longitudinal folds, as a result of studies with the scanning electron microscope, and suggest that these may undulate thus causing motion. Another type of movement, characteristic of some species, as well as of many other protozoa, consists of changes in body shape. There is enough similarity between this type of movement when exhibited by *Euglena* and gregarines so that the former is said to have been mistaken, on occasion, for the latter. Undoubtedly, this is due to the contraction of minute fibrils in the cytoplasm.

The belief that longitudinal epicytic folds play an important role in movement is supported by a recent study of Hildebrand and Vinckei (1975), although they suggest that ectoplasmic annular myonemes and cytoplasmic streaming are also involved. Their work was done on *Didymophes gigantea,* an especially favorable subject for research because of its large size (with a length sometimes exceeding 1 cm it is said to be the second largest gregarine known).

Didymophes gigantea is a parasite of the larvae of the scarabeid beetle *Oryctes nasicornis.* In part because the adult beetles are never infected, the authors think that the parasites' life cycle is in part determined by physiological events in the life cycle of the host.

Among the more recent studies is that of Stebbings *et al.* (1976). These authors subjected the pellicle of *Selenidium fallax* to examination by electron microscopy and concluded that, at least in this species, the pellicular fibrils are microtubular aggregates having a motile function (Fig. 5). As the animals move (and they are almost continuously active),

Fig. 5. A diagrammatic representation of *Selenidium fallax,* showing a cross section indicating the pellicular folds and arrangement of the subpellicular microtubules (see top). By courtesy of Dr. H. Stebbings and *Z. Zellforsch.* (Stebbings *et al.,* 1976).

the microtubules seem to change in length and are thought to slide on one another. However, the problem of whether all such fibrils are contractile remains unanswered; it is suggested that some may have a skeletal function.

3. Life Cycles

The life cycles of gregarines vary almost as much as their size and are often quite complex. *Filopodium ozakii,* a parasite of the sipunculid (annelid) worm, *Siphonosoma cumaense,* is a good example of a cephaline eugregarine. It was studied by Hukui (1939). After fertilization and the formation of a zygote (Fig. 6), sporoblasts develop within the oocyst, which is the infective stage. Sporozoites are formed from these and emerge after ingestion of the oocyst by the prospective host. They then attach themselves to epithelial cells of some portion of the midgut. After a period of growth, during which they become trophozoites, they leave the host cell, of which little remains, and as sporonts wander about the coelom looking for partners. If successful, they join in pairs and a period of division ensues; the resulting offspring, now gametes, unite to form zygotes. Each trophozoite is therefore a gametocyte.

The acephaline gregarine, *Monocystis,* of which there are a number of species, may be considered next. Unlike many of its kin, the trophozoite is intracellular and resides in the sperm mother cells of its host. Once inside its host cell, the parasite begins a period of growth which ends only when the cell is destroyed. By this time, it has become what is known as a sporadin and is ready to return to the gut where it seeks a partner; the two then form a gametocyte. What follows is similar to the cycle previously described. Repeated nuclear divisions are followed by cytoplasmic division, resulting in a crop of gametes. It is claimed that all the offspring of a given parent are of the same sex.

Union of male and female gametes results in the formation of a zygote and then of an oocyst. Eight sporozoites are eventually formed which, when released in the gut of the next host, are ready to repeat the cycle. As with coccidia and other gregarines, only the zygote is diploid, although there is disagreement about when meiosis occurs; it may vary in different species. The mature sporocysts remain viable in the soil for an uncertain length of time since they do not tolerate dessication well. They are only liberated when the host dies.

The genus *Porospora* presents an interesting variation from the one-host pattern of most gregarines. It has two hosts, a lamellibranch mollusc and a lobster. It is also remarkable for its very large size with sporadins reaching a length of almost 10 mm.

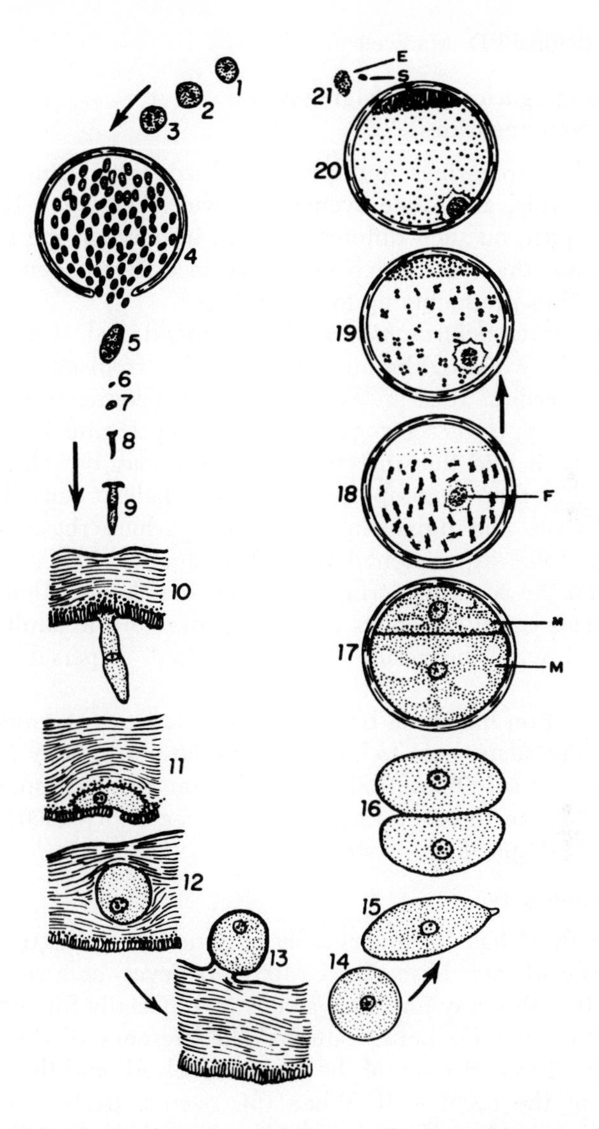

Fig. 6. Life cycle of the eugregarine *Filipodium ozakii,* from the sipunculid worm *Siphonosoma cumaense.* (1–3) Zygote; (4) formation of sporoblasts; (5) sporozoite; (6–9) formation of the trophozoite; (10) trophozoite attached to the intestinal wall; (11–12) penetration of wall by trophozoite; (13) trophozoite leaving the gut wall to enter coelomic cavity and become the sporont; (14) sporont; (15) change in shape of sporont to spindle form; (16) pseudoconjugation, resulting from the union of two sporonts upon mutual contact; (17) formation of cyst wall by pseudoconjugants; (18–20) initiation of a series of mitotic divisions, culminating in meiosis, and micro- and macrogametes; (21) fertilization. E, macrogamete or egg; S, microgamete of sperm; F, residual fat body; m, portion of encysted pair destined to produce microgametes; M, portion destined to produce macrogametes. (After Hukui, used by permission of Dover Publications and taken from Manwell, 1968.)

D. Schizogregarines (Archigregarines and Neogregarines)

The schizogregarines multiply by schizogony during some phase of their life cycle, and some even do it twice. Weiser's (1955) scheme is based, in part, on such differences, and, in part, on the number of divisions within the cyst, which may vary in different genera from one to fourteen, thus producing up to 8200 spores.

Ophryocystis elektroschirrha, already mentioned, is a species recently described by McLaughlin and Myers (1970) from monarch and Florida queen butterflies. It is an example of a schizogregarine (neogregarine) having two types of schizogony in its life cycle and is also a pathogen. Although it is said that infections in nature may be relatively light and, therefore, perhaps unimportant, it nevertheless caused many deaths among larvae and pupae in an insectary where these butterflies were being bred for experimental use. The infective spores in this case are carried on the scales covering the body. It is known that in some other host species larval infections are carried over to the adult through metamorphosis, with the result that widespread dispersal of the parasite occurs.

The question has often been asked to what extent gregarines are host specific. The facts seem to be that while host specificity is often strict, it is not always so. Susceptibility to infection by some species extends to hosts belonging to several genera. However, Ball (1960) states that, in general, host specificity is strict.

E. Pathogenicity

Although it has been generally assumed that eugregarines are not pathogenic to their hosts, Lipa (1957) believes, as a result of extensive studies, that this may not always be true, especially for species inhabiting the gastric ceca. He points out that the presence of the parasites often causes great hypertrophy of the tissues involved, and that this may cause rupture of the cecal wall. When this occurs, bacteria enter, often inducing septicemia and consequent death of the host. When all the ceca are parasitized, which frequently occurs, the risk is greatly increased. When only the intestine is invaded, the hazard is less because the damaged intestinal epithelium is quickly regenerated, but even here injury may be severe.

The mechanism of injury, Lipa suggests, is the loss of excessive amounts of fluid due to its absorption by the parasite in the course of pinocytosis, as evidenced by the occurrence of numerous tiny cavities in the surrounding tissue.

The probability of injury is often increased by the simultaneous presence of two or more species of gregarines in the same host. It appears that there is seldom any cross-immunity to such invasion.

That schizogregarines are likely to be lethal to their hosts has long been recognized, and Lipa believes that the mechanism of injury is the same in both groups.

There is also another complicating factor. Autoinfection is probably frequent in infections by some species of schizogregarines, as pointed out by Zizka (1972), who made an intensive study of *Farinocystis tribolii*, one of several species parasitizing the flour beetle. For gregarines residing in the Malpighian tubules or gut, such autoinfection would not occur, since oocysts are voided with the host's feces and must have a period of 2 weeks or so thereafter to complete maturation; in others, maturation is completed in the host.

It also appears that under certain conditions gregarine species not ordinarily pathogenic may become so. Such is the case with meal worm larvae infected with *Gregarina polymorpha*. They are usually not significantly affected by the parasite, but are harmed when fed a suboptimal diet (Harry, 1967).

It would be pleasant to be able to include gregarines among pathogens of pest insects. Perhaps the best that can be done is to mention *Lankesteria clarki*, a parasite of the tree-hole species, *Aedes sierrensis*. This gregarine destroys epithelial cells of the midgut and Malpighian tubules of its host (Sanders and Poinar, 1973). Unfortunately, the harm done may not be great since such cells are thought to regenerate rapidly.

F. Hyperparasitism

Few opportunities to secure food and shelter are ignored in the world of living things, and this is, of course, the key to their survival. It should, therefore, come as no surprise that gregarines themselves have their protozoan parasites, and doubtless others. Indeed, there is an entire family of Microsporidia, the Metchnikovellidae, all of which are gregarine parasites. [For many years this family was placed in the Haplosporidia, but Vivier (1965) states it should be classed with the Microsporidia.] Hosts of the latter are usually annelids, which may be further evidence of the antiquity of both gregarines and their hyperparasites. However, some are known from other hosts. Codreanu (1967) cites a microsporidian which parasitizes the gregarine *Enterocystis rhithrogenae*, the host of which is a mayfly, and Corbel (1967) mentions another microsporidian which resides in a gregarine (*Gregarina cousinae*), parasitic in the cricket (*Gryllus assimilis*). Parasitized gregarines from marine crabs and ascidians have also been reported; many others are probably still undiscovered.

G. Problems of Dispersal

Although it is clear that host infection results from the ingestion of cysts containing sporozoites, which may be produced in great numbers [two examples are mentioned by Corbel (1968) of species producing, in one case, about 512,000 sporozoites, and in another, 800,000], it seems likely that things are not always so simple. Ball (1960), for example, remarks "It is difficult to understand how (in marine forms) infections could be maintained by the relatively few cysts discharged into the sea." Yet perhaps one-third of the approximately 500 known species of gregarines parasitize ocean-dwelling hosts. Specifically, though most earthworms harbor *Monocystis,* it seems hard to account for this in terms of the necessary abundance of infective cysts in the soil. For this parasite, sexual union seems a likely means of transmission.

Watson (1916) explains the dispersal of gregarines in this way: ". . . the spores are liberated from the cyst through the spore ducts which are formed from the residual protoplasm of the cyst. They are scattered over the grass and ground by the wind and rain and are eaten by some host along with its food. Parasitism is thus accidental." Lipa (1967) remarks that gregarine infection is indeed more common among ground-dwelling hosts, such as centipedes and millipedes, and less frequently observed among flying insects. Coprophagous insects show a high incidence of infection due, perhaps, largely to the frequent density of their populations.

In species in which the food habits of larvae and adults are similar, gregarine infections naturally tend to be heavier, but when dietary requirements differ greatly from those of the adult, larval infections are often lost at metamorphosis. Even then (as in the monarch and Florida queen butterflies cited earlier) cysts may be carried over in or on the adult, and thus parasite dispersal is facilitated.

The spores and oocysts of gregarines, like those of coccidia, are often highly resistant and may survive months of exposure to ordinary climatic conditions, including those of temperate zone winters. This is an important factor in the persistence of a high level of infection in their hosts, few of which survive the cold season.

H. Evolutionary Origin

One can only speculate about the origin of any of the Sporozoa, since fossil remains are totally lacking. However, it seems likely that they originated from the flagellates. There is some evidence for this based on the fact that in both phytoflagellates (generally thought to be the most primitive protozoa) and Sporozoa, only the zygote is diploid. In addition, gametes are often flagellated, though this is true among the gregarines only for certain species in which the microgametes have a single flagellum; in others flagella are lacking. There is also some similarity in the mode of movement of gregarines and some phytoflagellates; this has already been discussed.

Among the gregarines, the most primitive forms are believed to be genera such as *Selenidium,* in which the sporozoites enter epithelial cells of the gut of their hosts (mostly polychaete annelids), become trophozoites ("cephalins") and undergo one generation of schizogony, from which arise several hundred spindle-shaped merozoites. These attach themselves to other epithelial cells, and then free themselves to become gamonts after a brief period of growth. They then associate themselves in pairs and form gametocysts in typical gregarine fashion.

That gregarines are of extreme antiquity is evidenced by their unique body form, and also by the complexity of life cycles of many species. They are also well adapted to life in their respective hosts, except for the schizogregarines, since, in general, they seem to cause relatively little injury. [The greater pathogenicity of the schizogregarines makes it difficult to believe that any of them have reacquired asexual schizogony, after descent from eugregarine ancestors which had lost it, as suggested by Grassé (1953). Asexual reproduction explains their pathogenicity, and pathogenicity is a disadvantage to any parasite.]

The great variety of body form among gregarines is further evidence of a long period of evolution. This is also reflected in the relatively large number of species known from marine hosts, since there is reason to believe that life originated in the ocean.

Insects (their most favored hosts) are of great antiquity, cockroaches and Paleodictyoptera (close relatives of dragonflies and damselflies) having appeared before the Carboniferous period, at least a quarter-billion years ago.

III. *Selenococcidium*

Before considering the haemogregarines, one additional genus remains to be discussed—*Selenococcidium.* It has a single species *S. intermedium.* It has no special importance, except that it is a parasite of the European lobster, which would doubtless be better off without its unwanted guest.

The species name is appropriate, since this coccidian appears to lie between the gregarines and the coccidia. The parasite lives as a growing form in the intestinal lumen, where it closely resembles a tiny nematode. This stage has several myonemes, like a gregarine, but the formation and behavior of the sexual stages are of the coccidian type.

Taxonomically, it is placed by Grassé in the subclass Prococcidia, family Selenococcidiidae, of which it is the sole member although Kudo places *Ovivora*, a parasite of an echiurid worm, in the same category. Others, however, regard this family simply as one in the Eimeriid group of coccidia.

Léger and Duboscq determined its life cycle in 1910, although certain gaps still remain. In view of the biological interest of the parasite, it is surprising that apparently it has not been studied since that time.

Mitosis is said to be of a very simple type—perhaps an indication of its relative primitiveness. Schizonts are small wormlike forms living free in the intestinal lumen. From them are derived eight merozoites which invade intestinal cells and, after a period of growth, divide as before, the resulting merozoites again entering the intestinal lumen. This cycle of events is repeated. Finally two types of schizonts are produced: from the smaller one a brood of eight microgametocytes arises and from the larger, four vermicular macrogametocytes. These, in turn, invade intestinal cells, fertilization occurs, and the zygotes form oocysts. Sporozoites must then develop, but how this occurs is still unknown.

IV. Haemogregarines

Despite the similarity of names, any relationship there may be between gregarines and haemogregarines is distant. It is probable that the latter were called haemogregarines by Danielewski (to whom we owe the name) in 1885 because of the gregarinelike movement exhibited by the parasites when observed free in blood films. The gamonts are also often seen in pairs (syzygy) in the invertebrate host, thus simulating true gregarines which are often present with them. Thus it may have been a matter of guilt by association. They are recognized as coccidia today.

A. Characteristics

The physical characteristics of these parasites so widely differ that it is very difficult to generalize. The chief variables are the preferred hosts and the life cycles, which involve schizogony in a vertebrate host and gamogony in an invertebrate. Infection in the former is usually the result of ingestion—often accidental—of the latter.

The best known and, in many cases, the only known stages are those in erythrocytes (Fig. 7) and leukocytes. They are usually elongate forms which stain with Romanowski stains much as do the malaria parasites (though often less intensely) but lack pigment. Although they are usually gametocytes there is little or no sexual dimorphism; macro- and microgametocytes look and stain alike. Occasionally, reproducing forms are also seen in the blood cells. If the invaded cells are erythrocytes

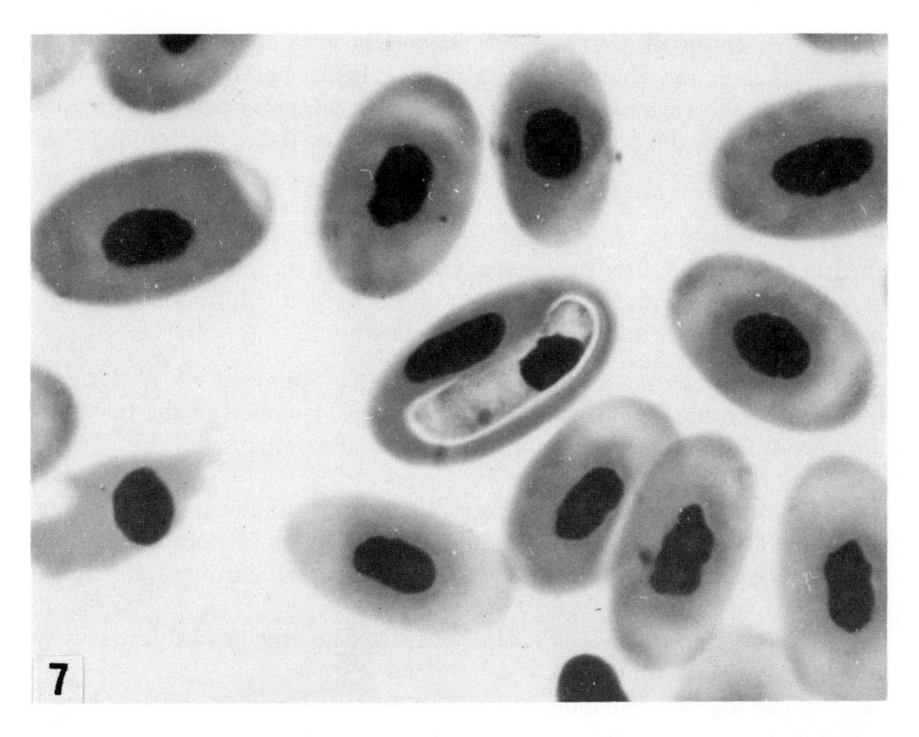

Fig. 7. A haemogregarine of unknown species (center) in a Giemsa-stained blood film from a Formosan "Beauty Snake" (*Elaphe taeniura*); prepared by the Navy Department Work Unit MF022.03.072003.

and nucleated, as they are in hosts other than mammals, the parasites lie beside the nucleus which they may displace, to some extent, laterally. If the invaded cell is a leukocyte, its nucleus may be indented as by a young *Leucocytozoon,* for which some species of these parasites have been mistaken.

Tissue stages occur in a variety of tissues and organs; in this respect, different species have their own preferences. The number of generations and duration of the schizogonic cycle are often limited, as in other coccidia.

Hosts are very varied, including many kinds of reptiles and a considerable number of amphibian, avian, and mammalian species, although man is probably not one of them. Vectors are almost equally varied. Leeches often serve as vectors for aquatic hosts; for terrestrial hosts there are ticks, mites, and even insects such as mosquitoes and tsetse flies.

In many cases transmission is mechanical but in others it is biological, and stages in the development of the parasite in the vector correspond roughly to those of the malaria plasmodia and their close relatives, *Leucocytozoon* and *Haemoproteus,* in their insect hosts. However, actual infection of the vertebrate host is usually by the more primitive method of ingestion of the vector, rather than by its bites, as pointed out previously.

B. Taxonomy

Although taxonomists vary considerably about the proper classification of the haemogregarines, they do not differ very much about the larger taxa. The system most often accepted follows:

Order: Coccidiomorphida Doflein 1901. Sporozoa of average or smaller size, typically intracellular in all stages. Gametes of the two sexes are very different ("anisogamous"). Male filamentous in form and often with two differing flagella.

Suborder: Coccidina Leuckart 1879. Intracellular parasites, most often of the digestive tract epithelium; reproduction in both asexual and sexual phases. Oocyst typically with sporocysts containing sporozoites. Wide host spectrum.

Superfamily: Adeleoidea Léger 1911. Gametocytes exhibit syzygy; few microgametes produced.

Family 1: Haemogregarinidae Neveu-Lemaire 1901. Oocysts without sporocysts. Sexual cycle in leeches. Asexual cycle in cold-blooded vertebrates, most often turtles. One genus: *Haemogregarina.*

Family 2: Hepatozooidae Wenyon 1926. Oocysts usually large; each with numerous sporocysts containing many sporozoites. Gametocytes in blood cells. Hosts mainly mammals and birds; vectors include ticks, mites, blood-sucking insects. One genus: *Hepatozoon.*

Family 3: Karyolysidae Wenyon 1926. Sporozoites penetrate egg of invertebrate host (mite), where they develop into sporocysts. Vertebrate host (usually lizard) harbors gametocytes in erythrocytes, with injury to their nuclei. One genus: *Karyolysus.*

Grassé (1953) prefers to group all three families into the single family Haemogregarinidae, calling these parasites "typical Adeleidea, which it would be better to term Haemococcidia." Reichenow was of the same opinion.

1. *Haemogregarina*

Haemogregarines of the genus *Haemogregarina* are parasites of turtles and other reptiles, and of some amphibia and fish. Although numerous species have been described it is almost certain that the list includes many which are invalid.

The life cycles of most fish haemogregarines remain incompletely or wholly unknown. Leeches are usually assumed to be the vectors of haemogregarine infections of aquatic vertebrates, but Davies and Johnston (1976) have adduced circumstantial evidence that hematophagous isopods, such as *Gnathia maxillaris,* may well be the vectors of at least one species, *Haemogregarina bigemina,* which has for its host the marine fish, *Blennius pholis.*

Haemogregarina stepanowi, a parasite of the common European freshwater turtle, *Emys orbicularis,* is probably the best known. Its life cycle was determined by Reichenow (1910). The slender crescentic sporozoites which have been liberated from the oocysts pass from the leech's intestine into the blood sinus and from there into its proboscis; it is then a short journey into the blood stream of the turtle, which they enter during feeding by the leech. They then invade erythrocytes in which they grow and eventually divide into 12–24 merozoites; these, in turn, invade other erythrocytes and a second generation of schizogony follows, but this time only six or fewer merozoites result. Such forms are said to occur especially in the bone marrow. One more schizogonic generation originates from these merozoïtes, the offspring of which are destined to become gametocytes, and are claimed to exhibit some sexual dimorphism. The male, according to Grassé, has a larger nucleus which stains more intensely than the smaller nucleus of the female; the latter also contains in its cytoplasm chromatoid granules which are lacking in the male.

Recently Patterson and Desser (1976) have made an extensive study of a haemogregarine, *Haemogregarina balli* n.sp., from the snapping turtle *Chelydra serpentina.* Although they were able to demonstrate development of the gametocytes in two species of leeches (*Placobella parasitica* and *P. ornata*), they were unable to achieve transmission.

Other species of *Haemogregarina* are known to have slightly different life cycles. *Haemogregarina mirabilis,* a parasite of the water snake, *Natrix piscator* (an Indian species studied by Ball, 1958), undergoes schizogony mainly in lung capillaries. Its vector is *Hirudinaria granulosa,* a leech. However, vectors are not always leeches. A haemogregarine of a lizard host is known which is transmitted by the assassin bug, *Triatoma rubrovaria.*

2. *Hepatozoon*

The first of the blood parasites included today in the genus *Hepatozoon* was seen by Balfour in jerboa leukocytes in 1905, though the genus itself was not created until 3 years later, when Miller proposed the name. By 1920, Reichenow could list twelve species in an equal number of hosts, almost all carnivores and rodents. Both known hosts and species have grown in number since then.

According to Miller (1908), who studied what is now known as *Hepatozoon muris* in a laboratory population of white rats, the parasites multiply only in the liver, where they undergo three generations of schizogony. Occasionally there may be a fourth or fifth generation. The offspring of the third generation are destined to become gametocytes and to enter mononuclear or, rarely, polymorphonuclear leukocytes. Whether gametocyte production is deferred when there are additional generations of schizogony is unknown. There is also some uncertainty about the type of hepatic cells parasitized. According to Wenyon (1926), Miller thought them to be glandular (parenchymal ?) cells, but Wenyon suggests that they may well have been endothelial or Kupffer cells.

When the blood of the infected rat is ingested by the spiny rat mite (*Laelaps echidninus*), the gametocytes leave their host monocytes and pair in typical haemogregarine fashion. Maturation then occurs, but there is some uncertainty about microgametogenesis.

Miller did not claim the production of more than one microgamete from a single microgametocyte, but Ball and Oda (1971) observed the formation of four biflagellated microgametes from a single microgametocyte of *Hepatozoon rarefaciens,* a parasite of snakes, and it seems likely that this may be true of other species.

Fertilization occurs after gametogenesis, and the zygote continues its development after penetrating the intestinal wall of the mite, where growth continues in the surrounding tissues until about 5 days after the initial blood meal; by this time it has become a cyst about 125 μm in diameter. Nuclear division ensues within the cyst and culminates in the production of a large number of sporocysts, each with 12 to 24 sporozoites. At this time the parent cyst may be as large as 250 μm in diameter.

Infection of a new host occurs when the infected mite is swallowed. The duodenal fluid of the rat is said to stimulate movement of the sporozoites, and thus probably triggers excystment. They then undoubtedly penetrate the intestinal mucosa and reach the liver via the portal circulation.

An important advance in our ability to study *Hepatozoon* in the laboratory has recently been made by Ball and Chao (1973), who achieved the development of the sporogonous stages of *Hepatozoon rarefaciens* in tissue culture, using a cell line derived from *Culex pipiens.* Mosquitoes of several species may act as vectors of this parasite.

The genus has a wide host spectrum. Not only is it known from many mammalian hosts, but also from some birds, as well as a number of cold-blooded vertebrates. Vectors are equally varied, including mites, ticks, tsetse flies, mosquitoes, midges (*Phlebotomus*), and probably many others. Infection of the vertebrate host is regularly accomplished by ingestion of the vector.

Although it is certain that many species of vertebrates harbor *Hepatozoon* and that numerous species of the parasite exist, it is quite likely that among the described species are a number of synonyms, one of the reasons being that little is known about host specificity in the genus. Some recent work by Ball and Oda (1971) suggests that at least among reptilian species of *Hepatozoon* it may not be very strict. However, there is also a likely possibility that numerous species of *Hepatozoon* remain to be discovered.

Hepatozoon muris is probably cosmopolitan, if only because its usual hosts have followed man around the world. It has been found in the blood of rats and mice, both wild and domestic, in Australia, England, continental Europe, Panama, and Japan. The author has seen it in the blood of rats captured in Taiwan.

Under laboratory conditions, Miller found *Hepatozoon muris* quite pathogenic, and, in recognition of the fact, gave it the species name *H. perniciosum*. However, in nature infections are said to be usually light and with little detectable harm done to their hosts. A variety of species names have been given to *Hepatozoon* from rats and mice, but it is generally thought that all are the same, with the possible exception of *H. musculi*. It is said that this species multiplies only in the bone marrow.

Another species, *Hepatozoon balfouri* (Laveran, 1905), has as its vertebrate host the jerboa (*Jaculus jaculus*), a small jumping rodent, in which it appears to be common. This species of jerboa is widely distributed in northern Africa. Apparently differing strains or subspecies of the parasite exist. Host specificity of Egyptian strains in both the jerboa and the vector mite (*Haemolaelaps aegyptius*) seems rather strict. Transmission via placenta may perhaps also occur (Hoogstraal, 1961). Perhaps the most interesting thing about *H. balfouri* is the pathogenicity of certain other strains for their mite vectors (*Haemolaelaps longipes* and *H. centrocarpus*) which are often killed (Furman, 1966).

Like *H. muris*, to which *H. balfouri* may be closely related, the latter undergoes schizogony in the liver, causing serious injury and the death of the host. Despite its pathogenicity to both vertebrate and invertebrate hosts in the laboratory, it is thought not to have much ecological significance in nature.

The life cycles of *H. muris* and *H. balfouri* are probably fairly typical; other species differ in minor ways, some preferring one organ and some another in which to undergo schizogony. They also may invade various types of blood cells.

A species of *Hepatozoon* which has considerable importance is *H. canis*, a parasite of dogs. It probably occurs also in cats, jackals, and hyenas, and perhaps in palm civets, although it has received a different species name in each of these hosts (*H. felis*, *H. rotundata*, and *H. chattoni*, respectively).

Hepatozoon canis apparently has a worldwide distribution, having been reported from dogs in Southeast Asia, North Africa, India, the Near East, and Europe.

Schizogony of *H. canis* occurs chiefly in the spleen and bone marrow, though it has been claimed to occur in the liver and even in the blood as well. The gametocytes (gamonts) invade polymorphonuclear leukocytes, which are infective for the brown dog tick (*Rhipicephalus sanguineus*). According to Levine (1961), both the nymph and the adult may transmit the infection, but transovarian infection of the tick does not occur as it does with some other tick-spread organisms. Oocysts occur in the tick hemocoel.

The parasite is a recognized pathogen of dogs and causes an infection which may be fatal in 1 or 2 months, although some dogs remain apparently healthy and become carriers. Fever, anemia, loss of weight, splenic enlargement, and sometimes paralysis of the hind quarters are the chief symptoms. No treatment is known.

Hepatozoon canis has also been reported in cats, but such infections must be very rare.

C. Haemogregarines in Man?

There is no clear evidence that haemogregarines ever infect man. Yet it would not be surprising if such infections occasionally occur. If they do, the parasite involved would undoubtedly be some species of *Hepatozoon*. Such infections are widespread in mammals, and host specificity (to the extent that anything is known about it) appears to be often limited. The vectors of some species of *Hepatozoon* may attack man, and while infection does not result from the bite, accidental ingestion is always possible.

There are a number of reports of human haemogregarine infections in the literature, several of them by respected and well-known investigators, such as Roubaud and the Sergents. Their figures suggest that they were actually dealing with haemogregarines in human blood, but these reports have been generally discounted. Wenyon (1923) concluded, after a careful review of the literature, that at best the evidence was equivocal, and that the bodies thought to be haemogregarines were more likely artifacts or "vegetable organisms." Others have reached a similar conclusion.

Karyolysus

The genus *Karyolysus* (or *Caryolysus*) was created by Labbé in 1894 for a parasite Danielewski had seen in the erythrocytes of a lizard (genus *Lacerta*) in 1886. Danielewski had initially called it *Haemocytozoon* and, a little later, *Haemogregarina*. Labbé coined the present generic name from two Greek words meaning "tissue dissolver," because of the destructive effect of the gamonts on the nuclei of their host cells.

Since then, numerous other reptilian hosts have been found with similar infections, and it now appears that *Karyolysus* rather commonly infects lizards and even snakes.

Reichenow (1921), the German protozoologist who first elucidated the life cycle of *Haemogregarina stepanowi* of the European water tortoise, was also the first to work out the life cycle of *Karyolysus lacertarum*. The host in this case was another species of lizard, *Lacerta muralis* (the "wall lizard").

Rather than being restricted to any one organ, this parasite undergoes schizogony in the endothelium of the capillaries generally. Each schizont gives rise to 8–30 morozoites, which for a time remain enclosed in a kind of cyst. Eventually they break out and invade other endothelial cells; this cycle may be repeated for a number of generations. After a time, some of the offspring, instead of again initiating another schizogonic cycle, invade erythrocytes in which they grow and become gametocytes. Except for types of host cells involved it is a quite typical coccidian life cycle. Aside from a small difference in size (the macrogametocyte is slightly larger) there is no discernible sexual dimorphism.

The sexual phase continues if these stages are fortunate enough to be ingested by the mite *Ophionyssus saurarum*. Maturation occurs in the mite's gut, beginning with the lateral pairing (syzygy) of the gametocytes. The pair penetrate an epithelial cell and become encysted within it. Maturation takes place which, for the microgametocyte, results in the production of two flagellated microgametes, one of which proceeds to fertilize the macrogamete. The zygote then begins a period of active growth culminating in schizogony and a brood of wormlike bodies to which Reichenow gave the name "sporokinetes." Some of these reach ova in the mite, in which they continue development which ends only in the nymph. Finally sporocysts are produced, each of which contains 20–30 sporozoites—the infective stage. Infection of a new host results when either the nymph or the adult developing from it is ingested; or it may even follow the ingestion of fecal droppings voided during the later stages of parasite development.

On the whole, little is known about most species of *Karyolysus*. No species is so far known from any warm-blooded host, and no one has cultivated the parasites. What is most needed is intensive laboratory study. It is certain that there are differences in life histories. It also seems likely that they may be pathogens, at least occasionally, and as such may play a significant role in the ecology of their vertebrate hosts. At present, we have no idea whether the lizard, snake, or even the mite with a case of *Karyolysus* is any the worse for it.

Much of interest might also be learned if there were a practical method of cultivating the parasites, either in tissue culture or in a synthetic medium. From a purely scientific point of view, it is unfortunate that unless living things are of some easily visible importance they are not likely to be intensively studied.

D. *Lankesterella* and Its Relatives

Like the Sporozoa (other than gregarines) considered earlier in this chapter, the Lankesterellidae are coccidia, but their life histories separate them from those previously examined. However, Wenyon (1926) remarks that they have long been "recognized as haemogregarines of cold-blooded animals."

The Lankesterellidae are also of great interest to anyone who has speculated about the evolution of the malaria parasites because they suggest how the coccidian ancestors of these most important disease-producing microorganisms may have become adapted to an alternating existence in a vertebrate and an insect host. As parasites, however, they have no known practical importance.

1. Characteristics

Stages seen in the erythrocytes are sporozoites having the appearance of typical haemogregarines. Its hosts are cold-blooded animals and birds.

Its life cycles are generally said to be of the "Eimeriid" type (see Chapter 2), in which there are typically a number of generations of schizogony, usually in the gut, culminating in the production of gamonts. From the male gamont, large numbers of microgametes originate, which then seek out the female gamete. After fertilization the zygote develops an oocyst about itself, in which repeated divisions result in varying numbers of sporozoites (the infective stage); these are often themselves enclosed in sporocysts. Both the number of sporocysts (absent in some species) and the number of sporozoites in each vary according to the genus. Most coccidia have only one host, but the Lankesterellidae are exceptions.

2. Taxonomy

Unfortunately the classification of the group is in a state of considerable confusion, partly because too little is known about its life cycles but more because taxonomists disagree about the relative importance of facts already known. Wenyon divides the family into two subfamilies, whereas Hoare (1933) recognizes them only as genera of the family Eimeriidae and subfamily Cryptosporidiinae. His scheme is logical, since it is based on the number of sporocysts within the oocyst, a characteristic which can be conveniently used in diagnosis. Grassé (1953) ranks Wenyon's subfamilies as families, while Levine (1961) follows Wenyon, but omits the subfamilies and places the genus *Tyzzeria* in the family Eimeriidae. We prefer Wenyon's scheme:

Family: Lankesterellidae Reichenow 1921. The oocyst lacks sporocysts, but contains sporozoites which, after liberation, make their way to the bloodstream, where they enter host cells, the type depending on the species. Transmission to another host depends on a blood-sucking vector.

Subfamily 1. Schellackinae Wenyon 1926. The entire development takes place in tissue of the intestinal wall. The final product, sporozoites, then enters erythrocytes. Hosts are reptiles (and birds, if Grassé's classification is accepted). One genus: *Schellackia.*

Subfamily 2. Lankesterellinae Wenyon 1926. Development is in the vascular endothelium. Sporozoites enter blood cells. Hosts are anuran amphibia and birds. One genus: *Lankesterella.*

Of the two subfamilies, the Schellackinae are the more primitive because schizogony occurs in the intestinal epithelium. Three species are known in the single genus of the family, all parasites of lizards. Reichenow (1919) worked out the life cycle of the first and best known of the three, *Schellackia bolivari.*

After the initial period of schizogony, merozoites which are to become gamonts are produced. Those destined to become macrogametocytes migrate to the submucosa where fertilization will take place, whereas those from which microgametocytes will develop remain in the intestinal epithelium. Numerous microgametes arise from the latter and seek out the female cells; fertilization ensues, an oocyst is formed, and within it three mitoses occur, with the production of eight sporozoites. These invade the blood stream after liberation from the confining walls of the oocyst, and seek out blood cells as hosts. Curiously, the type of cell selected depends on the species of lizard parasitized. In one lizard species (*Acanthodactylus vulgaris*) it is the erythrocytes, and in another (*Psammodromus hispanicus*) the lymphocytes. The vector of this species of *Schellackia* is the mite, *Liponyssus saurarum,* but transmission is purely mechanical; the parasite undergoes no development in it. Infection of a new host occurs when the mite is eaten.

Reichenow's study was done in Spain on lizards native to that country. A few years ago Bonorris and Ball (1955) described a species from California lizards of the genus *Sceloporus,* to which they gave the name *Schellackia occidentalis.* They determined that it had a similar life history, transmission occurring by a pterygosomid mite, *Geckobiella texana.* It, therefore, is probable that *Schellackia* has a wide distribution, which may be evidence of great antiquity.

If coccidia of vertebrates were ancestors of the malaria parasites (Manwell, 1955), a life cycle like that of *Schellackia* may represent an early stage in their evolution; schizogony occurs in the intestinal mucosa (the usual coccidian pattern), but the sporozoites invade the blood stream and parasitize blood cells. Specialization, however, has not proceeded far enough to ensure that such cells are always erythrocytes, and when they are, no pigment results from parasite metabolism. If culture methods could be devised it might be possible to discover which substances the parasites require from their host erythrocytes, and thus to compare their metabolic needs with those of the malaria plasmodia.

Transmission, however, is mechanical and the parasites, though able to live in erythrocytes are still unable to reproduce in them.

The disposition of the genus *Tyzzeria* is in dispute. As indicated above, Grassé, for reasons not quite clear, places it in the family Schellackiidae. There are several species of which one, *T. perniciosa,* is a serious pathogen of Pekin ducks. Another follows the usual life style of the family and has a reptile for its host, though in this case it is a snake rather than a lizard.

Lankesterella minima of frogs, for many years the only species known, exemplifies the type of life cycle typical of the Lankesterellinae. Although first seen by Chaussat in 1850, it remained for Nöller (1920) to work out the life cycle almost 70 years later. The parasites as seen in the erythrocytes of the frog are very slender, elongate organisms, lying beside the host cell's nucleus. For this reason Chaussat called it *Anguillula minima* which, freely translated, means the "very little worm" (if it is remembered that *Anguillula* is the tiny nematode commonly called the vinegar eel). Since then it has been given various other names, one of them being *Haemogregarina minima.* Another species is now known from toads.

The infection in the frog begins when it swallows an infected leech. Erythrocytes containing sporozoites acquired by the leech when feeding on an infected frog release their parasites during digestion, and somehow the latter make their way into the blood stream. (Why it is that the parasites can resist digestion when their host cells cannot is not known.) Once in the blood stream, the liberated sporozoites invade endothelial cells in which they grow and initiate schizogony. After a certain number of generations, multiplication ceases, and the merozoites produced develop into gametocytes. The microgametocytes mature and produce numerous microgametes, fertilization of the female cells follows, and each zygote secretes a wall about itself, thus forming an oocyst. Sporozoites develop within but there are no sporocysts. After liberation, the sporozoites penetrate erythrocytes, but they have no future unless picked up by a leech.

There are obviously some gaps in the story. The most important one concerns the problem of just how sporozoites from the infected leech reach the vascular endothelium after their ingestion by a new host.

Recently, Stehbins (1966), in a study of the ultrastructure of *Lankesterella hylae,* reported the finding of intracellular parasites in the intestinal submucosa. This may suggest a way station en route to the vascular endothelium after their ingestion in an infected leech. He also noted striking similarities in the ultrastructure of *Lankesterella, Plasmodium, Toxoplasma,* and *Sarcocystis.*

3. *Lankesterella* in Birds

Within the last few years the subject of the Lankesterellinae has become much more complicated, since a group of avian blood parasites long regarded simply as "of uncertain status" has been added to the subfamily. Typically, they appear as rather poorly defined, slightly elongated bodies in monocytes and lymphocytes; they often stain rather weakly with Romanowski stains. Nuclei of parasitized cells are often somewhat indented, as if by young leucocytozoa.

According to Lainson (1959), they were probably first seen many years ago by Laveran in the paddy bird (*Padda oryzivora*). Laveran thought them part of the life cycle of avian malaria parasites. Later, they were believed by other investigators to be haemogregarines, and still others regarded them as *Toxoplasma,* which they do indeed somewhat resemble. Marullaz called them *Toxoplasma avium.* Garnham (1950) realized that they were not *Toxoplasma* and called them, for lack of a better name, *Atoxoplasma;* he also separated them from haemogregarines, in part, because they lacked motility.

Then Lainson made a rather intensive study of these parasites in English sparrows, in which they are extremely common and probably cosmopolitan, and concluded that they belonged to the genus *Lankesterella.* For the parasites in the sparrow he proposed the name *Lankesterella garnhami* and for the very similar forms in the canary, *L. serini;* but he has since succeeded in transmitting the sparrow parasite to the canary and has, therefore, withdrawn the latter name (Lainson, 1960).

The life cycle he found involved schizogony in the lymphoid–macrophage cells of the spleen, bone marrow, and liver, with a later generation in the lungs, liver, and kidney; from the last generation gamonts developed which invaded lymphocytes and monocytes.

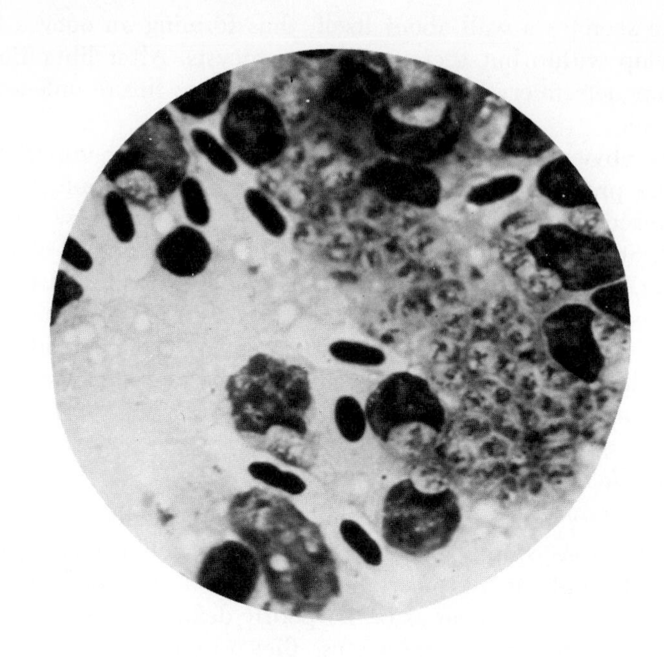

Fig. 8. A cluster of haemogregarine (*Lankesterella*) merozoites in the liver of an English sparrow (*Passer domesticus domesticus*). × 1500.

Many years ago, we also found similar forms (Fig. 8) in the liver and spleen of a fledgling English sparrow infected with what we then thought to be "avian *Toxoplasma.*" In retrospect, it seems certain that they were the same as what Garnham called *Atoxoplasma* and Lainson *Lankesterella.* In our infection, erythrocytes as well as monocytes and lymphocytes were invaded, though Lainson observed parasites only in the latter two cell types.

Although Lainson says nothing about pathogenicity it seems certain that *Lankesterella* infection may be serious. Our sparrow died soon after it was brought to the laboratory, and we have observed an apparent build-up in *Lankesterella* parasitemia in caged populations of English sparrows, and many such birds died.

Transmission is doubtless by the common bird mite, *Dermanyssus gallinae,* but Lainson found it difficult to prove this conclusively because natural infections are so common that to obtain uninfected birds is not easy. In later experiments he felt that he had succeeded not only in transmitting the infection by mites from canary to canary, but also from sparrows to canaries. In nature, mite infestation of nests and rookeries is virtually universal and these pests can overwinter. Hence, in susceptible avian species, it is easy to account for the near 100% incidence of *Lankesterella*. The birds probably ingest the mites by accident when they preen, and perhaps a few become mixed with the food the nestlings get. Lainson found some nestlings infected when they were less than a week old.

Lankesterella infections are common in many species of birds, but since the degree of host specificity is doubtful it is very uncertain whether infections in different host species represent the same or different species of *Lankesterella*.

Box (1973, 1975) studied this parasite in canaries and concluded that two species were involved: *Isospora serini* and *I. canaria*, n. sp. She found that a "disseminated infection of mononuclear phagocytes" is caused by *Isospora serini*, whereas *I. canaria* produces a "typical coccidian infection restricted to the intestinal epithelium." Asexual stages of parasites, she states, are indistinguishable from what others have called avian *Toxoplasma*, *Atoxoplasma*, and *Lankesterella*.

In an earlier study of the infection in English sparrows in Galveston, Texas, she discovered that 20% of them were infected, the highest incidence occurring in summer. Other species of birds were less often and less severely infected; in pigeons and ducks the parasite was entirely absent.

The ubiquity of this type of infection in birds is proof that the problem is a large one.

REFERENCES

Balfour, A. (1905). A haemogregarine of mammals, *H. jaculi* (*H. balfouri* Laveran). *J. Trop. Med. & Hyg.* 8, 240.

Ball, G. H. (1958). A haemogregarine from a water snake, *Natrix piscator*, taken in the vicinity of Bombay, India. *J. Protozool.* 5, 274–281.

Ball, G. H. (1960). Some considerations regarding the Sporozoa. *J. Protozool.* 7, 1–6.

Ball, G. H., and Chao, J. (1973). The complete development of the sporogonous stages of *Hepatozoon rarefaciens* cultured in a *Culex pipiens* cell line. *J. Parasitol.* 59, 513–515.

Ball, G. H., and Oda, S. N. (1971). Sexual stages in the life history of the haemogregarine *Hepatozoon rarefaciens* (Sambon and Seligman, 1907). *J. Protozool.* 18, 697–700.

Baudoin, J. (1971). Etude comparée de quelques Grégarines Acanthosporinae. *Jour. Protozool.* 18, 654–660.

Bonorris, J. S., and Ball, G. H. (1955). *Schellackia occidentalis* n. sp., a blood-inhabiting coccidian in lizards in Southern California. *J. Protozool.* 2, 31–34.

Box, E. D. (1973). Comparative development of two *Isospora* species in the canary. *J. Protozool., Suppl.* p. 510.

Box, E. D. (1975). Exogenous stages of *Isospora serini* (Aragao) and Isospora sp. n. in the canary (*Serinus canarius*) Linnaeus. *Jour. Protozool.* **22**, 165–169.

Codreanu, R. (1967). Sur une Microsporidie nouvelle hyperparasite d'une grégarine du genre *Enterocystis* et le statut systématique de la famille des *Enterocystidae* Margareta Codreanu 1940, propres aux Ephémères. *Protistologica* **3**, 351.

Corbel, J. C. (1967). *Metchnikovella martojai* n. sp. (Microsporidie, Metchnikovellidae) parasite d'une Grégarine de Gryllidae. *Protistologica* **3**, 365.

Corbel, J. C. (1968). Part of the cyst and spores in the frequency and intensity of gregarinian infestations in Orthoptera. *Protistologica* **4**, 19–26.

Davies, A. J. and Johnston, M. R. L. (1976). The biology of *Haemogregarina bigemina* Laveran & Mesnil, a parasite of the marine fish, *Blennius pholis Linaeus*. *J. Protozool.* **23**, 315–320.

Furman, D. P. (1966). *Hepatozoon balfouri;* (Laveran 1905) sporogonic cycle, pathogenesis, and transmission by mites to jerboa hosts. *J. Parasitol.* **52**, 373–382.

Garnham, P. C. C. (1950). Blood parasites of East African vertebrates, with a brief description of exoerythrocytic schizogony in *Plasmodium pitmani, Parasitology* **40**, 328–337.

Grassé, P. -P., ed. (1953). "Traité de Zoologie," Vol. 1, Part 2. Masson, Paris.

Harry, O. G. (1967). The effect of the eugregarine *Gregarina polymorpha* (Hammerschmidt) on the mealworm larva of *Tenebrio molitor* (L.). *J. Protozool.* **14**, 539–547.

Heller, G. and Weise, R. W. (1973). A scanning electron microscope study on *Gregarina* sp. from *Udeopsylla nigra*. *J. Protozool.* **20**, 61–64.

Hildebrand, H. F. and Vinckier, D. (1975). Nouvelles observations sur la grégarine *Didymophyes gigantea* Stein. *J. Protozool.* **22**, 200–213.

Hoare, C. A. (1933). Studies on some new ophidian and avian Coccidia from Uganda, with a revision of the classification of the Eimeriidae. *Parasitology* **25**, 359–396.

Hoogstrahl, H. (1961). The life cycle and incidence of *Hepatozoon balfouri* (Laveran, 1905) in Egyptian jerboas (*Jaculus* spp.) and mites (*Haemolaelaps aegyptius* Keegan, 1956). *J. Protozool.* **8**, 231–248.

Hukui, T. (1939). On the gregarines from *Siphonosoma cumanense*. *J. Sci. Hiroshima Univ., Ser. B,* Div. 1 **7**, Art. 1.

Kölliker, A. (1848). Die Lehre von den thierischen Zellen und den einfacheren thierischen Formelementen, nach den neusten Forschritten dargestellt. Z. *Wiss. Bot.* **1**, 1–37.

Lainson, R. (1959). *Atoxoplasma* (Garnham, 1950) as a synonym for *Lankesterella,* Labbé, 1889. Its life cycle in the English sparrow (*Passer domesticus domesticus,* Linn.). *J. Protozool.* **6**, 360–371.

Lainson, R. (1960). The transmission of *Lankesterella* (=*Atoxoplasma*) in birds by the mite, *Dermanyssus gallinae*. *J. Protozool.* **7**, 321–322.

Léger, L. (1907). Les schizogrégarines des trachéates. I. Le genre *Ophryocystis. Arch. Protistenkd.* **8**, 159–202.

Léger, L. (1909). II. Le genre *Schizocystis. Arch. Prostistenkd.* **18**, 83–100.

Léger, L., and Duboscq, O. (1910). *Selenococcidium intermedium* Léger et Duboscq, et la systématique des sporozoaires. *Arch. Zool. Exp.* **45**, 187.

Levine, N. D. (1961). "Protozoan Parasites of Domestic Animals and Man." Burgess, Minneapolis, Minnesota.

Lipa, J. J. (1967). Studies on gregarines (Gregarinomorpha) of Arthropods in Poland. *Acta Protozool.* **5**, 97–179.

Mackinnon, D. L., and Hawes, R. S. J. (1961). "An Introduction to the Study of Protozoa," Oxford Univ. Press (Clarendon), London and New York.

McLaughlin, R. E., and Myers, J. (1970). *Ophryocystis elektroschirrha* sp. n., a neogregarine pathogen of the Monarch Butterfly *Danaus plexippus* (L.) and the Florida Queen Butterfly *D. Gillippus berenice* Cramer. *J. Protozool* **17**, 300–305.

Manwell, R. D. (1955). Some evolutionary possibilities in the history of the malaria parasites. *Indian J. Malariol.* **9**, 247–253.

Manwell, R. D. (1968). "Introduction to Protozoology," Rev. ed. Dover, New York.

MacGregor, H. C. and Thomasson, P. A. (1965). The fine structure of two Archigregarines, *Selenidium fallax* and *Ditrypanocystis cirratuli. J. Protozool.* **12**, 438–443.

Miller, W. W. (1908). *Hepatozoon perniciosum* (n.g., n.sp.) a haemogregarine pathogenic for white rats; with a description of the sexual cycle in the intermediate host, a mite (*Laelaps echidninus*). *Bull. Hyg. Lab., Treas. Dep., Washington* **46.**

Nöller, W. (1920). Kleine Beobachtungen an parasitischen Protozoen (Zugleich verlaüfige Mitteilung über die Befruchtung und Sporogonie von *Lankesterella minima* Chaussat). *Arch. Protistenkd.* **41**, 169.

Ormieres, R. (1966). Grégarines parasites de myriapodes chilopodes. Observations sur les genres *Echinomera* Labbé 1899 et *Acutispora* Crawley 1903. *Protistologia* **1**, 15–21.

Patterson, W. B. and Desser, S. S. (1976). Observations on *Haemogregarina balli* n. sp. from the common snapping turtle, *Chelydra serpentina. J. Protozool.* **23**, 294–301.

Reichenow, E. (1910). Der Zeugungskreis der *Haemogregarina stepanowi. Sitzungsber. Ges. Nat. Freunde,* Berlin.

Reichenow, E. (1919). Der Entwicklungsgang der Hämococcidien *Karyolysus* und *Schellackia* nov. gen. *Stizungsber. Ges. Naturforsch. Freunde Berlin,* 440.

Reichenow, E. (1921). Die Hämococcidien der Eidechsen. Vor Bermerkungen und I. Theil. Die Entwicklungsgeschichte von *Karolysus. Arch. Protistenkd.* **42**, 180.

Roubaud, E. (1919). Un deuxième type d'hémogrégarine humaine. *Bull. Soc. Path. Exot.* **12**, 76.

Sanders, R. D., and Poinar, G. O. (1973). Fine structure and life cycle of *Lankesteria clarki,* sp. n. (Sporozoa, Eugregarinida) parasitic in the mosquito *Aedes sierrensis* (Ludlow). *J. Protozool.* **20**, 594–602.

Sergent, Ed. and Et., and Parrot, L. (1922). Sur un hémogrégarine de l'homme observée en Corse. *Bull. Soc. Exot.* **15**, 193.

Stebbings, H., Boe, G. S., and Garlick, P. R. (1976). Microtubules and movement in the archigregarine, *Selenidium fallax. Z. Zellforsch. Mikrosk. Anat.* (in press).

Stehbens, W. E. (1966). The ultrastructure of *Lankesterella hylae. J. Protozool.* **13**, 63–73.

Vavra, J., and Small, E. (1969). Scanning electron microscopy of gregarines (Protozoa, Sporozoa) and its contribution to the theory of gregarine movement. *J. Protozool.* **16**, 745–757.

Vivier, E. (1965). Étude au microscope électronique des spores de *Metchnikovella hovassei* n. sp.; appartenance des Metchnikovellidae aux Microsporidies. *C. R. Hebd. Seances Acad. Sci.* **260**, 6982–6984.

Walsh, R. D., and Callaway, C. S. (1969). The fine structure of the gregarine *Lankesteria culicis,* parasitic in the yellow-fever mosquito, *Aedes aegypti. J. Protozool.* **16**, 536–545.

Watson, M. E. (1916). Studies on gregarines. *Ill. Biol. Monogr.* **2**, 1–258.
Weiser, J. (1955). A new classification of the Schizogregarina. *J. Protozool.* **2**, 6–12.
Wenyon, C. M. (1923). "Haemogregarines" in man, with notes on some other supposed parasites. *Trop. Dis. Bull.* **20**, 527–550.
Wenyon, C. M. (1926). "Protozoology," Vol. 2, pp. 779–1563. Wm. Wood, New York.
Zizka, Z. (1972). An electron microscope study of auto-infection in Neogregarines (Sporozoa, Neogregarinida). *J. Protozool.* **19**, 275–280.

2

Avian Coccidia

M. D. Ruff and W. M. Reid

I. Introduction

Most species of wild and domestic birds host one or more species of coccidia. The widely dispersed nesting and feeding habits of wild birds often result in only light infections which produce no visible damage but induce some protective immunity for the host. The term coccidiasis is employed to emphasize the relatively nonpathogenic nature of these infections in contrast to coccidiosis where severe damage may be inflicted on the host. Severity of disease is directly related to the numbers of sporulated oocysts ingested with food or drink. Crowding of bird hosts before migration or by management practices may concentrate oocyst populations to the danger point.

Coccidiosis is more of a problem with domesticated birds. The risk increases as management practices call for concentration of bird populations. If the brooding hen whose scattered flock of 15 chicks is replaced by an artificial brooder containing 100 or more chicks, fecal debris may concentrate large numbers of oocysts which sporulate and become infective under warm moist conditions. The sudden outbreak of coccidiosis may wipe out 20% of the flock. The present trend is to further concentrate flocks by moving from out-of-doors pens or ranges to confinement rearing in long houses of 5000 or more birds with less than 1 square foot of floor space for each. Coccidiosis would certainly have prevented development of the vast broiler industry had anticoccidial drugs not been available to control this disease through continuous feed medication.

Coccidiosis constitutes the most economically important parasitic disease of chickens. Losses from mortality were serious before 1949 when preventive anticoccidials began to reduce or eliminate death from this disease. Losses due to morbidity from coccidiosis continue to persist, resulting from reduced weight gains, egg production, and decreased feed conversion. The cost of coccidiosis in broiler production has been estimated at ½ to 1 cent/lb. Based on an annual production of 3,000,000,000 broilers annually in the United States, losses would total between 60 and 120 million dollars. To this figure must be added the cost of anticoccidials estimated at 35 million. These impressive figures emphasize the importance of coccidiosis in chickens.

Interest in coccidiosis has resulted in an enormous and increasing volume of literature. Four successive annotated bibliographies abstract over 1800 selected papers (Merck & Company, Inc., 1953, 1961, 1965, 1970). In 1906, biological information on the Sporozoa was summarized by Hartog in 14 pages of "Cambridge Natural History" while recently several expansive volumes deal solely with the biology of the coccidia. These include "Coccidiosis" (Davies *et al.*, 1963), "The Coccidia" (Hammond and Long, 1973), "Coccidia and Coccidiosis" (Pellérdy, 1973), and "Life Cycles of Coccidia of Domestic Animals" (Kheysin, 1972). Levine (1973) has devoted 100 pages to coccidia with emphasis on genera and species found in domestic and wild host species.

II. Morphology

Eimeria tenella from the cecum of the chicken can be used to illustrate the life cycle and development (Fig. 1) of the avian coccidia, although some variations between species may occur in the prepatent period, number of schizont stages, and area of the digestive tract infected. The life cycle begins when the host digests the previously sporulated oocysts (Fig. 2). The outer wall of the oocyst is first removed through the mechanical grinding action of the gizzard which then ruptures the

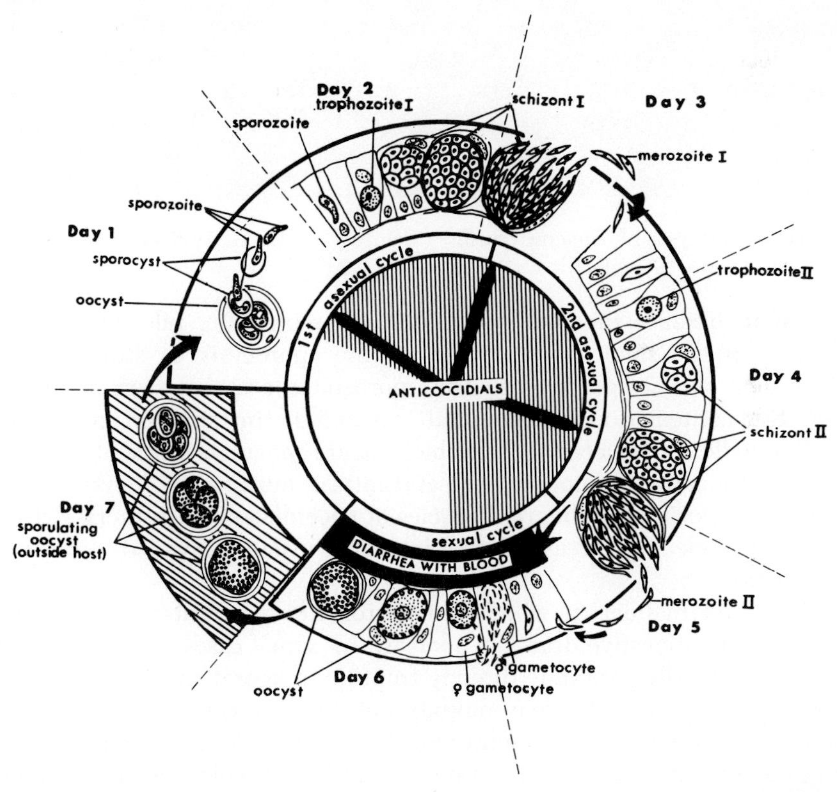

Fig. 1. Life cycle of *Eimeria tenella* from the ceca of the chicken. Reproduced from the *Proc. Symp. Coccidia Related Organisms,* p. 133. University of Guelph, Guelph, Ontario, 1973.

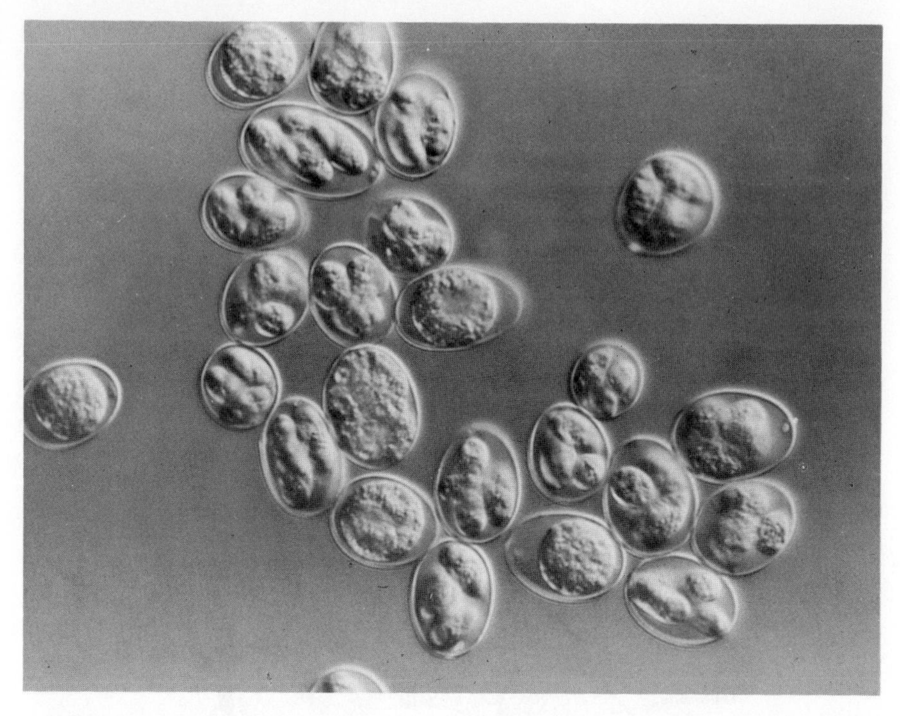

Fig. 2. Oocysts of *Eimeria tenella.* (Courtesy of D. R. Witlock, University of Georgia, Athens, Georgia.)

oocyst to liberate the four sporocysts. Chymotrypsin, bile, and carbon dioxide in the small intestine act on the sporocysts to stimulate the liberation of the sporozoites. The sporozoites are then carried in the digestive contents to the appropriate area of the intestine or, in the case of *E. tenella,* to the ceca. When chickens are inoculated subcutaneously, intraperitoneally, intramuscularly, and intravenously with oocysts of any of several species, the different species of coccidia establish themselves in the same selective areas of the intestine usually invaded after oral inoculation.

The exact manner in which the sporozoites penetrate the epithelial cells of the digestive tract is unclear. In some cases the sporozoites appear to rapidly penetrate directly from the mucosal surface of the cell. Sporozoites (*E. tenella*) may migrate between the cells as deeply as the lamina propria prior to penetration. The exact role of macrophages in migration of sporozoites is also unclear. In those species which are found deep in the intestinal tissue (*E. tenella* and *E. maxima*), sporozoites have been observed in macrophages which may carry the parasite to a deep location in the glands or crypts of Lieberkühn. Some workers, however, suggest that the ingestion by macrophages represents a protective mechanism of the host and thus is incidental to the life history of the parasite.

After penetration of the epithelial cell, the sporozoite rounds up and becomes a trophozoite. This stage of the parasite grows rapidly and undergoes the process of schizogony (asexual multiple fission). The nucleus divides to give rise to the schizont stage which develops merozoites. The schizont ruptures, releasing the first-generation merozoites which, in turn, enter other epithelial cells and undergo a second similar schizogonic development. Many of the life cycle variations between the different species of chicken coccidia occur in the schizont stage. The number of schizont generations is not known exactly for all species, but ranges from one to two generations (*E. tenella*) or up to four generations (*E. acervulina* and *E. mivati*). The size of the first-generation schizonts is generally small while schizonts from the later generations range from approximately 9 (*E. maxima*) up to 66 μm (*E. necatrix*). The location of the schizonts in relation to the surface of the epithelium may also vary. In the case of *E. acervulina*, although the sporozoites penetrate the glandular epithelium, the schizonts develop closer to the villar tips and there is relatively little change in the epithelial cells. A slightly different reaction is seen with the first-generation schizonts of *E. necatrix* which tend to push the nucleus of the host cell toward the lumen with resulting hypertrophy. The location of the parasite changes in the second-generation schizonts of such species as *E. necatrix* or *E. tenella*. With those species the large developing schizont stimulates such enlargement that the host cell can no longer fit in the row of epithelial cells and is squeezed down into the lamina propria underneath the epithelium.

The number of merozoites within the schizont is also variable and may range from as few as 8 merozoites per schizont (*E. maxima*) to as many as 200–350 merozoites per schizont in the second generation (*E. tenella*). The average number of merozoites per schizont for any specific generation for any one species is also variable (i.e., with the second generation of *E. necatrix*, there may be 32–132 merozoites per schizont).

The final development which occurs inside the host is the development of gametocytes and gametes. The male gametocyte (microgametocyte) produces a large number of biflagellate microgametes. These penetrate the female gametes (macrogametes). After fertilization, a thickened wall forms around the zygote which then becomes an oocyst. The oocyst is passed out of the host with the droppings. The minimum time, known as the prepatent period, required to complete this portion of the life cycle can be as short as 84 (*E. praecox*) or as long as 138 hours (*E. mitis* and *E. tenella*). The potential number of gametes formed is dependent upon the particular species involved. Calculating from the number of merozoites per schizont and the number of schizonts per generation, the potential number of gametes which might develop from a single oocyst if all sporozoites and merozoites penetrate and develop would range from a low of 512 with *E. mitis* up to a high of 1,344,000 with *E. mivati* and 1,800,000 with *E. tenella*. All sporozoites and merozoites do not, of course, survive and continue development, but the reproductive potential of coccidia may nevertheless be great.

Once outside the host, the oocyst undergoes both nuclear and cellular division resulting in the formation of four sporocysts, each containing two sporozoites. Actual sporulation is dependent upon environmental conditions such as temperature, oxygen, and moisture. The large oocysts of *E. maxima* require 30 hours to sporulate compared with 12 to 18 hours for the other species of avian coccidia.

III. Taxonomy and Description of Economically Important Species

The taxonomic relationships of the coccidia to other protozoa have been described in Chapter 3, this volume. Nine genera of coccidia (*Eimeria, Isospora, Caryospora, Dorisiella, Tyzzeria, Wenyonella, Sivatoshella, Cryptosporidium,* and *Lankesterella*) are known to infect birds (Table I). The life cycles of these organisms are all quite similar and the major distinctions are based on the number of sporocysts per oocyst and the number of sporozoites per sporocyst. Of the genera which infect birds, the genus *Eimeria* contains the most important species from an economic standpoint. Only the more economically important species will be described here. Other species of coccidia found in birds are reviewed by Levine (1973) and Pellérdy (1973).

Table I
The Genera of Coccidia Parasitic in Birds

Genus	Original description	No. sporocysts/ oocyst	No. sporozoites/ sporocyst
Eimeria	Schneider, 1875	4	2
Isospora	Schneider, 1881	2	4
Caryospora	Léger, 1904	1	8
Dorisiella	Ray, 1930	2	8
Tyzzeria	Allen, 1936	0	8
Wenyonella	Hoare, 1933	4	4
Sivatoshella	Ray and Sarkan, 1968	2	16
Cryptosporidium	Tyzzer, 1907	0	4
Lankesterella	Labbé, 1899	0	32

A. Chicken

For a number of years, nine species of *Eimeria* have been recognized in the United States (Fig. 3). Although they vary greatly in pathogenicity, all species are considered pathogenic to some degree. Recently, British workers have suggested subordinating up to four of these species to a subspecies of *Eimeria acervulina*. The critical test in species determination rests upon absence of cross immunogenicity when one species is tested against another in live birds. Five of the nine species establish themselves in the duodenum and upper small intestine. In the United States, Edgar (1958) includes eight of the nine species in his immunization programs to provide protection against the various strains or species of coccidia.

1. *Eimeria acervulina* Tyzzer, 1929

This species invades the epithelial cells of the anterior small intestine. The infection is most severe in the region of the duodenal loop and, to a lesser extent, in the midgut. Parasites may be found as far posterior as the ceca or rectum. In light infections the whitish lesions are oriented transversely in "ladderlike" bands which are visible from both the serosal and mucosal surfaces. In heavier infections the colonies of parasites are coalesced and are distinct only in the posterior areas of the intestine. Doran (1966) suggested that the sporozoites penetrate the tips of the villi and then proceed to the lamina propria. Macrophages then engulf the sporozoites and transfer them to the glandular epithelium. On the other hand, Pout (1967) has described the penetration of the glandular portion of the epithelium by sporozoites and the subsequent movement of "trains" of maturing parasites proceeding up the villi. As many as four schizogonic generations have been suggested for *E. acervulina*. The developing stages of this parasite lie supernuclearly in the epithelial cell. Because they observed no mortality in any but the heaviest of infections, several earlier workers including Tyzzer, who described the species, have suggested that *E. acervulina* is nonpathogenic. Mortality occurs only if several million or more oocysts are inoculated. Nevertheless, the severe weight depressions which have been demonstrated in infected chickens by numerous workers show that this species must be considered severely pathogenic.

2. *Eimeria mivati* Edgar and Siebold, 1964

This species, which parasitizes primarily the duodenal loop area, is characterized by a change in infection sites downward into the ileum, rectum, and ceca as infections progress. Oocysts average only 15.6 μm in length (18.3 for *E. acervulina*), and the individual plaques bearing nests

DIFFERENTIAL CHARACTERISTICS FOR 9 SPECIES OF CHICKEN COCCIDIA

DIAGNOSTIC CHARACTERISTICS

CHARACTERISTICS	*E. acervulina*	*E. brunetti*	*E. hagani*	*E. maxima*	*E. mivati*	*E. mitis*	*E. necatrix*	*E. praecox*	*E. tenella*
ZONE PARASITIZED	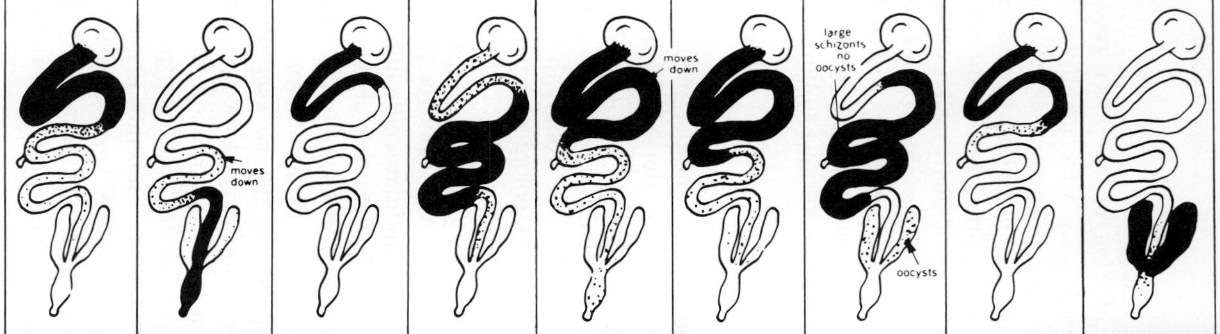	moves down			moves down		large schizonts no oocysts; oocysts		
MACROSCOPIC LESIONS	light infection: transverse, whitish bands of oocysts heavy infection: paques coalescing thickened wall	coagulation necrosis mucoid, bloody enteritis	pinhead hemorrhages petechiae	thickened walls, mucoid, blood-tinged exudate, petechiae	light infection: rounded plaques of oocysts heavy infection: thickened walls coalescing plaques	no lesions, mucoid exudate	ballooning, white spots (schizonts), petechiae, mucoid blood-filled exudate	no lesions, mucoid exudate	onset: hemorrhage into lumen later: thickening, whitish mucosa, cores clotted blood
MICROSCOPIC CHARACTERISTICS OOCYSTS REDRAWN FROM ORIGINALS			none available						
LENGTH x WIDTH	AV 18.3 x 14.6 (μ)	24.6 x 18.8	19.1 x 17.6	30.5 x 20.7	15.6 x 13.4	16.2 x 16.0	20.4 x 17.2	21.3 x 17.1	22.0 x 19.0
LENGTH WIDTH	17.7–20.2 13.7–16.3	20.7–30.3 18.1–24.2	15.8–20.9 14.3–19.5	21.5–42.5 16.5–29.8	11.1–19.9 10.5–16.2	14.3–19.6 13.0–17.0	13.2–22.7 11.3–18.3	19.8–24.7 15.7–19.8	19.5–26.0 16.5–22.8
OOCYST SHAPE AND INDEX—LENGTH / WIDTH	ovoid 1.25	ovoid 1.31	broadly ovoid 1.08	ovoid 1.47	ellipsoid to broadly ovoid 1.16	subspherical 1.01	oblong ovoid 1.19	ovoidal 1.24	ovoid 1.16
SPORULATION MINIMUM HR	17	18	18	30	12	18	18	12	18
SCHIZONT, MAX IN MICRONS	10.3	30.0		9.4	17.3	11.3	65.9	20	54.0
PARASITE LOCATION IN TISSUE SECTIONS	epithelial	2nd generation schizonts subepithelial	epithelial	gametocytes subepithelial	epithelial	epithelial	2nd generation schizonts subepithelial	epithelial	2nd generation schizonts subepithelial
PREPATENT PERIOD HR	97	120	99	123	93	138	99	84	138

of oocysts are more circular than are those characteristic of *E. acervulina.* Mortality seldom occurs, but since infection causes depression in weight gain, poor feed conversion, and reduced egg production, the species is classed as a pathogen. Although Edgar and Siebold (1964) reported an absence of reciprocal cross-immunity between *E. mivati* and the other eight recognized species of *Eimeria,* British investigators (Joyner and Long, 1974) have questioned this characteristic on the grounds that they found cross-immunity to be present in tests against *E. acervulina.* Because of their observations, Joyner and Long (1974) consider *E. mivati* to be a variant of *E. acervulina* and propose the name *E. acervulina* var. *mivati.* There remains some question as to whether or not the same or different strains are being tested in laboratories of Great Britain and the United States. Admittedly, species identification is difficult since distinguishing morphological characteristics may not be easily recognized and cross-immunity studies and length of life cycle tests may be impractical. Differences in response to anticoccidials of these two species and the proved need to include both species in inoculum prepared for planned immunization programs in the United States strengthen the case for recognition of *E. mivati* as a separate species.

3. *Eimeria mitis* Tyzzer, 1929

This species infects predominantly the anterior half of the small intestine and does not produce any gross lesions. Colonies are seldom seen in the epithelium but there is a uniform distribution of the parasite in the infected area. This coccidium was once regarded as nonpathogenic, but recent studies have shown mortality in young chicks. The original isolants no longer exist, and some attempts to reisolate this species have been unsuccessful. The validity of this species has been questioned by British investigators who suggest that it may be a subspecies of *E. acervulina.*

4. *Eimeria praecox* Johnson, 1930

Eimeria praecox infects the upper intestine and is distinguished by the short prepatent period (84 hours) between the time of inoculation and oocyst production. Although mortality does not occur, even with dosages as high as 10 million oocysts, some depression in rate of weight gain or loss of body weight has been found and the species must be considered mildly pathogenic. British investigators suggest that *E. praecox* may be a subspecies of *E. acervulina.*

Fig. 3. A diagnostic chart for nine species of fowl coccidia from Research Report 163, June 1973, College of Agriculture Experiment Stations, University of Georgia, Athens, Georgia 30602.

5. *Eimeria hagani* Levine, 1938

This coccidium, which is moderately pathogenic, was originally isolated from a chicken in New York State. Although N. D. Levine (personal communication, 1970) claims the species has disappeared from this State, Edgar has reported reisolation of this species. The characteristics of the species will remain doubtful until the endogenous stages and lesions have been more fully described. Some pinpoint hemorrhages have been observed in the mucosa following heavy experimental infections, but these lesions are not characteristic. Some British investigators now consider this parasite a subspecies of *E. acervulina*. However, Edgar has found necessary the inclusion of his isolant of *E. hagani* in inoculum used in his planned immunization program so this species would appear to be an antigenically unique organism.

6. *Eimeria necatrix* Johnson, 1930

Eimeria necatrix is one of the two important pathogenic species of *Eimeria* which invade chiefly the jejunum or midgut area in the zone near the yolk sac diverticulum. In severely infected chickens, lesions may be found in the anterior and the lower portion of the small intestine. Massive dilation or "ballooning" of the middle intestine also occurs in severely infected birds. The intestine may appear spotted, with white and dark reddish petechiae visible on the serosal surface. The whitish yellow plaquelike colonies of parasites, often invisible from the mucosal surface, are composed of groups of large second-generation schizonts. Freshly passed or clotted blood along with mucus and debris from the sloughing epithelial tissue can be found on the mucosal surface of the intestine. The oocysts of *E. necatrix* do not develop in the small intestine, but the third-generation merozoites migrate to and penetrate the epithelial cells of the ceca where they form gametes. *Eimeria necatrix* infection causes higher mortality than any other species with the exception of *E. tenella*. Mortality usually occurs when the parasites are still in the early stages of their life cycle. Mortality may be as high as 100% in experimentally infected chickens.

7. *Eimeria maxima* Tyzzer, 1929

Eimeria maxima, named for the distinguishing large oocysts, most severely parasitizes the middle portion of the intestine. Additional lesions may also be found in the duodenum and ileum. Infection causes a hemorrhagic enteritis which is associated with a thickening of the intestinal wall. There is now little question that *E. maxima* may cause severe pathological changes and mortality of up to 50% has been reported in experimentally infected chickens. Weight loss, poor feed conversion, reduced egg production, and decrease of yellow pigment in the skin are common. A single light infection usually produces protective immunity.

8. *Eimeria brunetti* Levine, 1942

This species establishes itself in the lower portion of the small and the large intestine. In severe infections the parasite may be found in the anterior portion of the small intestine and may also occur in the neck of the ceca. The mucosa may be destroyed by coagulation necrosis leaving a caseous eroded surface. In some infected birds, a complete blockage of the upper large intestine may occur. Sloughed mucosa and bloody exudate may be found inside the small intestine and sometimes soft cores extend into the ceca. In lightly infected birds, characteristic hemorrhagic streaks may occur in the mucosa. Severe weight loss may occur in infections which are overlooked by diagnosticians. This species is a severe pathogen producing heavy mortality.

9. *Eimeria tenella* (Railliet and Lucet, 1891) Fantham 1909

Eimeria tenella invades the epithelium and later the submucosa of the cecal pouches causing "bloody" or "cecal" coccidiosis. The protozoa will also parasitize adjacent areas of the digestive tract on rare occasions. Infection is characterized by bleeding and thickening of the cecal walls on the fifth and sixth day after inoculation and the later development of hardened cheesy cores. Death usually occurs on the fifth or sixth day of the infection at the time the second-generation schizonts mature. Anorexia is the first sign of infection. This species is extremely pathogenic and mortalities of 100% in laboratory and up to 20% in field infections have been reported.

Other species of coccidia have been described in the domestic fowl, although these are not of major economic importance. These include: *Eimeria tyzzeri, Isospora gallinae, Isospora gallinarum,* and *Wenyonella gallinae.*

B. Turkey

Seven species of *Eimeria* have been recognized in turkeys in the United States. They vary greatly in pathogenicity and in the site in the gut of the host in which they develop. Two additional species have been partially described by Edgar and colleagues, but until a fuller description is published they cannot be considered valid species. Two species, *Eimeria innocua* and *E. subrotunda,* are relatively nonpathogenic and parasitize the duodenum as well as sections of the intestine further down the digestive tract. *Eimeria innocua* may infect the cecal pouches. Neither species produces mortality nor causes formation of gross lesions.

1. *Eimeria meleagrimitis* Tyzzer, 1929

This species is the most pathogenic of the four which affect the duodenum and upper jejunum of turkeys. Although *E. meleagrimitis* was once considered to be nonvirulent, infection can cause mortality in as many as 90% of infected birds. The duodenum enlarges and petechial hemorrhages may be apparent in other regions of the mucosa. A reddish brown necrotic core sometimes forms in the intestine.

2. *Eimeria dispersa* Tyzzer, 1929

This species, which is but mildly pathogenic in turkeys, commonly infects bobwhite quail and has also been reported from grouse and partridges. The first sign of infection is a cream-colored change in the duodenum, followed by a progressive dilation and thickening of the entire intestine. A heavy infection can cause mild diarrhea and some depression of weight gain.

3. *Eimeria gallopavonis* Hawkins, 1952

The heaviest infection by this species is found in the large intestine although the lower portion of the small intestine and the ceca are also involved. Lesions are generally restricted to the area in the small intestine posterior to the yolk sac diverticulum. A milky white exudate consisting of oocysts and debris often fills the infected areas. This species should be classed as pathogenic since 100% mortality may follow artificial infection.

4. *Eimeria adenoeides* Moore and Brown, 1951

This species mainly infects the ceca but infection extends into the large intestine and lower small intestine. Caseous plugs are often formed in the ceca on the sixth to the eighth days after infection. *Eimeria adenoeides* is probably the most pathogenic of all the turkey coccidia. The schizonts which develop deep in the mucosa are primarily responsible for the lesions.

5. *Eimeria meleagridis* Tyzzer, 1929

This species infects the same regions of the digestive tract as *E. adenoeides*. Although cecal cores containing yellow caseous materials may form in infected turkeys, no mortality, depression in weight gain, or other economically important losses can be attributed to this species. Thus, this species may be said to produce coccidiasis and not coccidiosis.

6. *Other species*

Cryptosporidium meleagridis is a very small coccidium which occurs in the villar epithelium of the posterior third of the small intestine of turkeys. *Isospora heissini* (small intestine) has been reported from turkeys but its oocysts are similar to those of *I. lacazei* from the English sparrow, and this claim is of doubtful validity.

C. Duck

Although Pellérdy (1974) lists sixteen species of coccidia as occurring in domestic and wild ducks, most of these have been insufficiently described. Generally accepted species from the domestic duck include *Eimeria battakhi, E. schachdagica, Tyzzeria perniciosa, Wenyonella anatis, W. gagari,* and *W. philiplevinei.* Three species which parasitize the digestive tract, *E. danailovi, T. perniciosa,* and *W. philiplevinei,* have been shown to be distinctly pathogenic.

D. Goose

1. *Eimeria truncata* Railliet and Lucet, 1891

This species establishes itself in the kidneys of the domestic goose and may produce heavy mortality. The species appears to be worldwide in distribution. Infection occurs primarily in goslings although older birds may be carriers. The normal reddish brown color of the kidneys is altered to light grayish-yellow or red by infection.

2. *Eimeria anseris* Cotlan, 1933

This species generally infects the intestinal epithelium of geese which are 2–3 months old. Younger birds are not generally infected. The coccidium may produce a hemorrhagic enteritis with anorexia and diarrhea. The small intestine of infected geese becomes enlarged and filled with a reddish-brown fluid.

3. *Other species*

Three other species have been reported from the goose: *E. nocens, E. kotlani* (infecting the cloaca and causing rectal coccidiosis), and *Tyzzeria parvula.*

E. Guinea Fowl

Three species of *Eimeria* have been described as infecting the guinea fowl. These are *E. gorakhpuri, E. grenieri,* and *E. numidae.* The latter species produces a severe disease characterized by diarrhea which is frequently lethal.

IV. Parasite Physiology

A. Vitamin Requirements of Coccidia

The direct effects of vitamin deprivation on the coccidia must be separated from secondary effects brought about by vitamin deprivation of the host. Vitamin deficiencies in the host could create conditions unfavorable to the growth of the parasite, thus erroneously suggesting that the parasite required the vitamin. To avoid this problem, the role of vitamins for both the host and the coccidia together will be discussed. Warren (1968) reviewed the earlier work and gives data on the effect of nineteen different vitamins and growth factors on the severity of infection with *E. tenella* or *E. acervulina*.

There is little question that coccidial infection of the intestine, particularly by *E. acervulina,* can decrease the host's absorption of vitamin A. Vitamin A has a role in maintaining the epithelial tissues. Hosts receiving adequate vitamin A have lower morbidity and mortality from coccidial infection than hosts not receiving adequate levels. Chickens fed an adequate amount of vitamin A also recover more quickly from the disease and suffer less severe weight losses.

Vitamin K has a vital role in the production of prothrombin by the liver and deficiency leads to an increase in blood coagulation time. Thus, this vitamin is extremely important in determining the outcome of infection by those species of coccidia which produce hemorrhage (*E. tenella, E. necatrix, E. maxima,* and *E. brunetti*). Several workers have shown that elimination of vitamin K from the diet can increase mortality in chickens infected with *E. tenella* and *E. necatrix,* while the addition of vitamin K can reduce mortality. Most commercial poultry feed is now supplemented with added vitamin K.

The effects of vitamin C on the host appear to differ depending on the species of coccidium which the host is carrying. With *E. brunetti,* L-ascorbic acid seems to be beneficial. Conversely, with *E. tenella, E. necatrix, E. maxima,* and *E. acervulina,* no effect of feeding vitamin C has been shown.

Different species of coccidia require different B vitamins. There is some indication that *E. acervulina* and *E. tenella* require thiamine, riboflavin, biotin, nicotinic acid, and folic acid for normal development.

A deficiency of vitamin E restricts the growth of *E. tenella,* but not of *E. acervulina.* The mechanism by which this is brought about is unknown.

B. Cytochemical Studies

Only limited information is available on the composition of coccidia and on their absorption of nutrients. The intracellular nature of the parasites make such studies difficult. Advances in tissue culture techniques, and development of techniques for recovering intact parasites freed of host cells, will open up new avenues of investigation in the future. Ryley (1973) points out that cytochemical studies can provide useful information about the chemical nature and intracellular location of substances in coccidia, provided sufficient caution is exercised in interpreting results and if undue emphasis is not placed on stain specificity.

Pattillo and Becker (1955) detected DNA in sporozoites, schizonts, merozoites, microgametocytes, and unsporulated oocysts, but were unable to find any in macrogametocytes using the Feulgen reaction. DNA is probably present in the macrogametocyte but is difficult to demonstrate. Coccidia in tissue culture incorporate cystine, uridine, and adenosine, but not thymidine. RNA appears present in all stages of coccidia. Two common constituents of cells, protein and lipid, are also found in coccidia, although the latter appears limited to macrogametocytes, oocysts, and sporozoites.

Originally, coccidia were believed to use glycogen as their main energy storage substance. Later it was recognized that the polysaccharide differed from metazoal glycogen and such terms as "coccidien glykogen" and "parglycogen" were used. More recent studies have shown that the polysaccharide in coccidia is amylopectin, a material with a longer unit chain than glycogen. The asexual stages of *E. acervulina* and *E. necatrix* contain acid mucopolysaccharides, as do the sexual stages of *E. acervulina.* Coccidia also contain mucoprotein and glycoprotein.

Several studies have been made on the localization of enzymes within coccidial parasites. Strongly positive alkaline phosphatase activity has been described in the nuclei of all stages of *E. tenella.* Positive acid phosphatase activity has been shown in the schizont and macrogamete of *E. tenella.* Other histochemical reactions shown with *E. tenella* parasites include those for cytochrome oxidase and NADPH diaphorase.

C. Physiology of Coccidia

The intracellular nature of coccidia has made study of their physiology, metabolism, and biochemistry difficult. Metabolic studies by Smith and Herrick (1944) demonstrated that the respiratory rate increased in sporulating oocysts, and then decreased sharply after sporulation was complete. Numerous workers have shown sporulation to be an aerobic process. The respiratory rate of the sporozoites increases rapidly during the first 6 minutes of excystation, then decreases to a lower rate, presumably after cell penetration. Endogenous sources of energy for respiration are appreciable in the form of amylopectin. Under anaerobic conditions, metabolism produces predominately lactic acid with lesser amounts of CO_2 and glycerol. Exogenous sources of energy could be provided by glucose and lactose and, to a lesser extent, by fructose and mannose.

V. Cultivation

Considerable progress in cultivation has been made in the 10 years since the first avian species (*Eimeria tenella*) was grown in cell culture (Patton, 1965) and in the embryonated chicken egg (Long, 1965). Detailed reviews have been made by Doran (1973) and Long (1973b).

A. Cultivation in Avian Embryos

Numerous studies have involved the inoculation of sporozoites or second-generation merozoites into the allantoic cavity of chick embryos. Development of the parasites to oocysts which could subsequently be sporulated and used to infect chickens has been obtained with *E. brunetti, E. mitis, E. necatrix, E. praecox,* and *E. tenella.* In the case of *E. tenella,* hemorrhage can occur in the embryo and, if the dosage is large enough, death can result. Attempts to obtain development with *E. acervulina* and *E. maxima* have been unsuccessful.

Several factors influence cultivation of coccidia in chick embryos. Dosage is important as too large an inoculum of sporozoites can result in early embryo mortality. Differences in the virulence of strains has been reported. Temperature can play a role in the development of the coccidial infection in the embryo. *Eimeria tenella* appears to develop more rapidly at 41°C than at lower temperatures. The age, strain, and sex of the embryos are all factors influencing development.

B. Cultivation in Cell Culture

Cells from numerous species of animals and several different techniques have been used for the cultivation of the chicken coccidia (Doran, 1973). Development of *E. necatrix* and *E. tenella* progresses farthest when the parasites are cultured in cell lines from the chicken rather than from other animals. To date, of the avian coccidia, only *E. tenella* has been cultured through the complete life cycle from schizont to oocyst. Five species (*E. gallopavonis, E. mivati, E. praecox, E. maxima,* and *E. acervulina*) show little or no development in cell culture. Failure of these species to develop occurs when cultivation is attempted in the same host cell type as is used to cultivate *E. tenella,* suggesting that the former five species have different nutritional requirements than *E. tenella.*

Embryo and cell cultures currently have several applications beyond that of studying the life cycle of the parasites. Cell culture experiments may provide evidence to aid in species identification. This information would supplement information from cross-immunity studies in chickens. Cell cultures have been used to study the nutrient requirements of the parasites. The rate of incorporation of several nucleic acids into the nuclear material of the parasites has been measured in such systems. Perhaps one of the most rapidly expanding uses of cultivation of coccidia involves the screening for or testing of anticoccidial compounds. Although Ryley (1968) suggested that the chick embryo system was unsuitable for general screening because of the large quantity of sporozoites and oocysts needed for the test if mortality was used as a parameter, Long (1970) showed that a method based on counting the macroscopic lesions visible on the chorioallantoic membrane was satisfactory for drug screening. The cell culture system appears to present an accurate, rapid method for screening anticoccidial activity. This method requires only microgram quantities of material and is currently being used by some pharmaceutical companies for anticoccidial studies. A need exists for a method of culturing coccidia in a cell-free, chemically defined media.

VI. Host–Parasite Interactions

A. Factors Influencing Pathogenesis

Despite numerous studies, the mechanisms by which pathological changes are brought about in the host by coccidial infection are relatively unknown. Even the cause of death, which is the most severe effect observed, is subject to question. Some workers have concluded that death following a single large inoculation with *Eimeria necatrix* is due to the alarm reaction and not to any specific reaction of the parasite. Conversely, other workers have suggested that toxic products of the coccidia are responsible for many of the changes observed. With *E. tenella* some investigators are convinced that hemorrhage is the major cause of death.

Equal numbers of oocysts will not produce the same degree of damage in different birds. Several factors influence this response including (1) the condition and characteristics of the host (age, breed, nutritional condition, and physiological status), and (2) the state and nature of the parasite (the particular species involved, the age of the oocysts, the tissue site parasitized, and the nature and mode of inoculation).

The age of the host plays a major role in determining the severity of the infection. Although several earlier workers reported that resistance increased with age, many of these earlier experiments were clouded by immunity which developed from subclinical infections in birds which were not raised parasite-free.

Recent studies (reviewed by Long, 1973a) have indicated that susceptibility to infection increases with age. Oocyst yield following *E. acervulina* infection increases with the age of the host. Several studies also showed that the host susceptibility to coccidial infection may be linked in some way with genetic resistance.

The physiological and nutritional state of the host is important in determining the severity of the coccidial infection. A relationship be-

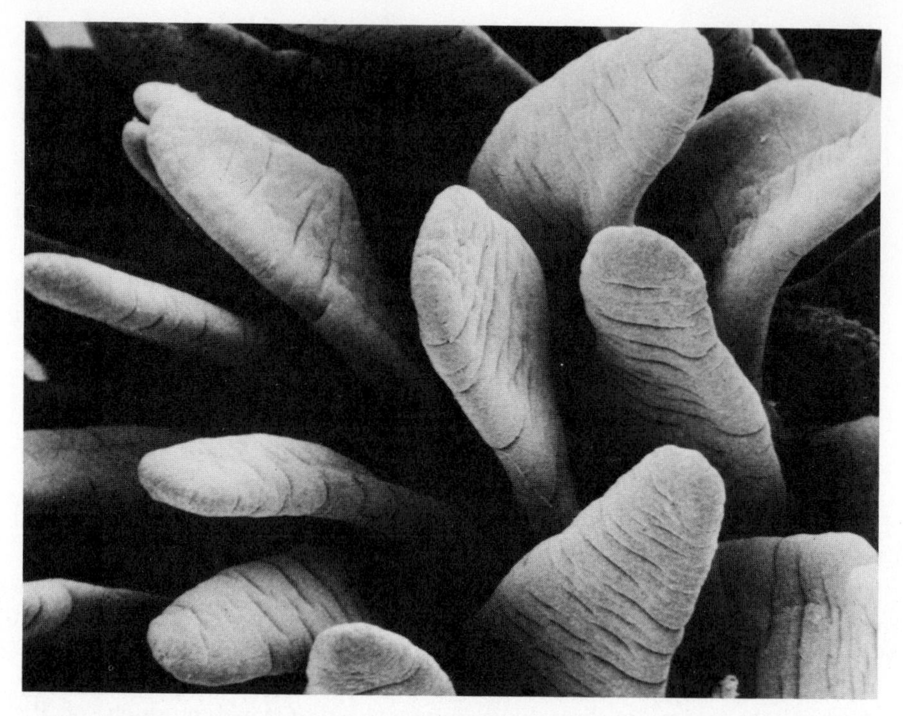

Fig. 4. The epithelial surface of the duodenum of an uninfected control chicken. (Courtesy of D. R. Witlock, University of Georgia, Athens, Georgia.)

tween dietary protein levels and severity of coccidiosis has been shown. High levels of dietary protein result in severe infections, probably because the high trypsin activity in the intestine caused by the protein diet results in efficient excystation. Factors such as low methionine level and the presence of aflatoxin in the diet can increase the severity of a coccidial infection. Seasonal differences in infection, with oocyst output being greater in winter, have also been reported to occur.

The species of coccidia involved in the infection undoubtedly plays a major role in the pathogenicity observed. Species such as *E. necatrix* or *E. tenella,* which have large stages in the life cycle, may produce extensive tissue damage from infection with only a few parasites. The site preference of the coccidia also seems to play a role since interference with nutrient absorption appears to be more severe by those coccidia which parasitize the upper intestinal tract than by those that parasitize the lower intestinal tract or cecum. Quite obvious differences in pathogenicity exist between those species which do and those which do not produce hemorrhage. There is no clear-cut explanation why 100,000 oocysts of *E. tenella* will frequently produce a severe weight loss and sometimes death, while a similar dose of *E. acervulina* may be relatively

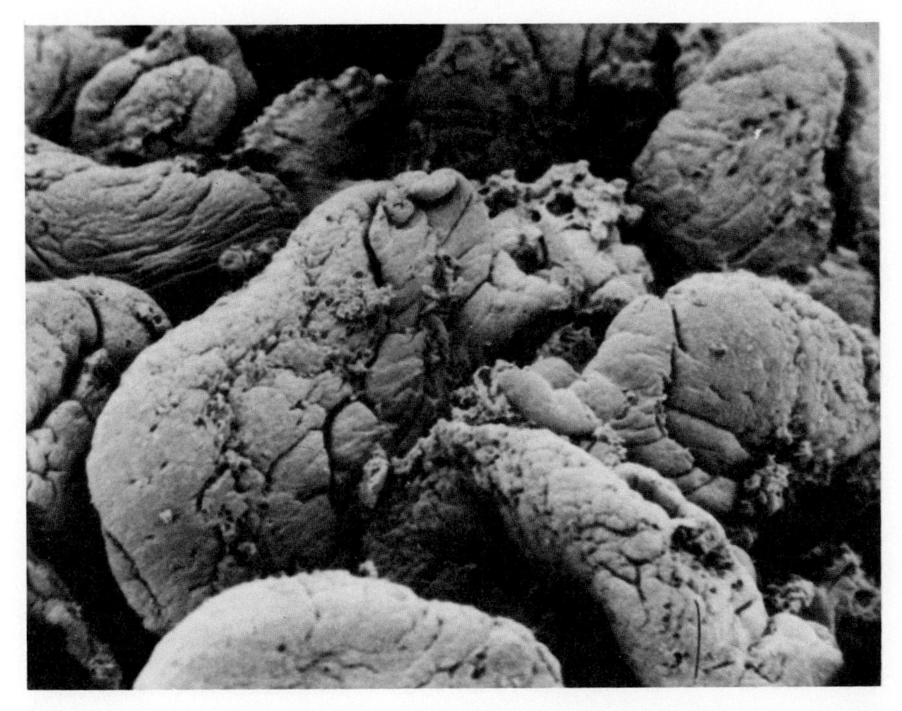

Fig. 5. The epithelial surface of the duodenum of a chicken infected with *Eimeria acervulina,* 7 days postinoculation with 1,000,000 sporulated oocysts. (Courtesy of D. R. Witlock, University of Georgia, Athens, Georgia.)

innocuous. Almost without exception, workers have reported that severity of the infection by a given coccidium (as measured by weight gain, feed conversion, lesion score, and mortality) increases as the number of oocysts given increases. This response, however, is not linear and large increases in the inoculation dose may produce a disproportionate response in the above parameters. The subsequent oocyst output from the infected chicken is dependent upon the oocyst number given. Oocyst output per oocyst given increases, depending upon the particular species, until a plateau is reached, after which the oocyst return per oocyst given decreases. This latter phenomenon is known as the "crowding effect" and may be related to the failure of merozoites to find a suitable unoccupied epithelial cell in heavy infections.

The viability of the oocysts is also important in determining a dose response. Storage of oocysts results in a progressive loss of viability which may be related to death of the sporozoites after the depletion of energy stores. Temperature, moisture, and oxygen levels play a definite role in the survival of oocysts. Sporulation is essential for infectivity.

The relationship between the coccidia, the bacterial flora of the host,

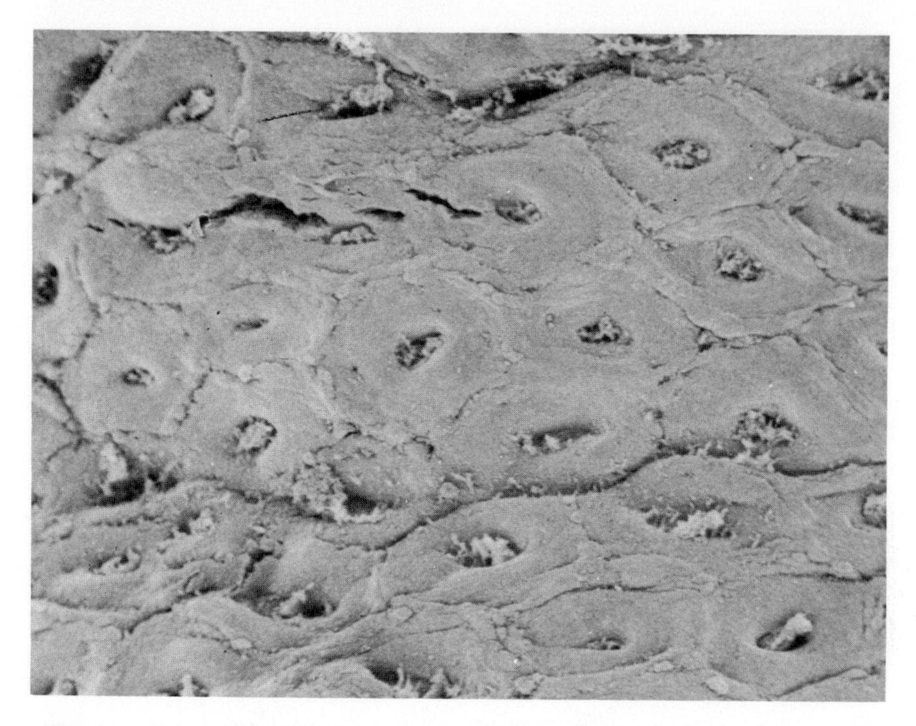

Fig. 6. The epithelial surface of the midcecal pouch of an uninfected control chicken.

and the subsequent severity of the infection is not clear. No differences could be shown in the weight gains or lesion scores between conventional and bacteria-free birds infected with *E. brunetti* or *E. necatrix*. Conversely, *E. tenella* infections were not as severe in bacteria-free chickens as in conventional birds. In the absence of a bacterial flora, no decrease in pH of the intestinal tract occurs in chickens with *E. brunetti* infection, yet a decrease in pH occurs in the duodenum of *E. acervulina*-infected chicks. Changes in the bacterial flora have been shown in the ceca of conventional birds infected with *E. tenella, E. brunetti, Eimeria brunetti* and *E. necatrix* have also been implicated along with certain species of bacteria in the production of ulcerative enteritis (a necrosis of unknown etiology in the lower part of the digestive tract) in chickens.

B. Histological Changes

The infection of the cells of the intestinal mucosa with coccidia brings about dramatic changes in the structure and appearance of the villi. These changes are especially apparent in the scanning electron microscope images of *E. acervulina*-infected duodenum (Figs. 4 and 5) and *E. tenella*-infected ceca (Figs. 6 and 7). Damage to the intestinal mucosa

Fig. 7. The epithelial surface of the midcecal pouch of a chicken infected with *Eimeria tenella,* 7 days postinoculation with 100,000 sporulated oocysts. (Courtesy of D. R. Witlock, University of Georgia, Athens, Georgia.)

can be shown by the low ratio of villous height to the total mucosal thickness in infected birds. The ratio of villous height to total mucosal thickness is lowest when maximum cell damage is present (Pout, 1967). The change in mucosal morphology may be influenced by changes in the cellular renewal rate of the intestinal epithelial cells in infected chickens.

In addition to the sloughing of the epithelial cells of the mucosa, there are also changes in the histology of the remaining tissue which progress as the coccidial infection develops (Long, 1973a). During development of the first-generation schizonts there is a progressive increase in the numbers of heterophil polynuclear leukocytes present in the submucosa surrounding the glandular tissue. Later there is an increase in numbers of granulocytic leukocytes and pyroninophilic cells in the submucosa and lamina propria. Infection with species such as *E. tenella* and *E. necatrix,* which develop in the lamina propria below the epithelial cells, cause lymphocyte and leukocyte infiltration accompanied by a thickening of the intestinal wall. Morphological changes in infected epithelial cells are also evident at the ultrastructural level. These changes include development of an enlarged and abnormal appearance of the principle cell mitochondrion in epithelial cells of chickens with *E. acervulina* infections.

C. Physiological Changes

Infection with coccidia produces numerous physiological changes in the host, including changes in almost all systems of the body. This is not surprising in view of the extensive intestinal damage during infection and the decrease in substrate absorption which results. A list of some of the major dietary nutrients which are malabsorbed by chickens with coccidial infection is shown in Table II. Infection in the intestine, particularly in the upper and midportion, affects absorption of nutrients, while infection in the ceca with *E. tenella* does not. In a series of papers, Turk and Stephens have suggested that the effect of the particular species of parasite on absorption is dependent on the area of the intestine infected and the area in which major absorption of the substrate occurs (Turk, 1974). The kinetics of methionine absorption shows that both diffusion and the mediated or selective portion of the transport system are decreased by infection. This appears to be associated with the damage done to the intestinal mucosa by the parasites and may be related to changes in the surface area of the mucosa of the villi.

One of the most striking effects of infection is the decrease in xanthophyll levels of infected chickens. The oxycarotenoid pigments, particularly xanthophyll, are responsible for the yellow color seen in the plasma, shanks, and skin of chickens and the yolk of eggs. In chickens with coccidial infections the level of plasma carotenoids is low and the skin loses its pigmentation (Yvoré and Mainguy, 1973). Infection with any of the six major species of chicken coccidia will reduce levels of carotenoids in the bloodstream of the host. The mechanism by which this decrease is brought about differs according to the species involved. With

Table II
Effect of Coccidial Infection on Intestinal Absorption by the Intestine of Chickens [a]

Substrate	*E. acervulina*	*E. necatrix*	*E. brunetti*	*E. tenella*
Zinc	+	+	−	−
Oleic acid	+	+	−	−
Methionine	+	+	ND	−
Histidine	+	ND	ND	ND
Calcium	+	+	+	−
Glucose	+	+	+	ND
Xanthophyll	+	ND	ND	−

[a] +, Decreased absorption; −, no effect; ND, not done.

E. acervulina in the upper intestine, interference of absorption of xanthophyll from the intestinal lumen causes the decrease; while with *E. tenella* in the ceca, the decreased carotenoid levels are a reflection of xanthophyll leakage through the damaged cecal wall. Other mechanisms which may contribute to this reduction in carotenoid levels include changes in the rate of loss of carotenoid with the feces and eggs, and interference with transfer substances in the bloodstream.

The marked decrease in intestinal pH caused by *E. acervulina* infection may be responsible for poor xanthophyll absorption (Kouwenhoven and van der Horst, 1969). The effect of infection on intestinal pH by all six major species of chicken coccidia has been studied (Ruff and Reid, 1975). The acidity level in the intestine varied with the particular species of parasite involved. The greatest and most consistent decrease in pH occurred in the region of the intestine where the particular species localized. Infection with *E. tenella* did not affect the intestinal pH. Therefore, at least with this species, a pH increase in the intestine does not contribute to the decrease seen in blood carotenoid levels.

Some workers have suggested that the decrease in nutrient absorption in the intestine may be related to the speed with which ingested feed passes through the digestive tract in infected birds. The overall passage time of food is increased with infection, but this is predominantly due to the retention of food in the crop. Upon entering the small intestine, food advances at the same rate in infected and uninfected birds. Chickens with cecal coccidiosis may, however, empty their ceca infrequently.

Extensive changes in renal function during the course to *E. tenella* infection have been reported. A decrease in renal clearance was noted during the first 3 days of infection followed by a rise in uric acid excretion during the fourth and fifth days. In addition, infection damaged the distal convoluted tubules, and caused dilation of Bowman's capsules. During infection there was an increase of cellular components of the glomeruli.

Coccidiosis can also influence the metabolic processes of the host. The respiratory quotient of muscle of chickens infected with *E. tenella* was less than that of uninoculated controls. Less glucose and ribonucleotides are present in the liver of *E. necatrix*-infected chickens than in livers of uninoculated controls. Coccidial infection decreases the ability of chickens to do muscular work. The rate of corticosterone secretion in chickens infected with *E. tenella* increased during the hemorrhagic phase of the infection. This increase was associated with the secretion of adrenal cholesterol and with an elevation of ascorbic acid and corticosterone levels of the adrenal tissue. The increase in adrenal ascorbic acid may result from an increased uptake from the circulation.

Coccidial infection brings about changes in the blood of the host. Numerous workers have reported a decrease in packed cell volume caused by the hemorrhage which occurs during infection. A hyperglycemia was reported in *E. tenella*-infected chickens. The mechanism for the hyperglycemia has not been determined. Increases in other blood constituents including chloride, sodium, protein, and potassium have been reported. Changes in the amino acid composition of muscle tissue also occur.

Eimeria tenella infection appears to cause a change in the blood clotting mechanism of chickens. Vitamin K is integrally involved in the blood clotting system and blood clotting may be affected by the change in the vitamin K requirement which occurs in the coccidia-infected birds. Prolonged blood clotting time may be responsible, at least in part, for the hemorrhage which occurs in infected birds. While the blood volume and percentage corpuscular volume change during infection, the plasma volume does not change even during the hemorrhagic phase of the disease.

D. Physical Changes

One of the most striking effects of a moderate to severe infection with avian coccidia is a decrease in the rate of weight gain, but decreases may be difficult to detect in lightly infected birds. Conversely, with heavy infections, an actual weight loss may be observed during and after the peak of the infection. This decrease in weight is proportional to the severity of the infection, although the relationship is not linear.

The use of "pair-fed" birds has shown that a decrease in feed consumption accounts for some but not all of the depression in weight gain. In such experiments uninoculated groups of birds are either pair-fed or restricted to the amount of feed consumed in the previous 24 hours by a corresponding group of inoculated birds. The weight depression in most of the pair-fed birds is statistically the same as that in the inoculated birds. Since 1 kg of feed consumed does not result in 1 kg of weight on the bird, a weight reduction equal to the reduction in feed consumption indicates the operation of many factors. Energy utilization from feed is less in infected birds and this may play a role in the decreased rate of weight gain. In addition, water intake decreases and water loss due to diarrhea increases in infected birds and this may also account for a portion of the weight loss.

Coupled with the failure to gain weight is a decrease in the chicken's ability to utilize the feed which is taken in as is shown by the increase in the feed conversion ratio (pounds of feed consumed/pound of gain) observed in infected birds. These two effects (failure to gain weight and increased feed conversion ratio) are of extreme economic importance to the poultry producers.

Although documentation is sparse, there are several reports showing that coccidiosis can decrease both the number and quality of eggs produced by laying hens. In chickens with heavy infections by such species as *E. acervulina, E. maxima,* and *E. tenella,* a complete cessation of egg production may occur. The thickness of the egg shell may decrease during the interval from initial infection until laying ceases, possibly as a result of malabsorption of calcium.

E. Immune Reactions

Birds show both innate and acquired immunity to coccidia, although the latter is unquestionably the most studied. Reviews of immunity have been made by Horton-Smith *et al.* (1963) and Rose (1973a,b). Some innate immunity, related to host specificity, age of the host, and breed of the bird has been shown. Certain nonspecific mechanisms of immunity, such as those mediated by interferon, act against coccidia.

Coccidia of the various species differ in their relative ability to stimulate immunity in the host. *Eimeria maxima* may stimulate complete immunity to subsequent challenge after a single immunizing dose. Other species, such as *E. acervulina,* may fail to stimulate complete immunity even after a series of inoculations. When tested by a standard procedure the species *E. maxima, E. tenella, E. acervulina,* and *E. necatrix* follow a pattern of descending immunogenicity.

A number of factors influence the immunological response following a coccidial infection. With some species of *Eimeria,* the immunizing dose is quite important and several studies have shown that satisfactory immunization can be obtained with small or moderate numbers of oocysts. Age of the host seems to play some role as older animals respond better than younger ones. The simultaneous association of other diseases may block or hinder development of the immunity resulting from a coccidial infection. Prior infection with Marek's disease results in increased oocyst production in the primary and in subsequent coccidial infections. Likewise, the presence of aflatoxin in the feed has been shown to interfere with development of immunity to *E. tenella.* The duration of immunity may vary from a few days to many weeks. Studies on the actual duration of immunity must be carried out with extreme care to avoid stimulation of immunity by accidental infection during the period of observation. The duration of immunity is dependent upon the method of immunization and the age of the host.

The life cycle stage which stimulates the immune response cannot be determined by oocyst inoculation since all stages subsequently develop.

Second-generation merozoites of *E. tenella* and *E. necatrix* recovered from infected birds and inoculated intrarectally into chickens did not produce immunity as there was no reduction in the numbers of oocysts produced after a later challenge with oocysts.

The stage in coccidial development affected by the immune response is not known. It is also not clear which stage is affected by the immune response. In partially immune hosts there is a reduction in the number of oocysts passed. However, these oocysts appear to be perfectly normal although the patent period is short. In hosts which are completely immune, some workers postulate that entry of the epithelial cells may be partially blocked or that those sporozoites which do penetrate do not undergo further development. Rose (1973a) points out that as the parasite's development proceeds in a challenged partially immunized host, the infection acts as a boosting antigenic stimulus. This results in an immune response which affects the terminal stages of the parasite's life cycle, thus shortening the patent period of the challenge infection.

Numerous serum antibodies have been demonstrated in chickens with coccidial infections. The techniques used to demonstrate antibody are complement fixation, precipitation, lytic agglutination, and indirect fluorescent antibody staining. Humeral antibodies can be demonstrated by precipitation or complement fixation 3–7 days after oocyst inoculation. The antibodies may immobilize the invasive stages of the parasite or induce slight changes in the morphology of the parasite. Some of the antibodies are directed against surface antigens of the parasite while others may be directed against enzymes or biochemical structures within the coccidia. The actual coccidial antigens eliciting these responses have not been fully identified, although both particulate and soluble antigens have been identified. Suspensions of free invasive stages such as sporozoites or merozoites have been used as antigens in some tests. Soluble antigens obtained by extracting parasites or infected tissues have been used in precipitation and in complement fixation tests. The suggestion has been made that the parasites may possess antigens with determinate groups antigenically similar to those of the host tissue. Rabbits immunized with a highly purified preparation of *E. tenella* sporozoites form antibodies which precipitate with normal fowl serum. Thus, some type of host "mimicry" may be found in coccidia similar to that of the helminth parasites. There are some indications that certain of the coccidial infections may trigger the development of autoantibodies although their role in causing disease symptoms is unclear. Antibodies which are produced and localized in the alimentary tract (copro-antibodies) may play a fairly important role in the immunological defense against coccidial infection. A strong reaction in immune birds results in a severe sloughing of epithelial tissue which occurs within a few hours of coccidial reinfection.

Both passive and maternal transfer of immunity to coccidia has been demonstrated, however, correct timing of the serum transfer is necessary if protection is to be observed. Inhibition of subsequent oocyst production is directly proportional to the volume of immune serum transferred and inversely proportional to the size of the challenge dose of oocysts.

VII. Epizootiology

Coccidia have a worldwide distribution being found wherever their hosts are. The six most economically important species from chickens have been reported from Europe, England, India, South America, and Australia. Coccidiosis is a "man-made" problem to the degree that the frequency and severity of infections increases as birds are raised in larger numbers and in more confined conditions. In the southern United States, 19% of the birds submitted to diagnostic laboratories were found to harbor coccidia.

VIII. Diagnosis

Scrapings from the mucosal surface of the intestine of a bird may reveal oocysts or endogenous stages of the parasite, but such a demonstration does not automatically indicate a flock diagnosis of coccidiosis. Mild infections are so common that considerable study may be required to distinguish between a serious coccidiosis outbreak and relatively harmless coccidiasis. The unwary diagnostician may be misled about the condition of the entire flock if he bases his diagnosis on a single cull bird with coccidial infection brought to the diagnostic laboratory. The skilled diagnostician will make a postmortem examination of several representative birds and will require a personal visit or serviceman's report of the condition and stage of disease of the entire flock. A differential diagnosis must consider other disease conditions which may produce cecal cores or bloody droppings such as blackhead. A definitive diagnosis of coccidial infection must include the demonstration of coccidian life cycle stages under the microscope.

At least a presumptive species diagnosis is usually desirable. Such identification is especially important in evaluation of coccidial infections

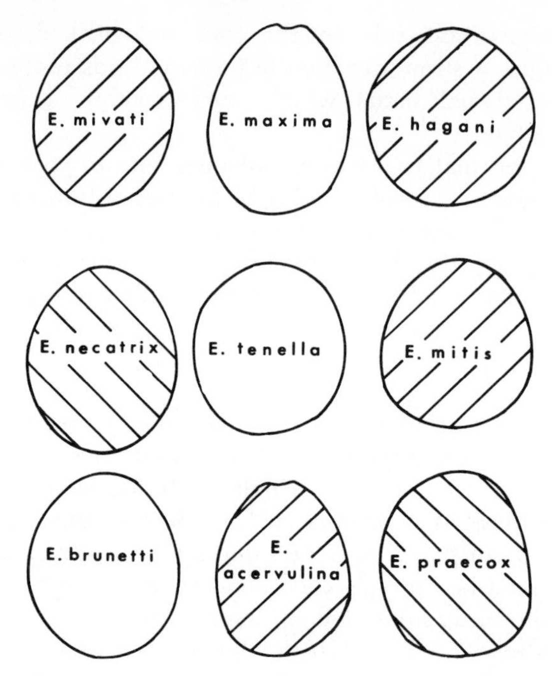

Fig. 8. The relative sizes of oocysts of nine species of *Eimeria* from chickens. Sizes shown are the minimum (no cross hatching), mean (\\\ ), or maximum (///) of the normal range for each species.

in turkeys. *Eimeria innocua, E. meleagridis,* and *E. subrotunda* may be considered practically nonpathogenic and if they are the only coccidia present, some other pathogenic agent should be considered. There is variability in degree of pathogenicity of the different coccidia of chickens so species identification aids prognosis. Identification is more readily accomplished by use of a chart than by use of the common dichotomous keys (Fig. 3). The chart will provide information on the zone of the intestine infected, appearance of macroscopic lesions, and other distinctive characteristics such as schizont size for each species of coccidia and thus may assist in identification of the species. Among chicken coccidia only *E. maxima* can be identified from the oocysts alone based on the irregular surface and distinctive color. Some information useful in identification can be obtained by determining length, width, and ranges in size of oocysts using an ocular micrometer. Overlapping size ranges (Fig. 8) may mislead even a careful technician who fails to make accurate measurements of at least ten oocysts. Other confirmatory tests to determine the prepatent period, minimum sporulation time, and cross-immunity tests may be required for positive species identification.

IX. Control of Coccidiosis

A. Management

Before the discovery of useful chemotherapeutic agents, poultry producers were forced to attempt to control coccidia by various management programs. The least successful of such programs were attempts at sanitation through disinfection, or by mechanical removal of litter. Despite these sanitation efforts, sufficient oocysts usually remained to initiate reinfection and build-up of the oocyst population to dangerous numbers. Since oocysts are unusually resistant, disinfection is usually impractical. If disinfection is needed in parasitology laboratories involved in working with coccidia, cresylic acid, ammonia, methyl bromide gas, or heat (30 minutes at 70°C) may be useful. Of older management measures, only the avoidance of wet spots from leaky roofs or waterers seems worth retaining.

A change from litter-covered floors to wire-floored pens greatly reduces exposure to coccidia. Although wire-flooded pens are used successfully with cage layers, problems including installation expense, production of breast blisters, difficulty in removal of droppings and dead birds, and difficulty in housefly control have prevented widespread adoption of this management system with broilers.

B. Immunization

Many poultry raisers are unaware of the protective flock immunity which often readily develops under field conditions. This immunity, which is brought about by the low level cycling of a coccidial infection produces an immunity of sufficient level to forestall active coccidiosis outbreaks in the majority of flocks. Some raisers of replacement pullets rely heavily on development of immunity after the introduction of young birds onto old litter.

A program of planned immunization has been developed by Edgar and his colleagues (Edgar, 1958). Under this program a mixture of a small number of oocysts of each of eight species of coccidia is given to chicks in the feed or drinking water during the first week of life. The coccidia become established in a "light" infection and subsequently produce daughter oocysts. These oocysts are, in turn, ingested by the birds and immunity develops following two or more generations of cycling. This method is not used for broilers which can more easily be protected for 7 to 8 weeks by feeding an anticoccidal drug. During immunization by this procedure, control of the oocyst exposure of individual birds is difficult and some birds may develop severe coccidiosis while others may remain susceptible. Severe coccidial breaks requiring treatment have occurred during immunization when oocyst numbers in the litter reached high levels. Since this procedure produces mixed infections, sometimes adequate immunity does not develop to all species given. Failure to immunize against *E. necatrix* is most common. This program is successful only if unusual care is taken in storing and administering oocysts and in maintaining moderate levels of litter moisture.

C. Treatment

Since the discovery that sulfanilamide has anticoccidial activity against six species of coccidia, the search for additional compounds for treatment of coccidiosis has continued. The initiation of treatment of the flock usually follows diagnosis of the disease in some birds and, therefore, medication is frequently started too late to be effective (Fig. 1 on life cycle). Individual birds showing coccidial symptoms will probably not respond to treatment. However, some benefit to other birds in the flock may be obtained.

Many different anticoccidials have been approved for treatment (Table III). Initiation of flock treatment should begin as early as possible after recognition of the disease. Numerous precautions and restrictions are placed on the use of anticoccidials for treatment of coccidiosis and the manufacturer's directions should be followed implicitly. The best method of administering anticoccidial drugs has been the subject of some debate. Initially, drugs for treatment were administered in the water on the assumption that water consumption during infection remained unchanged, while birds infected with coccidia show decreased feed consumption. However, several studies have shown that there is little difference in drug intake regardless of whether it is administered in feed or in water (Reid and Pitois, 1965). Medication in water may be the more convenient method of administration if bulk feeders are in use.

D. Preventive Medication

The introduction of the prophylactic medication in feed made possible the rapid increase in commercial broiler production in the United States. Since the advent of the sulfonamide drugs as the first group of compounds used in the prophylactic preventative medication against coccidia, over forty compounds have been marketed, although, currently, only approximately ten of these still have any widespread usage. ·

The selection of drugs for the prevention of coccidiosis has not been without problems. Edgar (1970) and Reid (1972) list 22 desirable characteristics of potential new anticoccidials. These characteristics mainly concern the mixing properties of the drug in the feed and the lack of toxic effects in the chicken. Other problems associated with preventive

Table III

Anticoccidial Drugs for Treatment of Coccidiosis of Chickens in the United States

Chemical name	Trade name	Manufacturer	Feed or water	Use level (%)	Treatment Duration	First commercial use
Roxarsone	3-Nitro	Salsbury Laboratories	Water	0.0076	Up to 10 Days	1945
Sulfamethazine	Sulmet	American Cyanamid	Water	1	2 Days	1947
				5	4 Days	
4,4′-Diaminodiphenylsulfone + N'-phenylsulfanilamide	Sulfa (Veterinary)	Salsbury Laboratories	Feed	0.1	2–3 Days on, 3 days off	1948
			Water	0.04	2 Days on, 3 off, 2 on	
Sulfamerazine	Numerous	American Cyanamid	Feed	0.4–0.5	2 Days	1948
				0.2–0.25	4 Days	
Nitrofurazone	nfz	Hess and Clark	Feed	0.11	5 Days	1948
	nfz soluble		Water	0.0082	5 Days	1956
2,4-Diamino-5-[p-chlorophenyl]-6-ethylpyrimidine + sulfaquinoxaline	Whitsyn-S	Whitmoyer Laboratories	Water	0.0015 Pyrimidine + 0.005 sulfaquinoxaline	2–3 Days on, 3 days off 2 days on	1957
Amprolium	Amprol	Merck	Water	0.012–0.024	3–5 Days	1960
				0.006	1–2 Weeks	
Sodium sulfachloropyrazine monohydrate	Esb_3	Gland-o-Lac Co.	Water	0.03	3 Days	1967
Sulfadimethoxine	Agribon	Hoffman-La Roche	Water	0.05	6 Days	1968

Table IV

Anticoccidials Approved for Use in Poultry Feeds in the United States [a]

Chemical name	Use level (ppm)	Trade name	Manufacturer	First commercial use	Premarketing withdrawal (days)
Sulfaquinoxaline	125	SQ	Merck	1948	10
Nitrofurazone	56	nfz	Hess & Clark	1948	5
		Amifur	Norwich		
Nicarbazin	125	Nicarb	Merck	1955	4
Furazolidone	55	nf-180	Hess & Clark	1957	5
		Furox	Norwich		
Amprolium	125	Amprol	Merck	1960	0
Chlorotetracycline	220	Aureomycin	Cyanamid	1960	0
Zoalene	125	Zoamix	Dow	1960	0
Nihydrazone	110	Zonifur	Norwich	1963	5
Aklomide	250	Novastat	Salsbury	1965	5
Buquinolate	82.5	Bonaid	Norwich	1967	0
Clopidol	125	Coyden	Dow	1968	0
Decoquinate	30	Deccox	May & Baker	1970	0
Monensin	100–121	Coban	Lilly	1971	3
Robenidine	33	Robenz	Cyanamid	1973	5
Lasalocid	75	Avatec	Hoffman-La Roche	1976	5

[a] From Feed Additive Compendium, May (1975).

medication include: anticoccidial contamination of layer flock feeds, inclusion of excessive anticoccidial drug in the feed causing toxicity in the birds, and omission of the anticoccidial from the feed resulting in coccidiosis breaks.

Drugs currently approved by the United States Food and Drug Administration (FDA) singly or in combination for inclusion in poultry feed for the control of coccidiosis are shown in Table IV. Several new drugs, which are not listed, are currently awaiting clearance by FDA. Strict adherence to the manufacturer's recommendations for use levels is essential. With certain of the anticoccidials the drug must be removed from the feed for a specified number of days prior to marketing to ensure that tissue residue levels do not exceed approved levels. Compliance with these withdrawal regulations is necessary since, if tolerance levels are exceeded, the entire flock may be condemned. Several combinations of anticoccidial drugs have also been approved (Table V).

Table V
Combinations of Drugs Approved for Use in Feed for Coccidiosis Control in Chickens in the United States [a]

Trade name	Chemical names of ingredients	Use level (ppm)	Manufacturer	Premarketing withdrawal
Aklomix-3	Aklomide	250	Salsbury	5
	Roxarsone	25–50		
Novastat	Aklomide	250	Salsbury	5
	Sulfanitran	200		
Novastat-3	Aklomide	250	Salsbury	5
	Sulfanitran	200	Salsbury	
	Roxarsone	25–50		
Amprol Plus	Amprolium	125	Merck	0
	Ethopabate	4		
Amprol Plus with 3-Nitro	Amprolium	125	Merck	5
	Ethopabate	4		
	Roxarsone	25–50		
Amprol Hi-E	Amprolium	125	Merck	0
	Ethopabate	40		
Polystat-3	Butynorate	200	Salsbury	5
	Sulfanitran	300		
	Dinsed	200		
	Roxarsone	25–50		
Unistat-2	Nitromide	250	Salsbury	5
	Sulfanitran	300		
Unistat-3	Nitromide	250	Salsbury	5
	Sulfanitran	300		
	Roxarsone	50		

[a] From Feed Additive Compendium, May (1975).

E. Mode of Action

Some confusion has arisen about the terminology used to describe anticoccidial drugs. Coccidiocidal drugs are ones which have the ability to kill the developing stages of the parasite, while coccidiostatic drugs only have the ability to retard or suspend development of the parasite for a period of time without killing the parasite. However, many compounds are coccidiocidal in varying degrees. Efficacy or field usefulness does not correlate with this mode of action. Care must be taken during the withdrawal period if birds have been heavily exposed to a coccidiostat, for delayed stages of the parasite may resume development and coccidial outbreaks can occur several days after drug withdrawal. Danger of such an outbreak has sometimes been exaggerated by competitive advertising.

The specific mode of action against the parasite is known for a limited number of anticoccidials. The three classes, sulfonamides, ethopabate, and pyrimidines, are known to block synthesis of tetrahydrofolic acid which is an essential vitamin for the development of the parasite. Ethopabate is a PABA (*p*-aminobenzoic acid) analog which competes with PABA and blocks dihydrofolate synthesis, although it does so at a later stage than sulfonamides. A number of the pyrimidines act as dihydrofolate reductase inhibitors. This enzyme converts folic acid into dihydrofolate which, in turn, becomes the active tetrahydrofolate vitamin.

Amprolium is structurally related to thiamine and acts by a reversible thiamine inhibition in the parasite. The 4-hydroxyquinolines are reported to inhibit mitochondrial respiration of coccidia (Wang, 1975). Of the newer anticoccidials only the mode of action of monensin is partially known. This compound affects ion transport and distribution in the cells, especially the potassium ion.

F. Drug Resistance

In 1954, 6 years after the introduction of sulfaquinoxaline as an anticoccidial in 1948, a strain of *E. tenella* was found which was resistant to this compound as well as to sulfamethazine. This was the first report of drug resistance which has been of increasing concern to those who use anticoccidials for preventive medication. Since 1954, numerous studies have demonstrated drug resistance to most, if not all, anticoccidials marketed. In recent work, Jeffers (1974a,b,c) has surveyed over 1000 litter samples collected from all major broiler producing regions in the United States. He reported finding strains of *E. acervulina, E. tenella*, and *E. maxima* which showed resistance to buquinolate, decoquinate, and clopidol. He also found nicarbazin-resistant *E. acervulina* and reported, for the first time, monensin- and robenidine-resistant strains of *E. maxima*.

The number of generations required for the appearance of drug resistance following exposure of a coccidial population to a specific medication varies depending upon the medication used. In some cases this may be a matter of only a few generations while with other anticoccidials resistance may not appear for several years.

One system which has been developed to deal with drug resistance is the "shuttle" program in which birds are shifted from one anticoccidial to another. As many as three different anticoccidials have been used during an 8-week period with a single flock of birds. The shuttle program may prevent development of resistant strains or it may merely delay their appearance. In selecting compounds for a shuttle program, care should be taken, however, to assure that the two anticoccidials selected for use are compatible in terms of their coccidiostatic and coccidiocidal actions, and in their requirement for withdrawal prior to slaughter.

The speed with which resistant strains emerge in the field may be roughly classified as follows: (1) glycomide, very rapid; (2) buquinolate and decoquinate, rapid; (3) clopidol, less rapid; (4) sulfonamides, nitrofurans, and robenidine, moderate; (5) amprolium, zoalene, and nitromide, slow; (6) nicarbazin and monensin, very slow (Reid, 1975). Although cross-resistance between drugs appears to be the exception rather than the rule, among exceptions reported are cross resistance between zoalene and nitrofurazone.

The intensive screening programs of the pharmaceutical companies have continued to provide new anticoccidials as the older ones become ineffective. The appearance of a few resistant strains does not necessarily nullify the usefulness of an otherwise successful anticoccidial. A controlled floor-pen comparison of the feed efficiency and performance of a number of field flocks by the feed manufacturer is still the best index of drug usefulness.

Although drug resistance continues to occupy the attention of drug users, continued exploration for new drugs by the pharmaceutical industry has kept pace by making replacement anticoccidials available. The poultry industry has been exploring methods of management other than feeding anticoccidials such as using wire or plastic floors with limited success. Layers and breeders maintained on wire do not suffer much from coccidiosis but to successfully raise most broiler stock on litter necessitates continuous use of large quantities of drug for feed medication.

REFERENCES

Davies, S. F. M., Joyner, L. P., and Kendall, S. B. (1963). "Coccidiosis." Oliver & Boyd, Edinburgh.

Doran, D. J. (1966). The migration of *Eimeria acervulina* sporozoites to the duodenal glands of Lieberkühn. *J. Protozool.* 13, 27–33.

Doran, D. J. (1973). Cultivation of coccidia in avian embryos and cell culture. *In* "The Coccidia" (D. M. Hammond and P. L. Long, eds.), pp. 183–252. Univ. Park Press, Baltimore, Maryland.

Edgar, S. A. (1958). Coccidiosis of chickens and turkeys and control by immunization. *World's Poult. Congr., Proc., 11th, 1958* pp. 1–19.

Edgar, S. A. (1970). Coccidiosis: Evaluations of coccidiostats under field conditions; statement of problem. *Exp. Parasitol.* **28**, 90–94.

Edgar, S. A. and C. T. Seibold (1964). A new coccidium of chickens, *Eimeria mivati* sp. n. (Protozoa: Eimeriidas) with details of its life history. *J. Parasitol.* **50**, 193–204.

Feed Additive Compendium (1975). May. Miller Publ. Co., Minneapolis.

Hammond, D. M., and Long, P. L., eds. (1973). "The Coccidia." Univ. Park Press, Baltimore, Maryland.

Horton-Smith, C., Long, P. L., Pierce, A. E., and Rose, M. E. (1963). Immunity to coccidia in domestic animals. *In* "Immunity to Protozoa" (P. C. C. Garnham, A. E. Pierce, and I. Roitt, eds.), pp. 273–295. Blackwell, Oxford.

Jeffers, T. K. (1974a). *Eimeria tenella:* Incidence, distribution, and anticoccidial drug resistance of isolants in major broiler-producing areas. *Avian Dis.* **18**, 74–84.

Jeffers, T. K. (1974b). Anticoccidial drug resistance: Differences between *Eimeria acervulina* and *E. tenella* strains within broiler houses. *Poult. Sci.* **53**, 1009–1013.

Jeffers, T. K. (1974c). *Eimeria acervulina* and *E. maxima:* Incidence and anticoccidial drug resistance of isolants in major broiler-producing areas. *Avian Dis.* **18**, 331–342.

Joyner, L. P., and Long, P. L. (1974). The specific characteristics of the *Eimeria,* with special reference to the coccidia of the fowl. *Avian Pathol.* **3**, 145–157.

Kheysin, Y. M. (1972). "Life Cycles of Coccidia of Domestic Animals." Univ. Park Press, Baltimore, Maryland.

Kouwenhoven, B., and van der Horst, C. J. G. (1969). Strongly acid intestinal content and lowered protein, carotene, and vitamin A blood levels in *Eimeria acervulina* infected chickens. *Z. Parasitenkd.* **32**, 347–353.

Levine, N. D. (1973). "Protozoan Parasites of Domestic Animals and of Man." Burgess, Minneapolis, Minnesota.

Long, P. L. (1965). Development of *Eimeria tenella* in avian embryos. *Nature (London)* **208**, 509–510.

Long, P. L. (1970). *Eimeria tenella:* Chemotherapeutic studies in chick embryos with a description of a new method (Chorioallantoic membrane foci counts) for evaluating infections. *Z. Parasitenkd.* **33**, 329–338.

Long, P. L. (1973a). Pathology and pathogenicity of coccidial infections. *In* "The Coccidia" (D. M. Hammond and P. L. Long, eds.), pp. 251–294. Univ. Park Press, Baltimore, Maryland.

Long, P. L. (1973b). The growth of *Eimeria* in cultured cells and in chicken embryos: A review. *In* "Proceedings of the Symposium on Coccidia and Related Organisms," pp. 57–82. University of Guelph, Guelph, Ontario.

Merck & Company, Inc. (1953). "Coccidiosis Annotated Bibliography." Merck, Sharp, & Dohme, Rahway, New Jersey.

Merck & Company, Inc. (1961). "Coccidiosis Annotated Bibliography." Merck, Sharp, & Dohme, Rahway, New Jersey.

Merck & Company, Inc. (1965). "Coccidiosis Annotated Bibliography." Merck, Sharp, & Dohme, Rahway, New Jersey.

Merck & Company, Inc. (1970). "Coccidiosis Annotated Bibliography." Merck, Sharp, & Dohme, Rahway, New Jersey.

Pattillo, W. H., and Becker, E. R. (1955). Cytochemistry of *Eimeria brunetti* and *E. acervulina* of the chicken. *J. Morphol.* **96,** 61–95.

Patton, W. H. (1965). *Eimeria tenella:* Cultivation of the asexual stages in cultured animal cells. *Science* **150,** 767–769.

Pellérdy, L. P. (1973). "Coccidia and Coccidiosis." Akadémiai Kiadó, Budapest.

Pout, D. D. (1967). Villous atrophy and coccidiosis. *Nature* (*London*) **213,** 306–307.

Reid, W. M. (1972). Coccidiosis. *In* "Diseases of Poultry" (M. S. Hofstad *et al.,* eds.), pp. 944–989. Iowa State Univ. Press, Ames.

Reid, W. M. (1975). Progress in the control of coccidiosis with anticoccidials and planned immunization. *Am. J. Vet. Res.* **36,** 593–596.

Reid, W. M., and Pitois, M. (1965). The influence of coccidiosis on feed and water intake of chickens. *Avian Dis.* **9,** 343–348.

Rose, M. E. (1973a). Immunity. *In* "The Coccidia" (D. M. Hammond and P. L. Long, eds.), pp. 293–341. Univ. Park Press, Baltimore, Maryland.

Rose, M. E. (1973b). Immune responses to the *Eimeria:* Recent observations. *In* "Proceedings of the Symposium on Coccidia and Related Organisms," (M. A. Fernando, B. M. McCraw, and H. C. Carlson, eds.) pp. 92–118. University of Guelph, Guelph, Ontario.

Ruff, M. D., and Reid, W. M. (1975). Coccidiosis and intestinal pH in chickens. *Avian Dis.* **19,** 52–58.

Ryley, J. F. (1968). Chick embryo infections for the evaluation of anticoccidial drugs. *Parasitology* **58,** 215–220.

Ryley, J. F. (1973). Cytochemistry, physiology, and biochemistry. *In* "The Coccidia" (D. M. Hammond and P. L. Long, eds.), pp. 145–181. Univ. Park Press, Baltimore, Maryland.

Smith, B. F., and Herrick, C. A. (1944). The respiration of the protozoan parasite, *Eimeria tenella. J. Parasitol.* **30,** 295–302.

Turk, D. E. (1974). Intestinal parasitism and nutrient absorption. *Fed. Proc., Fed. Am. Soc. Exp. Biol.* **33,** 106–111.

Wang, C. C. (1975). Studies of the mitochondria from *Eimeria tenella* and inhibition of the electron transport by quinolone coccidiostats. *Biochim. Biophys. Acta* **396,** 210–219.

Warren, E. W. (1968). Vitamin requirements of the coccidia of the chicken. *Parasitology* **58,** 137–148.

Yvoré, P., and Mainguy, P. (1973). "The Effect of Coccidiosis on the Metabolism of the Carotenoids," Roche Inf. Serv., Publ. No. 270–65482, 1423. F. Hoffman-La Roche & Company, Basel.

3

Coccidia of Mammals except Man*

Kenneth S. Todd, Jr. and John V. Ernst

I. Introduction

Coccidia are usually intracellular parasites of the intestinal tracts of vertebrates; however, they are found in invertebrates and in organs other than the intestinal tract. The name "coccidia" has varied meanings to different individuals. The use of the term in this chapter will be discussed in Section III on "Taxonomic Position." Coccidiosis is the disease caused by the coccidia. The most common genera of coccidia infecting mammals are *Eimeria* and *Isospora*. The names and hosts of the various species of coccidia mentioned in this chapter are presented in Table I.

Levine (1973a,b) has discussed the history of the coccidia. The first coccidium, and evidently the first protozoan, that was seen was *Eimeria stiedai,* which Leeuwenhoek observed in 1674 in the bile of a rabbit. The

* Mention of a trade name, proprietary product, or specific equipment does not constitute a guarantee or warranty by the U.S. Department of Agriculture and does not imply its approval to the exclusion of other products that may be suitable.

Table I

List of Coccidia and Hosts Mentioned in the Text

Parasite	Host
Cryptosporidium muris	Mouse (*Mus musculus*)
Eimeria species	
E. alabamensis	Ox (*Bos taurus*), zebu (*Bos indicus*), water buffalo (*Bubalus bubalis*)
E. arloingi	Sheep (*Ovis* species)
E. auburnensis	Ox (*Bos taurus*), zebu (*Bos indicus*), water buffalo (*Bubalus bubalis*)
E. bilamellata	Ground squirrels (*Spermophilus* species), white-tailed prairie dog (*Cynomys leucurus*)
E. bovis	Ox (*Bos taurus*), zebu (*Bos indicus*), water buffalo (*Bubalus bubalis*)
E. callospermophili	Ground squirrels (*Spermophilus* species), white-tailed prairie dog (*Cynomys leucurus*)
E. contorta	Rat (*Rattus norvegicus*), mouse (*Mus musculus*), multimammate rat (*Mastomys natalensis*)
E. chinchillae	Chinchilla (*Chinchilla laniger*), other rodents
E. falciformis	Mouse (*Mus musculus*)
E. larimerensis	Prairie dogs (*Cynomys ludovicianus* and *C. leucurus*), ground squirrels (*Spermophilus* species)
E. magna	Rabbits (*Oryctolagus cuniculus, Lepus* species, *Sylvilagus* species)
E. neitzi	Impala (*Aepyceros melampus*)
E. nieschulzi	Rats (*Rattus* species)
E. ninakohlyakimovae	Goats (*Capra* species), sheep (*Ovis* species), various other ruminants
E. scabra	Pig (*Sus scrofa*)
E. stiedai	Rabbits (*Oryctolagus cuniculus, Lepus* species, *Sylvilagus* species)
E. tenella	Chicken (*Gallus domesticus*)
E. vermiformis	Mouse (*Mus musculus*)
E. zuernii	Ox (*Bos taurus*), zebu (*Bos indicus*), water buffalo (*Bubalus bubalis*)
Isospora canis	Dog (*Canis familiaris*)

parasite was not named until 1865 and did not receive its present name until 1907. Only recently has the life cycle of the type species of the genus, *Eimeria falciformis,* been described (Haberkorn, 1970). Todd and Lepp (1971) believed that much of the early work on *E. falciformis* involved a complex of species from which other species could be separated only by studying their endogenous life cycles.

Most coccidian species have been described on the basis of the structure of their oocysts (Fig. 1). With information on the structures in the oocyst, Levine (1962) calculated that the genus *Eimeria* might contain

Table II
Numbers of Named Species of Coccidia for Mammals [a]

Order	Eimeria	Isospora	Others
Marsupialia	16	2	0
Insectivora	18	4	*Yakimovella,* 1; *Cyclospora,* 3
Chiroptera	10	0	0
Primates	8	7	0
Edentata	5	0	0
Lagomorpha	43	0	0
Rodentia	287	18	*Wenyonella,* 3; *Cryptosporidium,* 2; *Tyzzeria, Dorisiella, Caryospora,* and *Skrjabinella,* 1 each
Carnivora	33	30	*Hoareosporidium, Toxoplasma,* and *Cryptosporidium,* 1 each
Hyracoidea	1	0	0
Perissodactyla	3	0	0
Artiodactyla	119	7	*Wenyonella,* 1

[a] From Pellérdy (1974).

2,654,736 possible morphologically different oocysts. Some researchers believe that before a new coccidian species is named, both the oocyst and endogenous stages should be described. We know of no studies in which two structurally dissimilar oocysts have the same life cycles. However, the oocysts of some species that appeared to be structurally identical were found to be more than one species when the endogenous stages were studied. Numbers of named species of coccidia were listed for mammals by Pellérdy (1974) (Table II).

In addition, the oocysts of the genera *Sarcocystis, Besnoitia,* and *Hammondia* are now known to occur in carnivores. Thus, of the eighteen living orders of mammals, coccidia have not been described from seven (Monotremata, Dermaptera, Pholidota, Cetacea, Tubulidentata, Proboscidea, or Sirenia).

II. Morphology—Life Cycles

The coccidia have stages found within the host (endogenous stages) and outside the host (exogenous stages). The endogenous life cycle includes both asexual (schizogonic) and sexual (gametogonic) stages. The exogenous stages are oocysts. Various structures are present within the oocysts, depending on the genus and species present. A typical sporulated oocyst of the genus *Eimeria* is illustrated in Fig. 1.

A generalized coccidian life cycle is illustrated in Fig. 2. Unsporulated oocysts are passed in the feces. The protoplasm within the oocyst (the

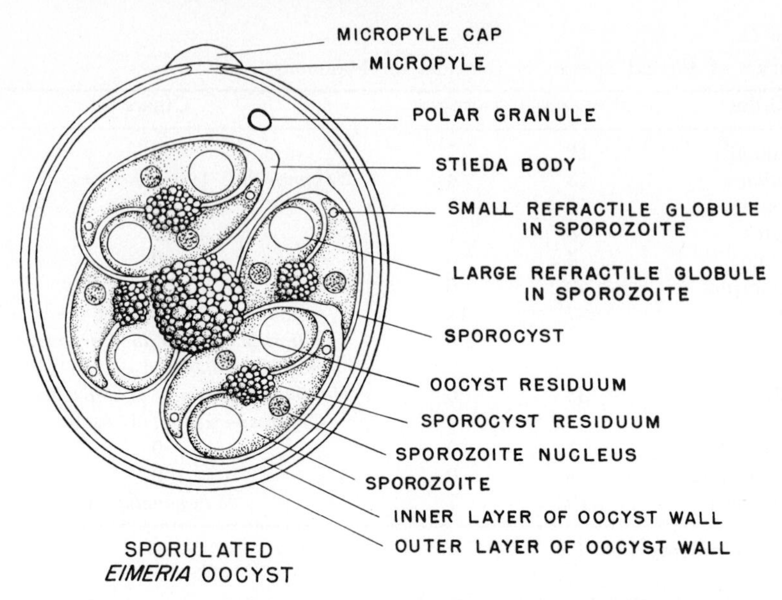

Fig. 1. Sporulated *Eimeria* oocyst. [From Levine (1973a).]

sporont) usually contracts after the oocyst is passed. With proper temperature and in the presence of moisture and oxygen, the oocysts sporulate to an infective state. When the sporulated oocysts are ingested, the motile infective stage (the sporozoite) is released. A sporozoite enters a cell and rounds up to form a trophozoite, which grows and becomes a schizont (schizogony). Growth of the schizont is accompanied by nuclear division. Elongate merozoites, each containing a nucleus from the schizont, are then formed. The number of merozoites in a schizont depends on the species of coccidia. The schizont ruptures and the merozoites enter new host cells to form a second generation of schizonts. Schizogony continues for a set number of generations, depending on the species, and eventually the merozoites form gamonts and the sexual cycle (gametogony) begins. Some merozoites become microgametocytes in which nuclear division takes place. During maturation of the microgametocytes, elongation of the nuclei and the growth of flagella result in the formation of microgametes. Other merozoites form macrogametes, each of which contains a nucleus that does not divide. After a period of growth and maturation, macrogametes are fertilized by microgametes to form zygotes. A wall of one or more layers develops around the zygote to form an oocyst. The host cell ruptures, and the oocyst is passed from the host into the external environment.

Many variations occur in the typical coccidian life cycle. One is the

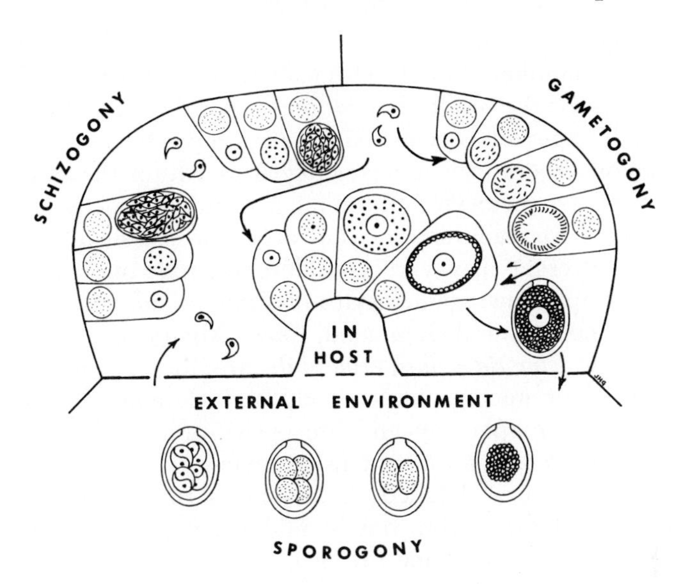

Fig. 2. Generalized coccidian life cycle. [Adapted from Todd (1975).]

location of the parasite in the host. Not all coccidia are intestinal parasites, and not all are found in host cell cytoplasm. *Eimeria alabamensis* and other coccidia have been reported to be intranuclear parasites (Davis *et al.*, 1955). Some coccidia, such as *Cryptosporidium muris,* are in a pericellular location (Tyzzer, 1910). Examples of coccidia found in organs other than the intestine are *E. stiedai* in the bile ducts of the liver (Metzner, 1903) and *Eimeria neitzi* in the uterus (McCully *et al.*, 1970). Kheysin (1972) has listed several other species of coccidia that are found in organs other than the intestine of various animals.

A single species of coccidium may have different endogenous stages in different parts of the intestinal tract. One example is *Eimeria bovis;* first-generation schizonts are in the small intestine, and second-generation schizonts and sexual stages are located in the large intestine (Hammond *et al.*, 1946, 1963).

The number of asexual generations of coccidian parasites is usually a genetic property of the parasite; however, influence by the host may play a role in determining the number of schizogonic generations. In normal mice, *Eimeria vermiformis* has two generations of schizonts (Todd and Lepp, 1971). However, in mice that have been injected with dexamethasone, the numbers of asexual generations can be increased (K. S. Todd, Jr. unpublished data). Although the size and number of merozoites may vary between generations, the size and number of merozoites in each generation is fairly constant for each species. Variations in the shape, size, and number of merozoites may be present for a single species. Variation might be the result of crowding or possibly sexual dimorphism.

The sexual stages of coccidia also vary, depending on the species. Little variation occurs in macrogametogenesis within the same species. The size of microgametocytes and number of microgametes produced is about the same for each species; however, the sizes and numbers of microgametes present may vary considerably from one species to another. For example, microgametocytes of *Eimeria auburnensis* are up to 79 × 80 μm (Chobotar and Hammond, 1969). Microgametocytes of other species, such as *Eimeria callospermophili*, are 13 × 10 μm and contain considerably fewer microgametes (Todd and Hammond, 1968a).

Details on the morphology and life cycles of coccidia of mammals would fill many pages, and an extensive review is not within the scope of this chapter. For a detailed discussion on morphology, including ultrastructure and life cycles, one should refer to Hammond (1973), Scholtyseck (1973), and Levine (1973c).

III. Taxonomic Position

What "coccidia" include depends on the classification scheme used. Some investigators consider that the coccidia include only those organisms in the family Eimeriidae, but limitation of the term to this family is too restrictive because it omits such genera as *Haemoproteus, Hepatocystis, Leucocytozoon,* and *Plasmodium.* As Scholtyseck (1974) has stated, "Most protozoologists doing research on Coccidia share the viewpoint that the Haemosporidia are Coccidia; however, not many malaria researchers agree." For the sake of clarity, we will limit the use of the term coccidia to the suborder Eimeriorina Leger, 1911 which, according to Levine (1973a), has the following characteristics: "Macrogamete and microgametocyte develop independently; syzygy absent, microgametocyte typically produces many microgametes; zygote not motile; sporozoites typically enclosed in a sporocyst; endodyogeny absent; monoxenous or heteroxenous." The classification used by Levine (1973a) is probably the most recent except, as noted by Scholtyseck (1974), the *Toxoplasma,* which Levine included in the family Sarcocystidae, should be included as a subfamily of Eimeriidae. Peteshev *et al.* (1974) and Wallace and Frenkel (1975) have found that oocysts are produced by *Besnoitia* species; thus, this genus probably belongs in the same subfamily. Because of the recent state of flux of the taxonomy of the genera *Isospora, Toxoplasma, Sarcocystis, Besnoitia,* and related organisms, it is probably best to wait until further research clarifies the taxonomic position of these organisms. The classification of the coccidia in Table III is that of

Table III

Classification of the Coccidia and Related Organisms [a]

Subphylum Apicomplexa Levine, 1970: Apical complex, generally consisting of polar ring, micronemes, rhoptries, subpellicular tubules, and conoid, present at some stage; micropore(s) generally present; single type of nucleus; cilia and flagella absent, except for flagellated microgametes in some groups; sexuality, when present, syngamy; cysts often present; all species parasitic

Class Sporozoasida Leuckart, 1879: Apical complex well developed; reproduction generally both sexual and asexual; oocysts (sometimes called "spores" in some forms) present; locomotion by body flexion, gliding or undulation of longitudinal ridges; microgametes flagellated in some groups; pseudopods ordinarily absent; if present used for feeding, not locomotion; monoxenous or heteroxenous

Subclass Gregarinasina Dufour, 1828: Mature gamonts extracellular, large; conoid modified into mucron or epimerite in mature organisms; endodyogeny absent; gametes similar (isogamous) or nearly so; equal numbers of male and female gametes produced by gamonts; zygotes form oocysts within gametocysts; parasites of digestive tract or body cavity of invertebrates or lower chordates; generally monoxenous

Subclass Coccidiasina Leuckart, 1879: Mature gamonts small, typically intracellular; conoid not modified into mucron or epimerite; endodyogeny present or absent; mostly in vertebrates, but a few in invertebrates

Order Protococcidiorida Kheisin, 1956: Merogony absent; in marine invertebrates

Order Eucoccidiorida Léger and Duboscq, 1910: Merogony present; in vertebrates or invertebrates

Suborder Adeleorina Léger, 1911: Macrogamete and microgametocyte associated in syzygy during development; microgametocyte usually produces from one to four microgametes; sporozoites enclosed in envelope, endodyogeny absent; monoxenous or heteroxenous

Suborder Eimeriorina Léger, 1911: Macrogamete and microgametocyte develop independently; syzygy absent; microgametocyte typically produces many microgametes; zygote not motile; sporozoites typically enclosed in a sporocyst; endodyogeny absent or present; monoxenous or heteroxenous

Family Selenococcidiidae Poche, 1913: Meronts develop as vermicules in host intestinal lumen; meronts with myonemes and nuclei in a row; in lobster

Genus *Selenococcidium* Léger and Duboscq, 1910: With the characters of the family

Family Aggregatidae Labbé, 1899: Development in host cell proper; oocysts typically with many sporocysts; mostly heteroxenous—merogony in one host, gamogony in another

Genus *Aggregata* Frenzel, 1885: Oocysts large, with many sporocysts; sporocysts with from 3 to 28 sporozoites; heteroxenous—merogony in a decapod crustacean and gamogony in a cephalopod mollusk

Genus *Angeiocystis* Brasil, 1909: Oocyst with four sporocysts, each with about thirty sporozoites; gamonts at first in the form of a large sausage; merogony unknown; presumably heteroxenous; known stages in heart of polychaete *Cirriformia* (syn., *Audouinia*)

Genus *Merocystis* Dakin, 1911: Oocyst with many sporocysts, each with two sporozoites; merogony unknown, presumably heteroxenous; known stages in kidney of whelk (prosobranch mollusk) *Buccinum*

Table III (Continued)

Genus *Myriospora* Lermantoff, 1913: Oocyst with eight to several hundred sporocysts, each with 24 or 32 sporozoites; male gamont helicoid; merogony unknown; presumably heteroxenous; known stages in polychaetes

Genus *Pseudoklossia* Léger and Duboscq, 1915: Oocyst with no or many sporocysts, each with two sporozoites (if sporocysts occur); merogony unknown; presumably heteroxenous; known stages in lamellibranch mollusks

Genus *Ovivora* Mackinnon and Ray, 1937: Oocyst with no(?) sporocysts, each presumably with twelve sporozoites; monoxenous; in eggs of echiuroid annelid *Thalessema*

Family Caryothrophidae Lühe, 1906: Oocysts large, without definite wall, with about twenty sporocysts each with twelve sporozoites; monoxenous; in polychaete annelids

Genus *Caryotropha* Siedlecki, 1902: With the characters of the family

Family Lankesterellidae Nöller, 1902: Development in host cell proper; oocysts without sporocysts, but with eight or more sporozoites; heteroxenous, with merogony, gametogony, and sporogony in the same vertebrate host; sporozoites in blood cells, transferred without developing by an invertebrate (mite or leech); microgametes with two flagella, so far as is known

Genus *Lankesterella* Labbé, 1899 (syn., *Atoxoplasma* Garnham, 1950): Oocyst produces 32 or more sporozoites; in amphibia and birds

Genus *Schellackia* Reinchenow, 1919: Oocyst produces eight sporozoites; in lizards

Family Eimeriidae Minchin, 1903: Development in host cell proper; oocysts and meronts (schizonts) without attachment organelle; oocysts with 0, 1, 2, 4, or many sporocysts, each with one or more sporozoites; monoxenous; merogony in the host, sporogony typically outside; microgametes with two or three flagella

Genus *Eimeria* Schneider, 1875: Oocysts with four sporocysts, each with two sporozoites

Genus *Isospora* Schneider, 1881: Oocysts with two sporocysts, each with four sporozoites

Genus *Barrouxia* Schneider, 1885: Oocysts with *n* sporocysts, each with one sporozoite

Genus *Caryospora* Léger, 1904: Oocysts with one sporocyst containing eight sporozoites

Genus *Cyclospora* Schneider, 1881: Oocysts with two sporocysts, each with two sporozoites

Genus *Dorisiella* Ray, 1930: Oocysts with two sporocysts, each with eight sporozoites

Genus *Tyzzeria* Allen, 1936: Oocysts with eight naked sporozoites, without sporocysts

Genus *Wenyonella* Hoare, 1933: Oocysts with four sporocysts, each with four sporozoites

Genus *Mantonella* Vincent, 1936: Oocysts with one sporocyst containing four sporozoites

Genus *Octosporella* Ray and Ragavachari, 1942: Oocysts with eight sporocysts, each with two sporozoites

Genus *Sivatoshellina* Ray and Sarkar, 1968: Oocysts with two sporocysts, each with sixteen sporozoites.

Table III (Continued)

Genus *Yakimovella* Gousseff, 1937: Oocysts with eight sporocysts, each with *n* sporozoites

Genus *Hoarella* Arcay de Peraza, 1963: Oocysts with sixteen sporocysts, each with two sporozoites

Genus *Pythonella* Ray and Das Gupta, 1937: Oocysts with sixteen sporocysts, each with four sporozoites

Family Cryptosporidiidae Léger, 1911: Development just under surface membrane of host cell or within its striated border and not in cell proper; oocysts and meronts with a knoblike attachment organelle at some point on their surface; oocysts without sporocysts, with four naked sporozoites; monoxenous; microgametes without flagella

Genus *Cryptosporidium* Tyzzer, 1907: With the characters of the family

Family Pfeifferinellidae Grassé, 1953: Oocyst without sporocysts, with eight naked sporozoites; fertilization of macrogamete through a "vaginal" tube; monoxenous; in mollusks

Genus *Pfeifferinella* Wasielewski, 1904: With the characters of the family

Family Sarcocystidae Poche, 1913: Syzygy apparently absent; endodyogeny present; cysts of pseudocysts containing zoites in parenteral cells of host; all parasitic in vertebrates; monoxenous

Subfamily Toxoplasmatinae Biocca, 1956: Zoitocysts with thin membranes and pseudocysts present in parenteral cells of host; meronts (schizonts) and gamonts in intestinal cells; gamonts produce oocysts

Genus *Frenkelia* Biocca, 1968: Cysts in brain lobulate and septate

Genus *Toxoplasma* Nicolle and Manceaux, 1908: Cysts in brain not lobulate or septate

Subfamily Besnoitiinae Garnham, 1966: Zoitocysts with thick, laminated, nucleated walls and pseudocysts present in parenteral cells; sexual reproduction unknown

Genus *Besnoitia* Henry, 1913: With the characters of the subfamily

Subfamily Sarcocystinae Poche, 1913: Zoitocysts elongated, often septate, with cytophaneres, in parenteral cells, pseudocysts unknown

Genus *Sarcocystis* Lankester, 1882: Zoites elongate; septa in cysts, if present, thin

Genus *Arthrocystis* Levine, Beamer and Simon, 1970: Zoites spherical; septa in cysts thick; cysts jointed like bamboo

Suborder Haemosporina Danilewsky, 1885: Macrogamete and microgametocyte develop independently; syzygy absent; microgametocyte produces about eight flagellated microgametes; zygote motile (ookinete); sporozoites naked; endodyogeny absent; heteroxenous; with merogony in vertebrate host and sporogony in invertebrate; pigment (hemozoin) may or may not be formed from host cell hemoglobin

Class Piroplasmasida Levine, 1961: Small, piriform, round, rod-shaped, or amoeboid; components of apical complex reduced; spores or sporocysts absent; flagella and cilia absent; pseudopods, if present, used for feeding, not locomotion; locomotion by body flexion or gliding; reproduction asexual by binary fission or schizogony; pigment not formed from host cell hemoglobin; heteroxenous; parasitic in vertebrate erythrocytes, leukocytes, other blood system cells or liver parenchymal cells; known vectors ticks

---

[a] From Levine (1973a).

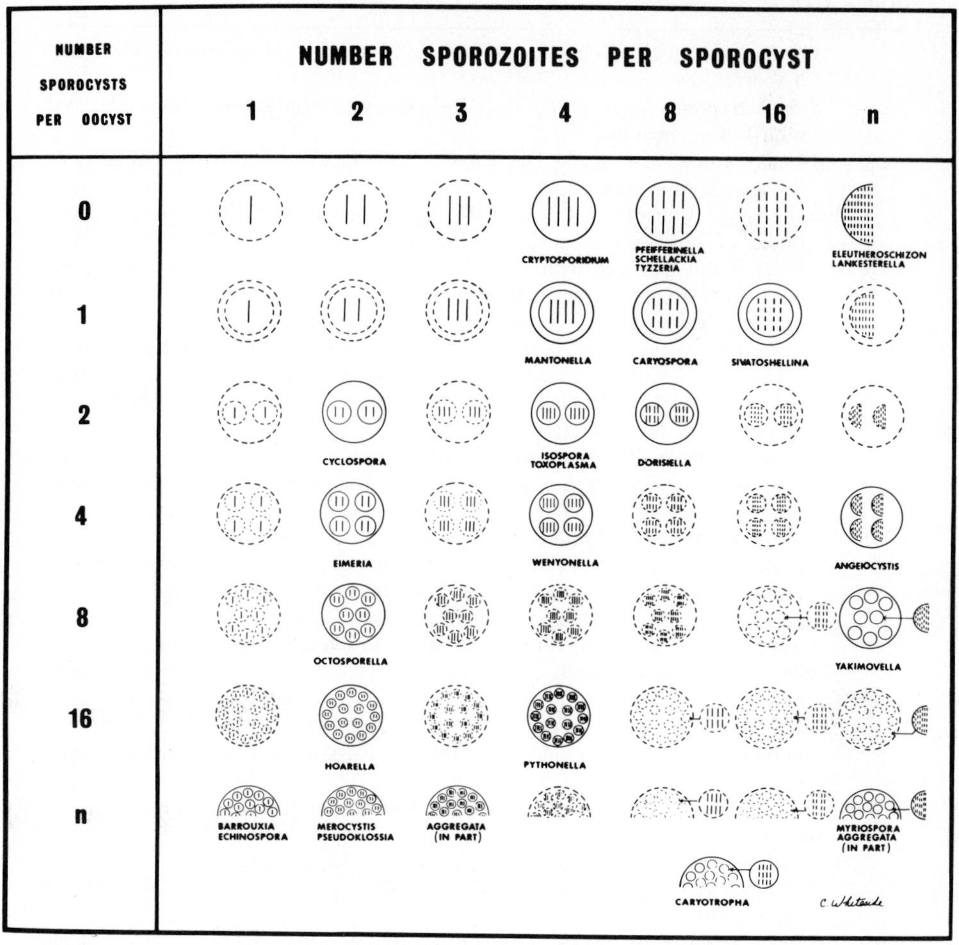

Fig. 3. Number of sporocysts per oocyst and of sporozoites per sporocyst in the genera of the suborder Eimeriorina. (In the genera without sporocysts, the numbers of sporozoites per oocyst are given.) [From Levine (1973a).]

Levine (1973a) and shows their relationship to other groups in the subphylum Apicomplexa.

The genera of coccidia in the suborder Eimeriorina are determined by the number of sporozoites per sporocyst and the number of sporocysts per oocyst. This "periodic table" of classification was first proposed by Hoare (1933). It has been updated by Levine (1973a) and is illustrated in Fig. 3.

IV. Metabolism and Biochemistry

The coccidia spend part of their life cycle as obligate intracellular parasites and are extremely difficult to study with classical metabolic and biochemical techniques. Most metabolic investigations have been restricted to sporulation and excystation of the oocysts. Studies on the intracellular stages have been restricted to histochemical and autoradiographic techniques that allow identification of the presence of enzymes and selected precursors but do not permit a detailed study of metabolism.

Dürr and Pellérdy (1969) summarized the literature on sporulation of oocysts, which is an aerobic process. *Eimeria stiedai* will not sporulate under anaerobic conditions. The total oxygen requirement was between 302 and 395 μl for 10 million oocysts to sporulate at 25°C. Wagenbach and Burns (1969) measured respiration of *E stiedai* oocysts during sporulation polarographically and compared the timing of events and respiratory rate during sporulation. An early increase in respiratory rate was followed by a depression in rate that correlated with the appearance of the early spindle stage. The rate again increased and then decreased toward a base rate during and after completion of sporulation.

Excystation of mammalian coccidian oocysts first requires alteration of the oocyst wall, followed by alteration of the sporocysts (Jackson, 1962). The oocyst walls are impermeable to all but a few substances and, thus, must be altered either mechanically or biochemically during the initial phase of excystation to allow the entry of trypsin and bile which affect the sporocyst walls. Jackson (1962) reported that when the oocysts of *Eimeria arloingi* were incubated under 1 atmosphere of CO_2, they became permeable to trypsin and bile; this permeability caused the sporozoites to excyst. The addition of a reducing agent increased the effect of CO_2. Subsequent studies on many other species have shown that CO_2 with reducing agents changes oocyst wall permeability; change of permeability may be a general phenomenon (Nyberg and Hammond, 1964; Nyberg *et al.*, 1968; Hibbert and Hammond, 1968; Bunch and Nyberg, 1970; Fayer and Leek, 1973). Jackson (1964) and later investigators all suggested that CO_2 acted to initiate an enzymatic alteration of the oocyst wall permeability.

After treating *E. stiedai* oocysts with $^{14}CO_2$, Jolley and Nyberg (1974) were unable to detect the radioisotope in oocyst walls, oocyst fluid, or sporocyst–sporozoite fragments. They noted that nearly all the ^{14}C was found in the excysting fluid and that CO_2 may have formed a complex with the reducing agent. They noted that the fluid within treated oocysts would not change the permeability of nontreated oocysts and concluded that an enzyme was probably not involved.

Recent findings by Jensen *et al.* (1976) have shown that changes in oocyst wall permeability of *E. bovis* and *E. stiedai* were induced when reducing gases were substituted for CO_2 and that some reducing agents destroyed the oocyst and its contents. They pointed out that in all previous experiments, reducing agents or rumen fluid, which may have contained natural reducing agents, were used and that the synergistic enhancement of mild reducing agents by CO_2 caused the change in oocyst wall permeability. Jolley *et al.* (1976) concluded that CO_2 enhanced the effect of the reducing agent on the oocyst wall without damaging the sporozoites inside.

Smetana (1933) found that the second stimulus required for excystation was trypsin. Lotze and Leek (1960) reported that bile, in combination with trypsin, greatly enhanced the process of excystation. Bile may facilitate the entry of enzymes through the altered micropyle or may alter the protein or lipoprotein surface of the Stieda body (Doran and Farr, 1962; Hibbert *et al.*, 1969). Bile seems to be essential for the excystation of some sheep, squirrel, and rabbit coccidia but not bovine coccidian species (Hibbert and Hammond, 1968). A number of bile acids or surface-active agents may be used instead of bile to induce excystation (Doran and Farr, 1962; Jackson, 1962; Hibbert *et al.*, 1969).

When trypsin and bile were added to CO_2-treated oocysts the sporozoites became motile within the sporocyst and the Stieda body became smaller and eventually disappeared. Activated sporozoites began to escape through the opening left by the disappearance of the Stieda body. After leaving the sporocysts, the sporozoites moved within the oocysts and then escaped through the micropyle (Nyberg and Hammond, 1964; Nyberg *et al.*, 1968; Hammond *et al.*, 1970; Roberts *et al.*, 1970).

Variations of the above described excystation process have been reported for some species. Fayer and Leek (1973) reported that sporocysts of *Sarcocystis fusiformis* from the feces of dogs must be treated with CO_2 prior to adding trypsin and bile before the sporocysts would excyst. Some *Eimeria* and *Isospora* oocysts have a substiedal body located beneath the Stieda body. The substiedal body must be expelled through the opening left by the dissolution of the Stieda body before the sporozoites can excyst (Hammond *et al.*, 1970; Duszynski and Brunson, 1973). Some *Isospora* species have sporocysts which do not have Stieda bodies, and after treatment with trypsin and bile the sporocyst walls collapse into fragments from which the sporozoites randomly escape (Speer *et al.*, 1973c; Duszynski and File, 1974).

Two fundamental problems are encountered in investigations of the biochemistry of the intracellular stages of mammalian coccidia: (1) The infected cells are widely dispersed in the host tissue, and their identification and removal for study are time-consuming. (2) The biochemical information obtained should be as valid for the parasite under natural conditions within its host cell as it is for the parasite removed from the host. Time spent searching for and removing infected cells is a time when the parasite is exposed to an abnormal environment—that of the dying or insulted host. Chemical and physical agents influencing the parasite must first pass through its host cell, and the interactions of the parasite with these agents must be distinguished from their previous interactions with the host cell.

Researchers have recently used two approaches to circumvent the problems listed above: (1) selection of *E. stiedai* as the source of intracellular stages because the rabbit liver is quickly accessible and often heavily infected and (2) cytochemical methodology. These approaches are not entirely adequate, however, because *E. stiedai* may not be representative physiologically of mammalian coccidia, in general, and because even the best controlled and designed cytochemical investigations produce quantitatively inaccurate results. Cytochemical methods are restricted in the compounds they can detect, and they are subject to a variety of factors that make even qualitative interpretation hazardous. The pertinent literature has been cited by Frandsen (1968) and reviewed by Ryley (1973). As mentioned in these references, T. V. Beier has conducted a series of cytochemical studies of the intracellular stages of coccidia infecting the intestinal epithelium of rabbits, and J. C. Frandsen has investigated the cytochemistry of the intracellular stages of *E. stiedai*. The development of host cell-free tissue culture techniques for the endogenous stages would open the door for rapid progress in biochemical research.

Oocysts are good subjects for biochemical investigations provided they can be obtained in adequate numbers and can be freed of contaminants. If intestinal inhabitants are studied, care must be exercised to ensure that the oocysts collected all belong to the same species. The enzymes lactate dehydrogenase (Frandsen and Cooper, 1972), fructose-1,6-diphosphate aldolase (Mitchell *et al.*, 1975), and glucose-6-phosphate dehydrogenase (Frandsen and Ennis, 1974; Frandsen, 1975a,b, 1976) have been isolated and described from unsporulated oocysts of *E. stiedai*. Other enzymes of mammalian coccidia have not been isolated. Ryley (1973) has reviewed the results of cytochemical investigations of oocysts.

V. Cultivation

Since the first description of the development of coccidia in cell cultures (Patton, 1965), there has been considerable research and interest in this area. *Eimeria tenella* is the only species that has been routinely cultured from sporozoites to oocysts. Most species develop only through the first-generation trophozoite or schizont stage (Doran, 1973). Merozoites of some mammalian coccidia taken from the host develop into sexual stages in cell cultures. Merozoites of *Eimeria magna* developed to gamonts and oocysts (Speer and Hammond, 1972), but merozoites of *Eimeria larimerensis* developed only to immature gamonts (Speer *et al.*, 1973a). First-generation merozoites of *E. bovis*, obtained from infected calves that had been inoculated 14 days earlier, were inoculated into bovine kidney cell cultures; second-generation schizonts, gamonts, and oocysts developed (Speer and Hammond, 1973). No attempts to sporulate mammalian coccidian oocysts produced in cell cultures have been made. First-generation merozoites of *E. bovis* from cell cultures produced gametocytes in the cecum of a calf (Hammond *et al.*, 1969).

Although there is not a good correlation between the sizes of schizonts in cell cultures and schizonts in the normal host, there is a general relationship between the two. For example, *E. bovis*, *E. auburnensis*, and *Eimeria ninakohlyakimovae* all have very large first-generation schizonts in their respective hosts (Hammond *et al.*, 1946; Chobotar *et al.*, 1969; Wacha *et al.*, 1971). The same species also produce large first-generation schizonts in cell cultures (Fayer and Hammond, 1967; Clark and Hammond, 1969; Kelley and Hammond, 1970).

The life cycle of *Eimeria zuernii* was described by Davis and Bowman (1957). With cell cultures, Speer *et al.* (1973b) discovered large first-generation schizonts for *E. zuernii* that were not found by Davis and Bowman (1957). However, large first-generation schizonts were later found by Stockdale (1975) using experimentally infected animals.

VI. Host–Parasite Interactions

In general, the coccidia have a narrow host range as well as a narrow organ specificity. In some species, the location within an organ and location within the host cell are also highly specific. As discussed above, some stages may be in one location and later stages may be in others. Host and site specificity of the coccidia have been reviewed by Marquardt (1973). Partial development of coccidia may occur in abnormal hosts without the formation of oocysts (Lotze *et al.*, 1961; Marquardt, 1966; Haberkorn, 1970; Todd *et al.*, 1971).

Rodent coccidia are generally assumed to be highly host specific. Levine and Ivens (1965) reported that of 47 attempts to transmit *Eimeria* species from one rodent genus to another, none were successful. Since then, Todd and Hammond (1968a,b) reported that two species of *Eimeria* were present in the closely related rodent genera *Spermophilus* and *Cynomys*. Mayberry and Marquardt (1973) and Mayberry *et al.*

(1975) reported that *Eimeria separata*, a parasite of *Rattus norvegicus*, could be transmitted to *Mus musculus*. Haberkorn (1971b) found that *Eimeria contorta* could be transmitted from *Rattus norvegicus* to *Mus musculus* and *Mastomys natalensis*. De Vos (1970) reported that *Eimeria chinchillae* from chinchillas could be experimentally transmitted to seven other species of rodents.

The same species of *Eimeria* are shared by the bovine genera *Bos* and *Bubalus*. Levine and Ivens (1970) reviewed the literature on cross-transmission studies of ruminant coccidia. Although 11 of 158 attempts to transmit coccidian species between genera were successful, the validity of some of the reports is questionable because adequate controls and coccidia-free animals were often not used.

Site specificity of the coccidia has been considered to be high. Todd *et al.* (1971) and Todd and Lepp (1972) reported that *E. vermiformis*, a coccidium of mice, could be transmitted to rats treated with dexamethasone. Although development took place in abnormal hosts, the endogenous stages were in the same location in the intestine and cells as in the normal host. In view of the findings that *Toxoplasma, Sarcocystis*, and related genera, which have asexual stages in various organs and tissues, have a coccidian life cycle, statements on host and site specificity of the coccidia will probably need to be revised as new data become available.

The nature of the host-parasite interaction is complex. In general, young nonimmune animals are more severely affected by coccidiosis than older immune ones. Early workers believed that coccidiosis was a disease of young animals because of age resistance. It is now known that age resistance is caused by prior exposure that results in immunity of older animals. As with other parasitic diseases, the general health, nutritional state, and prior exposure to the parasite are important factors in determining the degree of severity of the disease. Infections with coccidia may exacerbate infections by other agents.

The disease caused by coccidia varies not only with the state of the host, but also, and probably more important, with the species of coccidia present, the degree of parasitism, and the site of the infection. Other factors that may be involved are the breed or strain of the host and the viability and the virulence of the parasite.

Some species of coccidia are highly pathogenic whereas others are not. Therefore when coccidiosis is suspected and oocysts are found, the species present should be identified. When the disease is present, clinical signs may include diarrhea, and blood and intestinal tissue may be present in the feces. Enteritis is often complicated by the presence of secondary bacterial infections. Animals may have vague clinical signs of parasitism such as rough hair, poor weight gain or weight loss, weakness, and emaciation. Severely affected animals may die. With some species of coccidia, infected animals may have central nervous system disorders (Clayburg, 1970). The reasons for the nervous disorders are not known but may be related to electrolyte loss through the intestine.

Even with pathogenic species of coccidia, some endogenous stages may cause little, if any, harm. For example, schizonts of *E. bovis* do not affect the host even though the giant first-generation schizonts are located in endothelial cells of the lacteals of the small intestine (Hammond *et al.*, 1946). The pathological changes are associated with the sexual stages that are located in epithelial cells of the cecum and colon, and clinical signs first appear 18 days after ingestion of the oocysts (Hammond *et al.*, 1944).

Recent reviews on the pathology and pathogenesis of coccidial infections have been written by Long (1973) and Ryley (1975).

We have long known that a host develops a protective immunity to coccidia after an initial infection. This immunity is specific for the species of coccidium to which the animal has been exposed (Becker, 1934). Despite the ubiquitous nature of coccidia, most research on immunity has been limited to species that parasitize domestic animals; this emphasis reflects the economic importance of coccidia in domestic animals (Rose, 1973).

Little is known of the mechanisms of immunity to coccidia, and controversy exists as to whether the immune response is due to humoral or cellular factors. Although humoral antibodies to coccidia have been demonstrated by a variety of techniques in calves, rabbits, and mice (Rose, 1973), blood or serum from immune hosts given to susceptible hosts did not produce any noticeable protection against infection (Bachman, 1930; Becker and Hall, 1933; Senger *et al.*, 1959; Fitzgerald, 1964). Recently, protection by cells or cell fractions has, however, been shown. Rommel and Heydorn (1971) reported a highly significant reduction in numbers of oocysts excreted by rats given lymphocytes from mesenteric lymph nodes of immune animals. Lymphocytes from Peyer's patches did not transfer immunity. Negative results were obtained by these authors when they attempted to transfer immunity to a pure strain of *E. scabra* by transferring lymphocytes from regional lymph nodes from immune to nonimmune pigs. Liburd *et al.* (1973) reported that intravenous injections of thoracic duct lymphocytes from immunized rats elicited various degrees of adoptive immunity against primary infections of *E. nieschulzi.* Some of the rats were totally immune to a subsequent challenge of the parasite. Immunity against *E. nieschulzi* was passively transferred to uninfected rats with a dialyzable transfer factor made from lymphoid tissue from infected rats. The reduction in oocyst excretion by rats given transfer factor 48 hours before the primary infection was highly significant when compared with control groups (Liburd *et al.*, 1972).

Further evidence that immunity to coccidiosis is the result of cellular factors has recently been reported by Klesius and co-workers. Using a skin test antigen prepared from the walls of unsporulated oocysts, Klesius *et al.* (1975a) showed that delayed hypersensitivity, a recognized form of cell-mediated immunity, developed in rabbits infected with *E. stiedai.* Skin reactivity was passively transferred to noninfected rabbits with lymphocyte suspensions and cell-free transfer factor, but not with serum from infected skin-reactive rabbits. Later, an antigen prepared from unsporulated oocysts of *E. bovis* was used to demonstrate delayed hypersensitivity in calves infected with *E. bovis.* Both lymphocytes and transfer factor effectively transferred delayed hypersensitivity reactivity to oocyst antigen. Acquired immunity to *E. bovis* infection was found in some calves receiving bovine transfer factor but not in others (Klesius *et al.*, 1975b). The future use of lymphokines such as transfer factor should help to clarify the nature of the cellular immune response to coccidiosis.

Attempts to immunize animals against coccidiosis with nonviable parasite material has been largely unsuccessful. Bachman (1930) failed to immunize rabbits with suspensions of dried, pulverized oocysts, and Becker (1935) failed to immunize rats by injecting them with ground oocysts and infected intestine. Horton-Smith *et al.* (1963) and Rose (1961, 1963) attempted unsuccessfully to immunize rabbits against *E. stiedai* with antigens prepared from exudate of bile ducts containing a mixture of schizonts and gametocytes. Lowder (1966) reported that he immunized calves against *E. bovis* infections by oral inoculation, but not by intravenous inoculation, of a soluble extract of disrupted oocysts.

Use of attenuated oocysts for active immunization has been attempted in calves. Fitzgerald (1965) exposed sporulated oocysts of *E. bovis* to 60,000 rads of gamma radiation and found that the oocysts still produced infections in calves. In another study, Fitzgerald (1968) irradiated sporulated oocysts of *E. bovis* with cobalt-60 at levels of 10,000 to 100,000 rads. Oocysts irradiated in the unsporulated state were more resistant to the effects of gamma radiation than were oocysts irradiated in the sporulated state, and some unsporulated oocysts completed sporulation and caused mild infections in calves. Fitzgerald (1968) concluded that immunity to *E. bovis* depended upon development of infection and production of oocysts. Ingestion of large numbers of irradiated sporulated oocysts by the host had no advantage over ingestion of nonirradiated oocysts in producing immunity to coccidiosis.

VII. Epidemiology

Coccidia have worldwide distribution, and the geographical range of individual species is evidently limited only by the range of appropriate hosts. Limiting factors for survival of oocysts in the external environment include the presence of oxygen and water and suitable temperatures. Coccidiosis is transmitted in filth, and infection usually takes place by ingestion of sporulated oocysts in contaminated feed or water. Coccidiosis is much more likely to be a problem when animals are crowded than when they are not. For this reason, coccidiosis is generally less of a problem in wild animals than in domestic animals which are raised under intensive husbandry conditions. During coccidian infections, usually more than one species of coccidia are present. Because of the complex nature of bovine coccidiosis, the economic importance of the disease is difficult to estimate. All classes of domestic animals are affected by the parasites, but the losses due to mortality and morbidity are difficult to determine. Costs for special equipment, management practices, and drugs for prophylaxis and treatment must be considered when losses due to coccidiosis are calculated. The most extensive evaluation of the effect of coccidiosis on cattle was by Fitzgerald (1972a), who estimated that bovine coccidiosis caused losses of $472 million on a worldwide basis.

Coccidiosis cannot be prevented without a great deal of time and effort; however, the disease can be minimized by strict sanitation and good management. Proper construction of facilities to prevent contamination of feed and water with feces is important. Overcrowding enhances coccidiosis. None of the disinfectants commonly used are effective against coccidial oocysts.

Oocysts of *E. zuernii* can be killed with either 10^{-6} *M* mercuric chloride, 0.5% cresol, 0.05 *M* phenol, 0.25 *M* formaldehyde, or 4 hours of sunlight (Marquardt *et al.*, 1960). The effects of chemical and physical agents on coccidian oocysts were reviewed by Kheysin (1972) and Schneider *et al.* (1973). Although control of coccidiosis by killing oocysts seems to be impractical at the present time, the use of live steam and fumigation with NH_3 gas has been used in animal quarters under experimental conditions (Lepp and Todd, 1974).

The use of coccidiostats to prevent coccidiosis in mammals has received little attention. Amprolium (Corid) is recommended as a preventive for bovine coccidiosis on the manufacturer's label "during periods of exposure or when experience indicates that coccidiosis is likely to be a hazard." The use of drugs to control coccidiosis in domestic animals will be discussed in Section IX.

VIII. Diagnosis

The diagnosis of coccidiosis by clinical signs is often difficult. The symptoms of the disease are easily confused with those of other intestinal disorders.

In efforts to detect oocysts, feces can be examined by several methods; the most simple is a direct fecal smear. A small amount of feces is placed on a glass slide and mixed with a few drops of water. A cover slip is placed on the mixture, and the fecal suspension is examined with a microscope. Although this technique is fast and requires little equipment, oocysts may not be detected because of the small amount of feces examined.

Results are more satisfactory with methods that concentrate the oocysts in the feces. The most satisfactory method is with the use of Sheather's sugar solution, made by dissolving 500 gm of sucrose (cane or beet sugar) in 320 ml of distilled water. Melted phenol (6.5 gm) is added to the mixture as a preservative. Other flotation solutions such as zinc sulfate, magnesium sulfate, and sodium chloride may be used instead of Sheather's solution (Sloss, 1970). The sample can be processed as follows:

1. Place 2–5 gm of feces in a tea strainer and add 10–15 ml of water.
2. Stir the sample in a small bowl with a tongue depressor to make a heavy fecal suspension.
3. Add the suspension to a 15-ml centrifuge tube and centrifuge at 1500 rpm (ca. 500 *g*) for about 5 minutes.
4. Pour off the supernatant fluid and resuspend the sediment in Sheather's solution.
5. Add Sheather's solution with a wash bottle until the surface of the solution is just above the rim of the centrifuge tube.
6. Place a cover slip on the tube and centrifuge at 1500 rpm (ca. 500 *g*) for 5 minutes.
7. If desired, the preparation need not be centrifuged; instead let stand for 30 minutes to 1 hour.
8. Remove the cover slip, place it on a microscope slide, and examine with a microscope.

Quantitative techniques for counting oocysts are available (Davis, 1973), but they are of little use in clinical diagnosis. The most common quantitative technique used by researchers is the modified McMaster's technique (Whitlock, 1948). The McMaster's method we use in our laboratories is essentially that described by Levine *et al.* (1960).

Lesions caused by coccidia may be found at autopsy. Scrapings of the intestinal mucosa should be examined for the presence of endogenous stages.

At present, serodiagnosis of coccidian infections is of little, if any, value except for *Toxoplasma gondii*, which will be discussed in Chapter 4. Klesius *et al.* (1975, 1976) reported that infected calves and rabbits produced much larger dermal indurations than noninfected calves and rabbits when skin-tested with an antigen prepared from unsporulated oocysts of *E. bovis* and *E. stiedai*, respectively.

IX. Treatment*

Various compounds have been used to treat mammalian coccidiosis. The sulfonamides were the first group of drugs to be widely used. Coccidiosis in cattle caused by *E. bovis* was successfully controlled with sulfaguanidine (Boughton, 1943), sulfamethazine (Davis and Bowman, 1951, 1954; Horton-Smith, 1958; Hammond *et al.*, 1959, 1960; Peardon *et al.*, 1965), sulfamerazine (Davis and Bowman, 1951; Hammond *et al.*, 1956) and, somewhat less successfully, with sulfaquinoxaline (Davis and Bowman, 1951; Hammond *et al.*, 1956).

Dunlap (1949) reported that sulfaquinoxaline effectively controlled infections of *E. zuernii*. Peardon *et al.* (1963), however, found that sulfaguanidine and sulfamethazine were not effective in controlling natural infections of bovine coccidia.

Nitrofurazone was found to be effective in treating coccidiosis of sheep and goats (Deom and Mortelmans, 1956; Tarlatzis *et al.*, 1955, 1957). Sellier (1954) reported that nitrofurazone was effective against *E. zuernii* in calves; however, Hammond *et al.* (1960, 1965) reported that nitrofurazone was not effective against *E. bovis* in calves and was toxic to the animals at high doses. This finding was supported by the work of Peardon *et al.* (1963).

Amprolium was effective in controlling coccidiosis caused by *E. bovis* in calves (Casoroso and Zaraza, 1963; Peardon *et al.*, 1965; Hammond *et al.*, 1966; Newman *et al.*, 1968; Gretillat and Vassiliades, 1968; Slater *et al.*, 1970; Jolley *et al.*, 1971), *E. zuernii* in calves (Casoroso and Zaraza, 1963), and *E. ninakohlyakimovae* in lambs (Hammond *et al.*, 1967). The drug was most effective when administered to calves for the entire pre-

* Drugs mentioned in this review should be used only in accordance with the manufacturer's directions or as prescribed by a veterinarian.

patent period, but it also controlled infections when administered in large doses for 5 days, beginning on day 13 post-inoculation (Hammond *et al.*, 1966). Slater *et al.* (1970) found that amprolium was both coccidiostatic and coccidiocidal for first-generation schizonts of *E. bovis*. Baker *et al.* (1972) found amprolium to be very promising as a prophylactic drug for the prevention of coccidiosis in feedlot lambs. Although the mode of action of most drugs used for mammalian coccidiosis is unknown, amprolium is a thiamine antagonist (Ott *et al.*, 1965; Rindi *et al.*, 1966). Amprolium was effective against rabbit intestinal coccidia (Cvetković and Tomanović, 1967), but it did not control infection with *E. stiedai* (Fitzgerald, 1972b).

Some compounds other than those listed above have been tested for efficacy in controlling coccidial infections in mammals. Monensin is currently the most widely used anticoccidial in poultry and has been found to effectively control coccidiosis caused by *E. bovis* in calves (Fitzgerald and Mansfield, 1973) and hepatic coccidiosis in rabbits, although the rabbits ate only small amounts of feed containing the drug (Fitzgerald, 1972b). Decoquinate was effective against *E. bovis* and *E. zuernii* infections in calves (Miner and Jensen, 1976). Lincomycin was effective against *E. bovis* (Peardon *et al.*, 1965; Arakawa *et al.*, 1967, 1968). Todd and Thacher (1973) found that a mixture of chlorotetracycline and sulfamethazine inhibited the development of coccidiosis in young calves. Hammond *et al.* (1958) found that low levels of nicarbazine were not effective against *E. bovis* and that high concentrations were toxic. Other drugs not efficacious were glycarbylamide, framycetin sulfate, Zoalene, chloroquinine sulfate, and diphenthane-70 (Peardon *et al.*, 1963, 1965).

ACKNOWLEDGMENTS

The authors wish to express their appreciation to the following who have helped with the chapter: to John C. Frandsen for his assistance with the section on biochemistry; to James B. Jensen for his assistance with the section on metabolism and chemotherapy; to Peggy Holloway for assistance with the literature; to Eloise D. Graves for suggestions concerning the manuscript; to Sharon Harper and Marjorie Hildreth for typing the manuscript; and to Norman D. Levine and Frank D. Enzie for comments on portions of the manuscript.

REFERENCES

Arakawa, A., Kohls, R. E., and Todd, A. C. (1967). Effect of lincomysin hydrochloride upon experimental bovine coccidiosis. *Am. J. Vet. Res.* **28**, 653–657.

Arakawa, A., Kohls, R. E., and Todd, A. C. (1968). Lincomysin therapy for experimental coccidiosis in calves, with special reference to macroscopic and microscopic observation. *Am. J. Vet. Res.* **29**, 1195–1200.

Bachman, G. W. (1930). Immunity in experimental coccidiosis of rabbits. *Am. J. Hyg.* **12**, 641–649.

Baker, N. F., Walters, G. T., and Fisk, R. A. (1972). Amprolium for control of coccidiosis in feedlot lambs. *Am. J. Vet. Res.* **33**, 83–86.

Becker, E. R. (1934). "Coccidia and Coccidiosis of Domesticated, Game and Laboratory Animals, and of Man." Collegiate Press, Ames, Iowa.

Becker, E. R. (1935). The mechanism of immunity in murine coccidiosis. *Am. J. Hyg.* **21**, 389–404.

Becker, E. R., and Hall, P. R. (1933). Cross-immunity and correlation of oocyst production during immunization between *Eimeria miyarii* and *Eimeria separata* in the rat. *Am. J. Hyg.* **18**, 220–223.

Boughton, D. C. (1943). Sulfaguanidine therapy in experimental bovine coccidiosis. *Am. J. Vet. Res.* **4**, 66–72.

Bunch, T. D., and Nyberg, P. A. (1970). Effects of carbon dioxide on coccidian oocysts from 6 host species. *J. Protozool.* **17**, 364–369.

Casoroso, D. R., and Zaraza, H. (1963). Drug control of coccidiosis in the bovine—Amprol. *Proc. World Vet. Congr., 17th, 1963* p. 87.

Chobotar, B., and Hammond, D. M. (1969). Development of gametocytes and second asexual generation stages of *Eimeria auburnensis* in calves. *J. Parasitol.* **55**, 1218–1228.

Chobotar, B., Hammond, D. M., and Miner, M. L. (1969). Development of the first-generation schizonts of *Eimeria auburnensis*. *J. Parasitol.* **55**, 385–397.

Clark, W. N., and Hammond, D. M. (1969). Development of *Eimeria auburnensis* in cell cultures. *J. Protozool.* **16**, 646–654.

Clayburg, J. (1970). Neurological signs seen in coccidial infections. *Iowa State Univ. Vet.* **32**, 85.

Cvetković L., and Tomanović, B. (1967). Investigations of amprolium in prophylaxis of intestinal coccidiosis in rabbits. *Vet. Glas.* **7**, 607–612.

Davis, L. R. (1973). Techniques. *In* "The Coccidia" (D. M. Hammond and P. L. Long, eds.), pp. 411–458. Univ. Park Press, Baltimore, Maryland.

Davis, L. R., and Bowman, G. W. (1951). Coccidiosis in cattle. *Proc. U.S. Livestock Sanit. Assoc.* **55**, 39–50.

Davis, L. R., and Bowman, G. W. (1954). The use of sulfamethazine in experimental coccidiosis of dairy calves. *Cornell Vet.* **44**, 71–79.

Davis, L. R., and Bowman, G. W. (1957). The endogenous development of *Eimeria zurnii*, a pathogenic coccidium of cattle. *Am. J. Vet. Res.* **18**, 569–574.

Davis, L. R., Boughton, D. C., and Bowman, G. W. (1955). Biology and pathogenicity of *Eimeria alabamensis* (Christensen, 1941), an intranuclear coccidium of cattle. *Am. J. Vet. Res.* **16**, 274–281.

Deom, J., and Mortelmans, J. (1956). Observations sur la coccidiose du monton et de la cherve an Congo Belge. Essaus thérapeutiques. *Ann. Soc. Belge Med. Trop.* **36**, 47.

De Vos, A. J. (1970). Studies on the host range of *Eimeria chinchillae* De Vos and Van der Westhuizen, 1968. *Onderstepoort J. Vet. Res.* **37**, 29–36.

Doran, D. J. (1973). Cultivation of coccidia in avian embryos and cell culture. *In* "The Coccidia" (D. M. Hammond and P. L. Long, eds.), pp. 183–252. Univ. Park Press, Baltimore, Maryland.

Doran, D. J., and Farr, M. M. (1962). Excystation of the poultry coccidium, *Eimeria acervulina*. *J. Protozool.* **9**, 154–161.

Dunlap, R. E. (1949). Bovine coccidiosis. *Southwest. Vet.* **2**, 44–45.

Dürr, V., and Pellérdy, L. (1969). Zum Sauerstoffverbrauch der Kokzidienoocysten während der Sporulation. *Acta Vet. Acad. Sci. Hung.* **19**, 307–310.

Duszynski, D. W., and Brunson, J. T. (1973). Structure of the oocysts and excystation processes of four *Eimeria* spp. (Protozoa: Eimeriidae) from the Colorado pika, *Ochotona princeps*. *J. Parasitol.* **59**, 28–34.

Duszynski, D. W., and File, S. K. (1974). Structure of the oocyst and excystation of sporozoites of *Isospora endocallimici* n. sp., from the marmoset *Callimicol goeldii*. *Trans. Am. Microsc. Soc.* **93**, 403–408.

Fayer, R., and Hammond, D. M. (1967). Development of first-generation schizonts of *Eimeria bovis* in cultured bovine cells. *J. Protozool.* **14**, 764–772.

Fayer, R., and Leek, R. G. (1973). Excystation of *Sarcocystis fusiformis* from dogs. *Proc. Helminthol. Soc. Wash.* **40**, 294–296.

Fitzgerald, P. R. (1964). Attempted passive immunization of young calves against *Eimeria bovis*. *J. Protozool.* **11**, 46–51.

Fitzgerald, P. R. (1965). The results of parenteral injections of sporulated or unsporulated oocysts of *Eimeria bovis* in calves. *J. Protozool.* **12**, 215–221.

Fitzgerald, P. R. (1968). Effects of ionizing radiation from cobalt-60 on oocysts of *Eimeria bovis*. *J. Parasitol.* **54**, 233–240.

Fitzgerald, P. R. (1972a). The economics of bovine coccidiosis. *Feedstuffs* **44**, 28–29.

Fitzgerald, P. R. (1972b). Efficacy of monensin and amprolium in the prevention of hepatic coccidiosis in rabbits. *J. Protozool.* **19**, 332–334.

Fitzgerald, P. R., and Mansfield, M. E. (1973). Efficacy of monensin against bovine coccidiosis in young Holstein-Friesian calves. *J. Protozool.* **20**, 121–126.

Frandsen, J. C. (1968). *Eimeria stiedae:* Cytochemical identification of acid and alkaline phosphatases, carboxylic ester hydrolases, and succinate, lactate, and glucose-6-phosphate dehydrogenases in endogenous stages from rabbit tissues. *Exp. Parasitol.* **23**, 398–411.

Frandsen, J. C. (1975a). Glucose 6-phosphate dehydrogenase from *Eimeria stiedai* (Protozoa: Coccidia): Some properties of the semipurified enzyme. *J. Protozool.* **22**, 20A.

Frandsen, J. C. (1975b). Glucose 6-phosphate dehydrogenase from *Eimeria stiedai* (Protozoa: Coccidia): Some properties of the semipurified enzyme. II. *Proc. 50th Annu. Meet. Am. Soc. Parasitol.* p. 105.

Frandsen, J. C. (1976). Partial purification and some properties of glucose 6-phosphate dehydrogenase from *Eimeria stiedai* (Lindemann, 1865) Kisskalt & Hartman, 1907 (Protozoa: Coccidia). *Comp. Biochem. Physiol.* **45B**, 537–541.

Frandsen, J. C., and Cooper, J. A. (1972). Enzymes of coccidia: Purification and properties of L-lactate dehydrogenase from *Eimeria stiedae*. *Exp. Parasitol.* **32**, 390–402.

Frandsen, J. C., and Ennis, T. H. (1974). Properties of glucose 6-phosphate dehydrogenase from *Eimeria stiedai* (Protozoa: Coccidia). *J. Protozool.* **21**, 433–434.

Gretillat, S., and Vassiliades, G. (1968). Le traitement de la coccidiose des ruminants par L' "amprolium." *Rev. Elev. Med. Vet. Pays Trop.* **21**, 191–201.

Haberkorn, A. (1970). Die Entwicklung von *Eimeria falciformis* (Eimer, 1870) in der weissen Maus (*Mus musculus*). *Z. Parasitenkd.* **34**, 49–67.

Haberkorn, A. (1971). Zur Wirtsspezifität von *Eimeria contorta* n. sp. (Sporozoa: Eimeriidae). *Z. Parasitenkd.* **37**, 303–314.

Hammond, D. M. (1973). Life cycles and development of coccidia. *In* "The Coccidia" (D. M. Hammond and P. L. Long, eds.), pp. 45–79. Univ. Park Press, Baltimore, Maryland.

Hammond, D. M., Davis, L. R., and Bowman, G. W. (1944). Experimental infections with *Eimeria bovis* in calves. *Am. J. Vet. Res.* **5**, 303–311.

Hammond, D. M., Bowman, G. W., Davis, L. R., and Simms, B. T. (1946). The endogenous phase of the life cycle of *Eimeria bovis. J. Parasitol.* **32**, 409–427.

Hammond, D. M., Shupe, J. L., Johnson, A. E., Fitzgerald, P. R., and Thorne, J. L. (1956). Sulfaquinoxaline and sulfamerazine in the treatment of experimental infections with *Eimeria bovis* in calves. *Am. J. Vet. Res.* **17**, 463–470.

Hammond, D. M., Senger, C. M., Thorne, J. L., Shupe, J. G., Fitzgerald, P. R., and Johnson, A. E. (1958). Experience with nicarbazine in coccidiosis (*Eimeria bovis*) in cattle. *Cornell Vet.* **48**, 260–268.

Hammond, D. M., Clark, G. W., Miner, M. L., Trost, W. A., and Johnson, A. E. (1959). Treatment of experimental bovine coccidiosis with multiple small doses and single large doses of sulfamethazine and sulfabromomethazine. *Am. J. Vet. Res.* **20**, 708–713.

Hammond, D. M., Ferguson, D. L., and Miner, M. L. (1960). Results of experiments with nitrofurazone and sulfamethazine for controlling coccidiosis in calves. *Cornell Vet.* **50**, 351–362.

Hammond, D. M., Andersen, F. L., and Miner, M. L. (1963). The occurrence of a second asexual generation in the life cycle of *Eimeria bovis* in calves. *J. Parasitol.* **49**, 428–434.

Hammond, D. M., Sayin, F., and Miner, M. L. (1965). Nitrofurazone as a prophylactic agent against experimental bovine coccidiosis. *Am. J. Vet. Res.* **26**, 83–89.

Hammond, D. M., Fayer, R., and Miner, M. L. (1966). Amprolium for control of experimental coccidiosis in cattle. *Am. J. Vet. Res.* **27**, 199–206.

Hammond, D. M., Kuta, J. E., and Miner, M. L. (1967). Amprolium for control of experimental coccidiosis in lambs. *Cornell Vet.* **57**, 611–623.

Hammond, D. M., Fayer, R., and Miner, M. L. (1969). Further studies on *in vitro* development of *Eimeria bovis* and attempts to obtain second generation schizonts. *J. Protozool.* **16**, 298–302.

Hammond, D. M., Ernst, J. V., and Chobotar, B. (1970). Composition and function of the substiedal body in the sporocysts of *Eimeria utahensis. J. Parasitol.* **56**, 618–619.

Hibbert, L. E., and Hammond, D. M. (1968). Effects of temperature on *in vitro* excystation of various *Eimeria* species. *Exp. Parasitol.* **23**, 161–170.

Hibbert, L. E., Hammond, D. M., and Simmons, J. R. (1969). The effects of pH, buffers, bile and bile acids on excystation of sporozoites of various *Eimeria* species. *J. Protozool.* **16**, 441–444.

Hoare, C. A. (1933). Studies on some new ophidian and avian coccidia from Uganda, with a revision of the classification of the Eimeriidea. *Parasitology* **25**, 359–388.

Horton-Smith, C. (1958). Coccidiosis in domestic mammals. *Vet. Rec.* **70**, 256–262.

Horton-Smith, C., Long, P. L., Pierce, A. E., and Rose, M. E. (1963). Immunity to protozoa in domestic animals. *In* "Immunity to Protozoa" (P. C. C. Garnham, A. E. Pierce, and I. Roitt, eds.), pp. 273– 295. Blackwell, Oxford.

Jackson, A. R. B. (1962). Excystation of *Eimeria arloingi* (Marotel, 1905): Stimuli from the host sheep. *Nature (London)* **194**, 847–849.

Jackson, A. R. B. (1964). The isolation of viable coccidial sporozoites. *Parasitology* **54**, 87–93.

Jensen, J. B., Nyberg, P. A., Burton, S. D., and Jolley, W. R. (1976). The effects of selected gasses on excystation of coccidian oocysts. *J. Parasitol.* **62**, 195–198.

Jolley, W. R., and Nyberg, P. A. (1974). Formation of a carbon dioxide-cysteine complex in the incubation fluid used for excysting *Eimeria* species *in vitro*. *Proc. Helminthol. Soc. Wash.* **41**, 259–260.

Jolley, W. R., Hammond, D. M., and Miner, M. L. (1971). Amprolium treatment of six- to twelve-month-old calves experimentally infected with coccidia. *Proc. Helminthol. Soc. Wash.* **38**, 117–122.

Jolley, W. R., Burton, S. D., Nyberg, P. A., and Jensen, J. B. (1976). Formation of sulfhydryl groups in the walls of *Eimeria stiedai* and *E. tenella* oocysts subjected to in vitro excystation. *J. Parasitol.* **62**, 199–202.

Kelley, G. L., and Hammond, D. M. (1970). Development of *Eimeria ninakohlyakimovae* from sheep in cell cultures. *J. Protozool.* **17**, 340–349.

Kheysin, Y. M. (1972). "Life Cycles of Coccidia of Domestic Animals" (English translation by F. K. Plous, Jr.; K. S. Todd, Jr., ed.). Univ. Park Press, Baltimore, Maryland.

Klesius, P. H., Kramer, T., Burger, D., and Malley, A. (1975). Passive transfer of coccidian oocyst antigen and diphtheria toxoid hypersensitivity in calves across species barriers. *Trans. Proc.* **7**, 449–452.

Klesius, P. H., Kramer, T. T., and Frandsen, J. C. (1976). *Eimeria stiedai:* Delayed hypersensitivity response in rabbit coccidiosis. *Exp. Parasitol.* **39**, 59–68.

Lepp, D. L., and Todd, K. S., Jr. (1974). Life cycle of *Isospora canis* Nemeséri, 1959 in the dog. *J. Protozool.* **21**, 199–206.

Levine, N. D. (1962). Protozoology today. *J. Protozool.* **9**, 1–6.

Levine, N. D. (1973a). Introduction, history, and taxonomy. *In* "The Coccidia" (D. M. Hammond and P. L. Long, eds.), pp. 1–22. Univ. Park Press, Baltimore, Maryland.

Levine, N. D. (1973b). Historical aspects of research on coccidiosis. *In* "Proceedings of the Symposium on Coccidia and Related Organisms," pp. 1–10. University of Guelph, Guelph, Ontario.

Levine, N. D. (1973c). "Protozoan Parasites of Domestic Animals and of Man." Burgess, Minneapolis, Minnesota.

Levine, N. D., and Ivens, V. (1965). "The Coccidian Parasites (Protozoa, Sporozoa) of Rodents," Ill. Biol. Monogr. No. 33. Univ. of Illinois Press, Urbana.

Levine, N. D., and Ivens, V. (1970). "The Coccidian Parasites (Protozoa, Sporozoa) of Ruminants," Ill. Biol. Monogr. No. 44. Univ. of Illinois Press, Urbana.

Levine, N. D., Mehra, K. N., Clark, D. T., and Aves, I. J. (1960). A comparison of nematode egg counting techniques for cattle and sheep feces. *Am. J. Vet. Res.* **21**, 511–515.

Liburd, E. M., Pabst, H. F., and Armstrong, W. D. (1972). Transfer factor in rat coccidiosis. *Cell. Immunol.* **5**, 487–489.

Liburd, E. M., Armstrong, W. D., and Mahrt, J. L. (1973). Immunity to the protozoan parasite *Eimeria nieschulzi* in inbred CD-F rats. *Cell. Immunol.* **7**, 444–452.

Long, P. L. (1973). Pathology and pathogenicity of coccidial infections. *In* "The Coccidia" (D. M. Hammond and P. L. Long, eds.), pp. 253–294. Univ. Park Press, Baltimore, Maryland.

Lotze, J. C., and Leek, R. G. (1960). Some factors involved in the excystation of the sporozoites of three species of sheep coccidia. *J. Parasitol.* **46**, Suppl., 46.

Lotze, J. C., Leek, R. G., Shalkop, W. T., and Behin, R. (1961). Coccidial parasites in the "wrong host" animal. *J. Parasitol.* **47**, Suppl., 34.

Lowder, L. J. (1966). Artificial acquired immunity to *Eimeria bovis* infections in cattle. *Proc. Int. Congr. Parasitol., 1st, 1964* Vol. 1, pp. 106–107.

McCully, R. M., Basson, P. A., De Vos, V., and De Vos, A. J. (1970). Uterine coccidiosis of the impala caused by *Eimeria neitzi* spec. nov. *Onderstepoort J. Vet. Res.* **37**, 45–58.

Marquardt, W. C. (1966). Attempted transmission of the rat coccidian *Eimeria nieschulzi* to mice. *J. Parasitol.* **52**, 691–694.

Marquardt, W. C. (1973). Host and site specificity in the coccidia. *In* "The Coccidia" (D. M. Hammond and P. L. Long, eds.), pp. 23–43. Univ. Park Press, Baltimore, Maryland.

Marquardt, W. C., Senger, C. M., and Seghetti, L. (1960). The effect of physical and chemical agents on the oocyst of *Eimeria zurnii*. *J. Protozool.* **7**, 186–189.

Mayberry, L. F., and Marquardt, W. C. (1973). Transmission of *Eimeria separata* from the normal host, *Rattus*, to the mouse, *Mus musculus*. *J. Parasitol.* **59**, 198–199.

Mayberry, L. F., Plan, B., Nash, D. J., and Marquardt, W. C. (1975). Genetic dependence of coccidial transmission from rat to mouse. *J. Protozool.* **22**, 28A.

Metzner, R. (1903). Untersuchungen an *Coccidium cuniculi*. *Arch. Protistenkd.* **2**, 13–72.

Miner, M. L., and Jensen, J. B. (1976). Decoquinate in the control of experimental coccidiosis of calves. *Am. J. Vet. Res.* **37**, 1043–1045.

Mitchell, J. M., Daron, H. H., and Frandsen, J. C. (1975). Fructose-1, 6-diphosphate aldolase from *Eimeria stiedai*. *Abstr., Annu. Meet. Am. Soc. Microbiol.* p. 152.

Newman, A. J., MacKellar, J. C., and Davidson, J. B. (1968). Observations on the use of amprolium in the treatment of acute coccidiosis in calves. *Ir. Vet. J.* **22**, 142–145.

Nyberg, P. A., and Hammond, D. M. (1964). Excystation of *Eimeria bovis* and other species of bovine coccidia. *J. Protozool.* **11**, 474–480.

Nyberg, P. A., Bauer, D. H., and Knapp, S. E. (1968). Carbon dioxide as the initial stimulus for excystation of *Eimeria tenella* oocysts. *J. Protozool.* **15**, 144–149.

Ott, W. H., Dickson, A. M., and Van Iderstine, A. (1965). Amprolium. VIII. Comparison with oxythiamine and pyrithiamine as antagonists of thiamine in the chick. *Poult. Sci.* **44**, 920–925.

Patton, W. H. (1965). *Eimeria tenella:* Cultivation of the asexual stages in cultured animal cells. *Science* **150**, 767–769.

Peardon, D. L., Bilkovich, F. R., and Todd, A. C. (1963). Trials of candidate bovine coccidiostats. *Am. J. Vet. Res.* **24**, 743–748.

Peardon, D. L., Bilkovich, F. R., Todd, A. C., and Hoyt, H. H. (1965). Trials of candidate bovine coccidiostats: Efficacy of amprolium, lincomycin, sulfamethazine, chloroquine sulfate, and di-phenthane-70. *Am. J. Vet. Res.* **26**, 683–687.

Pellérdy, L. (1974). "Coccidia and Coccidiosis," 2nd ed. Akadémiai Kiadó, Budapest.

Peteshev, V. M., Galuzo, I. G., and Polomoshov, A. P. (1974). Cats-definitive hosts of *Besnoitia* (*Besnoitia besnoiti*). *Izv. Akad. Nauk. Kaz. SSR, Ser. Biol. 1974*(1), 33–38 (translated by V. Ivens, College of Veterinary Medicine, University of Illinois, Urbana).

Rindi, G., Ferrari, G., Ventura, U., and Trotta, A. (1966). Action of amprolium on the thiamine content of rat tissues. *J. Nutr.* **89**, 197–202.

Roberts, W. L., Speer, C. A., and Hammond, D. M. (1970). Electron and light microscope studies of the oocyst walls, sporocysts, and excysting sporozoites of *Eimeria callospermophili* and *E. larimerensis*. *J. Parasitol.* **56**, 918–926.

Rommel, M., and Heydorn, A. -O. (1971). Versuche der Übertragung der Immunität gegen *Eimeria*-Infektionen durch Lymphozyten. *Z. Parasitenkd.* **36**, 242–250.

Rose, M. E. (1961). The complement-fixation test in hepatic coccidiosis of rabbits. *Immunology* **4**, 346–353.

Rose, M. E. (1963). Some aspects of immunity to *Eimeria* infections. *Ann. N.Y. Acad. Sci.* **113**, 383–399.

Rose, M. E. (1973). Immunity. *In* "The Coccidia" (D. M. Hammond and P. L. Long, eds.), pp. 295–341. Univ. Park Press, Baltimore, Maryland.

Ryley, J. F. (1973). Cytochemistry, physiology, and biochemistry. *In* "The Coccidia" (D. M. Hammond and P. L. Long, eds.), pp. 145–181. Univ. Park Press, Baltimore, Maryland.

Ryley, J. F. (1975). Why and how are coccidia harmful to their hosts? *Pathog. Process. Parasitic Infect., Symp. Br. Soc. Parasitol., 13th, 1974* pp. 43–58.

Schneider, D., Ayeni, A. O., and Dürr, V. (1973). Zur Resistenz von Kokzidienoocysten gegen Chemikalien. *Dtsch. Tierarztl. Wochenschr.* **80**, 541–564.

Scholtyseck, E. (1973). Ultrastructure. *In* "The Coccidia" (D. M. Hammond and P. L. Long. eds.), pp. 84–144. Univ. Park Press, Baltimore, Maryland.

Scholtyseck, E. (1974). Toxonomy of coccidia. *Proc. Int. Congr. Parasitol., 3rd, 1974* Vol. 1, p. 9.

Sellier, R. (1954). Essais de traitement de la coccidiose bovine par le nitrofural. Thesis, Ecole Nationale Vétérinaire, Lyon.

Senger, C. M., Hammond, D. M., Thorne, J. L., Johnson, S. E., and Wells, G. M. (1959). Resistance of calves to reinfection with *Eimeria bovis. J. Protozool.* **6**, 51–58.

Slater, R. L., Hammond, D. M., and Miner, M. L. (1970). *Eimeria bovis:* Development in calves treated with thiamine metabolic antagonist (amprolium) in feed. *Trans. Am. Microsc. Soc.* **89**, 55–65.

Sloss, M. W. (1970). "Veterinary Clinical Parasitology," 4th ed. Iowa State Univ. Press, Ames.

Smetana, H. (1933). Coccidiosis of the liver in rabbits. I. Experimental study on the excystation of oocysts of *Eimeria stiedae. Arch. Pathol.* **15**, 175–192.

Speer, C. A., and Hammond, D. M. (1972). Development of gametocytes and oocysts of *Eimeria magna* from rabbits in cell culture. *Proc. Helminthol. Soc. Wash.* **39**, 114–118.

Speer, C. A., and Hammond, D. M. (1973). Development of second generation schizonts, gamonts and oocysts of *Eimeria bovis* in bovine kidney cells. *Z. Parasitenkd.* **42**, 105–113.

Speer, C. A., Hammond, D. M., and Elsner, Y. Y. (1973a). Development of second-generation schizonts and immature gamonts of *Eimeria larimerensis* in cultured cells. *Proc. Helminthol. Soc. Wash.* **40**, 147–153.

Speer, C. A., De Vos, A. J., and Hammond, D. M. (1973b). Development of *Eimeria zuernii* in cell cultures. *Proc. Helminthol. Soc. Wash.* **40**, 160–163.

Speer, C. A., Hammond, D. M., Mahrt, J. L., and Roberts, W. L. (1973c). Structure of the oocyst and sporocyst walls and excystation of sporozoites of *Isospora canis. J. Parasitol.* **59**, 35–40.

Stockdale, P. H. G. (1975). Life cycle of *Eimeria zuernii* (Rivolta, 1878) Martin, 1909. *Proc. 50th Annu. Meet. Am. Soc. Parasitol.* p. 102.

Tarlatzis, C., Panetsos, A., and Dragonas, P. (1955). Furacin in the treatment of ovine and caprine coccidiosis. *J. Am. Vet. Med. Assoc.* **126**, 391.

Tarlatzis, C., Panetsos, A., and Dragonas, P. (1957). Further experiences with furacin in treatment of ovine and caprine coccidiosis. *J. Am. Vet. Med. Assoc.* **131**, 474.

Todd, A. C., and Thacher, J. (1973). Suppression and control of bovine coccidiosis with Aureo-S-700[R] medication in the feed. *Vet. Med. & Small Anim. Clin.* **68**, 527–533.

Todd, K. S., Jr. (1975). Internal metazoal and protozoal diseases. *In* "Feline

Medicine and Surgery" (2nd ed.), p. 85–112. American Veterinary Publications, Santa Barbara, California.

Todd, K. S., Jr., and Hammond, D. M. (1968a). Life cycle and host specificity of *Eimeria callospermophili* Henry, 1932 from the Uinta ground squirrel *Spermophilus armatus*. *J. Protozool.* **15,** 1–8.

Todd, K. S., Jr., and Hammond, D. M. (1968b). Life cycle and host specificity of *Eimeria larimerensis* Vetterling, 1964, from the Uinta ground squirrel *Spermophilus armatus*. *J. Protozool.* **15,** 268–275.

Todd, K. S., Jr., and Lepp, D. L. (1971). The life cycle of *Eimeria vermiformis* Ernst, Chobotar and Hammond, 1971 in the mouse, *Mus musculus*. *J. Protozool.* **18,** 332–337.

Todd, K. S., Jr., and Lepp, D. L. (1972). Completion of the life cycle of *Eimeria vermiformis* Ernst, Chobotar, and Hammond, 1971, from the mouse, *Mus musculus,* in dexamethasone-treated rats, *Rattus norvegicus*. *J. Parasitol.* **58,** 400–401.

Todd, K. S., Jr., Lepp, D. L., and Trayser, C. V. (1971). Development of the asexual cycle of *Eimeria vermiformis* Ernst, Chobotar and Hammond, 1971, from the mouse, *Mus musculus,* in dexamethasone-treated rats, *Rattus norvegicus*. *J. Parasitol.* **57,** 1137–1138.

Tyzzer, E. E. (1910). An extracellular coccidium *Cryptosporidium muris* (gen. et sp. nov.), of the gastric glands of the common mouse. *J. Med. Res.* **23,** 487–509.

Wacha, R. S., Hammond, D. M., and Miner, M. L. (1971). The development of the endogenous stages of *Eimeria ninakohlyakimovae* (Yakimoff and Rastegaieff, 1930) in domestic sheep. *Proc. Helminthol. Soc. Wash.* **38,** 167–180.

Wagenbach, G. E., and Burns, W. C. (1969). Structure and respiration of sporulating *Eimeria stiedae* and *E. tenella* oocysts. *J. Protozool.* **16,** 257–263.

Wallace, G. D., and Frenkel, J. K. (1975). *Besnoita* species (Protozoa, Sporozoa, Toxoplasmatidae): Recognition of cyclic transmission by cats. *Science* **188,** 369–371.

Whitlock, H. V. (1948). Some modifications of McMaster helminth egg-counting technique and apparatus. *J. Counc. Sci. Ind. Res.* (*Aust.*) **21,** 177–180.

SUPPLEMENTARY REFERENCES

Becker, E. R. (1934). "Coccidia and Coccidiosis of Domesticated Game and Laboratory Animals and of Man." Collegiate Press, Ames, Iowa.

Becker, E. R. (1956). Catalog of eimeriidae in genera occurring in vertebrates and not requiring intermediate hosts. *Iowa State Coll. J. Sci.* **31,** 85–139.

Davies, S. F. M., Joyner, L. P., and Kendall, S. B. (1963). "Coccidiosis." Oliver & Boyd, Edinburgh.

Hammond, D. M. (1964). "Coccidiosis of Cattle. Some Unsolved Problems." Utah State University, Logan.

Hammond, D. M., and Long, P. L., eds. (1973). "The Coccidia." Univ. Park Press, Baltimore, Maryland.

Honigberg, B. M., Balamuth, W., Bovee, E. C., Corliss, J. O., Gojdics, M., Hall, R. P., Kudo, R. R., Levine, N. D., Loeblich, A., R., Jr., Weiser, J., and Wenrich, D. H. (1964). A revised classification of the phylum Protozoa. *J. Protozool.* **11,** 7–20.

Kheysin, E. M. (1967). "Zhiznennye Tsikly Koktsidii Domashnikh Zhivotnykh." Izd. Nauka, Leningrad. ["Life Cycles of Coccidia of Domestic Animals" (English translation by F. K. Plous, Jr.; K. S. Todd, Jr., ed. Univ. Park Press, Baltimore, Maryland, 1972.]

Kozar, Z. (1970). Toxoplasmosis and coccidiosis in mammalian hosts. *In* "Immunity to Parasitic Animals" (G. J. Jackson, R. Herman, and I. Singer, eds.), Vol. 2, pp. 871–912. Appleton, New York.

Kudo, R. R. (1966). "Protozoology," 5th ed. Thomas, Springfield, Illinois.

Levine, N. D. (1963). Coccidiosis. *Annu. Rev. Microbiol.* **17,** 179–198.

Levine, N. D. (1973). "Protozoan Parasites of Domestic Animals and of Man," 2nd ed. Burgess, Minneapolis, Minnesota.

Levine, N. D., and Ivens, V. (1965). "The Coccidian Parasites (Protozoa, Sporozoa) of Rodents," Ill. Biol. Mongr. No. 33. Univ. of Illinois Press, Urbana.

Levine, N. D. and Ivens, V. (1970). "The Coccidian Parasites (Protozoa, Sporozoa) of Ruminants," Ill. Biol. Monogr. No. 44. Univ. of Illinois Press, Urbana.

Miner, R. W., and Briggs, W., eds. (1949). Coccidiosis. *Ann. N.Y. Acad. Sci.* **52,** 429–624.

Musaev, M. A., and Veisov, A. M. (1965). "Koktsidii Gryzunov SSR" ("Coccidia of Rodents of USSR"). Izd. Acad. Nauk Az., S.S.R., Baku, U.S.S.R. (in Russian).

Orlov, N. P. (1956). "Koktsidiozy Sel'Skokhozyaistvennykh Zhivotnkh." Moscow. ["Coccidiosis of Farm Animals" (English translation by A. Ferber and A. Storfer; H. Mills, ed.), Publ. TT 70–50040. Israel Program for Scientific Translations, Jerusalem, 1970: U.S. Dept. of Commerce, Clearinghouse Fed. Sci. Tech. Inf., Springfield, Virginia, 1970]

Pellérdy, L. (1956). Catalogue of the genus *Eimeria* (Protozoa: Eimeriidae). *Acta Vet. Acad. Sci. Hung.* **6,** 75–102.

Pellérdy, L. (1957). Catalogue of the genus *Isospora* (Protozoa: Eimeriidae). *Acta Vet. Acad. Sci. Hung.* **7,** 209–220.

Pellérdy, L. (1963). "Catalogue of Eimeriidae (Protozoa: Sporozoa)." Akadémiai Kiadó, Budapest.

Pellérdy, L. (1969). "Catalogue of Eimeriidae (Protozoa, Sporozoa)," Suppl. I. Akadémiai Kiadó, Budapest.

Pellérdy, L. (1974). "Coccidia and Coccidiosis," 2nd ed. Parey, Berlin.

"Proceedings of the Symposium on Coccidia and Related Organisms." (1973). University of Guelph, Guelph, Ontario.

Wenyon, C. M. (1926). "Protozoology," 2 vols. Wm. Wood, New York.

Toxoplasma, *Hammondia*, *Besnoitia*, *Sarcocystis*, and Other Tissue Cyst-Forming Coccidia of Man and Animals

J. P. Dubey

I. *Toxoplasma gondii* (Nicolle and Manceaux, 1908) Nicolle and Manceaux, 1909

A. Introduction

Toxoplasma gondii is an intestinal coccidium of felids with an unusually wide range of intermediate hosts. Infection by this parasite is common in many warm-blooded animals including man. The name *Toxoplasma* (toxon = arc, plasma = form) is derived from its crescent shape. *Toxoplasma* was first discovered by Nicolle and Manceaux in 1908 in a rodent, *Ctenodactylus gundi*. At about the same time, Splendore (1908) independently described *Toxoplasma* in a laboratory rabbit in São Paulo, Brazil. Later it was found by Chatton and Blanc (1917) that the gundis were not infected naturally but acquired infection in captivity. Gundis live in the foothills and mountains of southern Tunisia.

They were used in research on leishmaniasis at the Pasteur Institute in Tunis. Chatton and Blanc (1917) suspected that *Toxoplasma* was transmitted by arthropods because it was found in the blood of the host. The authors in Tunis (Chatton and Blanc, 1917) and both Woke *et al.* (1953) and Frenkel (1965) in the United States investigated possible transmission by several species of arthropods with essentially unsuccessful results.

Wolf and Cowen (1937) were the first to report congenital *Toxoplasma* infection in man. Their report stimulated considerable interest in human toxoplasmosis and, within 5 years, Sabin (1942) had characterized the clinico-parasitological aspects of congenital toxoplasmosis in man. Pinkerton and Weinman (1940) reported the first known cases of fatal toxoplasmosis in adult human patients. The development of the "dye test" by Sabin and Feldman (1948) was and still is the key to much of our present knowledge of toxoplasmosis. It is the most specific serological test for toxoplasmosis available. Scientists using the dye test demonstrated that toxoplasmosis was a common infection in man throughout the world. While progress on the characterization of the disease in man and animals has been reported after Wolf and Cowen (1937) described the disease in man, the main routes of transmission remained a mystery. Congenital transmission occurred too rarely to explain widespread infection in man and animals. Weinman and Chandler (1954) suggested that transmission might occur through the ingestion of undercooked meat. Jacobs *et al.* (1960a) provided evidence to support this idea by the demonstration of the resistance of *Toxoplasma* derived from cysts to proteolytic enzymes. They found that the cyst wall was immediately dissolved by such enzymes but the released *Toxoplasma* survived long enough to infect the host. This hypothesis of transmission through the ingestion of infected meat was experimentally tested by Desmonts and associates (1965) in an ingenious experiment in children in a tuberculosis hospital in Paris. They compared the acquisition rates of *Toxoplasma* infection in children before and after admission to the sanitorium. The 10% yearly acquisition rate of *T. gondii* antibody rose to 50% after adding two helpings of barely cooked beef or horse meat and to a 100% yearly rate after the addition of lamb chops to the diet of children. Since the prevalence of *T. gondii* is much higher in sheep than in horses or cattle this illustrated the importance of carnivorism in transmission of *Toxoplasma.* Epidemiological evidence indicates that not only can *Toxoplasma* be transmitted by the ingestion of infected raw meat but it is common in man in some localities where raw meat is eaten (Wallace *et al.,* 1972). In Paris, where it is customary to eat raw meat, over 80% of the adult population has antibody to *Toxoplasma* (Desmonts *et al.,* 1965). Kean *et al.* (1969) described a small epidemic of toxoplasmosis in medical students at Cornell University who had eaten undercooked hamburgers.

While congenital transmission and carnivorism explain some transmission of *Toxoplasma,* they cannot explain the widespread *Toxoplasma* infection in vegetarians and herbivores. Prevalence rates for *Toxoplasma* in strict vegetarians were found to be similar to those in nonvegetarians (Rawal, 1959). Fresh excretions and secretions of animals which had even overwhelming infections proved essentially negative for *Toxoplasma* when tested in mice. Hutchison (1965) first discovered *Toxoplasma* infectivity in feline feces (Fig. 1). He suspected transmission of *Toxoplasma* through the eggs of the nematode *Toxocara cati,* like the transmission of the fragile flagellate *Histomonas* through *Heterakis* eggs. In a preliminary experiment, Hutchison (1965) fed *T. gondii* cysts to a cat infected with *T. cati* and collected feces containing nematode ova. Feces floated in 33% zinc sulfate solution and stored in tap water for 12 months induced toxoplasmosis in mice. A breakthrough had been obtained because, until then, both known forms of *Toxoplasma* were killed by water. Microscopic examination of feces revealed only *T. cati* eggs and *Isospora* oocysts. In his report, *Toxoplasma* infectivity was not attributed to oocysts or *T. cati* eggs. He repeated the experiment with two *T. cati*-in-

Fig. 1. William M. Hutchison in his laboratory in Strathclyde University, Scotland in 1975, 10 years after discovering the fecal transmission of *Toxoplasma gondii* in a cat.

fected and two *T. cati*-free cats. *Toxoplasma* was transmitted only in association with *T. cati* infection. On this basis, Hutchison (1967) hypothesized *Toxoplasma* was transmitted through nematode ova.

The report by Hutchison (1965) stimulated other investigators to examine fecal transmission of *Toxoplasma* through *T. cati* eggs (Dubey, 1966, 1967a, 1968b; Jacobs, 1967; Hutchison *et al.*, 1968; Frenkel *et al.*, 1969; Sheffield and Melton, 1969). The nematode egg theory of transmission was discarded after *Toxoplasma* infectivity was dissociated from *T. cati* eggs (Frenkel *et al.*, 1969) and *Toxoplasma* infectivity was found in feces of worm-free cats fed *T. gondii* (Frenkel *et al.*, 1969; Sheffield and Melton, 1969). Finally, in 1970, knowledge of the *Toxoplasma* life cycle was completed by the finding of the sexual phase of the parasite in the small intestine of the cat (Table I). *Toxoplasma gondii* oocysts, the product of schizogony and gametogony, were found in cat feces and characterized morphologically and biologically (Dubey *et al.*, 1970a,b). Of many species of animals experimentally infected with *T. gondii*, only felines shed *T. gondii* oocysts (Miller *et al.*, Jewell *et al.*, 1972; Janitschke and Werner, 1972). Several group of workers (Table I) independently and about the same time found *Toxoplasma* oocysts in cat feces. The discovery of the *T. gondii* oocyst in cat feces, and its aftermath, has been reviewed by Frenkel (1973a,b), Garnham (1971), and Jacobs (1973).

To finish the transmission story where it began, Ben Rachid (1970) fed *T. gondii* oocysts to *Ctenodactylus gundi.* The gundis died 6–7 days later from toxoplasmosis. This newer knowledge about the life cycle of *T. gondii* explains how gundis probably became infected in the laboratory of Nicolle (Fig. 2). At least one cat was present in the Pasteur laboratory in Tunis.

Other historical landmarks are given in Table I.

B. Structure and Life Cycle

Frenkel *et al.* (1970) and Dubey *et al.* (1970b) proposed a life cycle for *Toxoplasma* on the basis of the new information about the role of the cat in the infection (Fig. 3). Cats, not only the domestic but also wild Felidae, have been shown to be the definitive hosts and a variety of birds and mammals are now recognized to be intermediate hosts for *T. gondii.* The resistance of the oocyst to damage in the environment (Dubey *et al.*, 1970a; Frenkel *et al.*, 1975b) is the key to the understanding of the biology of *Toxoplasma* in nature. In the scheme of Frenkel *et al.* (1970), the three known modes of transmission of *Toxoplasma,* congenital, carnivorism, and fecal, are linked into a life cycle. The resistant oocyst in cat feces assures the dissemination of the parasite.

Cats can acquire infection by ingesting any of the three infectious

Fig. 2. Charles Nicolle in his laboratory in Pasteur Institute, Tunis in 1918 holding a cat 10 years after discovering *Toxoplasma gondii* (reproduced from a Bulletin of the Institute of Pasteur, Tunis, 1956 with permission of Director of the Institute).

stages of *Toxoplasma:* the tachyzoites (in pseudocysts), the bradyzoites (in cysts), and the sporozoites (in oocysts).

The term "Tachyzoite" (tachos = speed in Greek) was coined by Frenkel (1973b) to describe the rapidly dividing forms multiplying in any cell of the intermediate host and nonintestinal epithelial cell of the definitive host. The term tachyzoite relaces the previously used terms trophozoite (trophicos = feeding in Greek) and endodyozoite. Trophozoite is replaced by tachyzoite because the form of *Toxoplasma* which divides in the tissues of the intermediate host is not analogous to the trophozoite stage in the life cycle of other sporozoa. Endodyozoite is replaced because organisms within a cyst also divide by endodyogeny and it might, in addition, not be applicable to some of the forms in the cat's intestinal epithelial cells.

Table I
History of *Toxoplasma* and Toxoplasmosis [a]

Contributors and year	Contribution
Nicolle and Manceaux (1908)	Discovered in gundi
Splendore (1908)	Discovered in rabbit
Mello (1910)	Disease described in a domestic animal, dog
Janku (1923)	Identified in human eye at necropsy
Wolf and Cowen (1937)	Congenital transmission documented
Pinkerton and Weinman (1940)	Fatal disease described in adult humans
Sabin (1942)	Disease characterized in man
Sabin and Feldman (1948)	Dye test described
Siim (1952)	Glandular toxoplasmosis described in man
Hartley and Marshall (1957)	Abortions in sheep recognized
Beverley (1959)	Repeated congenital transmission observed in mice
Jacobs *et al.* (1960a)	Cysts characterized biologically
Hutchison, (1965)	Fecal transmission recognized
Hutchison *et al.* (1969; 1970; 1971); Frenkel *et al.* (1970); Dubey *et al.* (1970a,b); Sheffield and Melton (1970); Overdulve (1970); Weiland and Kühn (1970); Witte and Piekarski (1970); Zaman and Colley (1970)	Coccidian phase described
Frenkel *et al.* (1970); Miller *et al.* (1972)	Definitive and intermediate hosts defined
Dubey and Frenkel (1972a)	Five *Toxoplasma* types described from feline intestinal epithelium
Wallace (1969); Munday (1972a)	Confirmation of the epidemiological role of cats from studies on remote islands

[a] Modified from Frenkel (1973b).

Host cells containing numerous tachyzoites are called clones, "terminal colonies," groups, or pseudocysts. The word pseudocyst has been used because there is no well defined parasitic membrane surrounding a group of tachyzoites.

"Bradyzoites" (Brady = slow in Greek) are counterparts of tachyzoites. Whereas the tachyzoite multiplies rapidly, the bradyzoite is a similar organism multiplying slowly within a cyst. This term was also proposed by Frenkel (1973a) and to replace the term cystozoite used by Hoare (1972). Although cystozoite is a good term, bradyzoite provides a counterpart to tachyzoite.

A cyst is a collection of bradyzoites within a well defined parasitic

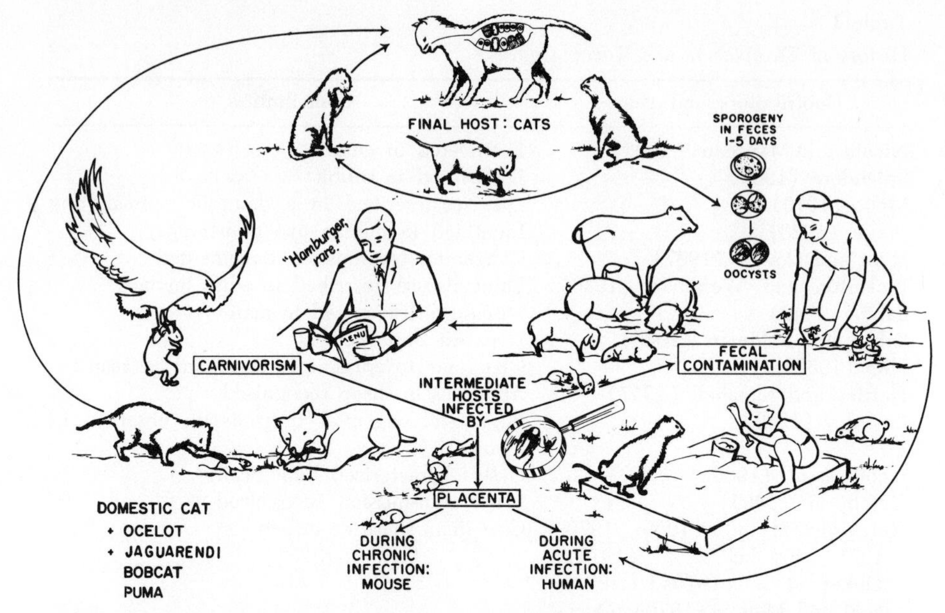

Fig. 3. Postulated transmission of *Toxoplasma gondii.* Cats are shown as definitive hosts and many birds and mammals as intermediate hosts. Oocysts are shed in feces of cats. After sporulation the oocysts are infectious to all warm-blooded hosts that have been experimented. Spread of *Toxoplasma* through oocysts, directly or indirectly, is important in nonfeline hosts. Cats probably acquire infection by preying on intermediate hosts. Man can become infected by ingesting undercooked meat (left) or congenitally (below). The congenital transmission was discovered in 1937, transmission through infected meat in 1965, and through oocysts in 1970. (From Frenkel, 1971a.)

membrane. This parasitic membrane differentiates it from clones or pseudocysts.

"Gametes" are the product of gametogony. The male gamete is called a microgamete while the female is called the macrogamete.

The "oocyst" is the developed zygote within a two-layered wall. The unsporulated oocyst contains a "sporont" or inner mass filling the oocyst. Upon sporulation the sporont divides into two round bodies called "sporoblasts" which later elongate to form sporocysts. Within each sporocyst develop four sporozoites. Confusion between the terms oocyst and cyst may occur, but one should remember that the cyst occurs following endodyogeny in the tissues whereas the oocyst is the result of sexual processes in the cat's intestine. The term tissue cyst is used here to avoid confusion between cyst and oocyst.

The tachyzoite is often crescent-shaped and is about 2×6 μm (Fig. 4). Its anterior end is pointed and its posterior end is round. It has a

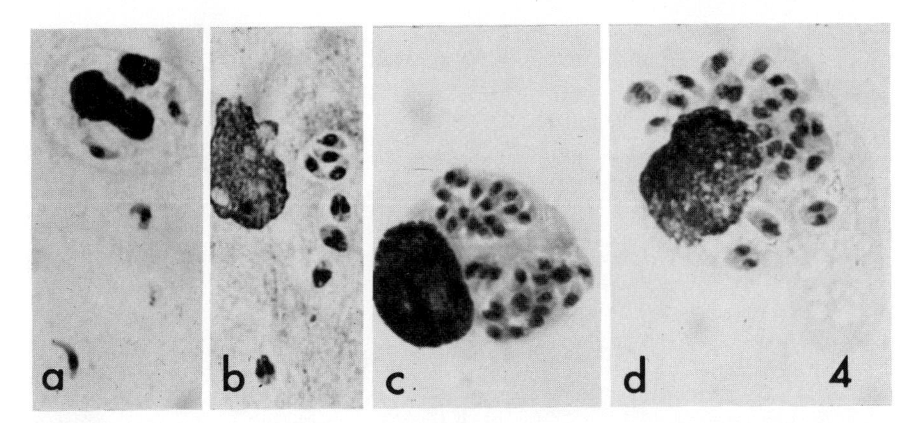

Fig. 4. Impression smears of *Toxoplasma* tachyzoites. Giemsa. × 1250. (a) An extracellular and two intracellular crescent-shaped tachyzoites. Cat lung. (b) Dividing tachyzoites. Note four paired organisms ready to separate from their mothers and four organisms inside the parasitophorous vacuole. Mouse brain. (c) Two separate groups totaling about 32 organisms almost filling the entire host cell cytoplasm. Mouse peritoneum. (d) Tachyzoites rupturing the host cell. Mouse peritoneum.

pellicle (outer covering), polar ring, conoid, rhoptries, micronemes, mitochondria, subpellicular microtubules, endoplasmic reticulum, Golgi apparatus, ribosomes, rough surface endoplasmic reticulum, micropore, and a well defined nucleus (Figs. 5–9) (Sheffield and Melton, 1968; de Souza, 1974). The nucleus is usually situated toward the posterior end or in the central area of the cell. Chromatin is distributed in clumps throughout the nucleus and the nucleolus is usually located centrally within the nucleus.

The pellicle consists of two membranes. The inner membrane is broken at three points: the anterior end (polar ring), on the lateral edge (micropore), and toward the posterior end. The polar ring is an osmiophilic thickening of the inner membrane at the anterior end of the tachyzoite. The polar ring encircles a cylindrical, truncated cone—the conoid (Fig. 5). The conoid consists of one or more convoluted rods or tubules wound like a compressed spring (Figs. 7 and 9). From the polar ring originate 22 subpellicular fibrils (microtubules) that run longitudinally almost the entire length of the cell. Terminating within the conoid are club-shaped organelles called rhoptries (Figs. 8 and 9). The rhoptries are glandlike structures with an anterior narrow neck (Figs. 6 and 7). Their saclike posterior end terminates anterior to the nucleus. They are 8–10 in number (Sulzer *et al.*, 1974). The micronemes (also called toxonemes) are convoluted tubelike structures at the anterior end, and possibly have a common origin with rhoptries. (Mehlhorn *et al.*, 1975).

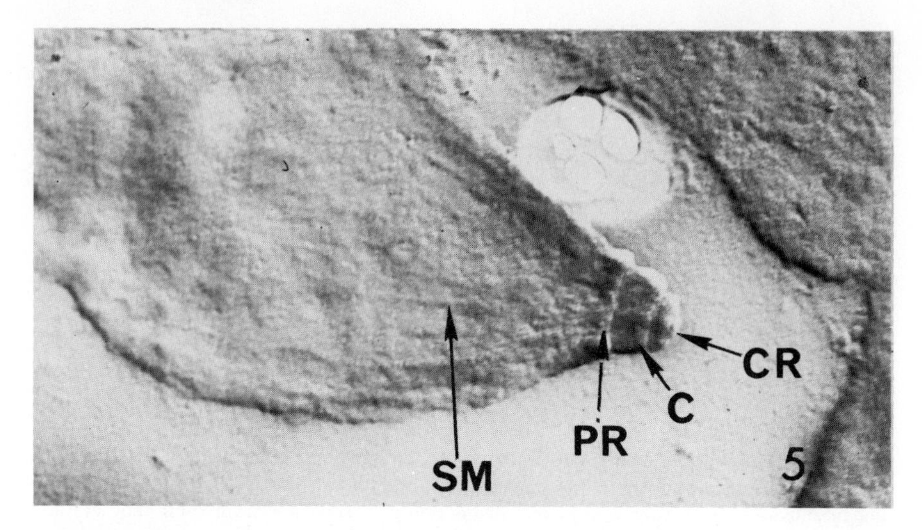

Figs. 5 and 6. Electron micrographs of *Toxoplasma gondii* tachyzoites. (From Sulzer *et al.*, 1974.)

Fig. 5. Carbon platinum replica of whole *Toxoplasma* from mouse peritoneal exudate. The conoid (C) and conoidal ring (CR) are shown protruding beyond the polar ring (PR). Subpellicular microtubules (SM) radiate from the polar ring. × 10,500.

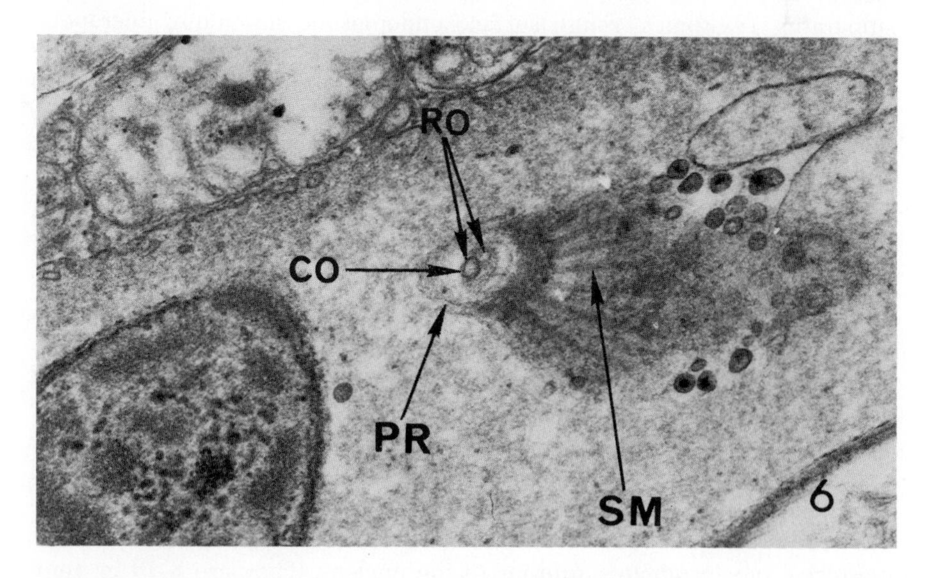

Fig. 6. Ultrathin section of anterior end of tachyzoite. Note a regular distribution of subpellicular microtubules (SM) and the polar ring (PR). Inside the polar ring the conoid appears to have a hollow center, (known as) the conoid opening (CO). Within the conoid are two circular objects that are believed to be openings of rhoptries (RO). × 48,000.

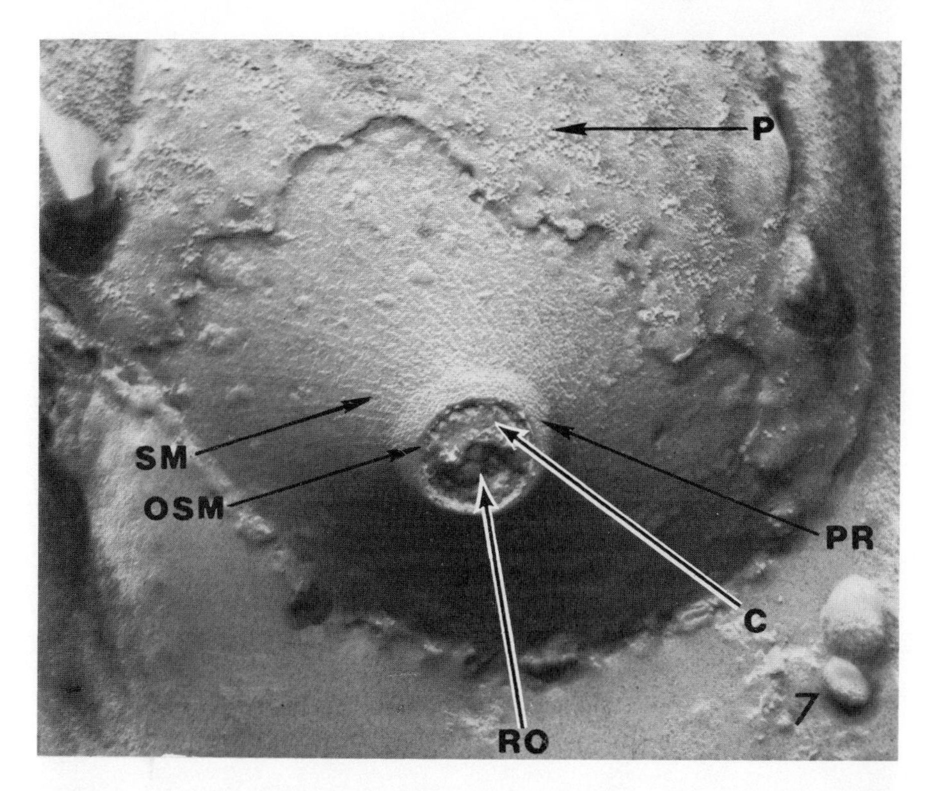

Fig. 7. Electron micrograph of a freeze-etch preparation of the anterior end of a tachyzoite. The pellicle (P) has separated from the anterior end. The subpellicular microtubules (SM) appear as double strands opening within the periphery of the polar ring (PR). (The openings of the subpellicular microtubules are designated OSM.) Rhoptries (RO) appear to open in the conoid (C). × 57,000 (From Sulzer *et al.*, 1974.)

The function of the conoid, rhoptries, and micronemes is not known. It is speculated that rhoptries have a secretory function, secreting their contents into the conoid. The conoid is probably associated with the penetration of the tachyzoite through the membrane of the host cell. The micropore is a cytostomelike structure formed by the invagination of the outer membrane of the pellicle (Fig. 8).

The tachyzoite enters the host cell by active penetration of the host cell membrane and possibly by phagocytosis (Bommer *et al.*, 1969; Jones *et al.*, 1972). Penetration is probably facilitated by secretions of hyaluronidase and lysozyme by rhoptries (Norrby, 1970). After entering the host cell the tachyzoite becomes ovoid in shape. The host cell isolates the tachyzoite by forming a parasitophorous vacuole. *Toxoplasma* can also invade in the host cell nucleus (Remington *et al.*, 1970b).

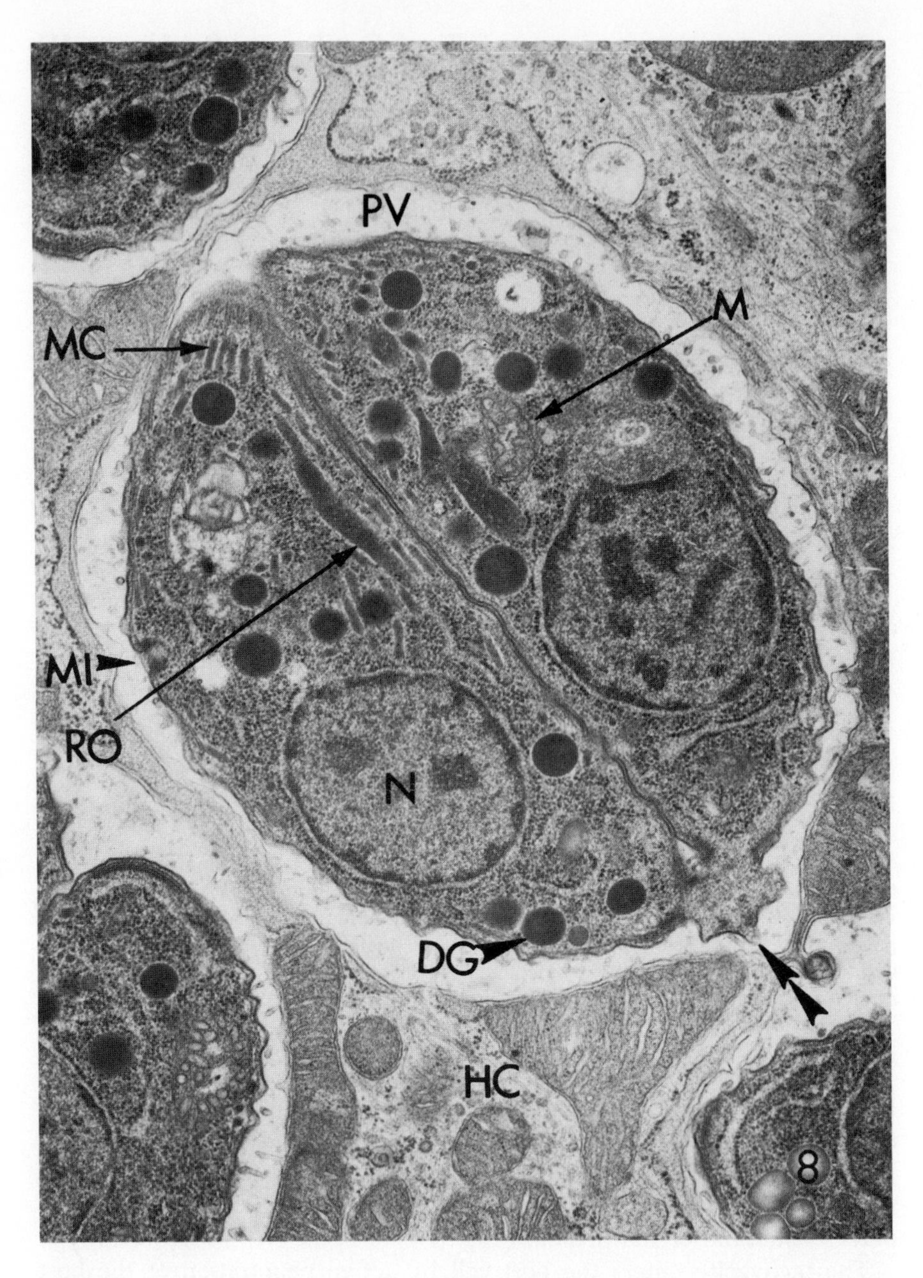

Fig. 8. Electron micrograph of a dividing tachyzoite. Note two fully formed daughter cells ready to separate lying in the parasitophorous vacuole (PV) within the host cell cytoplasm (HC). Each daughter cell has well formed anterior end, micronemes (MC), rhoptries (RO), micropore (MI), dense granules (DG), nucleus (N), and mitochondria (M). The daughter cells are still joined together (double arrow) at their posterior end. Cell culture. × 23,000 (Courtesy of Dr. H. G. Sheffield.)

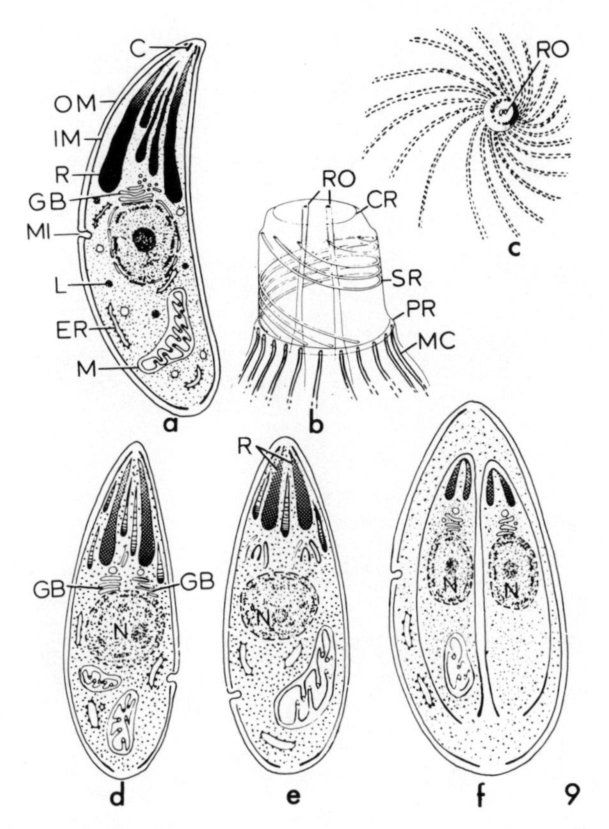

Fig. 9. Schematic drawings of tachyzoites. C, conoid; CR, conoidal ring; ER, endoplasmic reticulum; GB, Golgi body; IM, inner membrane; L, lysosomes; M, mitochondrium; MC, micronemes; MI, micropore; N, nucleus; OM, outer membrane; PR, polar ring; R, rhoptrie; RO, rhoptrie opening. SR, spiral ring. (a) Entire organism in longitudinal section. (b) Three-dimensional representation of the anterior end of the tachyzoite. The conoid is shown consisting of one or more convoluted tubules wound like a compressed spring. From the polar ring originate 22 subpellicular microtubules or micronemes. Terminating in the conoid are rhoptries. (c) Looking down on the anterior end showing the opening of the rhoptrie. (d, e, and f) Divisional (endodyogeny) stages of *Toxoplasma.* (d) First the Golgi complex divides into two at the anterior end of the nucleus. (e) Next the anterior membranes of the daughter cells appear as dome-shaped structures anteriorly. (f) Finally the mother cell nucleus becomes horseshoe-shaped and portions of the nucleus move into the dome-shaped anterior ends of the daughter organisms.

The tachyzoite multiplies asexually by repeated endodyogeny within the host cell. Endodyogeny is a specialized form of division in which two daughter cells are formed within the mother cell (Figs. 4, 8, and 10). The mother cell is destroyed during the formation of the daughter cells. This

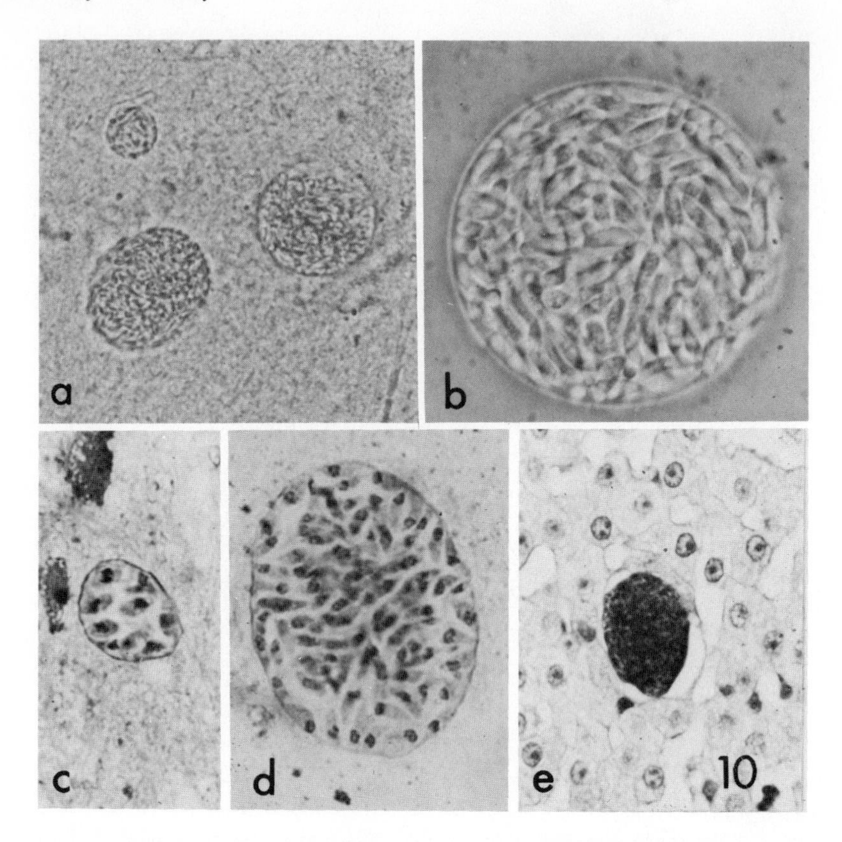

Fig. 10. Tissue cysts of *Toxoplasma*. (a) Squash preparation of an infected mouse brain. Note spherical cysts of different sizes. Unstrained. × 100. (b) A cyst freed from a mouse brain by grinding the brain in a pestle and mortar. Note the resilient cyst wall enclosing hundreds of bradyzoites. Unstained. × 1600. (From Dubey *et al.*, 1970b.) (c and d) Mouse brain preparations. Wilder's followed by Giemsa. × 1250. (c) A small cyst containing eight bradyzoites enclosed in a silver positive membrane. Most of the organisms have central nuclei. (d) A cyst containing many bradyzoites with terminal nuclei. (e) Section of a cyst in cat liver. The bradyzoites contain granules that stain bright red with PAS (periodic acid-Schiff). Here they appear black. Cysts persist in neural and muscular tissues for many months. This is the only cyst identified by the author in a visceral tissue (liver) of many (> 100) mice and cats infected with *Toxoplasma*. × 500. Periodic acid-Schiff hematoxylin.

divisional process has been studied in detail with the electron microscope (Sheffield and Melton, 1968; Galvin *et al.*, 1962; Vivier and Petitprez, 1969; Vivier, 1970). First, the Golgi complex divides into two at the anterior end of the nucleus. Next the anterior membranes of the daughter cells appear as dome-shaped structures anteriorly. The mother cell nucleus becomes horseshoe-shaped and portions of nucleus move into the dome shaped anterior ends of daughter organisms. The daughter cells continue to grow until they reach the surface of the mother cell. The inner membrane of the mother cell disappears and its outer membrane joins the inner membrane of the daughter cells. Tachyzoites continue to divide by endodyogeny until the host cell is filled with parasites (Fig. 4). Organisms within a group rarely divide simultaneously so the progeny are usually arranged at random. However, on occasion, rosettes may be formed by synchronous division of the daughter cells. A group of many tachyzoites surrounded by a parasitophorous vacuole is called a clone.

The tissue cyst is the "resting" stage of the parasite within the host. *Toxoplasma* cysts are usually subspherical (Fig. 10), or conform to the shape of the host cell. The cyst wall is elastic, argyrophilic (Fig. 10), and encloses hundreds or thousands of crescent-shaped bradyzoites. Initially the tissue cyst develops in the host cell cytoplasm and its wall is intimately associated with the host cell endoplasmic reticulum and mitochondria. The cyst wall is ultimately lined by granular material (Fig. 11) which also fills the space between bradyzoites (Wanko *et al.*, 1962).

Bradyzoites differ structurally only slightly from tachyzoites. They usually have a nucleus situated toward the posterior end whereas the nucleus in tachyzoites is more centrally located. They contain several glycogen granules which stain red with periodic acid-Schiff (PAS) reagent; such material is either indiscrete particles or absent in tachyzoites. Biologically, bradyzoites are less susceptible to destruction by proteolytic enzymes than tachyzoites, and infection with them in cats is associated with shorter prepatent periods than infection with tachyzoites (Fig. 12).

Tissue cysts grow intracellularly as the bradyzoites divide by endodyogeny. The tissue cysts vary in size. Young ones may be as small as 5 μm and contain only four bradyzoites, while older ones may have a diameter of 100 μm and contain hundreds of organisms (Fig. 10b). It is not certain whether tissue cysts become extracellular at any stage of development. Although tissue cysts may occur in visceral organs (Fig. 10e), persistent tissue cysts are found mainly in the neural and muscular tissues, including the brain, eye and skeletal and cardiac muscle. Intact tissue cysts probably do not cause any harm and may persist for the life of the host.

The factors influencing tissue cyst formation are not well known. Tissue cysts are more numerous in the chronic stage of infection after the host has acquired immunity. However, tissue cysts can occur in animals infected for only 3–4 days (Dubey and Frenkel, 1976) and in cells in culture systems devoid of known immune factors (Hoff *et al.*,

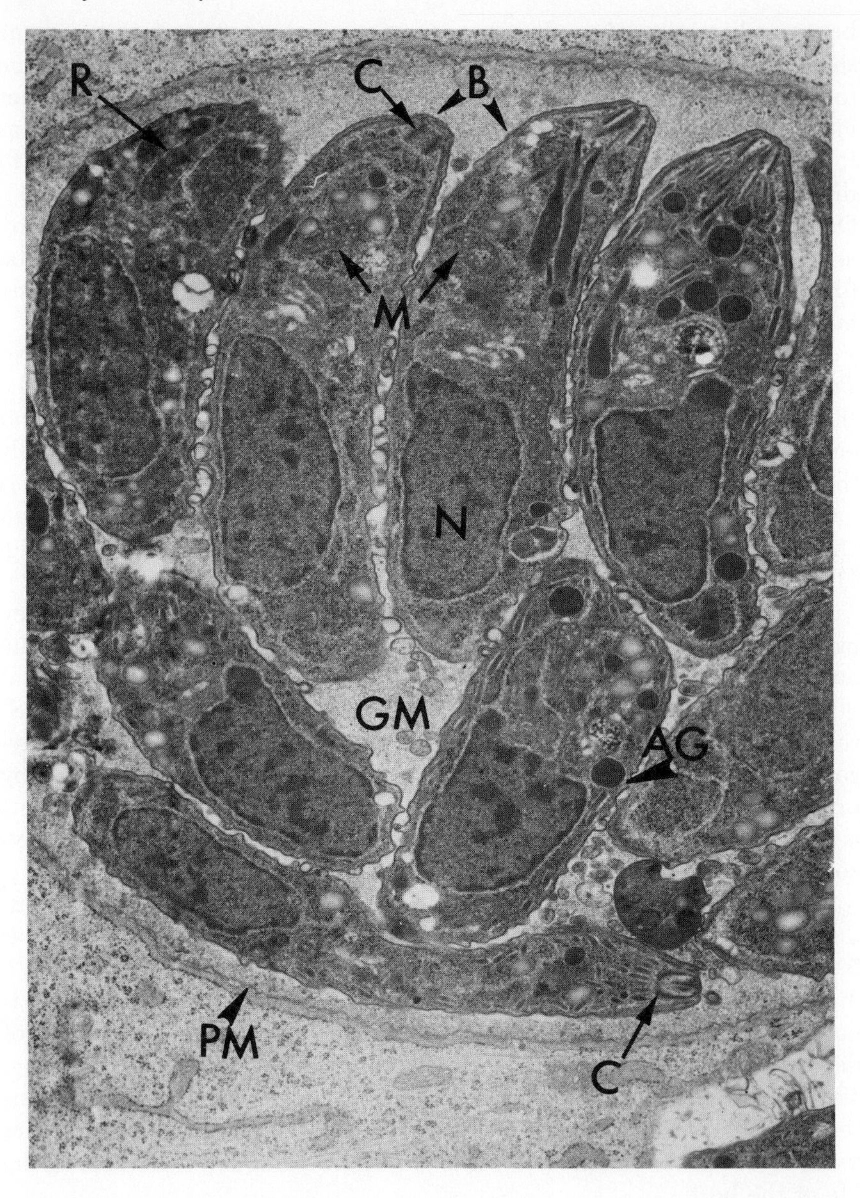

Fig. 11. Electron micrograph of a young tissue cyst. There are eleven bradyzoites (B) in this plane of section. They are enclosed within a well-defined parasitic membrane (PM). Bradyzoites are separated from each other by granular matrix (GM). Bradyzoites have a well-defined conoid (C), rhoptries (R), mitochondria (M), amylopectin granules (AG), and a posteriorly located nucleus (N). Mouse brain. × 14,000. (Courtesy of Dr. H. G. Sheffield.)

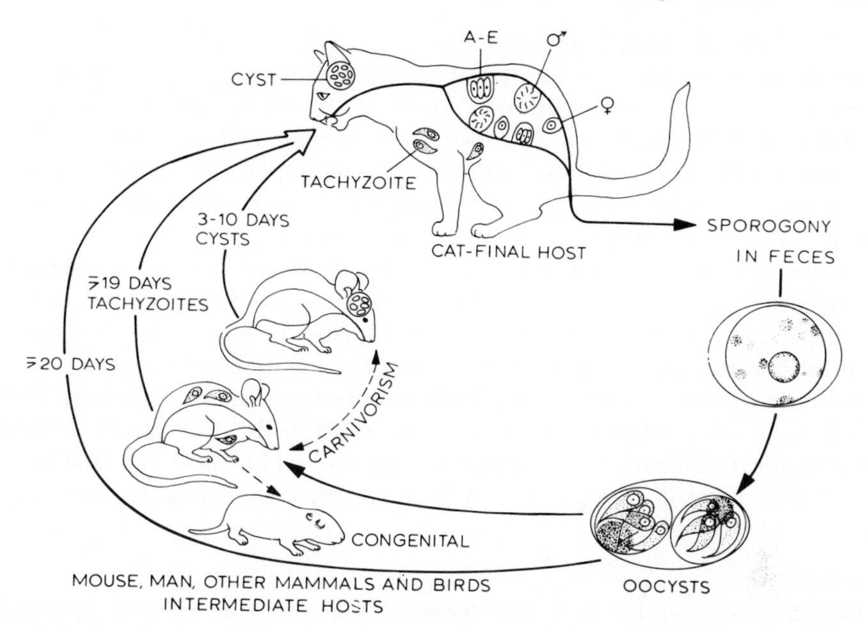

Fig. 12. Life cycle of Toxoplasma. Cats, the definitive hosts, can become infected by ingesting infected animals or sporulated oocysts. Prepatent periods to the shedding of oocysts vary with the stage of *T. gondii* ingested: 3–10 days after ingesting cysts, 19 days or longer after ingesting tachyzoites, and 20 days or longer after ingesting oocysts (Dubey and Frenkel, 1976). In nature it is the cyst-induced cycle in cats that is important because one-half or less numbers of cats ingesting tachyzoites or oocysts shed oocysts whereas almost every uninfected cat sheds oocysts after ingesting cysts (Dubey *et al.*, 1970b). Five types (A–E) of *Toxoplasma* and gametocytes occur in feline intestine after ingesting cysts in addition to the cycle that occurs in nonfeline hosts (Dubey and Frenkel, 1972a). The tachyzoite and oocyst-induced cycles in cats remain to be investigated (From Dubey, 1976b).

1977). Therefore, it is possible that functional immunity and the formation of tissue cysts are only coincidental. Tissue cysts appear to be an essential part of the life cycle of the parasite since cystless strains of *Toxoplasma* are not known. Even the "RH" strain of *Toxoplasma,* which has been subcultured as tachyzoites every few days for 36 years in nonimmune mice, will form tissue cysts under suitable conditions.

Cats shed oocysts after ingesting any of the three infectious stages of *Toxoplasma:* tachyzoites (in groups), bradyzoites (in tissue cysts), and sporozoites (in oocysts). Prepatent periods (time to the shedding of oocysts after initial infection) vary according to the stage of *Toxoplasma* ingested (Fig. 12). Less than 50% of cats shed oocysts after ingesting tachyzoites or oocysts whereas nearly all cats shed oocysts after ingesting cysts (Dubey and Frenkel, 1976; Wallace, 1973b).

The form of asexual reproduction of *Toxoplasma* differs from that of the *Eimeria*. In *Eimeria* the host becomes infected by ingesting sporulated oocysts. After excystation the sporozoites penetrate intestinal epithelial cells, round up, and lose some of their cell organelles, transforming into an intracellular parasite called a trophozoite or uninucleate schizont. The latter term is more descriptive and should replace trophozoite (Hammond, 1973). After two or more nuclear divisions, merozoite anlagen are formed and two merozoites are produced from each terminal nucleus (Scholtyseck, 1973). The released merozoites produce either another generation of schizonts or gametocytes. In most *Eimeria* species so far studied, there are fixed numbers of generations of schizonts, and each generation is morphologically distinct from the next generation. The schizogony is restricted mostly to epithelial cells. Intestinal *Eimeria* rarely form schizonts in extraintestinal locations (Lotze *et al.*, 1964).

Unlike *Eimeria*, the asexual cycle of *Toxoplasma* in feline intestinal

Table II

Characteristics of Types of Asexual Stages of *Toxoplasma* in Intestinal Epithelial Cells of Cats Fed Cysts [a]

	Type designation				
	A	B	C	D	E
1. Duration of occurrence	12–18 Hours	24–54 Hours	24–54 Hours	32 Hours–15 days	3–15 Days
2. No. of organisms in groups	2–3	2–30	16–40	2–35	4–24
3. Probable mode of division	Endodyogeny	Endodyogeny; endopolygeny	Schizogony	Schizogony; Endoploygeny; "splitting"	Schizogony
4. Presence of a residual body	No	No	Yes	No	Yes
5. Level of intestine parasitized	Jejunum	Jejunum and ileum	Jejunum and ileum	Jejunum, ileum, and colon [b]	Jejunum, ileum, and colon

[a] From Dubey and Frenkel (1972a).
[b] Rarely in bile duct and gall bladder.

epithelium can be initiated by tachyzoites and bradyzoites in addition to sporozoites. Only the bradyzoite-induced cycle has been studied in detail (Dubey and Frenkel, 1972a).

After the ingestion of cysts by the cat, the cyst wall is dissolved by the proteolytic enzymes in the stomach and small intestine. The released bradyzoites penetrate the epithelial cells of the small intestine and initiate the formation of numerous generations of *Toxoplasma.* Five morphologically distinct types of *Toxoplasma* develop in intestinal epithelial cells before gametogony begins (Dubey and Frenkel, 1972a) (Table II). These stages are designated types A–E instead of generations because there are several generations within one *Toxoplasma* type (Fig. 13), and the mode of division has not been studied in detail.

After entry into epithelial cells (Fig. 14a) of the upper part of the small intestine, the bradyzoites lose their PAS-positive granules and divide into two or three to form type A parasites (Fig. 14b). Type A is the smallest of all the five asexual intestinal *Toxoplasma* types. It occurs as collections of two or three organisms in the jejunum 12–18 hours after infection.

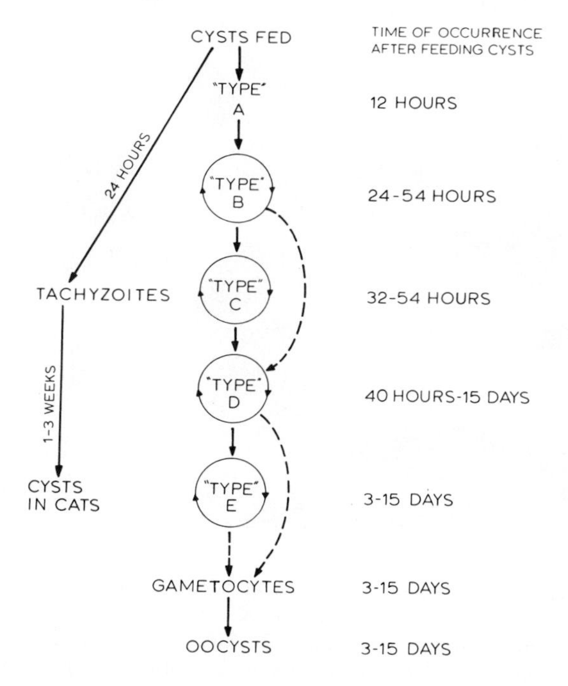

Fig. 13. Life cycle of *T. gondii* in the tissues of cats fed cysts. Areas of uncertainty are indicated by broken lines. Multiplication within a *Toxoplasma* type is indicated by circular arrows. (Modified from Dubey and Frenkel, 1972a.)

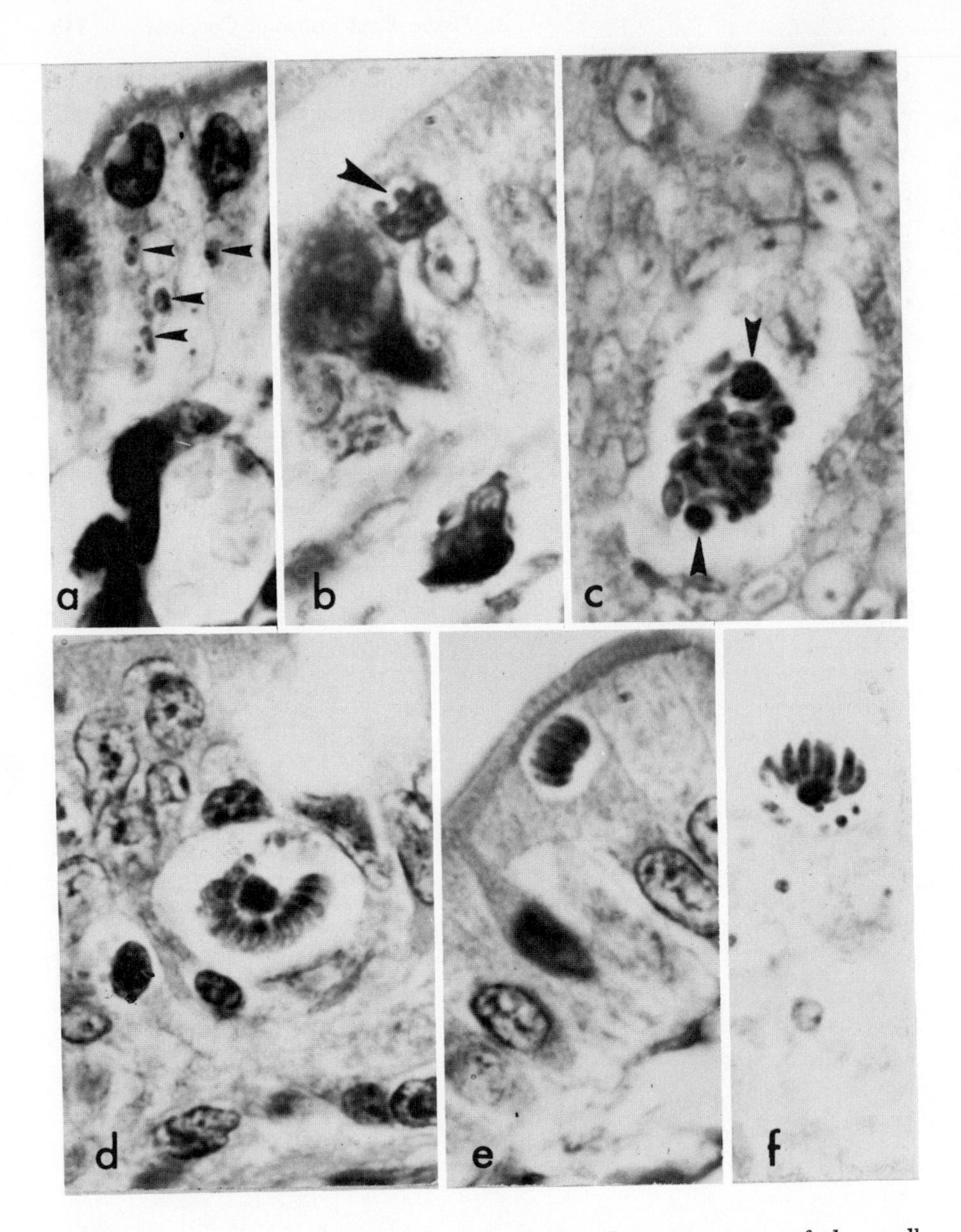

Fig. 14. Asexual enteroepithelial stages of *Toxoplasma* in section of the small intestine of cats after feeding cysts. All × 1600. (From Dubey and Frenkel, 1972a.) (a) Four bradyzoites (arrows) in jejunum. PAS-positive granules in bradyzoites are visible and were used to identify bradyzoites in epithelial cells; 8 hours. Periodic acid-Schiff hematoxylin. (b) A collection of two type A organisms (arrow) in a vacuole of an epithelial cell; 12 hours. Giemsa. (c) Type B. A vacuole with several free uninucleate organisms. At both ends are two undivided organisms (arrows)

Type B organisms are characterized by a centrally located nucleus and a prominent nucleolus. The cytoplasm appears dark blue after staining with Giemsa and the organism is bipolar (Fig. 14c). Type B organisms appear 12–54 hours after infection and presumably divide by simple endodyogeny and multiple endodyogeny (endopolygeny).

Type C organisms are elongated, have subterminal nuclei, and a strongly PAS-positive cytoplasm (Fig. 14d). They occur at 24 to 54 hours after infection and divide by schizogony. The ultrastructures of types A, B, and C organisms have not been described. Type D organisms are smaller than type C (Fig. 14e) and contain only a few PAS-positive granules. They occur from 32 hours after infection until oocysts are shed by the cat, usually for 12 days after infection. Type D organisms appear to divide by endopolygeny (Piekarski *et al.*, 1971), schizogony, and by "splitting" of their merozoites from the main nucleated mass without leaving a residual body. Type E organisms resemble that subtype of type D which divide by schizogony. However, after division of the type E organisms a residual body (Fig. 14f) remains. Type E organisms occur from 3 to 15 days after initial infection with cysts.

So far, only late stages, presumably type D, have been studied by electron microscopy. In these forms, the nucleus divides into two without any apparent cytoplasmic division (Sheffield, 1970; Ferguson *et al.*, 1974). A nuclear division is repeated and then merozoite formation is initiated by the development of an anterior membrane complex near the nucleus. Immature merozoites are formed by the inclusion of a nucleus in the membranes while they are still near the center of the parasite. In mature schizonts, merozoites lie near the outer surface of the schizont (Fig. 15). The outer membrane of the schizont invaginates around each merozoite and merozoites are released from the schizont.

The origin of gametocytes has not been determined. Probably the merozoites released from schizonts of types D and E initiate gamete formation. Gametocytes occur throughout the small intestine but most commonly in the ileum, 3–15 days after infection (Fig. 16). They occur distal to the host epithelial cell nucleus near the tips of villi of small

---

which contain several nuclei. These undivided organisms stain deeply when compared with already divided organisms. Type B frequently causes misorientation of infected host cells. It is not uncommon to find them in the lamina propria; 40 hours. Giemsa. (d) Type C. A mature schizont with merozoites arranged around a residual body. Infected host epithelial cell has probably sunk in the lamina propria. Type C is PAS positive and disappears by the time type E is recognized; 48 hours. Gomori-trichrome. (e) Type D in a vacuole of an epithelial cell. This type is located above the host cell nucleus. There is no residual body; 48 hours. Hematoxylin and eosin. (f) Type E. A schizont with a residual body and merozoites cut at an angle. Giemsa.

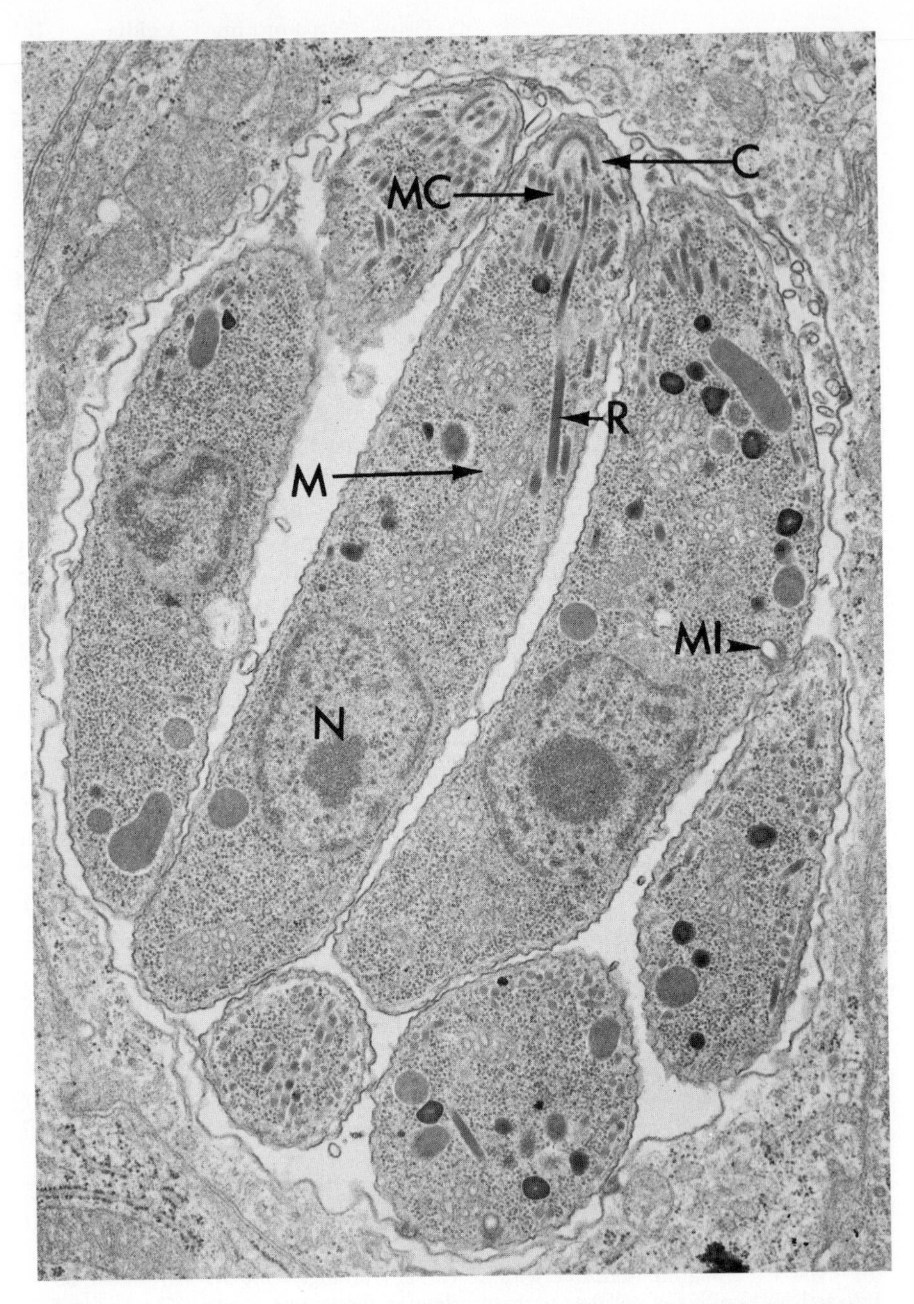

Fig. 15. Electron micrograph of a *Toxoplasma* schizont (probably type D) in the ileum of a cat fed cysts. There are seven mature merozoites. One longitudinally cut merozoite has a conoid (C), a convoluted mitochondrium (M), dense granules, a posteriorly located nucleus (N), rhoptries (R), and micronemes (MC). Micropores (MI) are visible in two other merozoites. × 19,000. (Courtesy of Dr. H. G. Sheffield.)

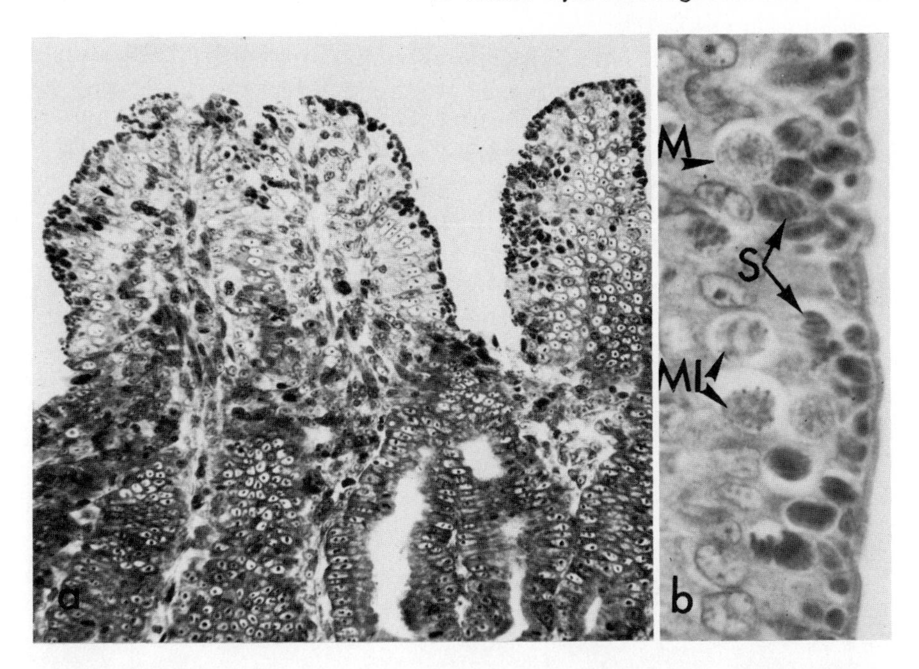

Fig. 16. *Toxoplasma* schizonts and gametocytes heavily parasitizing the villi of ileum of cats fed cysts 7 days previously. (From Dubey and Frenkel, 1972a.) (a) The entire length of villus is involved, but the glands of Lieberkühn are spared. *Toxoplasma* stages stain darker with Giemsa than the host tissue. × 240. (b) Higher magnification to show the degree of parasitism of epithelial cells by schizonts (S), macrogametocytes (M), and microgametocytes (MI). × 1000. Hematoxylin and eosin.

intestine. The female gamete is subspherical and contains a single centrally located nucleus and several PAS-positive granules (Fig. 17). Ultrastructurally, the mature female gamete contains a conoid (retained from the merozoite), several micropores, rough and smooth endoplasmic reticulum, numerous mitochondria, double-membraned vesicles, and wall-forming bodies; the latter two are typical of female gamete. The double-membraned bodies are located near the nucleus and are probably derived from it (Ferguson *et al.*, 1975). The wall-forming bodies (WFB) are of two types: I and II. WFB I are about 0.35 μm in diameter, osmiophilic, and appear before WFB II (Pelster and Piekarski, 1972; Ferguson *et al.*, 1975). The larger WFB II are fewer in number than WFB I (Fig. 17) and measure 1.2 μm in diameter (Ferguson *et al.*, 1975).

Mature male gametocytes are ovoid to ellipsoidal in shape (Figs. 18–21). When microgametogenesis takes place, the nucleus of the male gametocyte divides to produce 10–21 nuclei (Dubey and Frenkel, 1972a).

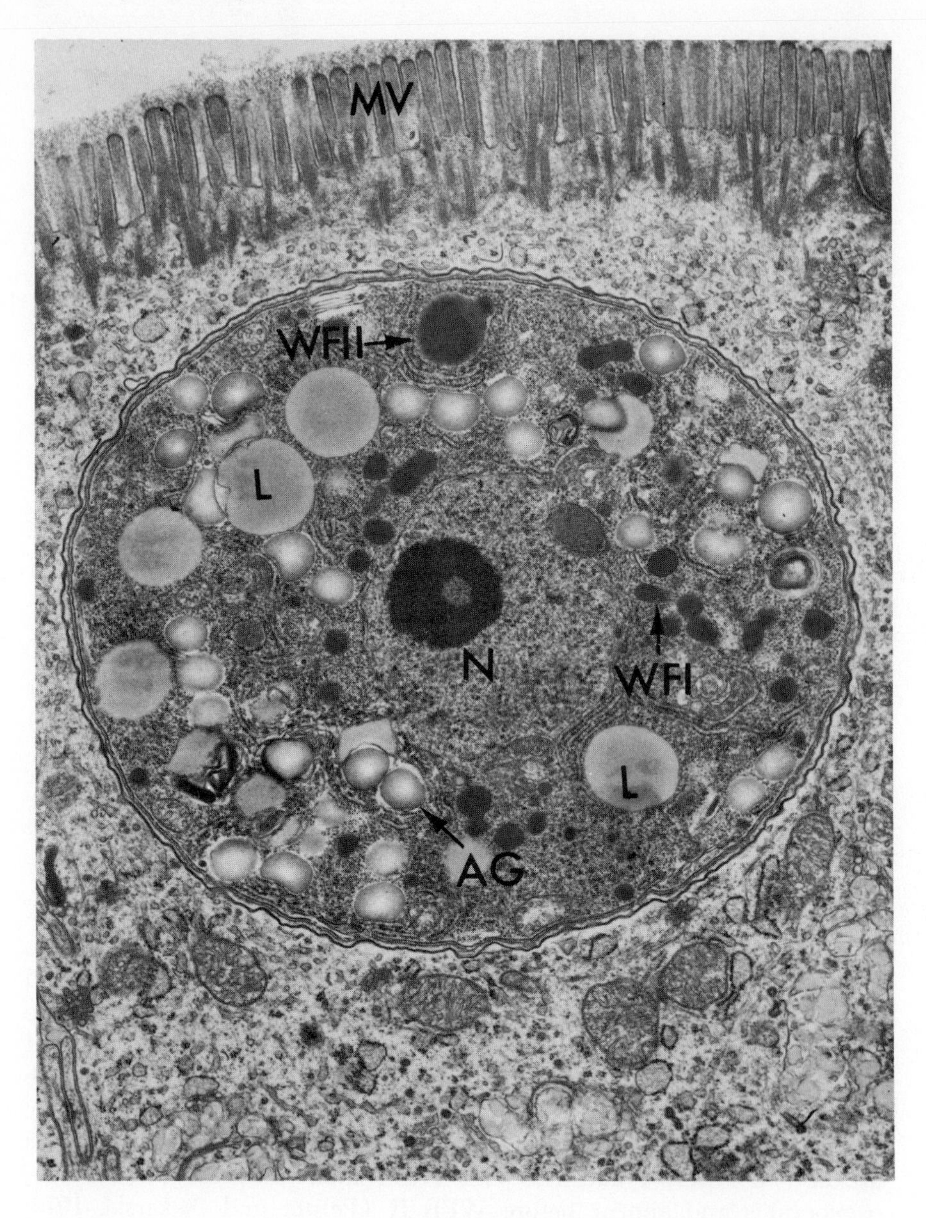

Fig. 17. Electron micrograph of a female gametocyte in the ileum of a cat. The gametocyte is located just below the host cell microvilli (MV) and above the host cell nucleus. The female gametocyte contains a centrally located nucleus (N) and amylopectin granules (AG). WFI, WFII, wall-forming bodies; L, lipid globules. × 12,000. (Courtesy of Dr. H. G. Sheffield.)

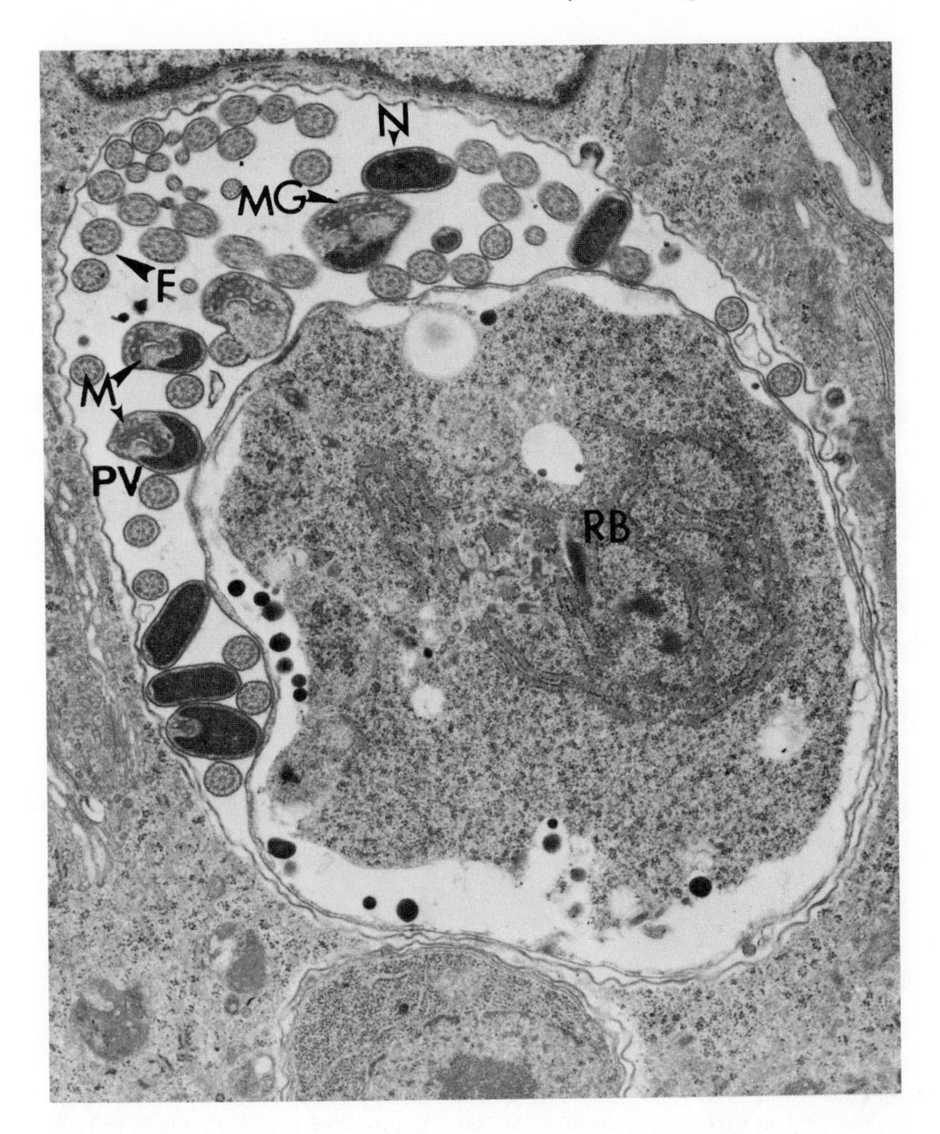

Fig. 18. Electron micrograph of a microgametocyte of *T. gondii*. This microgametocyte contains a residual body (RB) and cross sections of several microgametes (MG) that have extruded in the parasitophorous vacuole (PV). An arrow points to a cross section of a flagellum (F) showing characteristic 9 + 2 arrangements of microtubules. Also shown are the nucleus (N) and mitochondria (M) of individual microgametes. × 15,000. (Courtesy of Dr. H. G. Sheffield.)

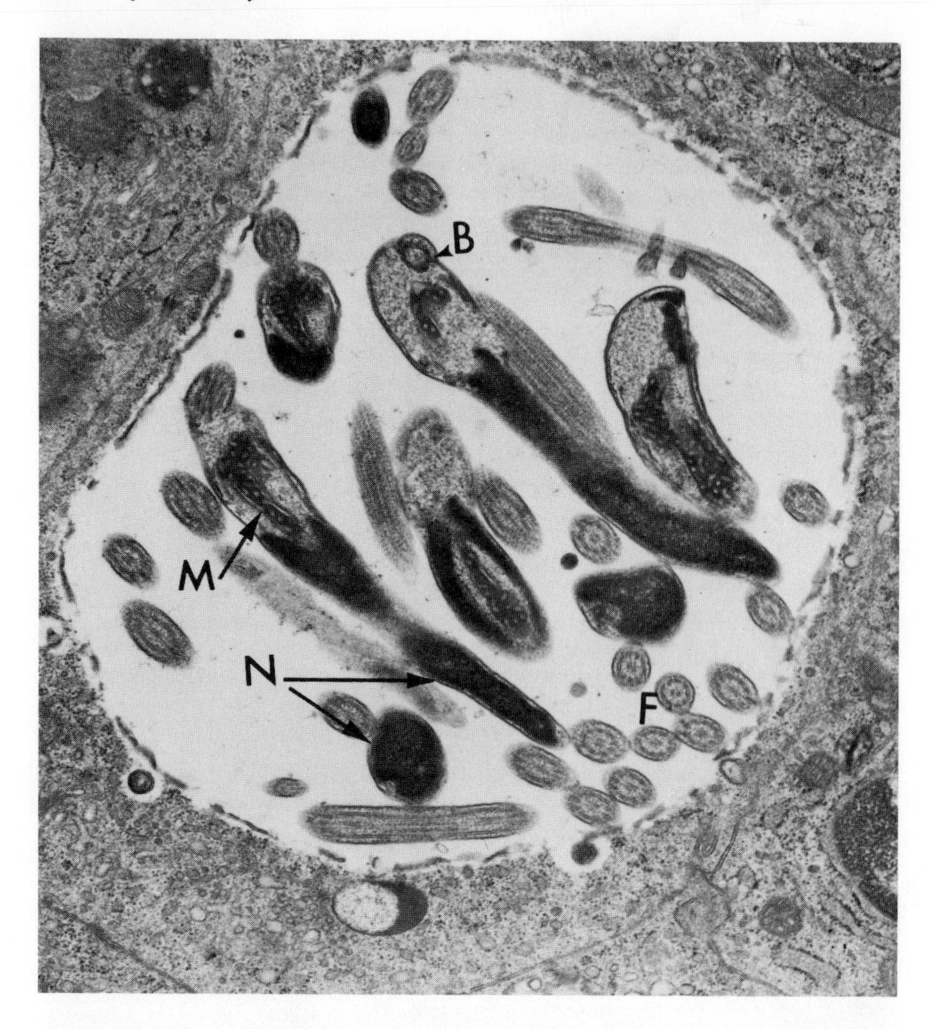

Fig. 19. Electron micrograph of a mature microgametocyte. Several microgametes lie free in the host cell. Each microgamete is biflagellate (F). Each flagellum arises from the basal body (B). There is a long mitochondrium (M) by the side of nucleus (N) in each microgamete. × 14,500. (Courtesy of Dr. H. G. Sheffield.)

The nuclei move toward the periphery of the parasite in protuberances formed in the pellicle of the mother parasite (Colley and Zaman, 1970; Pelster and Piekarski, 1971). One or two residual bodies are left in the microgametocyte after the division into microgametes (Fig. 18). Each microgamete is a biflagellate organism (Fig. 21). Ultrastructurally, microgametes appear to be laterally compressed. Microgametes consist mainly of nuclear material. At the anterior end is a pointed structure, the

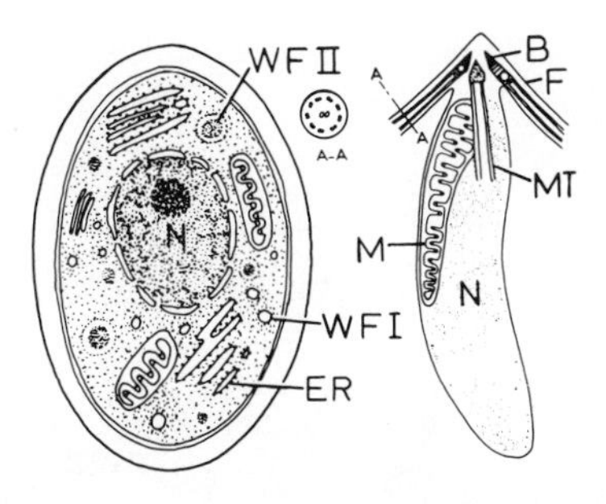

Fig. 20. Schematic drawings of a macrogamete (left) and a microgamete (right). Drawn after Pelster and Pickarski (1971, 1972). A-A cross section of flagellum; A, Basal body; ER, endoplasmic reticulum; F, flagellum; MT, microtubules; M, mitochondrium; N, nucleus; WFI, wall-forming bodies of the first type; WFII, wall-forming bodies of the second type.

perforatorium, in which lie the basal bodies. The two long free flagella originate from the basal bodies (Figs. 19 and 20). There is a large mitochondrium situated anterior to the nucleus. Five microtubules originate and extend posteriorly alongside the nucleus for a short distance. These microtubules may represent the rudiment of the third flagellum found in

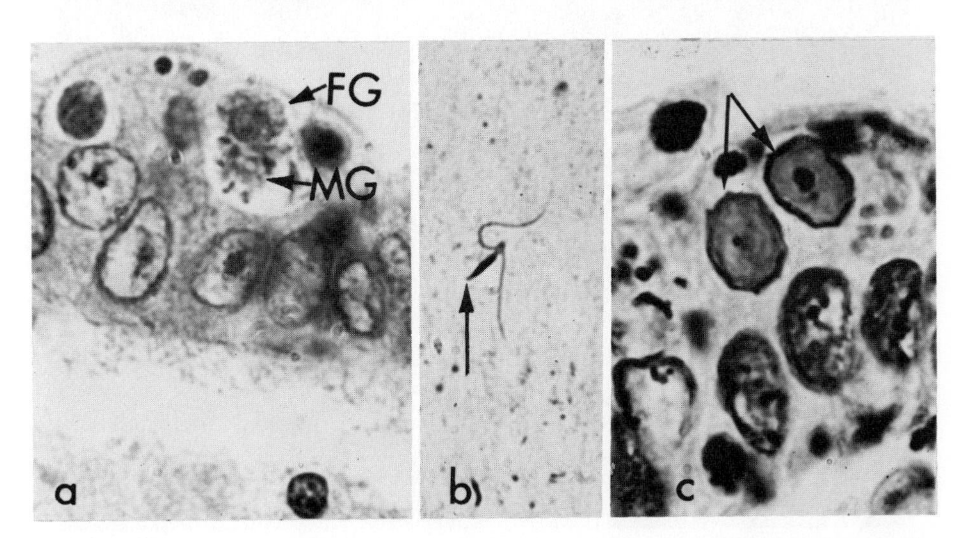

Fig. 21. Appearance of *Toxoplasma* gametocytes and oocysts in the intestine of cats using light microscope. All × 1600. (From Dubey and Frenkel, 1972a.) (a) A female (FG) and a male (MG) gametocyte presumably in a single epithelial cell. Hematoxylin and eosin. (b) A biflagellate microgamete. Impression smear. Giemsa. (c) Sections of two *T. gondii* oocysts in the ileum of a cat. The oocyst walls stain positively with a silver stain. Each oocyst has a centrally located nucleus. Wilder.

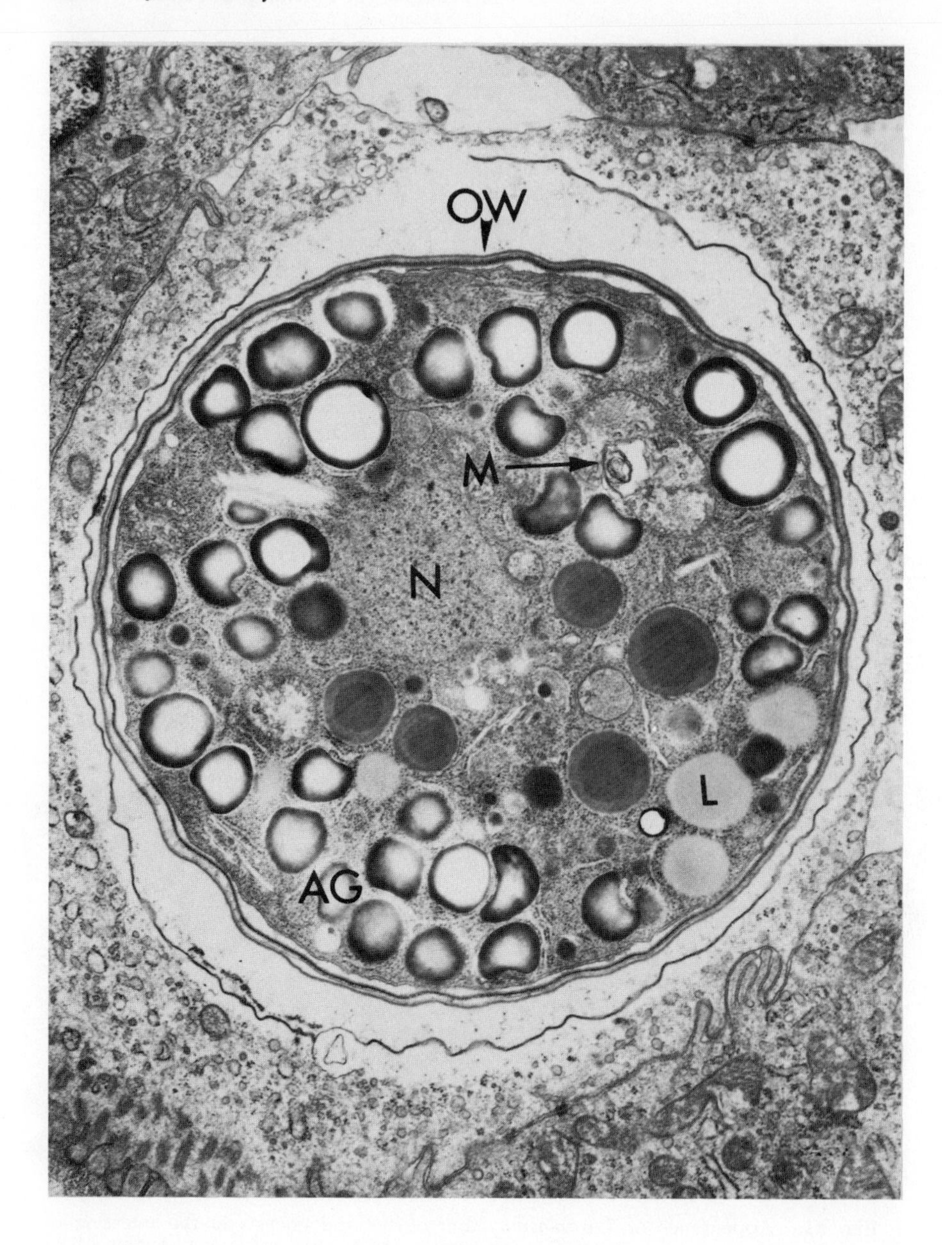

Fig. 22. Electron micrograph of a *Toxoplasma* oocyst. This electron micrograph shows the nucleus (N), amylopectin granules (AG), lipid globules (L), and oocyst wall complex (OW). M, mitochondrium. × 12,000. (Courtesy of Dr. H. G. Sheffield.)

several other coccidia (Scholtyseck, 1973). Male gametocytes are few in number and constitute 2–4% of the mature gametocyte population. The microgametes swim to and penetrate a mature macrogamete. After penetration, oocyst wall formation begins around the fertilized gamete (Fig. 22). According to Ferguson *et al.* (1975), five layers are formed around the pellicle of the gamete. No extensive cytoplasmic changes occur in female gametes while layers 1, 2, and 3 are formed. WFB I disappear with the formation of layer 4 and WFB II disappear when layer 5 is laid down (Ferguson *et al.*, 1975). Oocysts are discharged into the intestinal lumen by the rupture of intestinal epithelial cells when they are mature (Fig. 23).

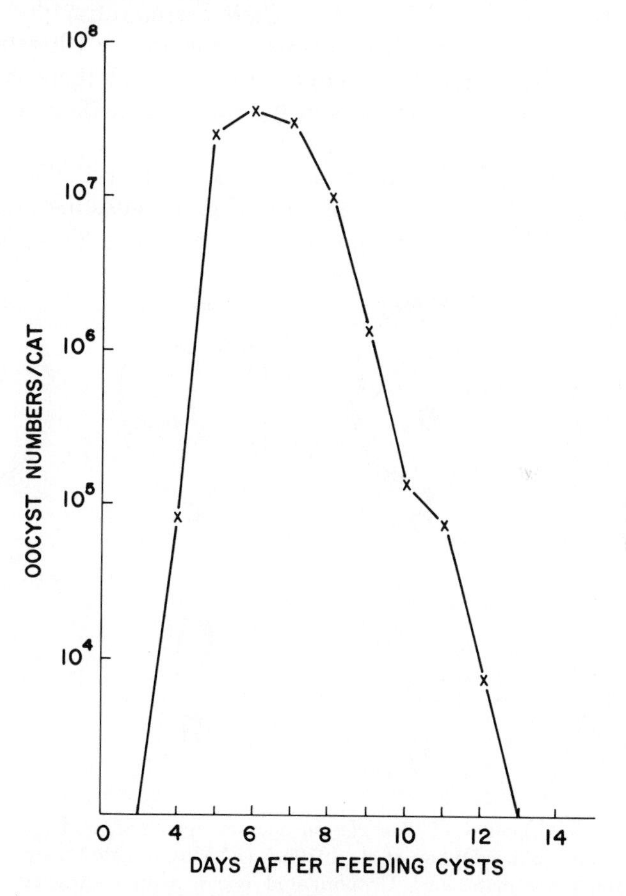

Fig. 23. Average oocyst numbers per day per cat from two weanling kittens each fed one mouse infected with *T. gondii*. Although oocysts are shed for a short period rather large numbers (over a billion) might be shed by each cat. (From Dubey and Frenkel, 1972a.)

Unsporulated oocysts are subspherical to spherical and are 10×12 μm in diameter (Figs. 24a,b). The oocyst wall contains two colorless layers. Micropyle and polar granules are absent. The sporont almost fills the oocyst, and sporulation occurs outside the cat within 1 to 5 days depending upon aeration and temperature.

Sporulated oocysts are subspherical to ellipsoidal and are 11×13 μm in diameter (Figs. 24c,d). Each sporulated oocyst contains two ellipsoidal sporocysts without a Stieda body. Sporocysts measure 6×8 μm (Fig. 24f). A sporocyst residuum is present. There is no oocyst residuum. Each sporocyst contains four sporozoites. The sporozoites are 2×6–8 μm in size with a subterminal to central nucleus and a few PAS-positive granules in the cytoplasm (Figs. 24g). There are no refractile bodies like those commonly found in other eimerian sporozoites. Ultrastructurally, the sporozoite is similar to the tachyzoite except that there is an abundance of micronemes and rhoptries in the former (Sheffield and Melton, 1970).

Simultaneously with the progress of the enteroepithelial cycle, the bradyzoites penetrate the lamina propria of feline intestine and multiply as tachyzoites. Within a few hours after infection, *Toxoplasma* may

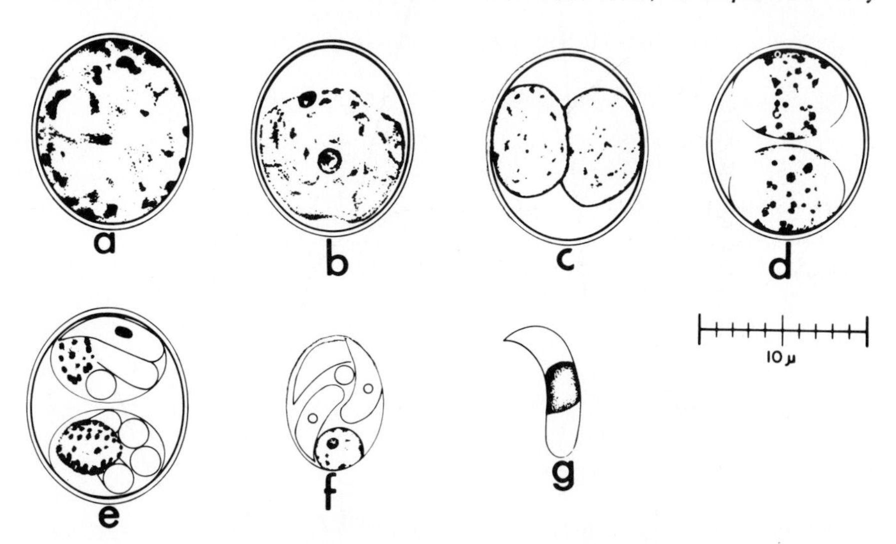

Fig. 24. Line drawings of *Toxoplasma* oocysts, sporocyst, and sporozoite drawn to the same scale. (From Dubey *et al.*, 1970b.) (a) Unsporulated oocyst with sporont occupying the inner mass. (b) Unsporulated oocyst with contracted sporont. (c) Oocyst with two sporoblasts. (d) Oocyst with two sporocysts. Note nuclei at both ends. (e) Sporulated oocyst with two sporocysts containing sporozoites. Note variation in sporocyst residua. (f) Sporocyst with sporozoites and a residual mass. (g) Sporozoite with a nucleus.

disseminate to extraintestinal tissues of cats. *Toxoplasma* were recovered from mesenteric lymph nodes of cats between 2 and 8 hours after feeding cysts (Dubey and Frenkel, 1972a). *Toxoplasma* persists in intestinal- and extraintestinal tissues of cats for at least several months, if not for the life of the cat (Dubey, 1967, 1977b; Dubey and Beverley, 1967). The extra-intestinal cycle of *Toxoplasma* in the cat is similar to that of nonfeline hosts with two exceptions: (1) tachyzoites have not been demonstrated in feline intestinal epithelial cells, whereas they do occur in nonfeline hosts (Dubey and Frenkel, 1973), and (2) *Toxoplasma* types D and E are noninfectious to mice by any route (Dubey and Frenkel, 1976). Therefore, it is concluded that feline enteroepithelial forms do not give rise directly to tachyzoites.

C. Cultivation

Toxoplasma has not been grown in cell-free media although *Toxoplasma* contains cytochromes and glycolytic and respiratory systems. The extracellular tachyzoites will synthesize their own RNA and DNA (Remington *et al.*, 1970c; Perrotto *et al.*, 1971) and they have been cultivated in tissue culture of anucleated cells in culture systems (Jones, 1973; Sethi *et al.*, 1973). These observations suggest that *Toxoplasma* can utilize preformed nucleic acids. *Toxoplasma* can be cultivated in laboratory animals, chick embryos, and cell cultures. Mice, hamsters, guinea pigs, and rabbits are all susceptible but mice are generally used as hosts because they are more susceptible than the others and are not naturally infected when raised in the laboratory on commercial dry food free of cat feces.

Tachyzoites grow in the peritoneal cavity, sometimes producing ascites, and also grow in most of the other host tissues after intraperitoneal inoculation with any of the three infectious stages of *T. gondii* in mice. Virulent strains usually produce illness in mice and sometimes kill them within 1 to 2 weeks. Tachyzoites of a virulent strain can be aspirated from the peritoneal cavity after light anesthesia. Avirulent strains grow slowly in mice and free tachyzoites of these strains are often difficult to obtain. However, tachyzoites of avirulent strains can be obtained from the peritoneal cavity or other organs of the mice immunosuppressed with corticosteroids (cortisone acetate, 2.5 mg, injected subcutaneously twice weekly).

Frequent rapid passage of tachyzoites of low virulence may increase the virulence. Strains become more virulent for mice after rapid intraperitoneal passage in mice (Table III). In this study the pathogenicity of tachyzoites derived from a strain maintained in the cyst stage was modified by twice weekly intraperitoneal passage in mice. Using tachyzoite inocula of the first passage, the ID_{50} was 2.56 log greater than the

Table III

Comparison of Pathogenicity of First and Sixty-Second Passages of Tachyzoites of M-7741 Strain with Tachyzoites of RH Strain in "CFW" Mice [a]

	Toxoplasma strain							
	M-7741				RH			
	"First" passage		62nd passage		> 1500 (30 years) (1971)		1956 [b]	
	Day of death [c]		Day of death [c]		Day of death [c]		Day of death [c]	
Minimal infecting dose	Individual	Mean	Individual	Mean	Individual	Mean	Individual	Mean
10^5	16,17,20,+,+,+	17.6	10,10,11,15	11.5	5,5,5,6,6,6	5.5	5,5,5,5,5,6	5.1
10^4	17,20,36,+,+,+	24.3	11,11,11,11	11.0	5,6,6,6,6,7	6.0	5,6,6,6,6,6	5.8
10^3	14,16,16,21,30	19.4	11,12,12,14	12.3	6,6,7,7,7,7	6.6	6,6,7,7,7,7	6.6
10^2	21,+,+,+,+,+	21	12,13,13,13	12.8	7,7,7,7,8,8	7.3	7,7,8,8,8,8	7.6
10^1	12,14,+,+,+,+	13	12,13,14,14	13.3	8,8,8,8,9,9	8.3	8,9,9,9,9,10	9.0
10^0	+,+,−,−,−,−		15,15,17,+	15.6	8,9,11,−,−,−	9.3	9,9,10,−,−,−	9.3
10^{-1}	−,−,−,−,−,−		−,−,−,−		−,−,−,−,−,−		−,−,−,−,−,−	
LD_{50}	$10^{-3.19}$		$10^{-5.0}$		$10^{-6.0}$		$10^{-6.0}$	
ID_{50}	$10^{-5.75}$		$10^{-5.0}$		$10^{-6.0}$		$10^{-6.0}$	

[a] Four to six mice per group, inoculated intraperitoneally. Dubey and Frenkel, unpublished.

[b] Frenkel (1956).

[c] +, Cysts or antibody; −, no antibody.

LD_{50}, and one-half of the mice inoculated with at least 10^4–10^5 infectious tachyzoites survived. By the time of the 30th mouse passage the difference between the ID_{50} and LD_{50} of the tachyzoites had all but disappeared, and the same was found in cyst- and oocyst-produced infections derived from these tachyzoites (unpublished). After the 62nd passage, the ID_{50} and LD_{50} were identical (Table III).

With the RH strain, which has been passed twice weekly for over 30 years in mice (Sabin, 1941), the ID_{50} and LD_{50} have been identical for years; time to death with the RH strain was ⅔ to ½ the time to death following administration of similar doses of the 62nd passage of parasites of the M-7741 strain. Survival time of mice infected with the RH strain was identical 15 years earlier (Frenkel, 1956).

Repeated frequent passages of a strain appear to modify some other biological characteristics of a strain. For example, old laboratory strains like the RH strain and some lines derived from the Beverley and M-7741 strain no longer produce oocysts after cysts are fed to cats (Frenkel *et al.*, 1969, 1976). Loss of oocyst production by mouse passage can perhaps be avoided by storing the oocysts at 4°C. I have observed oocysts to survive for 27 but not 29 months at 4°C. However, storage for longer than 1 year is not advised since the number of infectious oocysts decreases sharply between 12 and 18 months.

Toxoplasma tachyzoites will multiply in many cell lines in cell cultures. Mouse "virulent" strains rapidly destroy the cells while "avirulent" strains grow slowly causing minimal cell damage (Fig. 25). The mean generation time of *Toxoplasma* tachyzoites of the "RH" strain is 5 hours (Maloney and Kaufman, 1964; Jones, 1974). Unlike the situation which occurs after passage in mice, the passage of tachyzoites in cell culture is not known to alter the virulence of the organism.

Chick embryos can be infected by any route, but the chorioallantoic membrane route is most commonly used. *Toxoplasma* tachyzoites form lesions on the chorioallantoic membrane like those produced by poxviruses.

Tissue cysts are obtained by injecting tachyzoites, bradyzoites, or oocysts into mice. To obtain cysts from mice inoculated with a virulent strain, it is necessary to administer anti-*Toxoplasma* chemotherapy to prevent death from acute toxoplasmosis before cysts form. Sulfadiazine sodium is effective in controlling the acute stages of toxoplasmosis in mice. The effective dose of sulfadiazine varies with the virulence of the *Toxoplasma* strain. The administration of sulfadiazine sodium in the drinking water at a concentration of 15 mg in 100 ml from day 4 to 12 of infection is effective in controlling the M-7741 strain of *T. gondii* but will not control the "RH" strain. For the control of this strain, 125 mg of

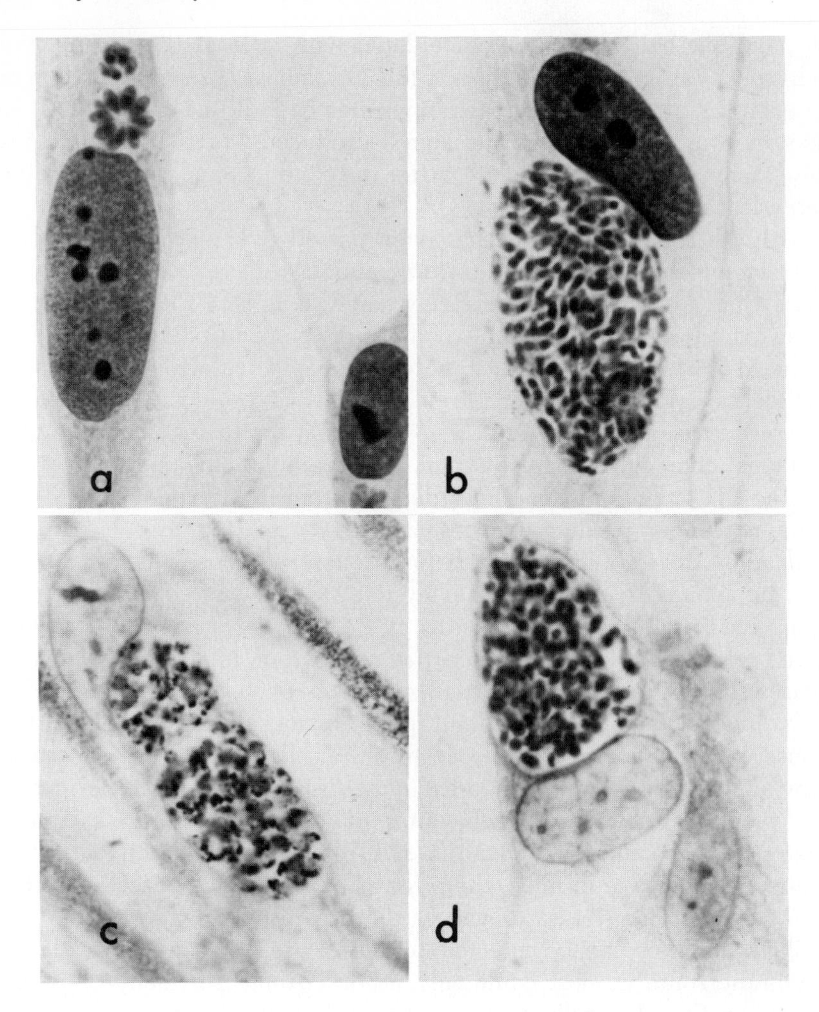

Fig. 25. *Toxoplasma* stages in cell culture (fibroblasts from the skin of a wooly monkey) after inoculation with tachyzoites of the BWM strain. The BWM strain was isolated from the skin of a naturally infected monkey. This cell line supports BWM *Toxoplasma* for many months. All × 1500. (From Hoff *et al.,* 1977.) (a) Two groups of tachyzoites, one in rosette form. Fourth day culture. Giemsa. (b) Cystlike structure. Fifth day culture. Giemsa. (c) Cystlike structure with prominent PAS-positive granules. Fifth day culture. PASH. (d) Cystlike structure with argyrophilic cyst wall. Wilder's. The formation of cysts in cell culture can vary with *Toxoplasma* strains, host cell growth rate, and selective manipulations. Cysts formed in the above cultures within 10 days of fibroblasts infection in the absence of all known immune factors. Addition of antibody enhanced the rate and speed of development of cysts in cell culture.

sulfadiazine sodium in each 100 ml of drinking water is needed from day 1 to 21 after infection. Tissue cysts are easily recognized in the mouse brain about 8 weeks after infection. They can be separated from the tissue by grinding the brain in a mortar and suspending the homogenate in saline. Many cysts remain intact after grinding the brain tissue.

The only report describing the development of schizontlike structures in cell culture is that of Azab and Beverley (1974). They grew schizonts in human embryonic lung cells bathed in Eagle's minimum essential medium supplemented with 10% or more fetal calf serum. With lower concentrations of calf serum, schizonts were absent but tachyzoites were numerous. Thus far feline enteroepithelial stages of *Toxoplasma* have not yet been cultivated *in vitro*. The schizontlike structures reported by Azab and Beverley (1974) do not appear to be *Toxoplasma* types in the cat (Dubey and Frenkel, 1972a).

Oocysts can be obtained by feeding tissue cysts from chronically infected mice to *Toxoplasma*-free cats. It is preferable to use recently weaned (10–12 weeks) kittens because they are easy to handle, their bowel movements are more regular, and they are less likely to be naturally infected at this age. As many as 30–60% of adult cats may be naturally infected with *Toxoplasma* (Dubey, 1968a,b). For feeding *Toxoplasma* to cats, a portion of the brain of an infected mouse should be crushed between a cover glass and glass slide and examined under the microscope to ascertain the presence of tissue cysts. The unhomogenized brain, which contains over 90% of the cysts present in the whole mouse (author's opinion), may be fed to the cat by placing it at the back of the tongue of the cat. The rest of the mouse carcass may be fed to the cat after cutting into small pieces or blending it in a chilled Waring blendor. Some cats will not eat mice or will vomit when "force fed" mice even after they have been starved for several days. Such cats can be lightly sedated using ketamine hydrochloride and mouse homogenate can then be "force fed" by placing the inocula on the back of the tongue. Swallowing reflexes remain unaffected by ketamine hydrochloride. Normally oocysts will appear in cat feces 3–8 days after the cat ingests tissue cysts.

D. Host–Parasite Relationships

Toxoplasma usually parasitizes the host (both definitive and intermediate) without producing clinical signs. Only rarely does it cause severe clinical manifestations. The majority of natural infections are probably acquired by ingestion of cysts in infected meat or oocysts from food contaminated with cat feces. The bradyzoites from the cysts or sporozoites from the oocyst penetrate the intestinal epithelial cells and multiply in the intestine (Fig. 26a). *Toxoplasma* may spread locally to

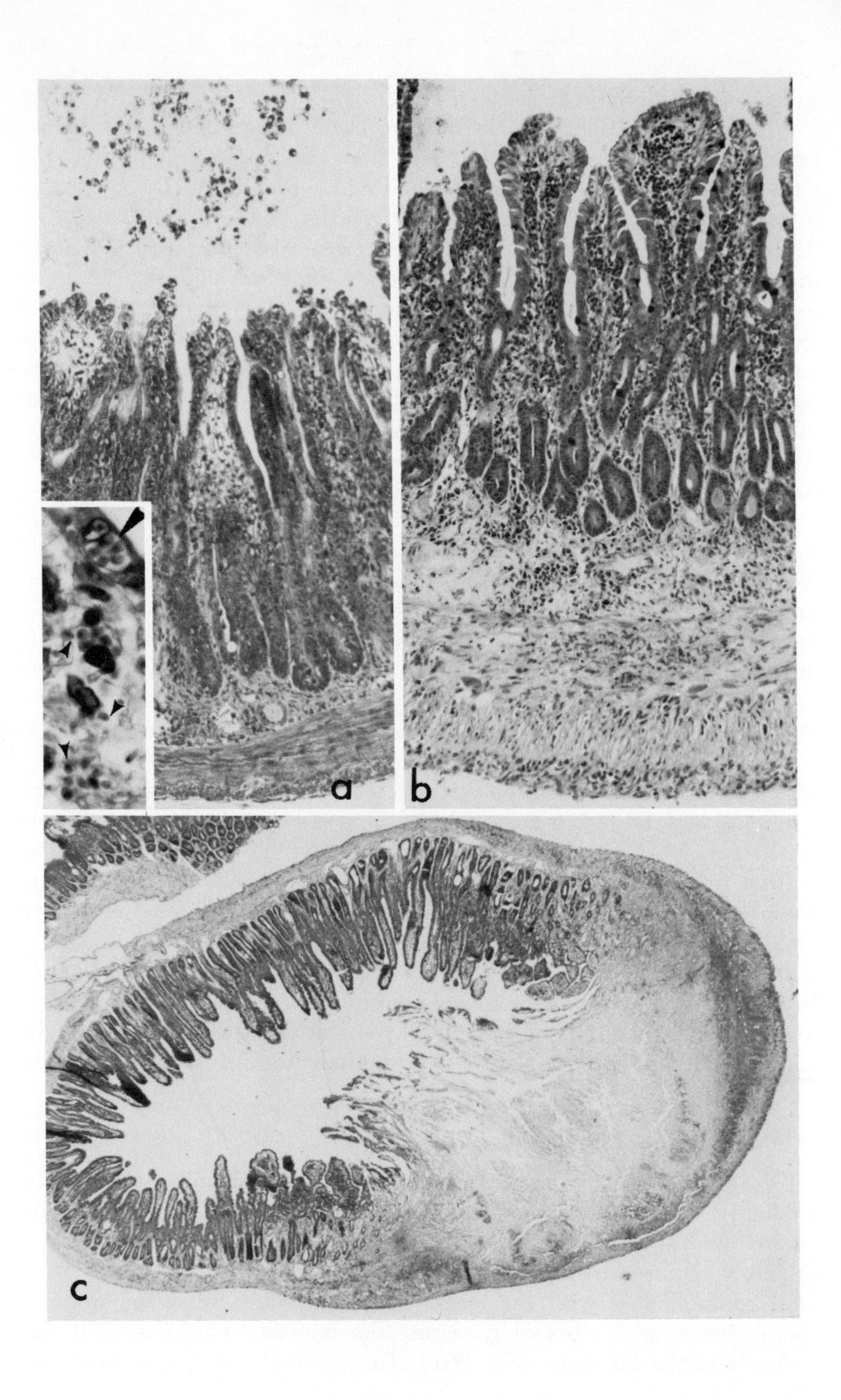
a
b
c

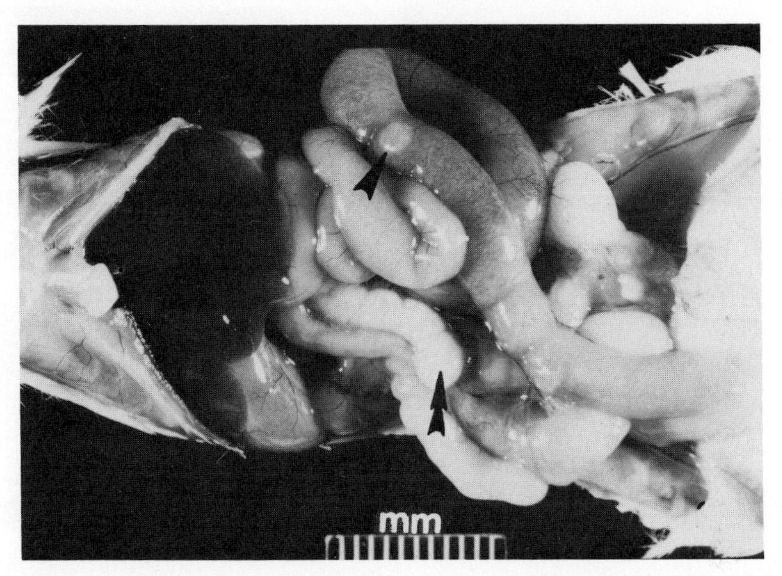

Fig. 27. Enteritis and mesenteric lymph node necrosis in a mouse 6 days after the ingestion of 10^4 oocysts of the M-7741 strain of *Toxoplasma*. Portions of the intestine are inflamed (prominent blood vessels). Peyer's patches (arrow) are prominent and mesenteric lymph nodes (double arrow) are necrosed. Tachyzoites are easily demonstrable in the smears of such a lymph node. Numerous tachyzoites also occur in the inflamed intestine but smears should be made when the animal is moribund or within 1 to 2 hours after death.

mesenteric lymph nodes (Fig. 27) and to distant organs by invasion of lymphatics and blood. A host may die because of necrosis of intestine and mesenteric lymph nodes (Fig. 28) before other organs are severely damaged (Dubey and Frenkel, 1973; Miller *et al.*, 1972). Focal areas of necrosis may develop in many organs. The clinical picture is determined by the extent of injury to these organs, especially vital organs such as the

Fig. 26. Toxoplasma enteritis in mice or hamsters following the ingestion of oocysts. (a) Tachyzoites destroying the superficial epithelium and cells in the lamina propria. Exudate cells are in the lumen. Mouse after feding 10^5 oocysts of the M-7741 strain. × 125. Hematoxylin and eosin. Inset shows tachyzoites in an epithelial cell (large arrow) and in the lamina propria (small arrows) of another mouse. 6 days. × 800. Periodic acid-Schiff-hematoxylin. (b) Extension of tachyzoites to the intestinal submucosa and muscle. Superficial epithelium appaears normal. Hamster, 8 days after feeding 10^5 of the M-7741 strain. × 125. Hematoxylin and eosin. (c) Ulcer in small intestine due to toxoplasmosis. There are numerous tachyzoites (too small to be seen at this magnification) in the necrotic tissue bordering healthy tissue. Hamster, 8 days after feeding 10^5 oocysts. × 125. Hemotoxylin and eosin. Now that the natural mode of *Toxoplasma* infection is known these intestinal lesions should be looked for in human toxoplasmosis.

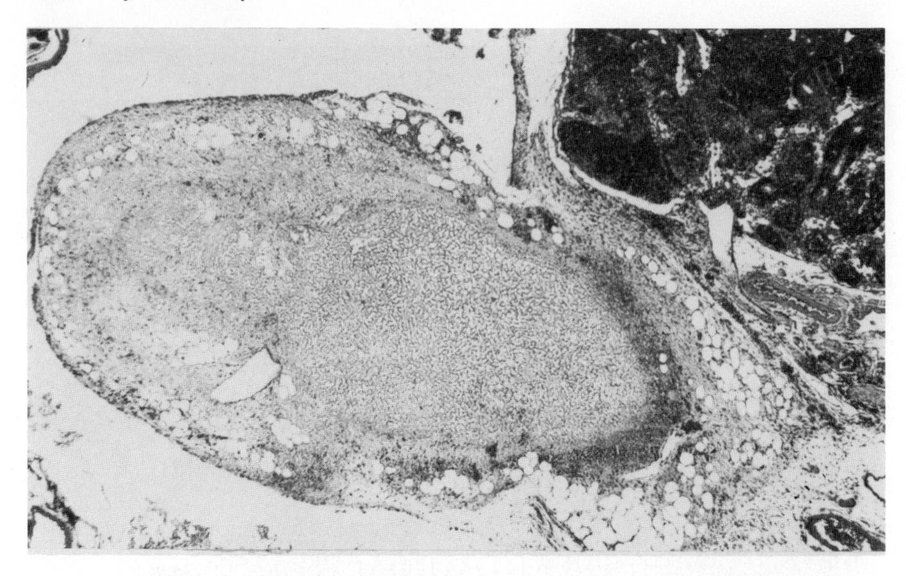

Fig. 28. Toxoplasmic coagulative necrosis in the mesenteric lymph node of a hamster, 12 days after the ingestion of cysts of the M-7741 strain. All cellular elements have been destroyed. An adjacent lymph node (upper right) appears normal. × 32. Hematoxylin and eosin.

eye (Fig. 29), heart, and adrenals (Fig. 30). Necrosis is caused by the intracellular growth of tachyzoites. *Toxoplasma* does not produce a toxin (Nozik and O'Connor, 1969; Pettersen, 1971).

The host may die due to acute toxoplasmosis but much more often recovers with the acquisition of immunity and the appearance of humoral antibodies. Inflammation usually follows the initial necrosis (Fig. 31). By about the third week after infection, *Toxoplasma* tachyzoites begin to disappear from visceral tissues and may localize as cysts in neural and muscular tissues. *Toxoplasma* tachyzoites may persist longer in the spinal cord and brain (Figs. 32 and 33) because immunity is less effective in neural organs than in visceral tissues. How *Toxoplasma* are destroyed in immune cells is not completely known (Frenkel, 1967; Jones *et al.*, 1975). All extracellular forms of the parasite are directly affected by antibody but intracellular forms are not (Sabin and Feldman, 1948). Cellular factors are more important than the humoral in mediation of effective immunity against *Toxoplasma* (Frenkel, 1967). Under experimental conditions, infection with avirulent strains protects the host from damage but does not prevent infection with more virulent strains (de Roever-Bonnet, 1964). Immunization with killed parasites is only par-

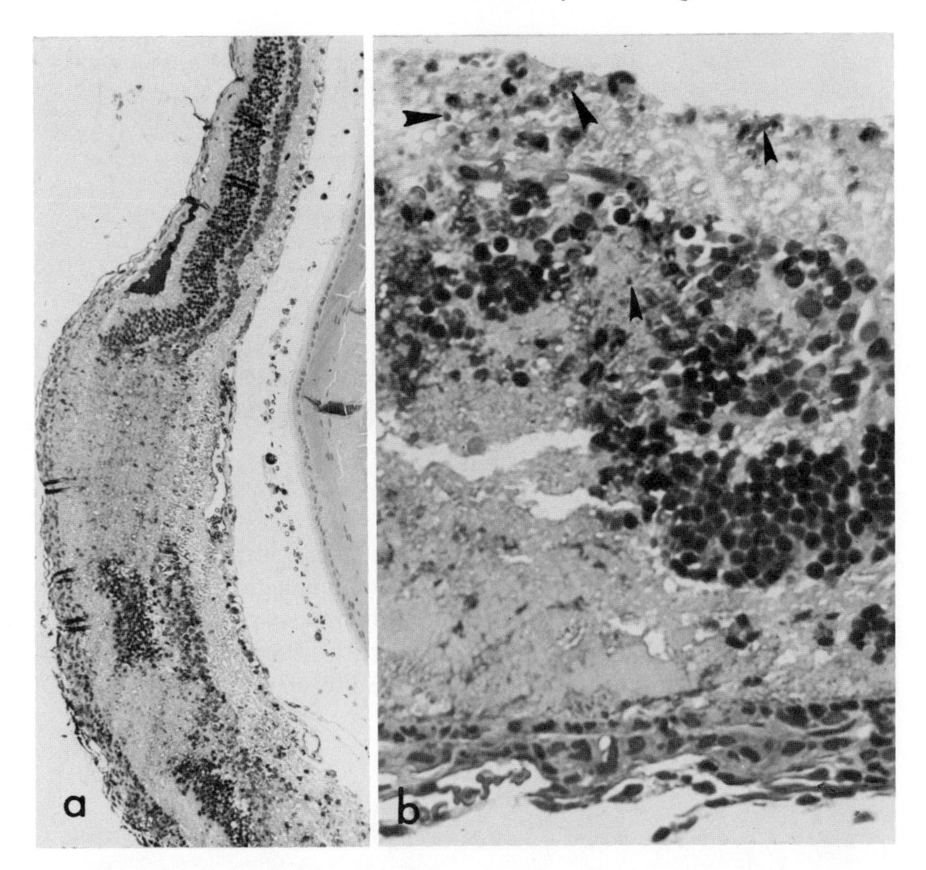

Fig. 29. Toxoplasmic retinal necrosis in a hamster, 20 days after feeding cysts of the M-7741 strain. (a) Lower power showing normal and destroyed portions of the retina. × 125. Hematoxylin and eosin. (b) Higher magnification of (a) showing tachyzoites (arrows) causing retinal necrosis. × 500. Hematoxylin and eosin.

tially protective (Cutchins and Warren, 1956; Ruskin and Remington, 1971; Beverley *et al.*, 1971b; Krahenbuhl *et al.*, 1972c).

Several interferon producing agents and certain viruses and bacteria can enhance the acquired resistance against experimental *Toxoplasma* infection in cell cultures or mice (Remington and Merigan, 1968; Ruskin *et al.*, 1969; Remington, 1970; Swartzberg *et al.*, 1975; Tabbara *et al.*, 1975). Toxoplasmosis has been reported to protect mice against a variety of unrelated organisms (Mahmoud *et al.*, 1976). This effect has been attributed largely to non-specific activation of macrophages (Swartzberg *et al.*, 1975). Whether this non-specific acquired resistance plays any important role in mediation of effective immunity against toxoplas-

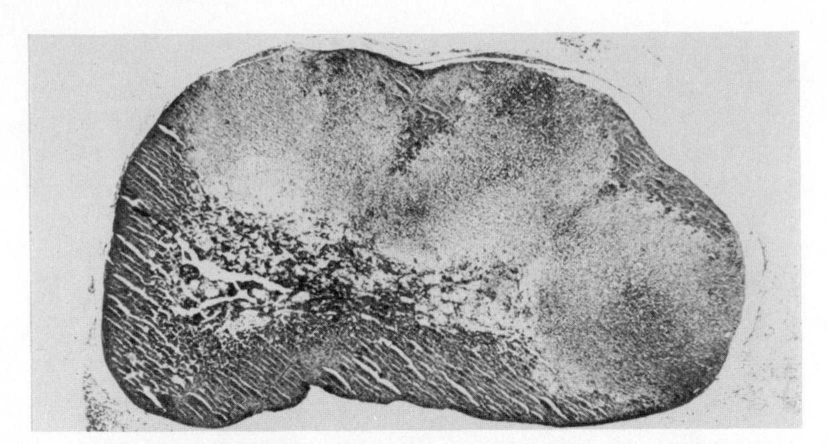

Fig. 30. Almost total destruction of the adrenal gland of a hamster, 14 days after feeding the cysts of the M-7741 strain. × 32. Hematoxylin and eosin.

mosis is open to question (Frenkel, 1973a). *Toxoplasma* infection can also produce long-lasting depression of immunological reactivity by suppressing immunocompetent B and T lymphocytes (Huldt *et al.*, 1973; Strickland *et al.*, 1975). In most instances, immunity following a natural *Toxoplasma* infection persists for the life of the host.

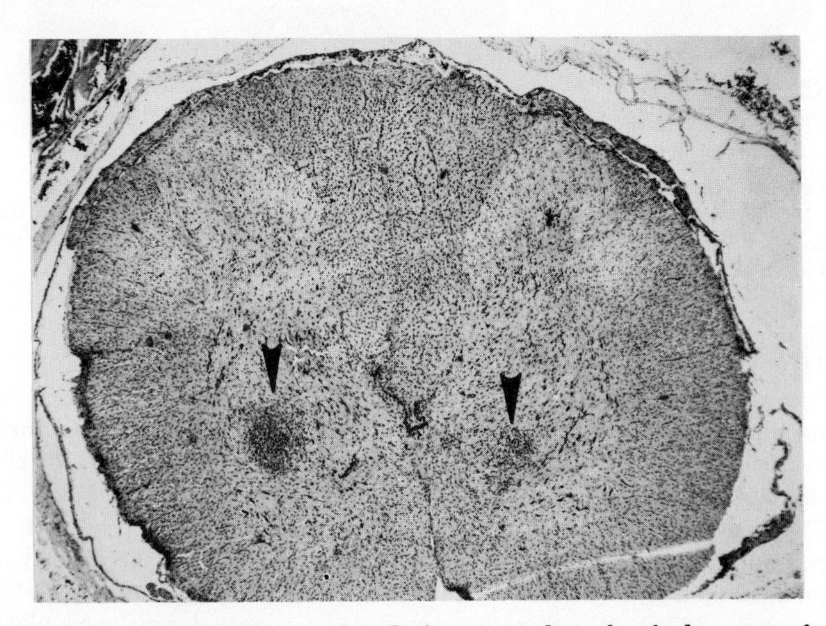

Fig. 31. A section of the spinal cord of a cat, 11 days after feeding cysts of the M-7741 strain. There are two microglial nodules. *Toxoplasma* were identified in one of the lesions. × 35. Hematoxylin and eosin. (From Dubey and Frenkel, 1972a.)

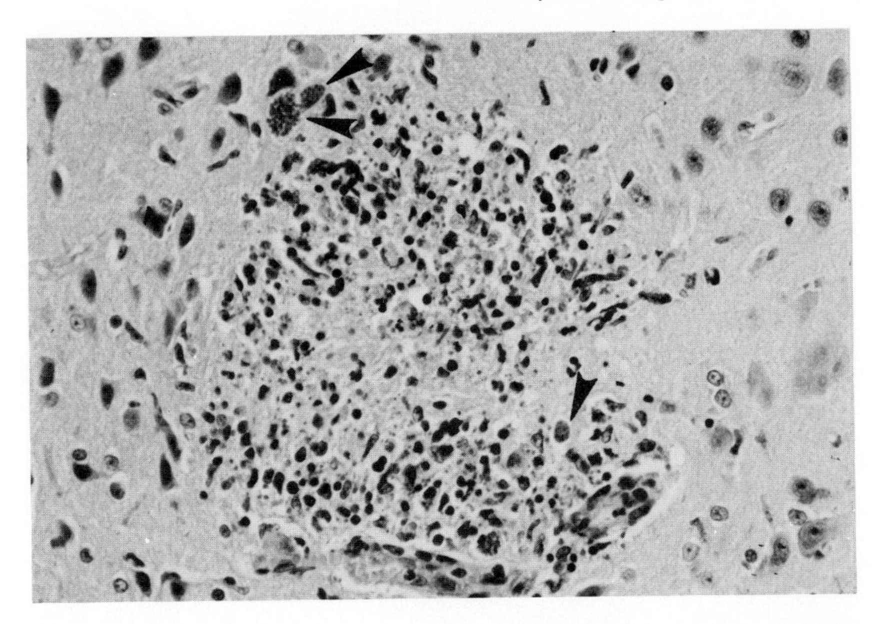

Fig. 32. Relapsing toxoplasmosis in the brain of a cat. The cat had been injected with 6-methylprednisolone acetate (40 mg/kg body wt/week) for 25 days starting 206 days after feeding *Toxoplasma* cysts. Numerous tachyzoites and three cysts (arrows) are seen in the periphery of the brain. × 400. Hematoxylin and eosin. (From Dubey and Frenkel, 1974.)

Immunity, however, does not eradicate infection. *Toxoplasma* persist in the cyst stage in musculature and neural organs for several years after acute infection. Immunity seems to be associated with the presence of living *Toxoplasma* (premunition). The fate of fully developed cysts is not known. It has been proposed (Frenkel, 1971b) that occasionally reactivation of chronic disease takes place. Chronic infections may be reactivated locally, for example, in the eye. Reactivation possibly results from the rupture of a cyst or by seepage of bradyzoites through the cyst wall. Probably many tissue cysts rupture at different times during the life of the host and the released bradyzoites are destroyed by the host's immune responses. This reaction may cause local necrosis accompanied by inflammation. Hypersensitivity plays a major role in such reactions. However, infection usually subsides with no local renewed multiplication of *Toxoplasma* in the tissue, but occasionally there may be formation of new tissue cysts (Frenkel, 1971b).

In immunosuppressed patients, such as those given large doses of immunosuppressive reagents in preparation for organ transplants, rupture of a cyst may result in renewed multiplication of bradyzoites into tachyzoites and the host may die from toxoplasmosis unless treated (Frenkel,

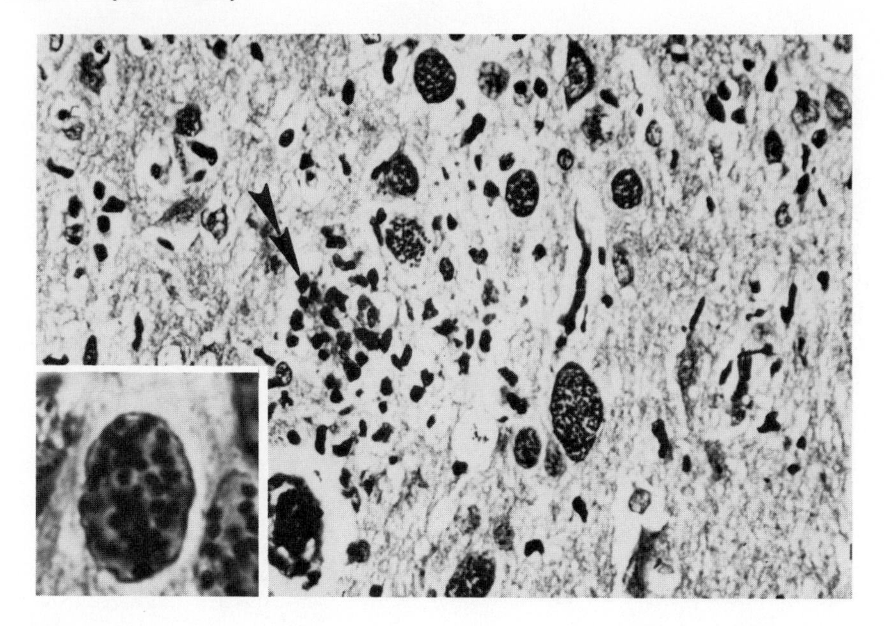

Fig. 33. Numerous *Toxoplasma* cysts around a focal microglial nodule in a cat after 30 days of prednisone (40 mg/kg body wt/day) feedings starting 14 days after primary infection. × 400. Wilder's silver impregnation. Inset shows silver positive cyst wall. × 1000. Formation of numerous cysts around a microglial nodule suggests partial suppression of immunity. (From Dubey and Frenkel, 1974.)

1971b; Frenkel *et al.,* 1975a). It is not known how corticosteroids cause relapse but it is unlikely that they directly cause rupture of the tissue cysts.

Pathogenicity of *Toxoplasma* is closely related to the virulence of the strain and host species. *Toxoplasma* strains may vary in their pathogenicity to a given host (Table III). Ten oocysts of the M-7741 or the Aldrin strain were lethal to mice whereas as many as 1000 oocysts of the S-1 strain caused inapparent infection in mice, and it took 5000 infectious oocysts of the BWM strain to kill one-half of the inoculated mice (Dubey and Frenkel, 1973).

Such differences are most marked in primary isolations since pathogenicity may be increased by frequent passages in a given host. Although this has been known for years (Jacobs and Melton, 1954), pathogenicity has not always been related to number of organisms inoculated until recently. Studies reported in Table III indicate that by frequent passages in mice not only the tachyzoites become more pathogenic, but that the cysts and oocysts derived from these lines become more pathogenic as well.

Adaptation to a given host may, however, differ with different strains of *Toxoplasma*. For example, the RH strain killed 4 out of 5 infected mice on first passage in 17 to 21 days and after only three intraperitoneal passages regularly killed all inoculated mice in 3 to 5 days (Sabin, 1941). The pathogenicity of this strain has been stable for the last 15 years (Table III). On the other hand, with the M-7741 strain, even after 62 passages, mice injected with 10^5 trophozoites survived for more than 9 days (Table III). Until recently, the virulence of strains has been measured by inoculation of *T. gondii* into mice by artificial parenteral routes. Titration of oocysts and cysts of isolates in mice by the oral route may provide a better measure of the virulence of the strain in nature than inoculation by parenteral routes. The pathogenicity of different isolates can be studied only in susceptible hosts like mice. Certain host species are genetically resistant to all strains of *Toxoplasma*. For example, adult rats are resistant while the young rats are susceptible. Mice of any age are susceptible to *Toxoplasma* infection. Adult dogs like adult rats are resistant whereas puppies are fully susceptible.

Ten oocysts of the M-7741 strain kill a mouse whereas circumstantial evidence indicates that this strain is not pathogenic for man (Miller *et al.*, 1972; Dubey and Frenkel, 1973). Individuals working with *Toxoplasma* should, however, consider all isolates as pathogenic and handle them with extreme care. Other strains avirulent for mice may be virulent for man.

Other unknown factors vaguely classified as stress may affect the *Toxoplasma* infection in a host. More severe infections are found in pregnant or lactating mice than in nonlactating mice (Beverley, 1961). Concomitant infection may make the host more susceptible or resistant to *Toxoplasma* infection (Ruskin and Remington, 1968a,b). Clinical toxoplasmosis in dogs is often associated with canine distemper virus infection (Campbell *et al.*, 1955; Capen and Cole, 1966).

1. Infection in Man

Most infections in man are asymptomatic but at times the parasite can produce devastating disease. Infection may be congenitally or postnatally acquired.

Congenital infection occurs only when a woman becomes infected during pregnancy. While the mother rarely has symptoms of infection she does have a temporary parasitemia. Focal lesions develop in the placenta and the fetus may become infected. At first there is generalized infection in the fetus. Later, infection is cleared from visceral tissues and may localize in the central nervous system. A wide spectrum of clinical disease occurs in congenitally infected children. Mild disease may consist of slightly diminished vision whereas severely diseased children may have the full tetrad of signs: retinochoroiditis, hydrocephalus, convulsions, and intracerebral calcification. Of these, hydrocephalus is the least common (about 3% of congenital infections) but most dramatic lesion of toxoplasmosis (Beverley, 1974). This lesion is unique to congenitally acquired toxoplasmosis in man. *Toxoplasma* proliferates in the walls of ventricles causing necrosis. If the flakes of necrotic material block the aqueduct of Sylvius, the drainage of ventricular fluid is blocked. As *Toxoplasma* destroys ependymal cells lining the ventricles the antigen seeps out of ventricles and reacts with antibody from within the blood vessels. This results in vasculitis and thrombosis as a result of the antigen–antibody reaction (Frenkel and Friedlander, 1951). The necrotic subependymal tissue may become calcified. This intracerebral calcification can be detected by X-ray examination.

By far the most common sequel of congenital toxoplasmosis is ocular disease. Except for an occasional involvement of an entire eye, virtually in all cases the disease is confined to the posterior chamber (Perkins, 1973). *Toxoplasma* proliferates in the retina (Wilder, 1952) and this leads to inflammation in the choroid. Therefore, the disease is correctly designated as retinochoroiditis. The lesions of ocular toxoplasmosis are fairly characteristic in man. Such lesions in the acute or subacute stage of inflammation appear as yellowish-white, cottonlike patches in the fundus (O'Connor, 1974). The lesions may be single or multiple and may involve either one or both eyes. During the acute stage, inflammatory exudate may cloud the vitreous and may be so dense as to preclude visualization of the fundus by an ophthalmoscope. As the inflammation subsides, the vitreous clears and the diseased retina and choroid can be seen through the ophthalmoscope. Retinal lesions may be single or multifocal small gray areas of active retinitis with minimal edema and reaction in the vitreous humor (Friedmann and Knox, 1969). The punctate lesions are usually harmless unless they are located in a macular area. Although severe infections may be detected at birth, milder infections may not flare up until adulthood (Hogan *et al.*, 1958). The cause of recurrent toxoplasmic retinochoroiditis is not known. It is surmised that tissue cysts present in the retina rupture from time to time reactivating the lesion (Frenkel, 1971b).

Postnatally acquired infection may be localized or generalized. The most common clinical manifestation of acquired toxoplasmosis is lymphadenopathy. Out of 1637 cases of toxoplasmosis reviewed by Perkins (1973), 89.2% had lymphadenopathy, 4.3% had central nervous system involvement, 1.4% had myocarditis, 0.8% had pulmonary involvement, and

4.2% had miscellaneous conditions. Typical toxoplasmic retinochoroiditis in postnatally acquired toxoplasmosis is rare (Perkins, 1973).

Lymphadenitis is the most frequently observed clinical form of toxoplasmosis in man. Although any nodes may be involved, the most frequently involved are deep cervical nodes. These nodes when infected are tender, discrete but not painful, and the infections resolve spontaneously in weeks or months. Lymphadenopathy may be associated with fever, malaise, and fatigue in about one-half of patients, muscle pains in one-third, and sore throat and headache one one-fifth (Beverley, 1974). Although the condition may be benign, its diagnosis is vital in pregnant women because of the risk to the fetus. In toxoplasmic lymphadenopathy, the Paul Bunnel test is negative. Histologically, there is reticular cell hyperplasia, necrosis and fibrosis are absent, and the node architecture is preserved (Frenkel, 1971b; Gray *et al.*, 1972). The absence of Reed-Sternberg cells in toxoplasmic lymphadenitis differentiates it from Hodgkin's disease. A search for *Toxoplasma* in the sections of lymph nodes is usually futile because *Toxoplasma* are too few to be detected in sections. The diagnosis can be confirmed by inoculation of lymph node material into mice.

Encephalitis is another important manifestation of toxoplasmosis in immunosuppressed patients (Vietzke *et al.*, 1968; Carey *et al.*, 1973; Cheever *et al.*, 1965; Roth *et al.*, 1971; Siegel *et al.*, 1971; Frenkel *et al.*, 1975a; Gleason and Hamlin, 1974). Clinically, patients may have headache, disorientation, drowsiness, hemiparesis, reflex changes, convulsions, and many may become comatose (Frenkel, 1971b). Diagnosis is aided by serological examination. However, in immunosuppressed patients both inflammatory signs and antibody production may be suppressed, thus making the diagnosis even more difficult. Encephalitis is now recognized with great frequency in patients treated with immunosuppressive agents (for review see Frenkel *et al.*, 1975a).

2. Infection in Lower Animals

Toxoplasma is capable of causing severe disease in animals other than man. Losses in lambs due to *Toxoplasma* infection are known to be of considerable economic importance in New Zealand (Hartley and Marshall, 1957; Hartley and Moyle, 1968) and Great Britain (Beverley and Watson, 1961; 1971) where a high proportion of all abortions in sheep are due to toxoplasmosis. *Toxoplasma*-induced abortions of lambs are known to occur in several other countries, but the relative incidence is not known. Congenital infection occurs when sheep become infected during pregnancy (Beverley and Watson, 1961; Jacobs and Hartley, 1964). The severity of disease depends upon the time of infection.

Infection at 45 to 55 days of pregnancy causes death of the fetus, at 90 days the fetus may be born alive but usually with clinical signs, and at 120 days (the gestation period in sheep is 145 days) lambs may be infected but rarely clinical (Beverley, 1974). Although disease is common in lambs, it is virtually unknown in adult sheep. Isolated reports of encephalitis in adult sheep (McErlean, 1974) are probably due to another organism structurally similar to *T. gondii* (Hartley and Blackmore, 1974).

The diagnosis of ovine abortion due to *T. gondii* is of paramount importance to the sheep industry. The most striking lesions are found in placental cotyledons. Foci of necrosis occur in the villous part of the cotyledon. (Fig. 34). These foci may be macroscopic and can be seen by immersing the cotyledon in physiological saline solution and looking

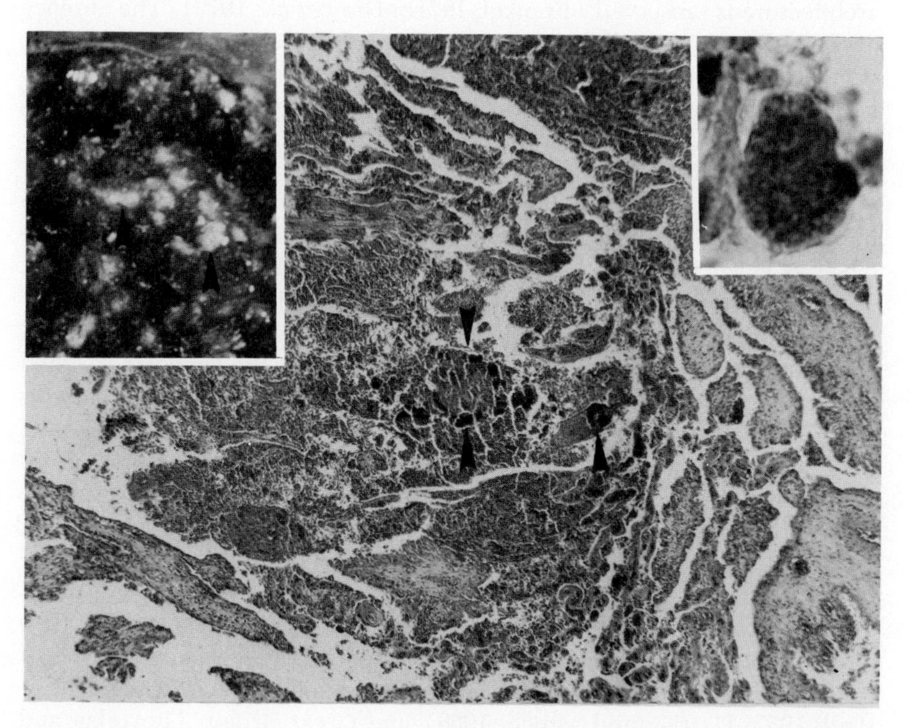

Fig. 34. Placental sheep cotyledon naturally infected with *Toxoplasma*. Multiple foci of necrosis (arrows) are present in the villus part of the cotyledon. Recognition of this lesion is useful in the diagnosis of toxoplasmic abortion in sheep because *T. gondii* are rarely identified with certainty in these necroti foci. × 50, hematoxylin and eosin. Inset (upper left) shows macroscopic necroti foci. × 4, unstained. Inset (upper right) shows tachyzoites. × 1250, hematoxylin and eosin. (Courtesy of Dr. W. J. Hartley.)

for white necrotic foci in the villi (Beverley *et al.*, 1971a,b). Leukoencephalomalacia is another common finding in congenitally infected lambs (Hartley and Kater, 1963).

The only report dealing with disease in cattle is that of Sanger *et al.* (1953) from the United States. Sanger *et al.* (1953) isolated *T. gondii* from the internal organs of two spontaneously infected calves and from the colostrum of one cow. *Toxoplasma*-like organisms were found in the sections of internal organs of three cows, one bull, and two calves from four widely separated herds in Ohio. A large number of cattle from these farms had nervous and respiratory signs of undiagnosed illness. Although one might presume that these epizootics of acute disease were caused by *T. gondii,* direct evidence is lacking. Successful attempts to produce acute fatal disease in cattle by inoculation with virulent *Toxoplasma* by several parenteral routes simultaneously is not representative of a natural process. Toxoplasmosis in cattle requires further study.

Outbreaks of toxoplasmosis in pigs have been reported from several countries (Farrell *et al.*, 1952; Folkers, 1962; Nobuto *et al.*, 1960; Harding *et al.*, 1961). Pneumonia, encephalitis, and abortions were found in affected animals (Koestner and Cole, 1970a; Møller *et al.*, 1970; Work *et al.*, 1970). Clinical disease is probably related to the number of oocysts or cysts and strain of *T. gondii* ingested. In recent experimental studies, the ingestion of 1000 oocysts produced clinical disease (Durfee *et al.*, 1974). In another study, the ingestion of 170 or more oocysts produced fatal toxoplasmosis in pigs. The pigs died from necrosis and hemorrhage in several visceral organs and lungs (Ito *et al.*, 1974a,b; Saski *et al.*, 1974).

In dogs, the disease is most severe in puppies. Common clinical manifestations of canine toxoplasmosis are: respiratory distress, ataxia, and diarrhea (Cole *et al.*, 1954; Koestner and Cole, 1960a). In most cases, pneumonia is caused by a combination of *Toxoplasma* and distemper virus (Campbell *et al.*, 1955). Distemper virus produces immunodepression (Krakowka *et al.*, 1975) and *Toxoplasma* proliferates probably as an opportunistic pathogen. However, undoubted cases of toxoplasmic pneumonia have been reported (Capen and Cole, 1966). *Toxoplasma* produces focal necrotic areas in the lung, liver, and brain. Occasionally, extensive necrosis occurs due to infarction of a blood vessel (Fig. 35). Toxoplasmosis must be considered in the differential diagnosis when focal necrotic lesions are found together in the liver, lung, and brain. Several cases of myositis have been reported (Hartley *et al.*, 1958).

Cats, like nonfeline hosts, also suffer from clinical toxoplasmosis. Pneumonia is the most important clinical manifestation of feline toxoplasmosis. Uncomplicated lesions consist of focal necrotic areas with

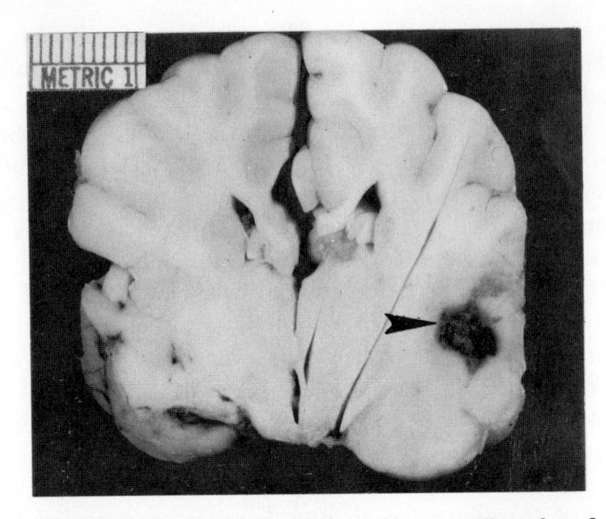

Fig. 35. Cross section through the cerebrum of a naturally infected dog showing an infarct in the left Sylvian gyrus. The dog registered fever and convulsive seizures before death. *Toxoplasma gondii* tachyzoites and cysts were demonstrated in the lesions. (From Koestner and Cole, 1960b.)

edema (Figs. 36 and 37). Affected cats may appear depressed, anorexic, and die suddenly with no obvious clinical signs. Other common clinical manifestations are hepatitis, pancreatic necrosis, myositis, and myocarditis (Meier *et al.*, 1957; Petrak and Carpenter, 1965; Dubey and Frenkel, 1974). Occasionally, *Toxoplasma* proliferate in the gall bladder and bile ducts of cats producing chronic inflammation (Smart *et al.*,

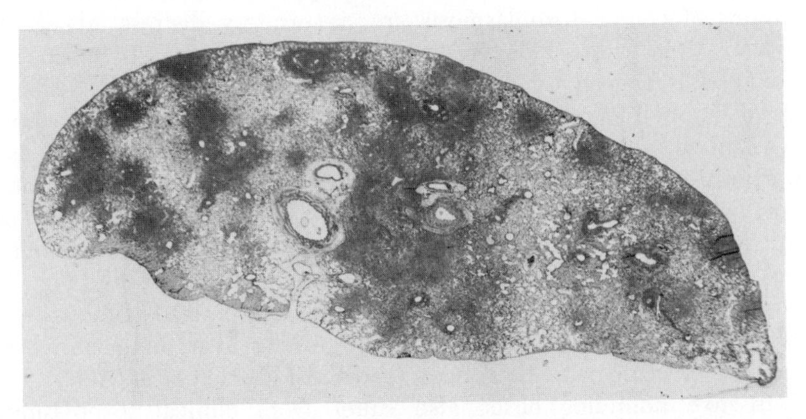

Fig. 36. (Figs. 36 and 37.) Primary and relapsing toxoplasmic pneumonia in cat. Cross section of the caudal lobe of the lung showing necrosis and edema 11 days after ingesting *Toxoplasma* cysts. × 3. Hematoxylin and eosin.

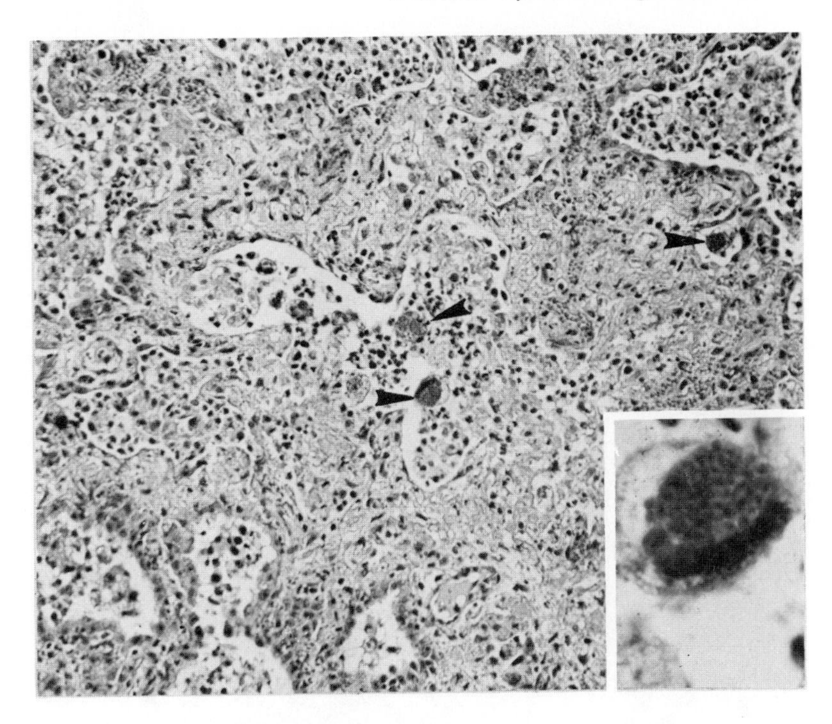

Fig. 37. Relapsing toxoplasmic pneumonia 30 days after feeding prednisone in a cat infected with *T. gondii* for 14 days (same cat as in Fig. 33). × 200. Hematoxylin and eosin. There is necrosis, edema, fibrin, and macrophages (arrow) are occluding the alveoli. Inset shows tachyzoites in a macrophage. × 1000. Hematoxylin and eosin. (From Dubey and Frenkel, 1974.)

1973; Neufeld and Brandt, 1974; Dubey and Frenkel, 1974). Although *T. gondii* multiplies extensively in intestine (Fig. 38), the affected cats rarely have diarrhea. Clinically apparent encephalitis is probably uncommon in cats.

Sporadic and widespread outbreaks of toxoplasmosis occur in rabbits, mink, and other domesticated and wild animals. These are summarized by Siim *et al.* (1963).

E. Diagnosis

Diagnosis can be aided by serological or histopathological examination. Clinical signs of toxoplasmosis are nonspecific and cannot be depended upon for a definite diagnosis because toxoplasmosis mimics several other infectious diseases clinically.

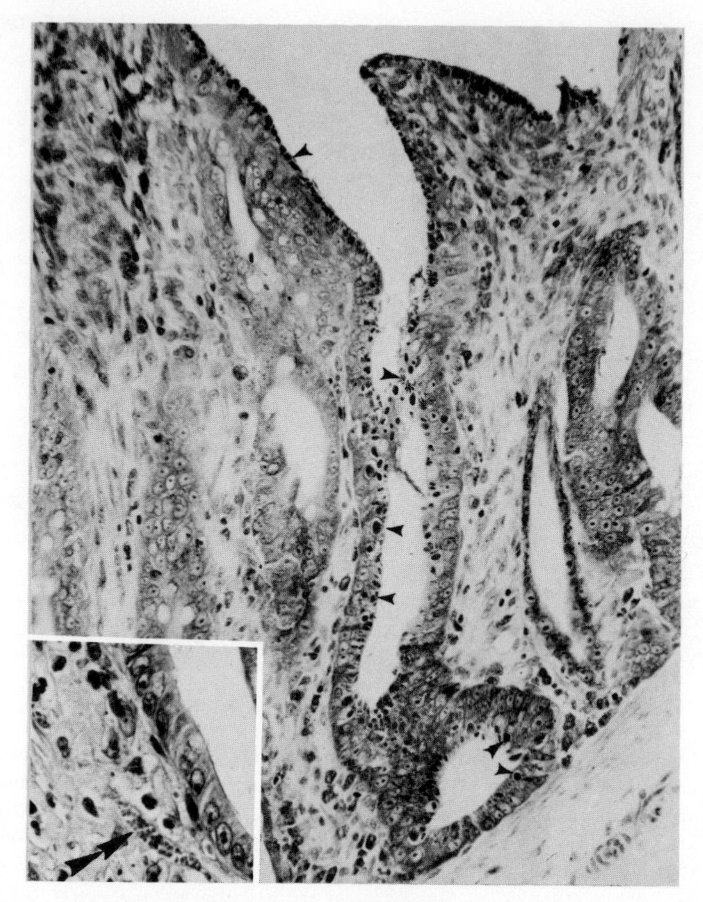

Fig. 38. Relapsing toxoplasmosis in the ileum of a cat, 12 days after injections of methylprednisolone acetate (80 mg/kg body wt/week) and 231 days after primary infection. Both the superficial epithelium and the glands of Lieberkühn are parasitized with *Toxoplasma* types D and E and gametocytes. The *Toxoplasma* organisms stain darkly and occupy the space between nucleus and free cell border. × 225. Giemsa stain. Inset shows tachyzoites (arrow) in the subepithelium. × 400. Hematoxylin and eosin. (From Dubey and Frenkel, 1974.)

1. Isolation

Toxoplasma can be isolated from patients by inoculation of laboratory animals. The choice of inoculum will depend upon the circumstances. Secretions, excretions, body fluids, and tissues taken by biopsy, such as lymph nodes or muscle tissue, are possible specimens from which to attempt isolation. Cerebral spinal fluid from a child with possible congenital infection and encephalitis or lymph node material from a person with lymphadenopathy are good sources of *Toxoplasma.* At necropsy, the tissues selected will be determined on the basis of macroscopic pathological changes, e.g., if there is pneumonia, lung tissue will be selected.

For the isolation of *Toxoplasma* from heparinized blood or cerebrospinal fluid, the samples are centrifuged for 10 minutes at 2000 rpm, the sediment is suspended in saline, and injected intraperitoneally into mice. Whole heparinized blood may be inoculated directly. For the isolation of *Toxoplasma* from tissues, the tissues are homogenized in a mortar, possibly with sterilized sand or Alundum, and are then suspended in saline. If bacterial contamination is suspected, tissues or fluids can be mixed with antibiotics in physiological saline solution to give a concentration of 1000 units of penicillin and 100 μg of streptomycin per milliliter without reducing the infectivity. Contaminated specimens suspended in antibiotics are allowed to stand at room temperature for 1 hour. Just before inoculation, the suspension is shaken; when the heavy particles have settled to the bottom the supernatant fluid is injected into mice.

Mice are more sensitive to infection than are chick embryos or cell cultures (Abbas, 1967). Inoculation may be by the subcutaneous, intraperitoneal, or oral routes. Each of these three has its advantages. The intraperitoneal route is most useful for isolating *Toxoplasma* from specimens not contaminated by bacteria. The subcutaneous route is useful for isolating *Toxoplasma* from specimens contaminated by bacteria. The oral route is best for isolating *Toxoplasma* from unpreserved feces of cats. However, it is often preferable to use combinations of these routes to get maximum numbers of isolations. About 0.5 to 1 ml of fluid can be inoculated by the intraperitoneal or subcutaneous route into a 20- to 25-gm mouse. To minimize overflow of the inoculum, the needle should be inserted subcutaneously for at least 1 cm before entry into the peritoneal cavity. Tincture of iodine (7% iodine in 95% alcohol) should be applied at the site of inoculation to kill *Toxoplasma* that may escape from the inoculation site. Extra precautions are necessary when handling oocysts because they survive well in the environment.

Depending upon the virulence of the strain, ascites might develop in the mice in 7 to 14 days after intraperitoneal inoculation. Tachyzoites may be found in the peritoneal fluid or in imprints made from the mesenteric lymph nodes. After oral inoculation the mice may die within 10 days due to enteritis and mesenteric lymphadenitis (Dubey and Frenkel, 1973). Imprints of intestinal and mesenteric lymph nodes should be made soon after the mice die or, which is even better, from mice killed in the moribund stage. When mice die in the second week they usually do so from penumonia and encephalitis, irrespective of the route of inoculation and imprints of the lungs and brain should be examined for *Toxoplasma* tachyzoites. After 14 days, organisms begin to disappear from the visceral organs and *Toxoplasma* are most likely to be demonstrated in the brain. A diagnosis is made by finding *Toxoplasma* in films of body fluids or tissue imprints stained by one of the Romanowsky methods.

Tissue cysts should be sought in survivors 6–8 weeks after inoculation even though *Toxoplasma* cysts may appear earlier since early cysts are usually small and may be missed. Examination of the brain for cysts is carried out following grinding with a pestle and mortar, adding saline (1 ml per mouse brain) slowly as grinding proceeds. One or two drops of

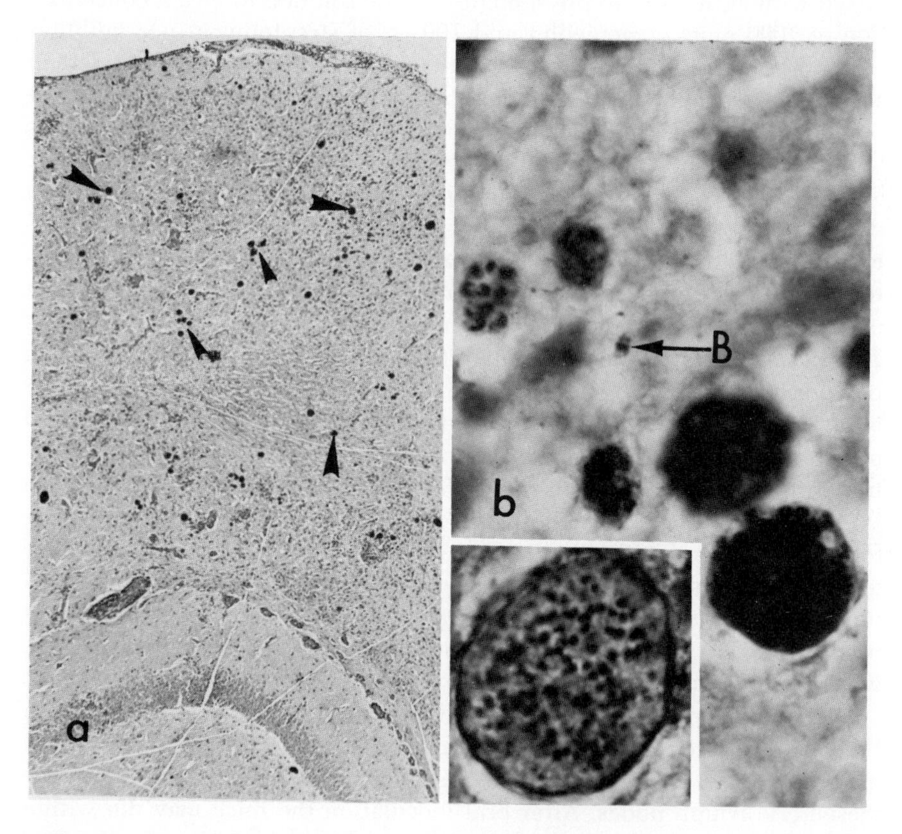

Fig. 39. *Toxoplasma* cysts in sections of mouse brain. (a) Numerous cysts (arrows) in the cerebrum. Bradyzoites in cysts stain bright red with PAS and cysts are easily seen under low magnification. × 50. Periodic acid Schiff-hematoxylin. (b) A collection of cysts with different staining intensities for PAS reagents. More mature cysts stain darker than younger cysts. Arrow points to PAS-positive granules in a bradyzoite (B). × 1250. Periodic acid Schiff-hematoxylin. Inset shows a cyst wall positively stained with Wilder's ammoniacal silver.

this saline–brain suspension are examined between a glass slide and cover slip using the low magnification lens of the microscope. *Toxoplasma* cysts are darker than the brain tissue and are easily seen at 100 × magnification (Fig. 10). It is best to bleed the mice out completely before removing the brains for examination. This prevents contamination of brain with blood, makes the identification of cysts easier, and the blood is available for serological examination. A portion of brain should be fixed for sectioning. Certain pollen grains structurally resemble *Toxoplasma* cysts. To avoid erroneous diagnosis, cover glasses, slides, and pestle and mortars should be kept covered when not in use and identification should always be confirmed on stained smears or sections (Fig. 39).

Failure to demonstrate *Toxoplasma* in mice indicates lack of infection but does not exclude it. Antibodies to *T. gondii* should be sought in the sera of inoculated mice to detect inapparent infection. Antibody tests may be done as early as 2 weeks after infection. To obtain blood for serological examination, anesthetized mice are bled from the orbital sinus. About 0.5 ml of blood can be easily obtained from the orbital sinus without killing the mice. Blood can also be obtained from the tail vein but the yield is poor and the method is more time-consuming. Mice develop antibody to *T. gondii* between 3 and 16 days after infection depending upon the strain of *Toxoplasma*, dose, and stage of the parasite inoculated. Mice inoculated with even one infectious organism of an avirulent strain usually develop antibody by the 16th day (Dubey and Frenkel, 1973).

2. Serology

Many serological tests have been used for the detection of antibodies of *T. gondii*. Of these, the most reliable is the cytoplasm modifying or "dye" test of Sabin and Feldman (1948). This is essentially a neutralizing type of antigen–antibody reaction. It is based on the principle that when antibody acts upon the parasite, it modifies it so that under suitable conditions *Toxoplasma* do not stain uniformly with alkaline methylene blue. Live virulent *Toxoplasma* are used as antigen and are exposed to dilutions of the test serum and a complementlike "accessory-factor" obtained from *Toxoplasma* antibody-free human serum. This test is sensitive and so far it is the most specific test for toxoplasmosis. Its main disadvantages are its high cost and the human hazard due to the use of live organisms. The indirect fluorescent antibody test (IFAT) overcomes some of the disadvantages of the dye test. In this test, killed *Toxoplasma*, which are available commercially, are used as antigen. Titers obtained by IFAT are comparable to those in the dye test. The disadvantages of the IFAT are that a microscope with UV light is required and fluorescent antispecies globulin is required for each species to be tested and false positive titers may occur in patients with antinuclear antibodies (Araujo *et al.*, 1971). Its suitability in animal diagnostic work is, therefore, limited but it has proved useful in the diagnosis of human toxoplasmosis. Another test, IgM-IFAT, has been used as an aid in the diagnosis of congenital toxoplasmosis in children (Remington, 1969). Heavier IgM antibody is not transferred across the intact placenta whereas IgG is passively transferred. Other serological tests, namely, the indirect hemagglutination test (IHA), the agglutination test, and the complement fixation test, each have some advantages. IHA (Jacobs and Lunde, 1957) and agglutination (Fulton and Turk, 1959) tests are easy to perform but they need further evaluation for specificity of titers. Antigens for IHA tests are now commercially available in several countries including the United States. Although the IHA test is easy to perform, it usually does not detect antibodies during the acute phase of toxoplasmosis. The agglutination test has disadvantages similar to those of the IHA test, but a recent modification of the agglutination test shows great promise. In the modified agglutination test, the test serum is treated with 2-mercaptoethanol to eliminate nonspecific agglutinins (Desmonts *et al.*, 1974; H. A. Feldman, 1976, personal communication). The complement fixation test is useful in diagnosing acute infection because antibody measured in this test appears during acute infection and usually disappears soon after the acute infection ends, but this depends upon the antigenic preparation.

The result of examining one positive serum sample only establishes that the host has been infected at some time in the past. A four- to eight-fold higher antibody titer in a serum taken 2–4 weeks after the first serum was collected more certainly indicates an acute acquired infection. A titer of 1:1000 or higher in the "dye" or IFA test indicates acute infection (Frenkel, 1971b). A high antibody titer sometimes persists for months and a rise may not be associated with clinical symptoms. As indicated earlier, most acquired infections in man are asymptomatic.

The significance of an antibody titer in relation to clinical toxoplasmosis is not easily evaluated. For example, even a titer of 1:2 in the dye test may be significant in the diagnosis of clinical retinochoroiditis. The diagnosis of ocular toxoplasmosis is difficult because the antibody titers are usually low and a biopsy obviously is not possible for the isolation of *T. gondii* from the eye. Desmonts (1966) has developed a test for the diagnosis of ocular toxoplasmosis. In this test, determination of a higher antibody protein concentration in the anterior chamber fluid of an affected eye than in the serum confirms ocular toxoplasmosis. However, there is the risk of damaging the eye while aspirating anterior chamber fluid and one needs specific equipment to determine antibody protein concentration.

3. Histology

Diagnosis can be made by finding *Toxoplasma* in host tissue removed by biopsy or at necropsy. A rapid diagnosis may be made by making impression smears of lesions on glass slides. After drying for 10 to 30 minutes, the smears are fixed in methyl alcohol and stained with Giemsa. Well preserved *Toxoplasma* are crescent-shaped and stain well with any of the Romanowsky stains (Fig. 4). However, degenerating organisms, which are commonly found in lesions, usually appear oval and their cytoplasm stains poorly as compared to their nuclei. Diagnosis should not be made unless organisms with typical structure are located because degenerating host cells may resemble degenerating *Toxoplasma*. In sections, the tachyzoites usually do not stain distinctively from host cells. In sections stained with hematoxylin and eosin, tachyzoites may be easily located by deeply staining the host tissue. Occasionally, cysts might be found in areas with lesions. Tissue cysts are usually spherical, have silver positive walls, and the bradyzoites are strongly PAS positive (Fig. 39). In properly stained sections, *T. gondii* tissue cysts can be easily located at 100 × magnification (Fig. 39). However, toxoplasmosis should not be diagnosed unless tachyzoites are demonstrated in the lesions.

F. Treatment

1. In Man and Other Intermediate Hosts

Sulfonamides and pyrimethamine (Daraprim) are two drugs widely used for therapy of toxoplasmosis. These two drugs act synergistically by blocking the metabolic pathway involving *p*-aminobenzoic acid and the folic–folinic acid cycle, respectively (Eyles and Coleman, 1953; Cook and Jacobs, 1958; Sheffield and Melton, 1975). These two drugs are usually well tolerated but sometimes thrombocytopenia and/or leukopenia may develop. These effects can be overcome by administering patients folinic acid and yeast without interfering with treatment because the vertebrate host can utilize presynthesized folinic acid while *Toxoplasma* cannot (Frenkel and Hitchings, 1957). The commonly used sulfonamides, sulfadiazine, sulfamethazine, and sulfamerazine are all effective against toxoplasmosis. Generally, any sulfonamide that diffuses across the host cell membrane is useful in antitoxoplasma therapy. While these drugs have beneficial action when given in the acute stage of the disease process when there is active multiplication of the parasite, they will not usually eradicate infection. It is believed that these drugs have little effect on subclinical infections, but the growth of tissue cysts in mice has been suppressed with sulfonamides (Beverley, 1958). Sulfa compounds are excreted within a few hours of administration, therefore, treatment has to be administered in daily divided doses (four doses of 500 mg each) usually for several weeks or months. A loading dose (75 mg) of pyrimethamine during the first 3 days has been recommended because it is absorbed slowly and binds to tissues. From the fourth day, the dose of pyrimethamine is reduced to 25 mg, and 2–10 mg of folinic acid and 5–10 gm of Bakers' yeast is added (Frenkel, 1971b).

Because pyrimethamine is toxic some authors use a combination of trimethoprim and sulfamethoxazole as possible alternatives to pyrimethamine and sulfadiazine (Norrby *et al.*, 1975). Although trimethoprim, like pyrimethamine, is a folic acid antagonist it has no synergistic effect in combination with sulfamethoxazole against murine toxoplasmosis (Feldman, 1973; Remington, 1976). Therefore, trimethoprim is not recommended against toxoplasmosis.

Certain other drugs, diaminodiphenylsulfone (SDDS), spiramycin, chlorinated lincomycin analogs, lasalocid, and monensin have been found effective in experimentally induced *Toxoplasma* infection in animals or cell cultures (Ohshima *et al.*, 1967; Garin and Eyles, 1958; Shimizu *et al.*, 1968; Beverley *et al.*, 1973; McMaster *et al.*, 1973; Araujo and Remington, 1974; Tabbara *et al.*, 1974; Melton and Sheffield, 1975). Spiramycin produces high tissue concentrations, particularly in placenta without crossing the placental barrier, and has been used in man without harmful effects (Desmonts and Couvreur, 1974a). It is used as an alternative therapy to sulfadiazine and pyrimethamine (SP) during pregnancy because spiramycin may produce hematological and tetratogenic effects. Cases of congenital toxoplasmosis were fewer (44%) among the offspring of mothers treated with spiramycin (4 × 500–750 mg daily for 3 weeks) than of untreated (76%) mothers (Desmonts and Couvreur, 1974a).

2. In Cats

The primary objective of treating cats is to suppress oocyst shedding because the diagnosis of acute illness due to toxoplasmosis is rarely made in cat. Sulfadiazine and pyrimethamine are 2 recommended drugs (Frenkel, 1975; Sheffield and Melton, 1976). Frenkel (1975) recommended 100 mg/kg/day of sulfadiazine and 1mg/kg/day of pyrimethamine for treating toxoplasmosis in cats. In his experiments, even 15 mg/kg/day sulfadiazine combined with 0.2 mg/kg/day pyrimethamine in food reduced oocyst numbers in feline feces by 1000-fold (Frenkel, 1975). Search for better anticoccidial drugs should continue because even toxic doses of sulfadiazine (240 mg/kg/day) and pyrimethamine

(1 mg/kg/day) did not completely suppress *Toxoplasma* oocyst production by cats (Dubey and Yeary, 1977). Furthermore, these drugs may interfere with acquisition of immunity to toxoplasmosis (judged by reshedding of oocysts). Two other antitoxoplasmic drugs, 2-sulfamoyl 1-4 diaminodiphenylsulfone (SDDS) at the rate of 100–1000 mg/kg body weight and clindamycin at the rate of 100 or 250 mg/kg body weight, suppressed but did not completely stop oocyst shedding (Dubey and Yeary, 1977).

G. Epidemiology

Toxoplasma infection in man is widespread throughout the world. Approximately one-half billion of the human population has antibody to *T. gondii* (Kean, 1972). Infection in man and animals varies in different geographical areas of a country (Walton *et al.*, 1966). Causes for these variations are not yet known. Environmental conditions, cultural habits of the people, and animal fauna are some of the factors that may determine the degree of natural spread of *Toxoplasma*. Infection is more prevalent in hot and humid areas than in dry and cold climates (Gibson and Coleman, 1958). Only a small proportion (less than 1%) of people acquire infection congenitally. A child may be born infected *in utero* but without symptoms of clinical disease. In a study of over 20,000 pregnant women in Paris, 6.3% of pregnant women acquired infection during pregnancy. Of 180 children whose mothers contracted toxoplasmosis during pregnancy, 110 (61%) were uninfected, 6 died neonatally, and 64 (36%) were infected. Of the 64 infected children, 46 (26%) had subclinical infection, 11 had mild diseases, and 7 were severely diseased (Desmonts and Couvreur, 1974b). Koppe *et al.* (1974) found 12 (0.65%) children congenitally infected in a prospective study of 1821 pregnant women in Holland. During a seven year follow-up of these 12 children, three were found to have mild retinal scars, 1 child was ill, and the other 8 were asymptomatic (Koppe *et al.*, 1974). In the United States, rates for congenital infections were 1:682 in Alabama (Alford *et al.*, 1969, 1974) and 1:1350 in New York City (Kimball *et al.*, 1971).

Congenital infection occurs only once in women because mothers of congenitally infected children have not been known to give birth to infected children in subsequent pregnancies (de Roever-Bonnet, 1961; Desmonts and Couvreur, 1974a). Unlike man, repeated congenital infection can occur in mice, rats, guinea pigs, and hamsters. Several litters may be infected from an infected mouse or hamster (Beverley, 1959; de Roever-Bonnet, 1969) without reinfection from outside sources. Congenitally infected mice can themselves produce congenitally infected mice for ten generations (Beverley, 1959). Although congenital infection occurs in guinea pigs (Huldt, 1960; Wright, 1972), rabbits (Uhliková and Hübner, 1973), dogs (Chamberlain *et al.*, 1953; Koestner and Cole, 1960a), pigs (Mariasaki *et al.*, 1976; Moller *et al.*, 1970), and perhaps other animals it is most important in sheep due to "storms" of abortions produced in New Zealand and England (Hartley and Marshall, 1957; Beverley and Watson, 1961). In sheep, like man, congenital infection occurs only when the ewe acquires infection during pregnancy. Normal lambs are born during subsequent pregnancies from infected ewes (Munday, 1972b). That there is no contact infection between infected and noninfected sheep is shown by an excellent study reported by Hartley and Moyle (1974). These authors studied by serological means a flock of congenitally infected and noninfected sheep running together on the same pasture. During a period of 7 years, infected sheep remained serologically positive and uninfected lambs remained serologically negative for 6 years (Hartley and Moyle, 1974). Although medically important, congenital infection is too rare to account for widespread infection in man and animals.

The relative frequency of acquisition of postnatal toxoplasmosis between that due to eating raw meat and that due to ingestion of food contaminated by oocysts from cat feces is not known and is difficult to investigate. *Toxoplasma* infection is common in many animals used for food. Sheep, pigs and rabbits are commonly infected throughout the world. *Toxoplasma gondii* has been isolated from 8 to 35% of apparently healthy sheep in several countries (Rawal, 1959; Jacobs *et al.*, 1960b; Jantischke *et al.*, 1957), and 24–43% of pigs (Jacobs *et al.*, 1960; Čatár, 1969; Work, 1967). Infection in cattle is less prevalent than in sheep or pigs at least in western Europe, the United States, New Zealand and Denmark (Jacobs *et al.*, 1960b, 1963; Remington, 1968; Work, 1967; Dubey and Streitel, 1976d). However, Čatár *et al.* (1969) isolated *T. gondii* from 8 out of 86 cattle in Yugoslavia. Infection is common in rabbits throughout the world. Such a high prevalence of *T. gondii* in these animals is important from the viewpoint of human infection. As stated earlier, humans can acquire infection by eating raw or undercooked meat (Desmonts *et al.*, 1965; Kean *et al.*, 1969). *Toxoplasma gondii* organisms are susceptible to extremes of temperatures. Tissue cysts are killed by heating the meat throughout to 150°F. *Toxoplasma gondii* in meat are killed by freezing (Jacobs *et al.*, 1960a; Work, 1968). However, there are also discrepant findings. On two occasions *T. gondii* survived freezing, once in a mouse frozen at −6°C for one day and in the skeletal muscle of a monkey frozen at −20°C for 16 days (Frenkel and Dubey, 1972c; Dubey and Frenkel, 1973). In further experiments, *T. gondii* tissue cysts were killed by freezing (Dubey, 1974). Cultural habits of people may play a role. For example, in France the prevalence of *Toxoplasma* antibody is very high. Whereas 84% of pregnant women in Paris have antibodies, only 32% in New York City and 22% in London have antibodies (Desmonts and Couvreur, 1974a). The higher incidence in France appears to be related in part to the French habit of eating some of their meat raw. It is of interest that the prevalence of *Toxoplasma* antibodies is much lower in immigrants to Paris than women of French origin (Desmonts and Couvreur, 1974a). To avoid *Toxoplasma* infection, hands should be washed after handling meat, and meat should be cooked to 140°F before eating. *Toxoplasma* tachyzoites and tissue cysts are killed even by tap water so surface cleaning of contaminated hands is easily achieved.

Oocysts are shed by cats, not only the domestic cat but other members of the Felidae like ocelots, marguays, Jagurundi, bob cats, and bengal tigers (Miller *et al.*, 1972; Jewell, *et al.*, 1972; Janitschke and Warner, 1972). Oocyst formation, however, is greatest in the domestic cat. Widespread natural infection is possible since a cat may excrete millions of oocysts after ingesting one infected mouse (Fig. 23). Oocysts are resistant to most ordinary environmental conditions (Table IV) and can survive in moist conditions for months and even years (Dubey *et al.*, 1970a; Yilmaz and Hopkins, 1972; Frenkel *et al.*, 1975b). Invertebrates like flies, cockroaches and earthworms can mechanically spread oocysts (Dubey *et al.*, 1970; Wallace, 1971b; 1972b).

Only a few cats may be involved in the spread of *Toxoplasma* since, at any given time, as little as 1% of the domestic cat population may be shedding oocysts (Wallace, 1973b; Janitschke and Kuhn, 1972; Pampiglione *et al.*, 1973; Knoch *et al.*, 1974; Christie *et al.*, 1976). One study reported finding *Toxoplasma* oocysts in 41.3% of 213 stray cats in Egypt (Rifaat *et al.*, 1976). Whether cats shed oocysts only once or several times in nature is not known. Under experimental conditions, cats usually did not reshed oocysts after reinoculation of *Toxoplasma* cysts (Piekarski and Witte, 1971; Dubey and Frenkel, 1974; Sheffield and Melton, 1974; Overdulve, 1974; Dubey *et al.*, 1977b,c). Immunity to *T. gondii* in cats may wane with time, and cats may reshed oocysts in nature. Under experimental conditions, cats shed large numbers of *T. gondii* oocysts after feeding *Isospora felis* (Chessum, 1972; Campana-Rouget *et al.*, 1974; Dubey, 1976a). Although these chronically *T. gondii* infected cats reshed large numbers of *T. gondii* oocysts after challenge with *I. felis,* they did not show any clinical signs, suggesting the relapse was confined to intestine (J. P. Dubey, unpublished). Chronically infected cats can also reshed oocysts after the administration of 40 mg or more of methyl prednisolone acetate (Depomedrol, Upjohn) per kg body weight of the cats.

Table IV
Resistance of *T. gondii* Oocysts

	Concentration (%)	Viability of oocysts	Reference
Reagent			
Formalin	10	24 Hours	Dubey *et al.* (1970a)
Sulfuric acid + dichromate	63 7	30 Minutes	Dubey *et al.* (1970a)
Ethanol	95	1 Hour	Dubey *et al.* (1970a)
Ethanol	20	7 Days	Dubey *et al.* (1970b)
Ethanol + ether	95	1 Hour	Dubey *et al.* (1970a)
Sodium hydroxide	6.0	24 Hours	Dubey *et al.* (1970a)
Urea		48 Hours	Dubey *et al.* (1970a)
Glycerine	50	5 Hours	J. P. Dubey, unpublished
Ammonium hydroxide	5.0	10 Minutes	Dubey *et al.* (1970a)
Ammonia, liquid	5.5	1 Hour	Frenkel and Dubey (1972b)
Tincture of Iodine	2.0	3 Hours	Frenkel and Dubey (1972b)
Tincture of Iodine	7.0	10 Minutes	Frenkel and Dubey (1972b)
Drying, at relative	80.0	18 Days	Frenkel and Dubey (1972b)
humidity	0	2 Days	Frenkel and Dubey (1972b)
Temperature (°C)			
Indoors			
20–22	In water	548 Days	Hutchison (1967)
4	In 2% H_2SO_4	578 Days	J. P. Dubey (unpublished)
−21	In water	28 Days	Dubey and Frenkel (1972c)
45	In water	1 Hour	Dubey *et al.* (1970b)
50	In water	30 Minutes	Dubey *et al.* (1970a)
Outdoors			
Texas: −6.5–37.5	Native feces	334–410 Days	Yilmaz and Hopkins (1972)
Kansas: −20–35	Native feces	548 Days	Frenkel *et al.* (1975b)
Costa Rica: 15–30	Native feces	56–357 Days	Frenkel *et al.* (1975b)

Unlike relapse with *I. felis,* the hypercorticoid cats shed fewer oocysts and developed clinical toxoplasmosis (Dubey and Frenkel, 1974).

How cats become infected in nature is an important consideration. In the laboratory, nearly all *Toxoplasma*-free cats shed oocysts after ingesting tissue cysts whereas far fewer (33% or less) do so after ingesting oocysts (Dubey *et al.*, 1970; Dubey and Frenkel, 1976; Wallace, 1973b). Even when they do, the numbers of oocysts shed are smaller. Carnivorism is, therefore, a more effective cause of oocyst formation in cats than is fecal contamination. In a survey in Kansas City cats, none of the 302 4½- to 10-week-old cats were infected with *T. gondii*, and only 16% of 80 11- to 26-week-old cats were infected, whereas 38% of adult cats were infected (Table V). These data indicate that few infections take place until the kitten is old enough to hunt for some of its food (Wallace, 1971a). Also, differences in the prevalences in domestic and stray cats from the same area suggest that cats which have to hunt for food are more likely to acquire infection. Cats probably become infected by preying on intermediate hosts. All evidence indicates that cats do not acquire infection congenitally (Dubey, 1977a; Dubey and Hoover, 1977). Up to 64% of cats in nature have antibody to *T. gondii* (Dubey, 1968a,c, 1973, 1976b). Infection rates of cats probably vary with the rate of infection in local avian and rodent populations.

Only a few surveys have been made of *Toxoplasma* infection in rodents (Wallace, 1973c). These report a low incidence of *Toxoplasma* antibody in mice and rats but *Toxoplasma* infection may be more prevalent than indicated by these surveys. Many infected rodents are not detected in these surveys because they become sick and die or become easy prey for other animals shortly after infection. Furthermore, not all infected ro-

Table V
Antibody Titers to *T. gondii* by Age and Type of Cat [a]

Age	Type	No. of cats	Percentage sero-positive 1:2 or more	No. of positive cats with antibody titers of: 1:2 to 1:32	1:64 or More
4½–10 Weeks	Domestic	302	8.6	26 [b]	0
11–26 Weeks	Domestic	80	16.2	1 [b]	12
6 Months	Domestic	128	37.5	27	21
6 Months	Stray	157	57.9	42	49

[a] From Dubey (1973).
[b] Probably passively transferred antibody, disappeared after 12 to 14 weeks of age.

dents become or remain serologically positive for long periods of time (Jacobs, 1964). More work is required to determine the exact prevalence of *Toxoplasma* in small rodents (Wallace, 1973c).

Proper identification of *Toxoplasma* oocysts in feces of cats is important from a public health viewpoint. Detection of *Toxoplasma* oocysts is possible using any of the standard fecal flotation techniques. However, the use of salt solutions over 1.18 specific gravity is not recommended because the distortion produced in the oocyst makes identification difficult. The following is an outline of a procedure now used for floating *Toxoplasma* oocysts from cat feces (Dubey *et al.*, 1972).

Materials Needed

A. Plastic, paper, or glass container (250 ml), tongue depressors, gauze or cheese cloth, Pasteur pipettes, rubber bulb, razor blade, scissors, centrifuge tubes, centrifuge (floating head type), microscope, cover glasses, slides, container to hold discarded pipettes.

B. Sucrose solution of 1.15 specific gravity (sugar 53 gm, water 100 ml, liquid phenol 0.8 ml).

Procedure

1. Collect the fecal specimen in a container, preferably, in a disposable cup. Add enough water to wet the feces. Leave the cup covered for 1 to 2 hours, until feces have softened.
2. Carefully drain the excess water into another container. This excess can either be carefully discarded down a drain, or sterilized by boiling. Emulsify the feces thoroughly before adding the sucrose flotation solution. Slowly add about 5 to 10 volumes of sucrose solution, while emulsifying the feces with a tongue depressor.
3. Strain feces through one to two layers of cheese cloth or gauze held over a container by means of a rubber band. Discard the gauze and the feces retained on it carefully so as not to spill the feces. The safest way to remove the gauze is to cut the rubber band with scissors or blade while the other hand holds the gauze and container firmly above the band. Any other method can cause the rubber band to catapult the feces.
4. Pour the strained feces into a 10- to 15-ml or larger centrifuge tube until the tube is filled to within 1 to 2 cm of the top.
5. After balancing the tube, centrifuge at about 1000 to 3000 rpm. Top speed in a clinical centrifuge is satisfactory.
6. After 5 to 10 minutes centrifugation, sample 1 to 2 drops from the very top of the tube using a pipette and a rubber bulb or a wire loop, mount the drop on a glass slide, and cover with a cover slip.
7. Let the slide sit on a flat surface for 1 to 5 minutes so that the fecal

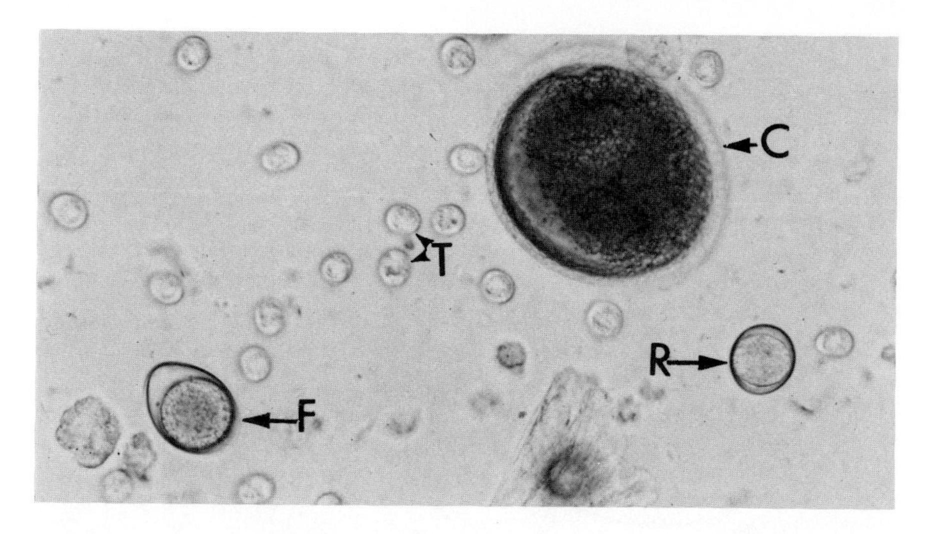

Fig. 40. Microphotographs of unsporulated oocysts of *Toxoplasma gondii* (T), *Levineia felis* (F), and *Levineia rivolta* (R) in a fecal sample. Centrifugal sucrose flotation was used. *Levineia felis* and *L. rivolta* are common coccidia of cats. *Toxoplasma gondii* oocysts are the smallest of all feline coccidia averaging 10×12 μm. *Levineia felis* oocysts are pear-shaped and average 40×30 μm. *Levineia rivolta* oocysts are ovoid and measure about 25×20 μm. All three coccidian oocysts are compared to the size of a *Toxocara cati* (C) egg, the common feline round worm (From Dubey, 1976b).

particles are immobilized. Examine under 100 × magnification and look for coccidian oocysts, and then further examine under 400 × magnification (Fig. 40). Unsporulated *Toxoplasma* oocysts measure 10 x 12 μm and are about one-quarter the size of *Levineia felis* (Fig. 41), one-sixth to one-eighth the size of a *Toxocara* egg, and about twice the size of red blood cells. Because of the small size of *Toxoplasma* oocysts, they usually lie in a different plane of focus than those of *L. rivolta, L. felis,* and *Toxocara* eggs. It is unusual to find both large and small structures in one focus as shown in Fig. 40. In some cats only a few *Toxoplasma* oocysts are shed, so a careful search may be required to find them. It is not unusual to find only one to five oocysts in the entire 22 x 22 mm cover glass preparation.

8. Discard all disposable material used for floating feces by burning. Alternatively, boil or autoclave all the material.

9. Fecal flotations should be done the same day feces are collected from cats or the feces should be stored at 4°C. *Toxoplasma* oocysts are unsporulated and not infectious in freshly passed feces, but sporulate

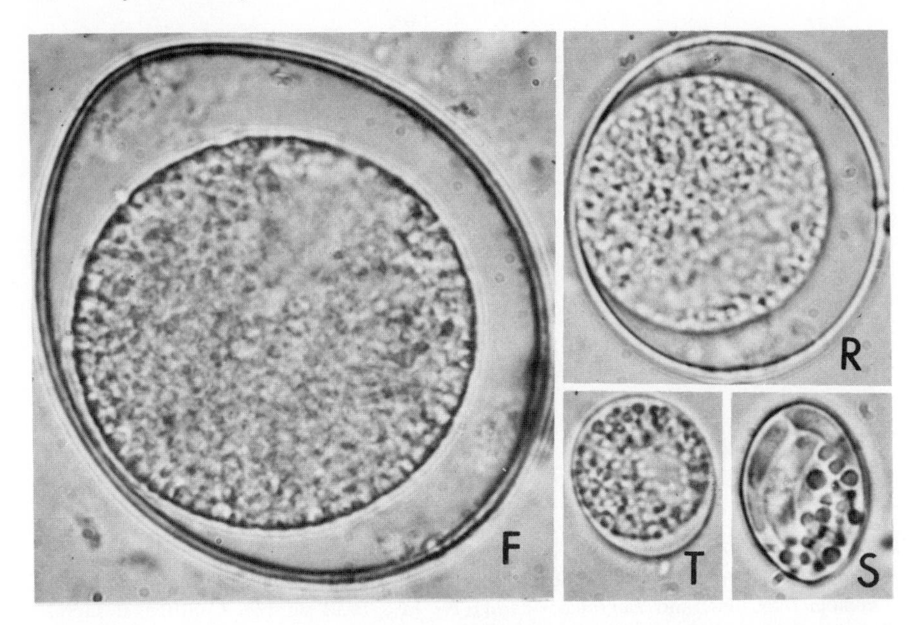

Fig. 41. Unsporulated oocysts of *Levineia felis* (F), *L. rivolta* (R), *T. gondii* (T), and *Sarcocystis* sp. (S) (from Dubey, 1973). Eleven species of feline coccidia are currently recognized. Of these, *L. felis* and *L. rivolta* can be easily distinguished by their larger size. *Besnoitia* oocysts are between the size of *T. gondii* and *L. rivolta*. *Besnoitia besnoiti* oocysts measure 13 × 16 μm (Peteshev *et al.*, 1974) and the murine *Besnoitia* oocysts average 12 × 17 μm (Wallace and Frenkel, 1975). *Toxoplasma* (Dubey *et al.*, 1970b) and *Hammondia hammondi* oocysts (Frenkel and Dubey, 1975) are structurally similar and measure about 10 × 12 μm. All these six species are shed unsporulated in feline feces. *Sarcocystis* is shed fully sporulated, therefore, it can be easily distinguished from other coccidia. There are at least five species of *Sarcocystis* in feline feces, involving mouse, cattle, sheep, cotton tail rabbits, and pigs as intermediate hosts.

and become infectious within 1 day at room temperature (22°–26°C) and aerobic conditions.

10. A presumptive diagnosis of toxoplasmosis can be made if oocysts measuring 10 x 12 μm are found in cat feces. However, for a definitive diagnosis, feces should be inoculated in mice to differentiate them from *Hammondia* oocysts. For this, 0.5 ml of the supernatant from the very top of the centrifuge fecal suspension is mixed with 4.5 ml of 2% sulfuric acid in a 30-ml bottle. After capping, the bottle is left at room temperature for 3 to 7 days. Shaking the bottle helps to aerate the oocysts (Dubey *et al.*, 1972).

11. For inoculation into mice, the sulfuric acid can be removed from stored oocysts by dilution in water and centrifugation or can be neutral-

ized with 3.3% sodium hydroxide (NaOH) containing 2% phenol red as indicator. The NaOH solution should be added drop by drop until the end point is reached at which time the color changes from yellow to orange.

12. Because *T. gondii* oocysts are pathogenic to man, mice should be inoculated with care. After intraperitoneal inoculation, apply 7% alcoholic tincture of iodine at the inoculation site to minimize contamination of the mouse cage or bedding by oocysts leaking from inoculation site. If oocysts are fed to mice, bedding should be replaced 1 day after the mice are fed since most of the unexcysted oocysts appear in the mouse feces within 24 hours.

Other possible means of transmission of toxoplasmosis are venereal, ingestion of milk or saliva, eating of eggs, through blood transfusions, organ transplants, and through arthropods (Frenkel, 1973a; Cathie, 1954; Janitschke and Nürnberger, 1975). These modes of transmission of *Toxoplasma* are not likely to occur in nature. *Toxoplasma* has been isolated from the milk of naturally and experimentally infected animals (Sanger and Cole, 1955; Sanger *et al.*, 1953) and rarely from chicken eggs (Jacobs and Melton, 1966). It is unlikely that either of these modes of transmission are significant in man. Parasitemia occurs during acute infection and may also occur during chronic infection. However, the risk of acquiring toxoplasmosis by blood transfusion is negligible (Janitschke *et al.*, 1974). Patients receiving organ transplants from *Toxoplasma*-infected donors may develop toxoplasmosis because of immunosuppressive therapy.

The incidence of *Toxoplasma* infection increases with age (Feldman and Miller, 1956; Feldman, 1965). In central France and El Salvador, over 87% of adult population has antibody to *T. gondii* (Desmonts *et al.*, 1965; Remington *et al.*, 1970a). The proportion of infections acquired through contact with soil containing *T. gondii* oocysts or eating infected meat needs to be determined by carefully conducted surveys. Probably both modes of infections are important. Cats are probably essential in the cycle of *Toxoplasma* (Wallace, 1969; Wallace *et al.*, 1972; Munday, 1972). These authors found no incidence or low incidence of *Toxoplasma* in isolated populations of man and animals on islands in the absence of cats and high incidence in association with cats.

For the prevention of *Toxoplasma* infection, the following precautions should be taken: (1) heat meat to 150°F throughout before eating; (2) wash hands with soap and water after handling meat; (3) never feed raw meat to cats; feed only dry, canned food or cooked meat; (4) keep cats indoors; (5) change litter boxes daily; flush cat feces down the toilet or burn them; clean litter pans by immersion in boiling water; (6) use gloves while working in garden; (7) cover children's sandboxes when not in use.

II. *Hammondia* Frenkel and Dubey, 1975

This genus was proposed for an intestinal coccidium of cats, *Hammondia hammondi.*

A. *Hammondia hammondi*

1. Introduction

Hammondia hammondi is an intestinal coccidium of cats. It is structurally similar to *Taxoplasma gondii.* Unlike *T. gondii*, it has an obligatory two-host cycle (Fig. 42). Cats are the definitive hosts and white mice, deer mice (*Peromyscus*), multimammate rats (*Mastomys*), guinea

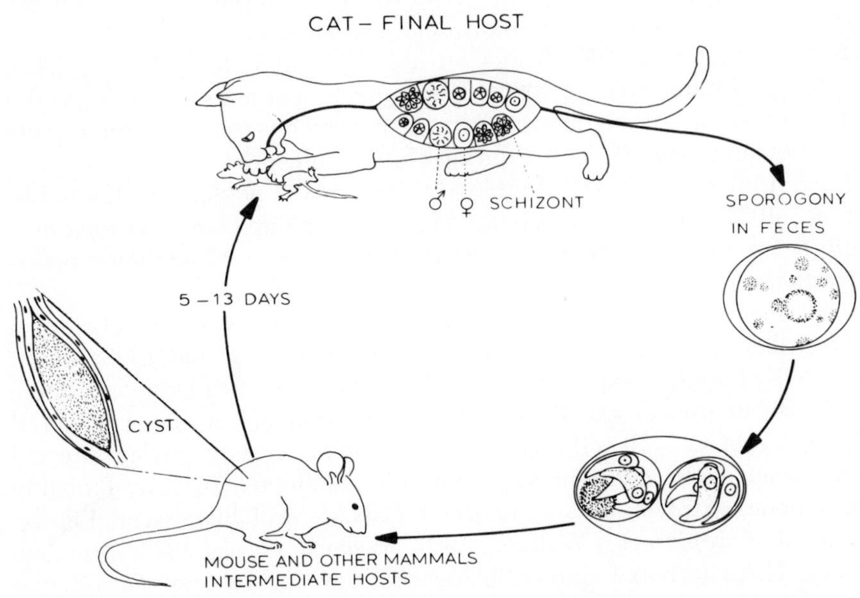

Fig. 42. Postulated life cycle of *Hammondia hammondi.* It has an obligatory two-host cycle. Cats are the definitive host and mice, rats, hamsters, guinea pigs, and dogs can act as experimental intermediate hosts. Unsporulated oocysts are shed in feline feces. After sporulation outside the host, the oocysts are infectious to intermediate hosts but not to definitive hosts, cats. Generalized infection occurs in the intermediate hosts. Cysts occur mainly in the musculature. Cats acquire infection by ingesting an intermediate host. *Hammondia hammondi* does not invade extraintestinal organs of the cat. There is no congenital infection in either definitive or intermediate hosts.

pigs, rats, hamsters, and dogs can act as experimental intermediate hosts (Frenkel and Dubey, 1975; Wallace, 1975; Dubey, 1975b; Dubey and Streitel, 1976b). Its natural intermediate hosts are not known.

2. History

A *Toxoplasma*-like organism, CR-4, was isolated by Frenkel and Dubey (1975) from the feces of a naturally infected cat in December, 1971. The cat started shedding *Toxoplasma*-like oocysts nine days after corticosteroid administration. At first it was thought that these oocysts were a species of *Toxoplasma*, but detailed life cycle studies indicated that unlike *T. gondii* the CR-4 isolate had an obligatory two-host cycle. This led Frenkel and Dubey (1975) to create a new genus and a new species, *Hammondia hammondi*, for this organism. While this paper was in press, Frenkel (1974) mentioned this organism in his review paper. Because Frenkel's (1974) review paper was published before the publication of the proper description of *H. hammondi* it has been said that Frenkel (1974) discovered *Hammondia hammondi* (Sheffield *et al.*, 1976; Tadros and Laarman, 1976). However, this is improper since the organism was discovered jointly. Therefore, this organism should be properly referred to as *Hammondia hammondi* Frenkel and Dubey, 1975.

Wallace (1973a) found a similar oocyst, WC-1170, in the feces of a nautrally infected cat in Hawaii. However, the oocyst was misidentified as being related to *Sarcocystis muris* which was also observed. Wallace (1975) later showed by cross protection experiments that WC–1170 was indeed a strain of *H. hammondi* and that the original cat feces contained *Sarcocystis* that had been overlooked.

3. Structure and Life Cycle

Unsporulated oocysts are shed in feline feces and measure 11 × 13 μm. After two to three days at 22°–26°C oocysts sporulate and contain two sporocysts each with four sporozoites (Fig. 43). The oocysts are impossible to distinguish structurally from those of *Toxoplasma gondii*.

After the oral administration of sporulated oocysts to intermediate hosts, the sporozoites are released in the intestine. Sporozoites penetrate the intestine, divide, and become tachyzoites. The tachyzoites multiply intracellularly by endodyogeny (Sheffield *et al.*, 1976) in the intestinal lamina propria, muscularis, Peyer's patches and mesenteric lymph nodes for up to 11 days after infection causing cell necrosis (Figs. 44 and 45). By day 10-11 cysts form in skeletal (Figs. 46 and 47) and cardiac muscle of mice (Frenkel and Dubey, 1975; Dubey and Streitel, 1976b). Cysts grow to the size of 300 x 95 μm in skeletal muscle and usually conform to the shape of the cell parasitized. They are rarely detected histologically

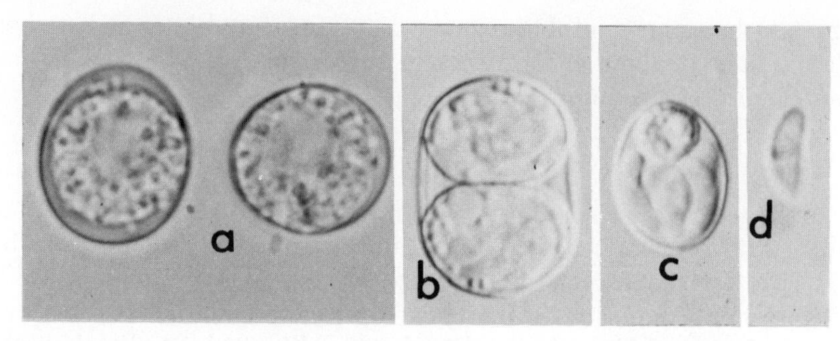

Fig. 43. Oocysts, sporocyst, and sporozoite of *H. hammondi.* All × 1664. Unstained. (a) Unsporulated oocysts. *Hammondia hammondi* oocysts are spherical to ovoid in shape. (b) A sporulated oocyst containing two sporocysts. (c) A sporocyst released fom the oocyst by pressure. Note the sporozoites and a residual body. (d) A sporozoite released by pressure from a sporocyst.

in the brain. Unlike cysts in muscle, the cysts in the brain are usually small (20–25 μm) (Fig. 48). Necrosis and inflammation may occur in the affected muscles. These lesions probably result because of the rupture of cysts without renewed multiplication; tachyzoites were not seen in lesions of myositis (Frenkel and Dubey, 1975).

Cats become infected by ingesting cysts found in the tissues of infected intermediate hosts. After ingestion, bradyzoites are released from cysts and schizogony is initiated in the epithelium of the small intestine of cats (Fig. 49). The schizonts occurring at 96 hours and thereafter resemble *Toxoplasma* types D and E; earlier stages have not been determined. Merozoites released from schizonts produce male and female gametocytes (Fig. 50). The gametocytes and oocysts of *H. hammondi* resemble those of *T. gondii* in structure and location.

Oocysts are shed 5–14 days after the ingestion of cysts and peak oocyst shedding occurs 1–2 days later (Fig. 51). Cats shed oocysts for 1–3 weeks. After the cessation of oocyst production, schizonts and gametocytes are not seen in intestinal epithelium. However, infection persists in the intestine in unknown stages as long as 3 months after oocysts have been shed (Frenkel and Dubey, 1975). These chronically infected cats spontaneously shed oocysts from time to time (Dubey, 1975a). Extraintestinal organs of the cat are not parasitized at any stage of the infection (Dubey, 1975a). Orally, oocysts or tachyzoites are noninfectious to cats (Frenkel and Dubey, 1975; Dubey and Streitel, 1976b). Unlike *T. gondii,* definitive hosts can become infected usually by ingesting cysts from intermediate hosts. Intermediate hosts can be infected with *H. hammondi* only through oocysts and not by inoculating tachyzoites and bradyzoites either parenterally or orally.

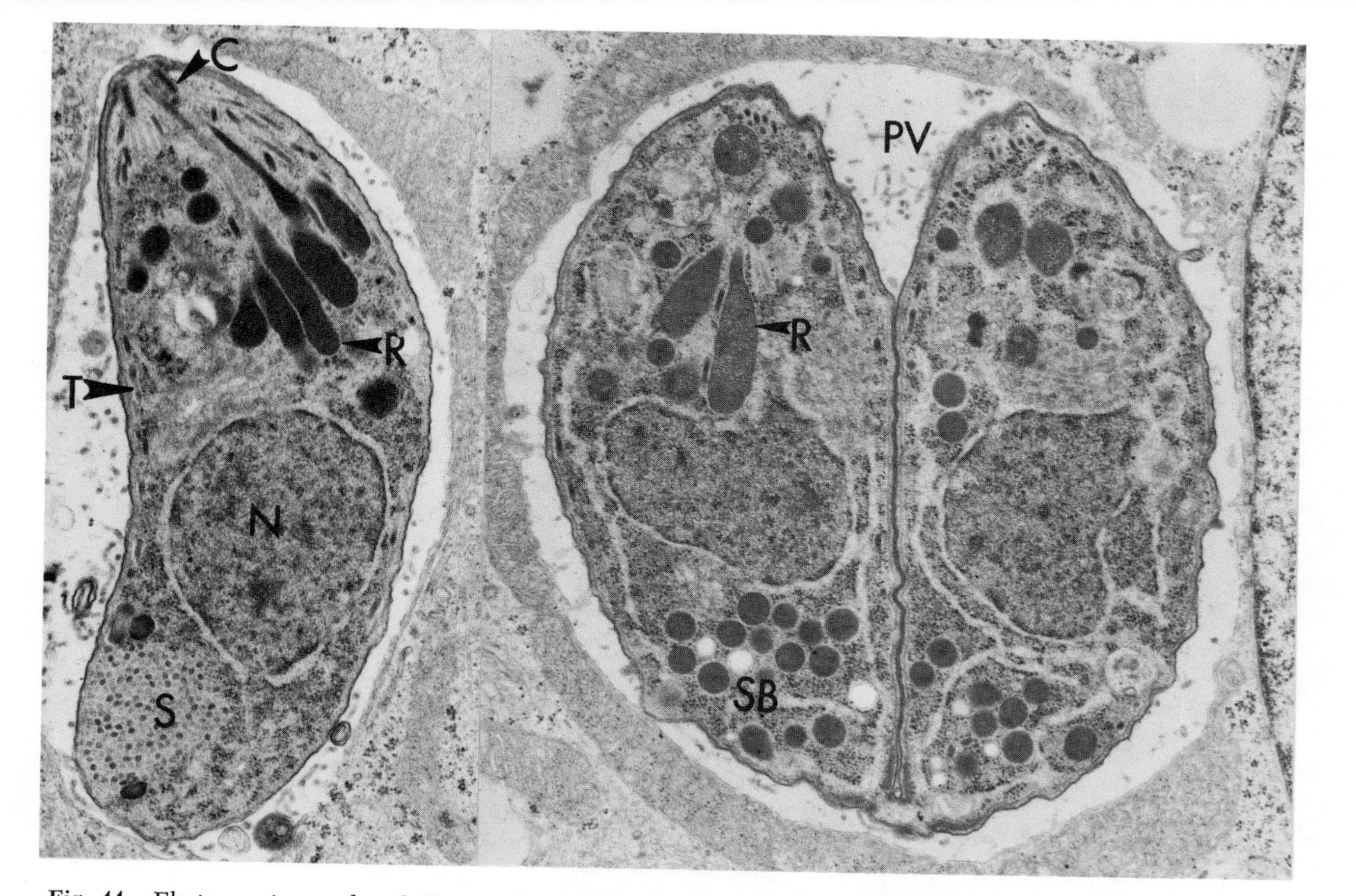

Fig. 44. Electron micrographs of *Hammondia hammondi.* Longitudinal section of organisms in parasitophorous vacuole (PV). A single organism on the left and a pair of dividing organisms on the right. C, conoid; N, nucleus; R, rhoptry; S, storage granules; SB, spherical bodies; T, subpellicular microtubules. Left: 12,570 ×. Right: 10,831 ×. (From Sheffield *et al.* 1976).

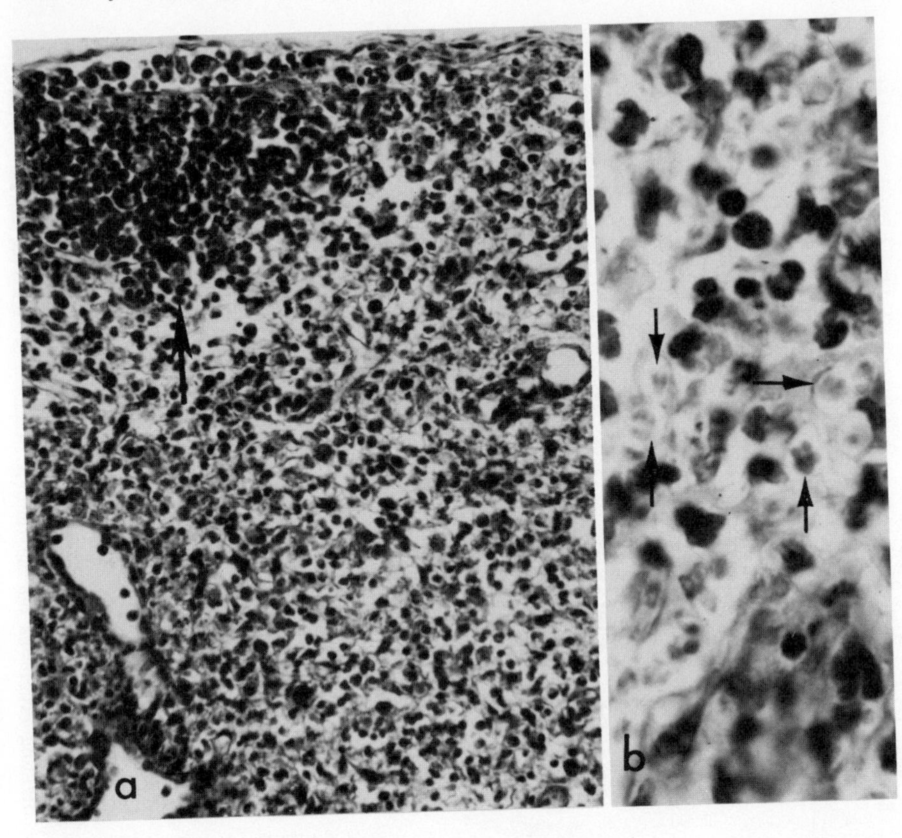

Fig. 45. A section of mesenteric lymph node of a mouse, 5 days postinfection. There is necrosis and depletion of lymphocytes. (a) A few lymphocytes are visible in the upper left corner (arrow). (b) Arrow points to tachyzoites. × 1250. Periodic acid Schiff-hematoxylin.

4. Pathogenicity

The pathogenicity of *Hammondia hammondi* varies with dose, host species and the route of inoculation. The CR-4 strain of *H. hammondi* is moderately pathogenic to mice. About 30% of mice fed 10^6 oocysts die after ingesting sporulated oocysts whereas 10^4 to 10^5 oocysts cause asymptomatic infection. Ooocysts are infectious but nonpathogenic to mice by the intraperitoneal or subcutaneous route. None of the seven strains of *H. hammondi* that we tested were pathogenic to hamsters (Christie and Dubey, 1977a).

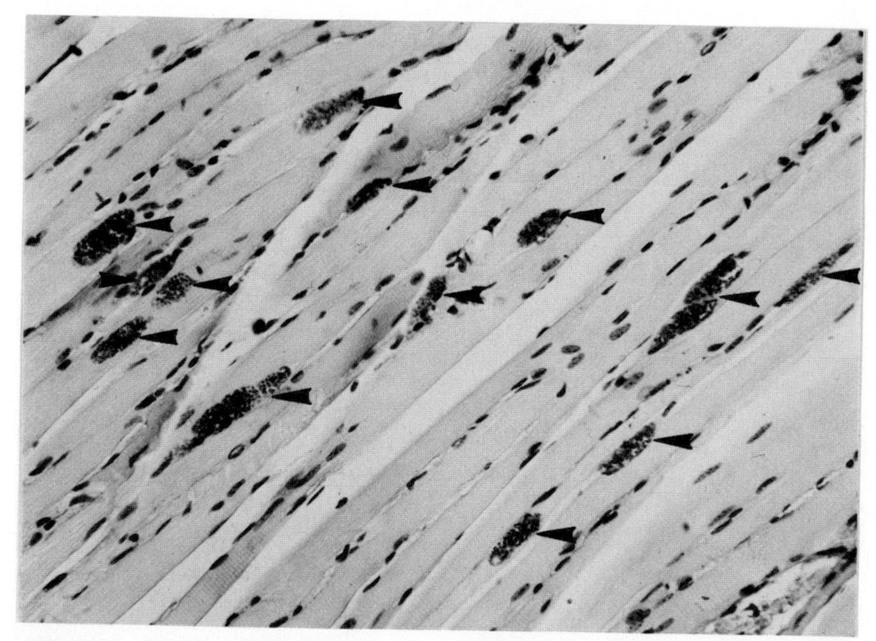

Fig. 46. Numerous *H. hammondi* cysts in the skeletal muscle of a mouse, 17 days postinfection. Inflammation is absent. × 200. Periodic acid Schiff-hematoxylin.

5. Immunological Relationship to *Toxoplasma gondii*

Mice, rats, hamsters, guinea pigs and dogs infected with *H. hammondi* develop antibody to *T. gondii* antigen (Frenkel and Dubey, 1975; Wallace, 1975). Antibody to *H. hammondi* antigen has also been detected in patients with *T. gondii* infection (Wallace, 1975). However, it is not known at the present time if *H. hammondi* infects human beings. Mice and hamsters immunized with *H. hammondi* became resistant to a fatal challenge with *T. gondii* (Frenkel and Dubey, 1975; Christie and Dubey, 1977a). However, unlike intermediate hosts, cats infected with *H. hammondi* neither develop antibody nor become immune to *T. gondii* (Frenkel and Dubey, 1975; Wallace, 1975; Christie and Dubey, 1977b).

B. *Hammondia heydorni* (Tadros and Laarman, 1976) nov. comb.

Synonyms: (Small *Isospora bigemina; Isospora wallacei* Dubey, 1976)

1. History

Isospora bigemina was first named *Coccidium begeminum*, based on finding oocysts with two sporocysts in the intestinal villus of a dog (Stiles, 1891; 1892). Oocysts structurally resembling those of *I. bigemina*

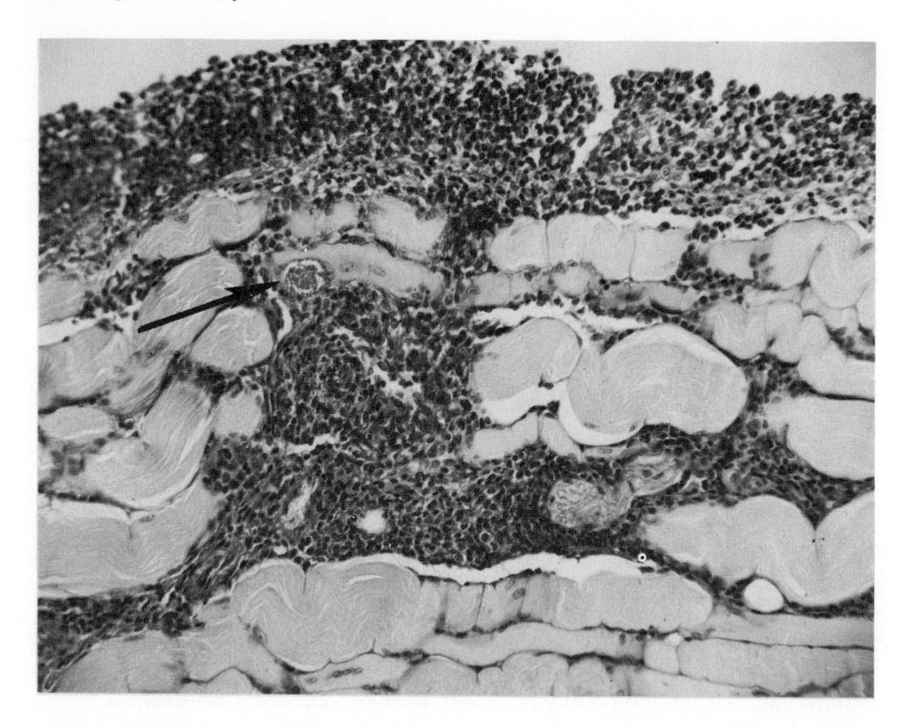

Fig. 47. Severe inflammation in abdominal muscle of a mouse, 39 days post-infection. Arrow points to a cyst. Tachyzoites have not been demonstrated in muscular lesions. × 250 Hematoxylin and eosin.

were later reported from several other hosts including cats and man (Wenyon, 1923). However, valid cross transmission experiments were not performed.

Two races of *I. bigemina*, different in size, were reported from dogs (Wenyon, 1926a; Wenyon and Sheather, 1925). The smaller race developed in the epithelial cells of the small intestine; its oocysts measured 10–14 × 7.5–9 μm and were shed unsporulated. The larger race developed in the lamina propria; its oocysts measured 18–20 × 14–16 μm and were shed sporulated.

It became clear from recent studies that the small and large races of *I. bigemina* were actually at least two distinct organisms (Heydorn, 1973; Heydorn *et al.*, 1975c). Because the name *I. bigemina* was first used for the large race developing in the lamina propria of the dog (Stiles, 1891), which is now *Sarcocystis bigemina* (Levine, 1977a), a new name, *Isospora heydorni*, was proposed by Tadros and Laarman (1976), for the small race of canine *I. bigemina*, because it had no name. Dubey (1976b)

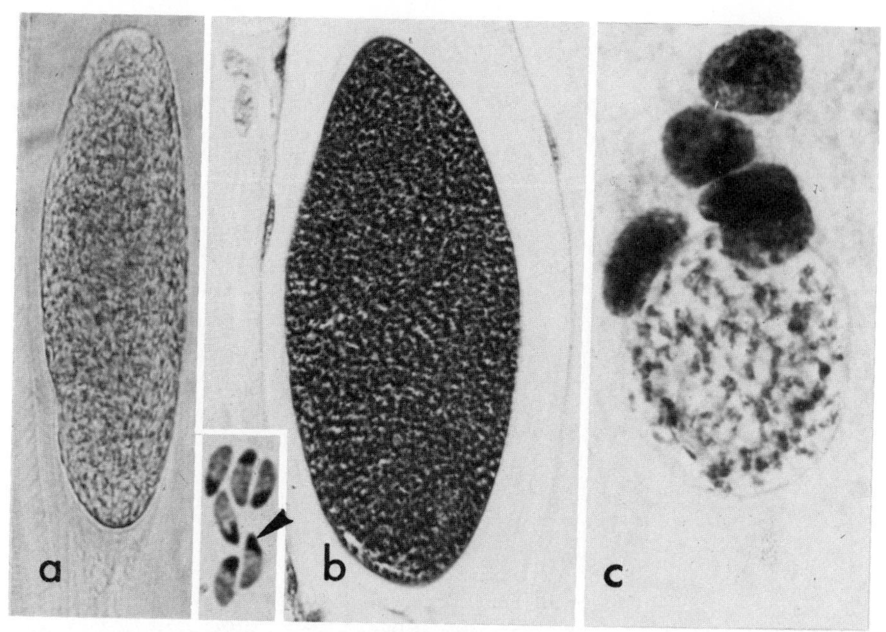

Fig. 48. *Hammondia hammondi* cysts in the muscle and brain of mice. (a) A cyst in abdominal muscle, 104 days postinfection. × 100. Unstained. (b) A cyst in skeletal muscle, 218 days postinfection. Bradyzoites stain bright red with PAS. Here they appear black. Inset shows terminal nuclei of bradyzoites, released from a cyst, 178 days postinfection. × 1600. Giemsa stain. (c) A cyst in mouse brain smear, 178 days postinfection. *Hammondia hammondi* cysts are rarely identified in brain. When cysts are found they are spherical and much smaller in size than in the muscle. × 1250. Giemsa stain.

called it *I. wallacei* for the same reason but the Tadros and Laarman name appeared a month earlier and has priority.

The classification of *H. heydorni* is at the present uncertain. Because its oocysts do not directly induce oocyst formation in dogs, it is not a true *Isospora* because oocysts of all known *Isospora* species induce oocysts in their definitive hosts. This organism is temporarily placed under the genus *Hammondia* because of structural and life cycle similarities to *Hammondia hammondi.*

2. Structure and Life Cycle

Cattle and dogs are two known hosts. Unsporulated oocysts are shed in canine feces and measure 11 × 13 μm (Levine and Ivens, 1975; Dubey and Fayer, 1976). After 12–48 hours, oocysts sporulate and contain two sporocysts each with four sporozoites. Sporulated oocysts are infectious

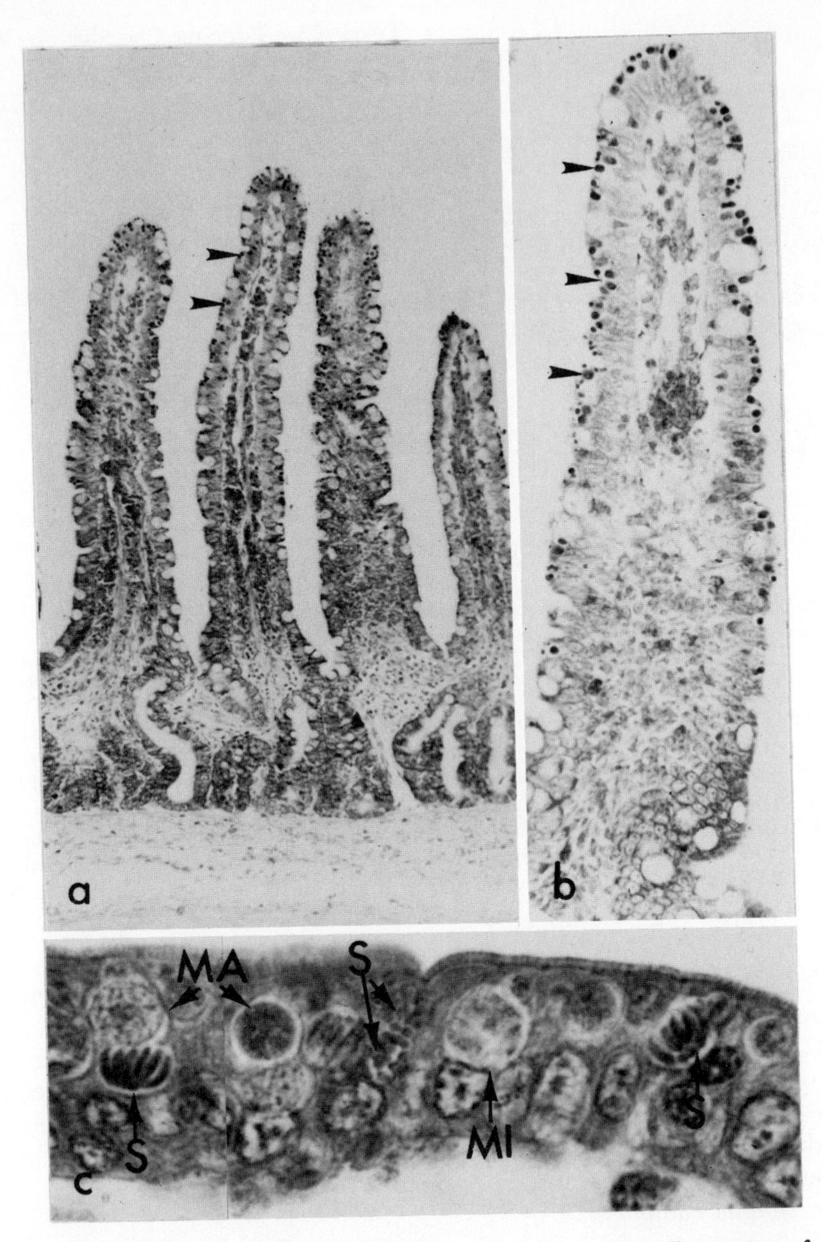

Fig. 49. *Hammondia hammondi* stages in the section of small intestine of cats fed cysts. *Hammondia hammondi* parasitizes the epithelial cells of the villus, whereas glands of Lieberkühn are usually not parasitized. (a) × 125. (b) × 315. Giemsa stain. *Hammondia hammondi* stains darker than the host tissue. Almost all epithelial cells are parasitized. (c) Schizonts (S), macrogametocytes (MA), and microgametocytes (MI) in epithelial cells. × 1250. Hematoxylin and eosin. Oocyst shedding by this cat is shown in Fig. 57.

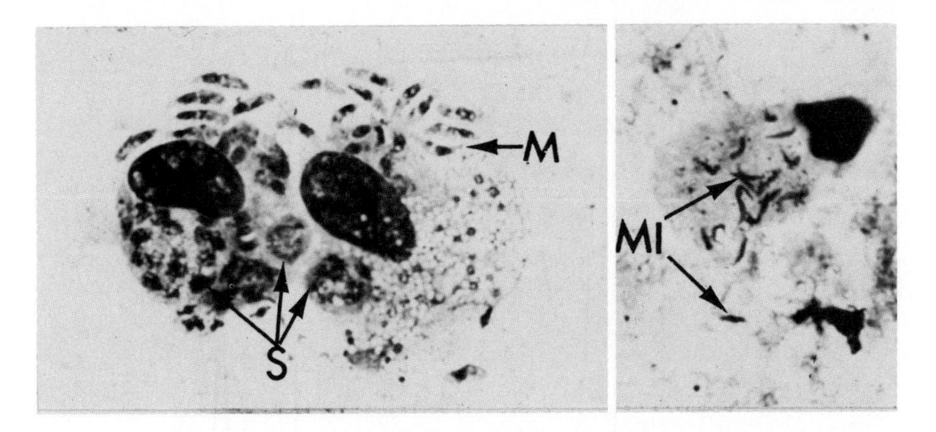

Fig. 50. Schizonts (S), merozoites (M), and microgametes (MI) of *H. hammondi* in intestinal smear of a cat, 10 days postinfection. × 1250. Giemsa stain.

to dogs but do not induce oocyst formation in canines (Heydorn, 1973; Dubey and Fayer, 1976). Stages that occur in canine tissue after the ingestion of oocysts have not been identified. After the ingestion of infected canine tissues, dogs excrete oocysts with prepatent periods of 7–15 days. Schizonts and gametocytes occur throughout the small intestinal epithelium after the ingestion of infected canine tissues (Heydorn *et al.*, 1975b; Dubey and Fayer, 1976). They occur distal to the host cell nucleus at the tips of villi. Mature schizonts are 5–8 μm and contain 3–12 merozoites (Dubey and Fayer, 1976). Male gametocytes lack a residual body. Stages were not found in extraintestinal organs of dogs (Dubey and Fayer, 1976).

Hammondia heydorni does not infect mice, rabbits and cats (Heydorn, 1973; Dubey and Fayer, 1976).

Dogs shed *H. heydorni*-like oocysts after the ingestion of naturally infected bovine musculature (Heydorn, 1973; Dubey and Fayer, 1976). However, convincing evidence is lacking at the present time that *H. heydorni* oocysts are infectious to cattle (Dubey and Fayer, 1976). Althought the life cycle of *H. heydorni* can be completed in "one" host (dog) in the laboratory it is unlikely to be the natural mode of infection. A canine-bovine-canine cycle observed by Heydorn (1973) would appear to be the natural mode of infection because the dogs shed oocysts after ingesting beef.

3. Pathogenicity

Hammondia heydorni is nonpathogenic to dogs (Dubey and Fayer, 1976).

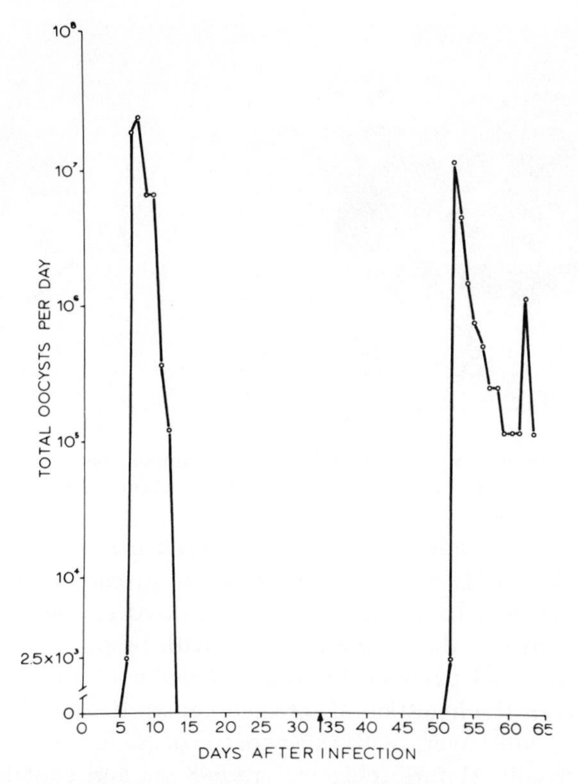

Fig. 51. Oocyst shedding by a *H. hammondi*-infected cat during primary infection and relapse. This cat shed millions of oocysts between the sixth and thirteenth day after feeding cysts. The cat started reshedding oocysts after weekly injections of 6-methylprednisolone acetate (40 mg/kg body wt/week) starting day 34 (arrow) after feeding cysts. It continued shedding oocysts until necropsy on day 64. The cat was asymptomatic throughout the course of infection. (From Dubey, 1975a.)

III. *Sarcocystis* Lankester, 1888

A. History

Sarcocystis miescheriana (Kuhn, 1865) Lankester, 1882, the type species of *Sarcocystis*, was originally described from the muscle of the pig. Later, in the late nineteenth and early twentieth century, *Sarcocystis* was recognized as a common parasite in the musculature of herbivores. However, its life cycle remained unknown until Fayer (1972) inoculated cells in culture with bradyzoites released from *Sarcocystis* cysts from a grackle, resulting in development of coccidian gametes and oocyst-like stages. In the same year, Rommel *et al.* (1972) independently found oocysts in feces of cats fed *Sarcocystis* cysts from sheep. Wallace (1973a) induced *Sarcocystis* in mice by feeding them coccidia from the feces of a naturally infected cat.

B. Structure and Life Cycle

Sarcocystis has an obligatory two-host life cycle (Fig. 52). Intermediate hosts (herbivores) acquire infection by ingesting sporocysts or oocysts that are shed in feces of infected definitive hosts (carnivores). Definitive hosts become infected with *Sarcocystis* by ingesting the encysted form of the parasite in the musculature of intermediate hosts.

Sporozoites are released from the sporocysts in the intestine of intermediate hosts and invade many tissues. Shizogony occurs in endothelial cells of blood vessels (Figs. 53 and 54) in most organs of the intermediate host (Fayer and Johnson, 1973), preceding the development of typical cysts in the muscle (Figs. 55–59).

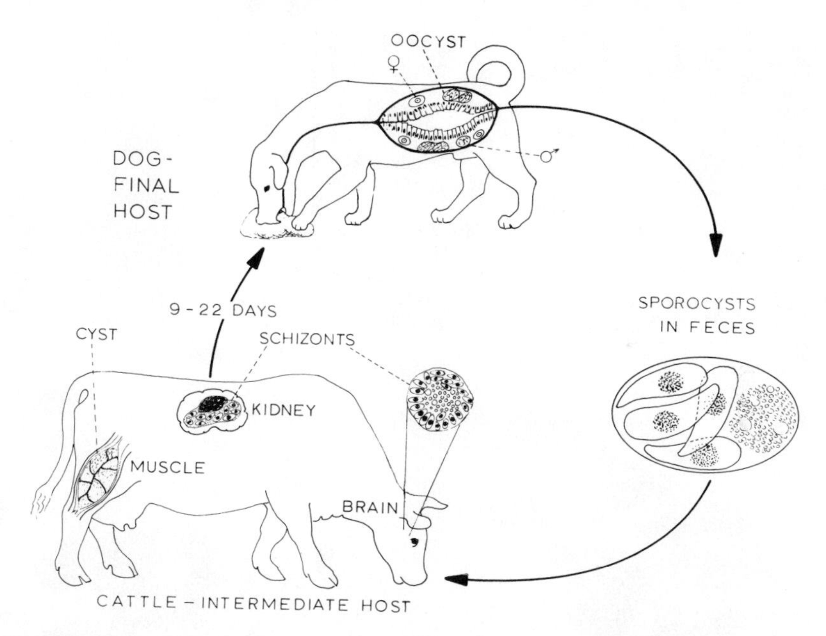

Fig. 52. Life cycle of *Sarcocystis cruzi* of cattle with the dog as an example of the definitive host. Dogs, wolves, coyotes, raccoons, and foxes shed sporulated oocysts or sporocysts in their feces after eating infected bovine musculature. Cattle became infected by ingesting sporocysts from the feces of carnivores. Generalized infection occurs in bovine tissues and schizonts are formed in many tissues especially in the kidneys and brain. After schizogonic cycles, cysts are formed in the musculature in 2 months. Current evidence indicates that canines become infected by ingesting only mature cysts. Sporocysts are noninfectious to definitive hosts. (From Dubey, 1976b.)

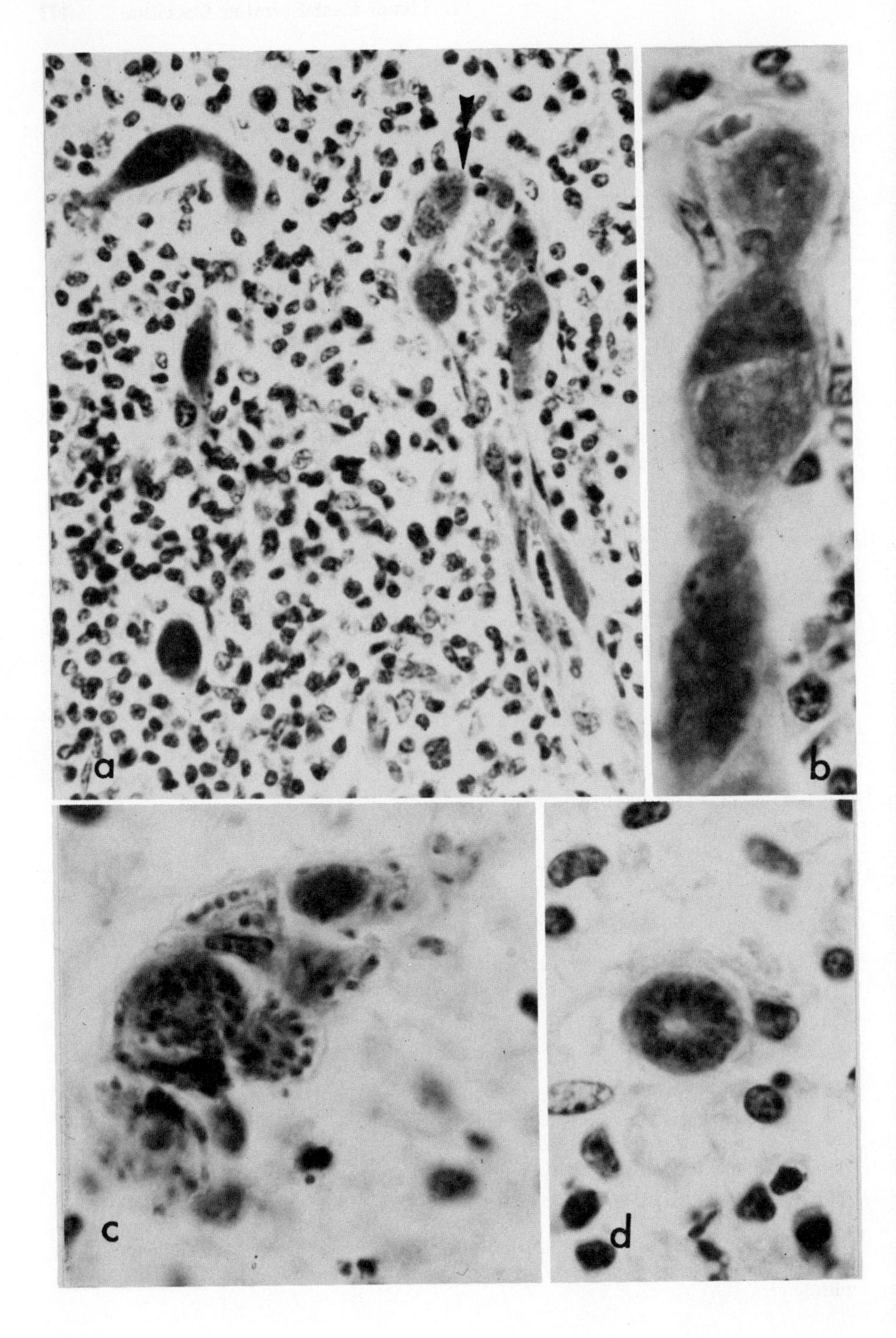
a
b
c
d

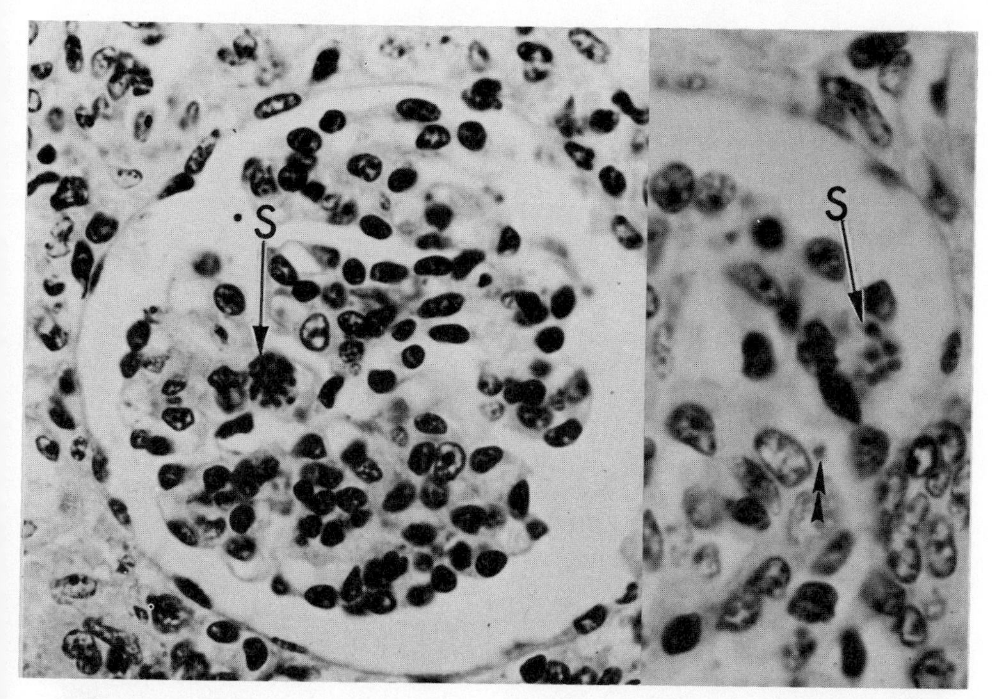

Fig. 54. Schizonts (S) of *Sarcocystis* in the kidney of a calf 33 days after feeding sporocysts from canine feces. (Courtesy of R. Fayer and A. J. Johnson.) (left) × 660. (right) × 1000. Hematoxylin and eosin. Double arrow points to single merozoites.

As the name implies, *Sarcocystis* forms cysts in the muscles of the intermediate hosts—reptiles, birds (Fig. 58), small rodents, and herbivores. Cysts are occasionally found in the brain as well (Hartley and Blackmore, 1974; Hilgenfeld and Punke, 1974). Cysts vary in length from a few micrometers to several centimeters, depending on the host and species. They are usually elongated and divided into compartments (Figs. 55 and 56). The thickness and structure of the cyst walls may vary with each *Sarcocystis* species and within each species as the cyst matures.

Fig. 53. *Sarcocystis* stages in the mesenteric lymph node of a naturally infected bovine from Kentucky. (Slide courtesy of Dr. J. K. Frenkel; from Dubey, 1976b). (a) Numerous irregularly shaped schizonts in endothelial cells. One capillary (arrow) has six schizonts. × 500. Hematoxylin and eosin. (b) Higher magnification of one of the schizontlike structures shown in (a). × 1250. Hematoxylin and eosin. (c) Numerous intracellular and extracellular merozoites. This stage appears similar to *Toxoplasma* tachyzoites. × 1250. Hematoxylin and eosin. (d) A subspherical schizont with peripherally arranged merozoites. × 1250. Hematoxylin and eosin.

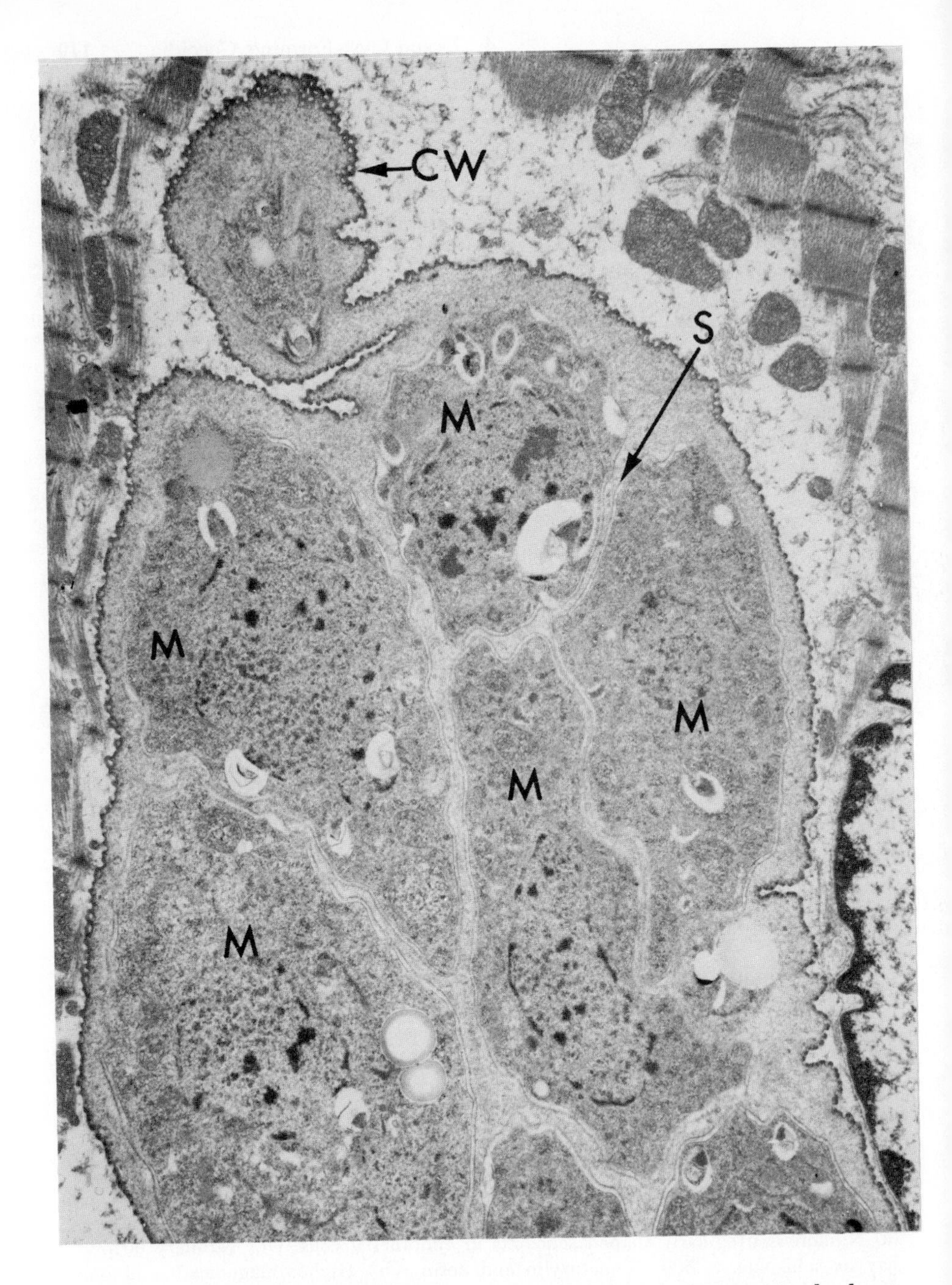

Fig. 55. Electron micrograph of a young cyst of *Sarcocystis cruzi* in the heart of a calf 58 days after feeding sporocysts from canine feces. There are several metrocytes (M) enclosed in the cyst wall (CW). The metrocytes are separated by septa (S). × 12000. (Courtesy of N. D. Pacheco and R. Fayer.) (From Dubey, 1976b.)

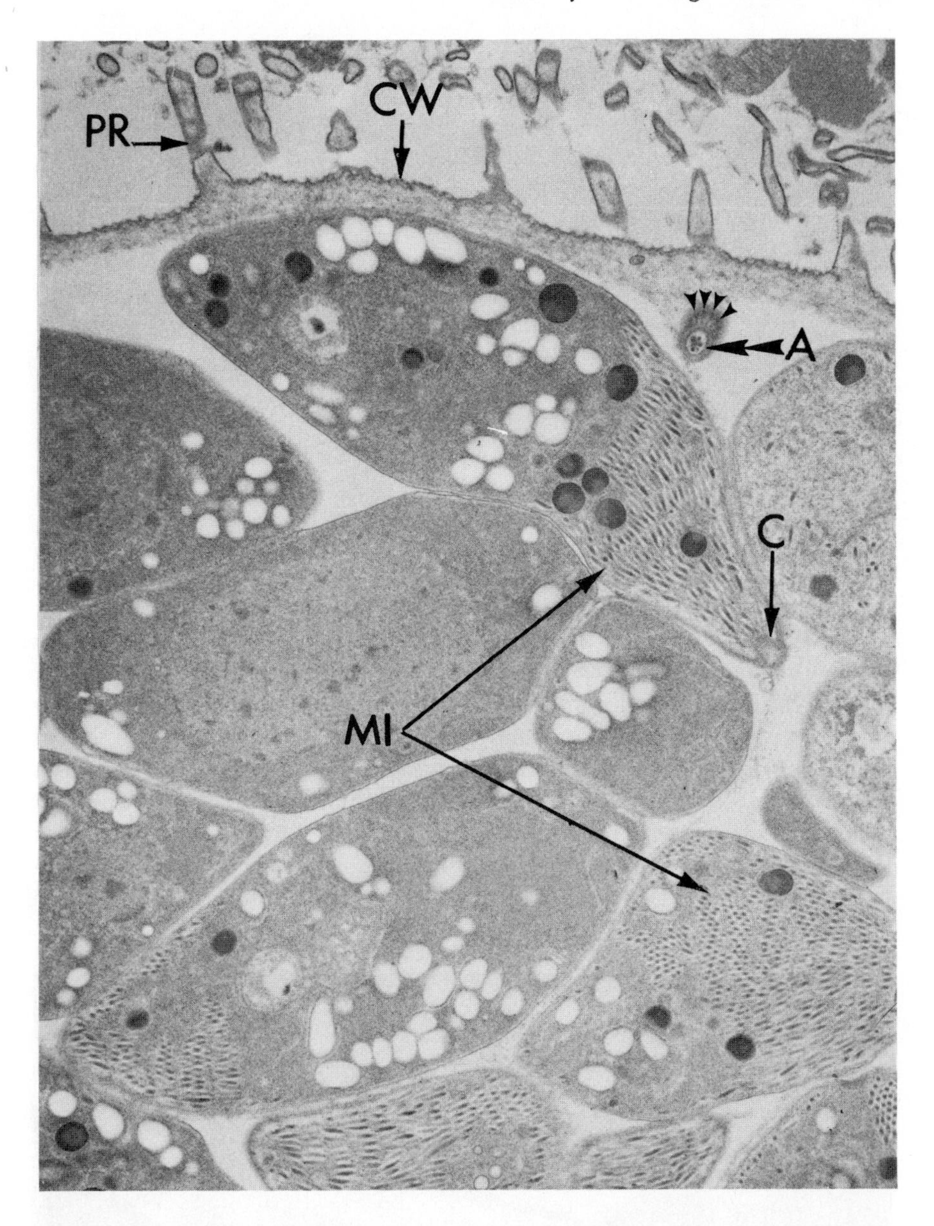

Fig. 56. Electron micrograph of a mature cyst of *Sarcocystis cruzi* in the heart of a calf 124 days after ingesting sporocysts from canine feces. The cyst wall (CW) is thin and has several projections (PR). The bradyzoites contain numerous micronemes (MI) that are absent in metrocytes shown in Fig. 64. The conoid (C) is visible in a longitudinally cut bradyzoite. Cross section of the anterior end (A) of a bradyzoite shows the microtubules (small arrows) originating from the polar ring. × 1000. (Courtesy of N. D. Pacheco and R. Fayer.) (From Dubey, 1976b.)

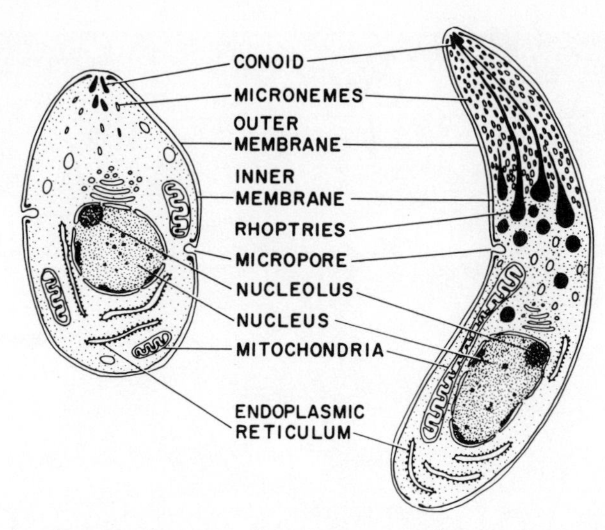

Fig. 57. A schematic drawing of a metrocyte and a bradyzoite of *Sarcocystis cruzi.* (Drawn after Heydorn *et al.*, 1975a.) (From Dubey, 1976b.)

Ultrastructurally, the sarcocyst has a primary cyst wall composed of a single unit membrane. The unit membrane is strengthened by an underlying layer of osmiophilic material. The primary cyst wall is folded into protrusions on the outside of the cyst. The protrusions vary in length and structure with each species of *Sarcocystis*. In some species the protrusions contain microtubules but not in others (Mehlhorn *et al.*, 1976). The unit membrane of cyst wall has numerous invaginations toward the interior of the cyst. These invaginations probably represent absorptive areas (Mehlhorn *et al.*, 1976). Beneath the primary cyst wall there is a zone of fine

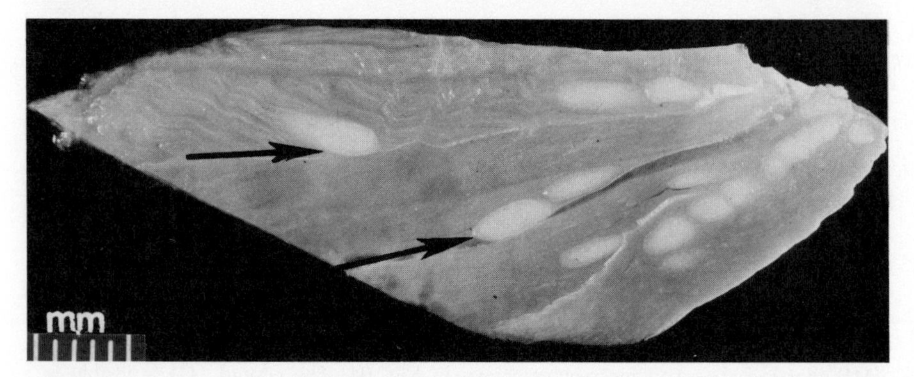

Fig. 58. *Sarcocystis* in a skeletal muscle section of a naturally infected duck. × 2. Unstained. (From Dubey, 1976b.)

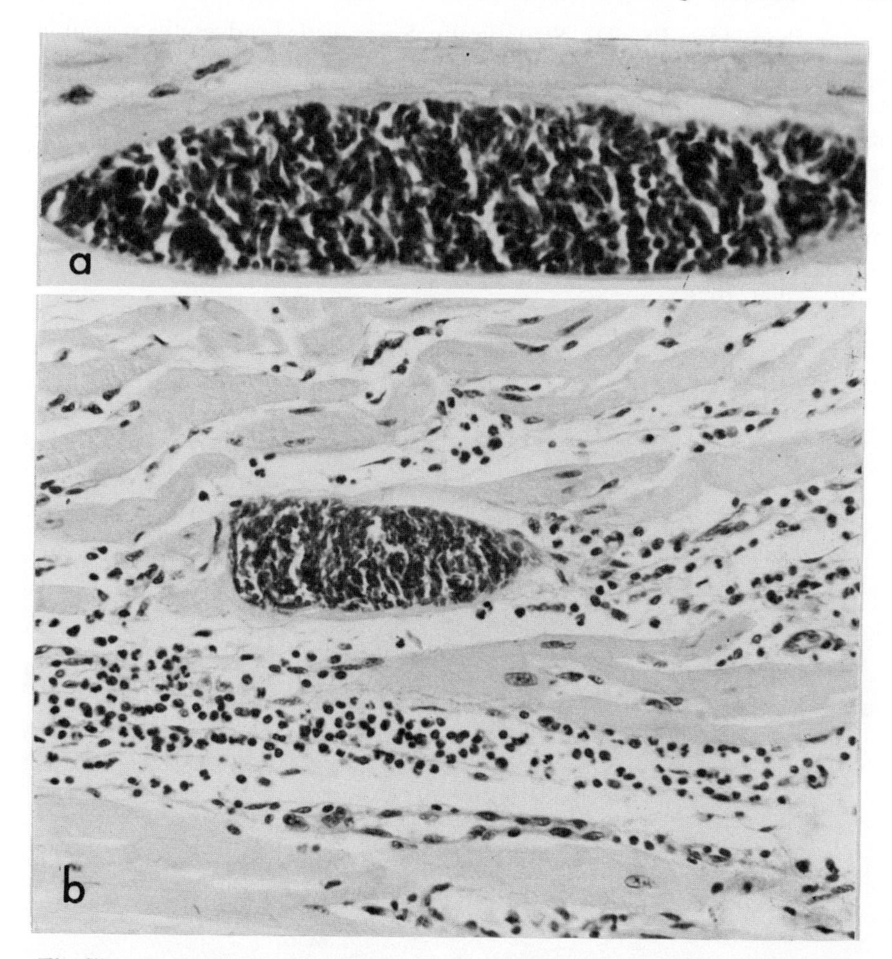

Fig. 59. *Sarcocystis* in the heart of a naturally infected cow. (a) A longitudinally cut cyst. There are hundreds of PAS-positive bradyzoites enclosed in the cyst wall. × 500. Periodic acid-Schiff hematoxylin. (b) Myositis in association with *Sarcocystis* cyst. These inflammatory lesions probably result due to rupture of cysts in sensitized myocardium. × 315. Periodic acid-Schiff hemotoxylin.

granular ground substance that extends into the interior of the cyst, forming compartments. In one species of *Sarcocystis* (*Sarcocystis tenella*) the entire parasitized host cell is surrounded by a secondary cyst wall that is composed of connective tissue (Mehlhorn *et al.*, 1976).

Two distinct regions can be recognized within the cyst proper. The peripheral region contains globular parasites called metrocytes (Figs. 55 and 57). Two daughter cells are formed by endodyogeny within each metrocyte. After several divisions, the globular metrocytes give rise to banana-shaped bradyzoites. Structurally, *Sarcocystis* bradyzoites resemble coccidian merozoites except that they have more (≥400) micronemes. A conoid and rhoptries are located at the more pointed end, a nucleus is situated toward the rounded end, and several PAS-positive cytoplasmic granules and mitochondria are present. Metrocytes are structurally like bradyzoites except that they lack rhoptries and micronemes (Mehlhorn and Scholtyseck, 1973) (Figs. 55–57).

Schizonts and metrocytes are noninfectious to definitive hosts (Ruiz and Frenkel, 1976).

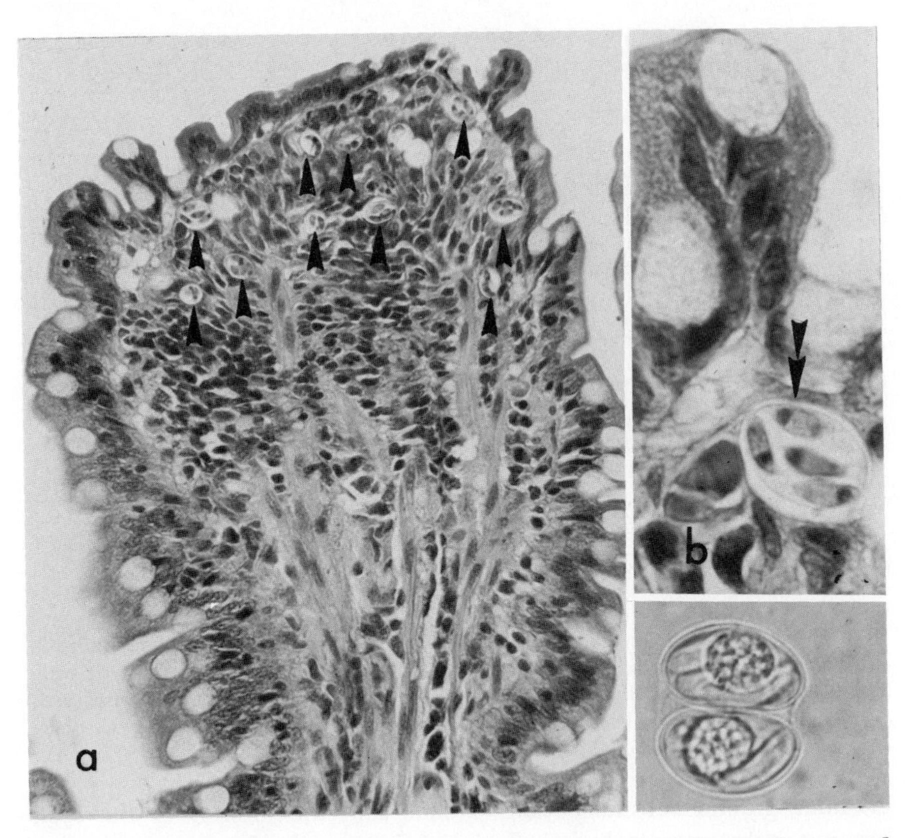

Fig. 60. *Sarcocystis cruzi* sporocysts in the ileum of a dog fed naturally infected bovine musculature. (From Dubey, 1976b.) (a) Numerous fully sporulated sporocysts (arrows) are visible in the lamina propria, just beneath the epithelium. × 315. Hematoxylin and eosin. (b) Higher magnification of an area from (a). A sporocyst with four sporozoites (double arrows) is seen in one plane of focus. It is rare to find all sporozoites in one focus. × 1250. Hematoxylin and eosin.

Fig. 61. (Lower right.) A sporulated oocyst of *Sarcocystis* in the feces of a dog. × 1250. Unstained.

The definitive host becomes infected by ingesting mature intramuscular cysts (containing bradyzoites) from infected intermediate hosts. The cyst wall is disrupted by proteolytic enzymes and the bradyzoites are released. The bradyzoites penetrate the lamina propria of the small intestine and form gametes without producing schizonts (Heydorn and Rommel, 1972a; Fayer, 1974; Munday *et al.*, 1975). The male gametes fertilize females gametes and unsporulated oocysts are produced in the lamina propria. Oocysts sporulate in the lamina propria, two sporocysts being produced in each oocyst. Four sporozoites develop within each sporocyst (Figs. 60 and 61). The oocyst wall surrounding two sporocysts is fragile and thin (0.1 μm) and often breaks releasing the sporocysts in the lamina propria (Mehlhorn and Scholtyseck, 1975). Thus, fully sporulated sporocysts are usually shed in feces (Fig. 62).

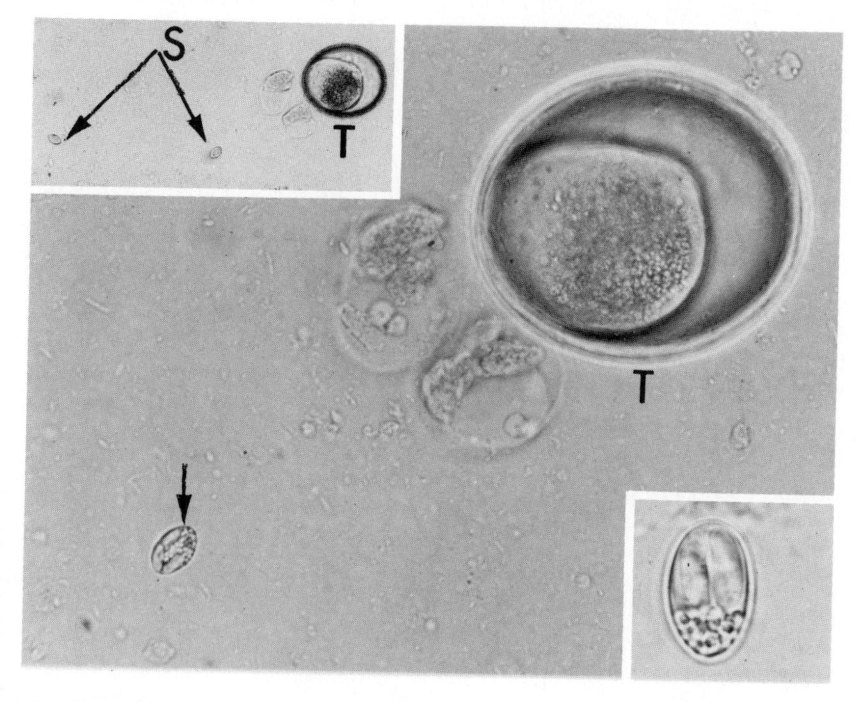

Fig. 62. *Sarcocystis* sporocyst (arrow) in fecal preparation, prepared using a sucrose centrifugal flotation method from a naturally infected dog. × 500. Unstained. Inset (upper left): × 125; inset (lower right): × 1250. *Toxascaris leonina* (T) eggs are shown for size comparison. Naturally infected dogs may shed only a few sporocysts (S) and it is not unusual to find only one or two sporocysts in an entire 22 × 22 mm cover slip area. Their detection is facilitated by leaving the slide–cover slip preparation for a period of 1 to 2 days. (From Streitel and Dubey, 1976.)

Sporocysts are infectious to intermediate hosts but not to the definitive hosts (Fishle, 1973; Ruiz and Frenkel, 1976).

C. Nomenclature

Considerable confusion exists in the literature regarding the species of *Sarcocystis*. Originally, a single species was thought to parasitize each herbivore host. Recent studies have shown that more than one species may parasitize a single intermediate host and one species may develop in several definitive hosts (Heydorn *et al.*, 1975c). A new classification scheme has been proposed to name the species of *Sarcocystis*, combining the generic names of the intermediate host and the definitive host (Heydorn *et al.*, 1975c). This scheme is simple and useful because it summarizes the life cycle of *Sarcocystis* in a single word. However, according to the rules of zoological nomenclature and in fairness to those who originally described the species *Sarcocystis*, it is not permissible to change the name of a valid existing species (Levine, 1977a). Therefore, old, valid species names must be retained.

D. *Sarcocystis* in Cattle

Until recently, only one species of *Sarcocystis* (*S. fusiformis*) was recognized to parasitize both cattle and buffaloes. Three species names of *Sarcocystis* were then proposed in the ox: *S. bovifelis*, *S. bovicanis*, and *S. bovihominis*, depending on their structural and life cycle differences (Heydorn *et al.*, 1975c). Prior to identification of these new species, several investigators had described separate species of *Sarcocystis* from the ox: *S. hirsuta* by Moulé, 1888; *S. cruzi*, by Hasselman, 1926; *S. iturbei* by Vogelsang, 1938, and *S. marcovi* by Vershinin, 1975. In addition, *S. fusiformis* was described from a water buffalo (*Bubalus bubalis*). The taxonomy of the species found in the ox has been reviewed recently by Levine (1977a). Three valid species are: *S. cruzi*, *S. hirsuta*, and *S. hominis* (Table VI).

The prevalence of bovine species varies from country to country and within the country (Fayer *et al.*, 1976; Dubey and Streitel, 1976b; Streitel and Dubey, 1976).

1. *Sarcocystis cruzi*

A. LIFE CYCLE. Dogs, wolves, coyotes, raccoons, and foxes (Fayer and Johnson, 1975; Fayer *et al.*, 1976a; Rommel *et al.*, 1974; Dubey and Streitel, 1976b shed sporulated oocysts or sporocysts (Fig. 52) in their feces after ingesting infected bovine musculature. After ingestion by cattle, sporozoites are released presumably in the small intestine. Probably two or more generations of schizonts (Figs. 53 and 54) develop in extra-

Table VI
Biological Features of *Sarcocystis* Species of Ox, Sheep, Pig, and Horse

Intermediate hosts / Species < Old names	Ox *S. fusiformis*			Sheep *S. tenella*		Pig *S. miescheriana*			Horse *S. bertrami*	
Current names	*S. cruzi*	*S. hirsuta*	*S. hominis*	*S. ovicanis*	*S. tenella*	*S. miescheriana*	*S. porcifelis*	*S. suihominis*	*S. bertrami*	*S. fayeri*
Cyst wall	Thin (0.5 μm)	Thick (6.0 μm) striated	Thick (5.9 μm) striated	Thick, radially striated	Thin	Not known	Not known	Not known	Thin	Thin
Pathogenicity	Pathogenic	Nonpathogenic or slightly pathogenic	Nonpathogenic or slightly pathogenic	Pathogenic	Nonpathogenic	Not known	Pathogenic	Not known	Not known	Not known
Definitive hosts	Dog Coyote Wolf Fox Raccoon	Cat	Man Rhesus monkey Baboon	Dog	Cat	Dog	Cat	Man	Dog	Dog
Prepatent period (days)	9–10	7–9	9–10	8–9	11–14	9–10	5–10	10–17	8	12–15
Sporocysts (μm)[a]	16 × 11	12 × 8	15 × 9	15 × 10	12 × 8	13 × 10	13 × 8	13 × 9	15 × 10	12 × 8

[a] Mean dimensions (Dubey, 1976b).

intestinal organs, primarily in vascular endothelial cells (Fayer and Johnson, 1973). Young cysts with metrocytes develop in skeletal and cardiac muscle (Fig. 55) as early as one month after the ingestion of sporocysts but are not commonly found until 2 months. Fully formed cysts (Figs. 56 and 59) containing only bradyzoites develop within 76 days (Heydorn *et al.*, 1975a). Metrocytes measure 7 × 5 μm and bradyzoites are 11–14 × 2.5–3.5 μm in size (Heydorn *et al.*, 1975a). Female gametocytes develop in the lamina propria of small intestine of dogs 2–6 days after the ingestion of infected bovine musculature. They measure 5–6 × 8.5–11.5 μm, depending upon the stage of development (Fayer, 1974). Unsporulated oocysts develop by the seventh day and sporulated oocysts or sporocysts may be shed on the ninth day of infection (Figs. 60–62). Sporocysts measure on an average 16 × 10 μm (Fayer, 1974).

B. PATHOGENICITY. This is the most pathogenic species in cattle. Until recently, sarcosporidia were considered of doubtful pathogenicity. Acute disease occurred in calves experimentally infected with sporocysts from canine feces (Fayer and Johnson, 1973; Johnson *et al.*, 1975). Clinical signs included: anorexia, pyrexia, anemia, cachexia, and weight loss. Affected calves became moribund or died within 33 days after ingesting 10^5–10^6 sporocysts. On necropsy, generalized lymphadenopathy and petechiae of serosal membranes were found. Schizonts were found in endothelial cells of blood vessels. These signs and necropsy findings were similar to those reported for Dalmeny diseases of cattle in Canada (Corner *et al.*, 1960).

Dalmeny disease was described in December, 1961, from a closed Holstein herd on a farm in Dalmeny in Ontario, Canada (Corner *et al.*, 1960). Twenty cows, 3 heifers, 2 steers, a bull and 10 calves were housed in one stable. Twenty-five animals became ill. The signs included: intermittent pyrexia, decreased milk yield, rapid loss of condition, and dyspnea. One animal had diarrhea, 13 had hemorrhagic vaginitis, and 9 salivated excessively. Of 17 pregnant cows, 10 aborted. Those that aborted were in the last trimester of pregnancy and only one dam survived. Chronic illness (up to 8 weeks) was characterized by extreme emaciation, pale mucous membranes, mandibular edema, exophthalmia, and sloughing of the tip of the tail. Of the 25 animals affected, eight eventually recovered. Of the 17 that died, 16 were necropsied. Ecchymotic hemorrhages were found in several organs, particularly in the myocardium. Protozoan bodies (schizonts) were found in endothelial cells of almost all organs examined. A similar disease has now been reproduced experimentally by feeding *Sarcocystis* sporocysts to cattle in Beltsville, Maryland (Fayer and Johnson, 1973; Johnson *et al.*, 1975; Fayer *et al.*,

1976b). Thus, Dalmeny disease appears synonymous with acute sarcocystosis. Two additional episodes of acute sarcocystosis in cattle in Canada (Meads, 1976) and the United States (Frelier *et al.*, 1977) have been reported. A similar infection was reported in a calf from England (Lainson, 1972).

Sarcocystis hirsuta and *S. hominis* are nonpathogenic for cattle.

Information on *S. cruzi* as well as other species of *Sarcocystis* of domestic animals is given in Table VI.

C. DIAGNOSIS. A presumptive diagnosis of acute bovine sarcocystosis may be made by clinical signs: anemia, swollen lymph nodes, excessive salivation, abortion, and loss of the hair at the tip of the tail. Determination of serum glutamic oxalacetic transaminase (SGOT), lactic dehydrogenase (LDH), and creatinine phosphokinase (CPK) and of hematologic data may aid clinical diagnosis. In experimentally infected calves there was an oligocythemic anemia, a leukocytic shift to the left, and an elevation of SGOT, LDH, and CPK levels in the serum (Mahrt and Fayer, 1975). An indirect hemagglutination (IHA) test has been recently developed for the diganosis of bovine sarcocystosis (Lunde and Fayer, 1977). In this test the antigen is prepared from *Sarcocystis cruzi* bradyzoites. Serum titers of more than 1:162 are considered specific for *Sarcocystis* infection. IHA titers began to rise in calves 30 to 45 days after inoculation with *S. cruzi* sporocysts and reached levels up to 1:39,000 90 days after inoculation. There was no cross reaction between *Sarcocystis* and *Toxoplasma gondii* in this test.

D. CONTROL. Shedding of *Sarcocystis* sporocysts in the feces of carnivores is the key factor in the spread of *Sarcocystis* infection. Therefore, control efforts must be based primarily on measures designed to break the cycle. Under certain conditions carnivores will eat live or dead livestock. Therefore, efforts should be made to bury or incinerate dead livestock. Stored grain used for feeding must be kept covered and carnivores should not be allowed inside animal housing facilities. Over 50% of adult swine, sheep, and cattle are estimated to be infected with *Sarcocystis* spp. Therefore, the ingestion of infected meat by carnivores is likely to result in their shedding of sporocysts and oocysts. There is virtually no immunity to sarcocystosis in carnivores, so repeated reinfection can occur (Fayer, 1974). Meat should be cooked well before being fed to animals (Fayer, 1975).

The use of anti-coccidial drugs may prevent or minimize *Sarcocystis* infection in the intermediate host. Under experimental conditions, feeding of amprolium at the rate of 100 mg/kg body wt reduced damage in cattle inoculated with *S. cruzi* (Fayer and Johnson, 1975).

E. *Sarcocystis* in Sheep

Sarcocystis infection is common in sheep throughout the world (Levine, 1973). In a recent survey involving sheep slaughtered in a Detroit, Michigan abbatoir, 73.3% of 789 adults and 10.8% of 108 lambs were infected (Seneviratana *et al.*, 1975). There are two species: S. *ovicanis* involving a dog–sheep cycle, and S. *tenella* involving a cat–sheep cycle.

Sarcocystis tenella is nonpathogenic whereas S. *ovicanis* is highly pathogenic to lambs. After ingesting S. *ovicanis* sporocysts, lambs became anorexic, weak, and died (Gestrich *et al.*, 1974; Heydorn and Gestrich, 1976; Leek and Fayer, 1976; Munday *et al.*, 1975). Schizonts were found in endothelial cells of almost all organs examined. Pregnant ewes fed 10^5 to 10^6 sporocysts of S. *ovicanis* developed pyrexia (42°C), became ataxic, and aborted. Schizonts were demonstrated in almost all organs of ewes but not in fetal tissues (Leek and Fayer, 1976).

A naturally occurring sporozoan induced encephalomyelitis in sheep has been reported from the United States (Olafson and Monlux, 1942), Ireland (McErlean, 1974), and Australia (Hartley and Blackmore, 1974). Affected sheep had myelomalacia. Schizont-like structures were found in astrocytes. It appears that these reports were of acute sarcocystosis, not toxoplasmosis.

F. *Sarcocystis* in Pigs

Sarcocystis infection is considered common in pigs (Levine, 1973). In a recent survey involving pigs killed in a Detroit, Michigan abbatoir, cysts were found in 12.7% of 55 pigs older than one year but not in 48 pigs younger than one year of age (Seneviratana *et al.*, 1975). There are three species: S. *miescheriana* (Kuhn, 1865) Lankester, 1882 with a pig–dog cycle, S. *porcifelis*, Dubey, 1976 with a pig–cat cycle, and S. *suihominis* (Tadros and Laarman, 1976) nov. comb. with a pig–man cycle.

Of these three species, the species with the pig–cat cycle (S. *porcifelis*) was reported to be highly pathogenic for pigs (Golubkovan and Kisliakova, 1974). Pigs grew poorly and developed diarrhea, myositis, and lameness after the ingestion of unknown numbers of sporocysts from feline feces. The pathogenicity of S. *miescheriana* and S. *suihominis* for pigs is not known.

G. *Sarcocystis* in Horses

Sarcocystis infection is considered common in horses throughout the world (Levine, 1973). The dog is the definitive host for *Sarcocystis* species in the horse (Rommel and Geisel, 1975; Dubey *et al.*, 1977). There are two species in the horse based on differences in sporocyst size and prepatent periods (Table VI). The pathogenicity of *Sarcocystis* infection in horses is not known.

H. *Sarcocystis* in Man

There are three species: *S. hominis* (Railliet and Lucet, 1891) Dubey, 1976, *S. suihominis* (Tadros and Laarman, 1976) nov. comb., and *S. lindemanni* (Lindemann, 1868) Rivolta, 1878. *Sarcocystis hominis* infection is acquired by eating meat of infected cattle and *S. suihominis* by ingesting infected pork (Table VI).

For *S. lindemanni*, man is the intermediate host; its definitive host is not known. Evidence of infection in man is usually an accidental finding during histologic examination; only 28 cases have been reported (Jeffery, 1974). Infection with *S. lindemanni* has not been associated with any definite symptoms (Jeffery, 1974). Anorexia, nausea, abdominal pain and diarrhea have been reported in patients naturally infected with *S. hominis* and *S. suihominis* (previously called *Isospora hominis*) (Barksdale and Routh, 1948). However, under controlled conditions none of the 5 of 12 people that excreted sporocysts of *S. hominis* after ingesting naturally infected beef in Germany became ill (Aryeetey and Pierkarski, 1976). They shed sporocysts for 9–179 days without reinfection.

I. *Sarcocystis* in Mule Deer

In wild deer large numbers of *Sarcocystis* sp. were found in the musculature of fawns that were killed and examined from Harney County, Oregon between March, 1974 and April, 1975 (Hudkins-Vivion *et al.*, 1976). The authors suspect that *Sarcocystis* infection might be a cause of declining deer populations in Oregon because *Sarcocystis* can cause acute fatal disease in experimentally infected mule deer (Hudkins and Kistner, 1977). Dogs and coyotes shed sporocysts in their feces 11 or more days after ingesting tissues of naturally infected mule deer. The sporocysts measured 13.8–16.1 × 9.2–11.5 μm (mean 14.4 × 9.3 μm). Nine of 11 fawns fed 10^5–10^6 sporocysts from coyotes' feces died of acute sarcocystosis between 27 and 63 days post inoculation. Schizonts were found in various tissues, especially around blood vessels in skeletal muscles. Cysts in muscles were found in the remaining two fawns when killed at 60 to 88 days after inoculation (Hudkins and Kistner, 1977).

Because sporocysts of the bovine species are larger (16.3 × 10.7 μm) than those derived from ingested mule deer tissues (14.4 × 9.3 μm) and the organisms would not develop in experimentally inoculated cattle and sheep, a new name (*S. hemionilatrantis*) was proposed by Hudkins and Kistner (1977) for the species from the mule deer.

IV. *Frenkelia* Biocca, 1968

A. Introduction

Frenkelia species form thin-walled cysts in the brain of field voles, meadow mice, chinchillas, muskrats, and bank voles (Bell *et al.*, 1964; Karstad, 1963; Frenkel, 1956; Rommel and Krampitz, 1975; Burtscher and Meingassner, 1976; Tadros and Laarman, 1976). Structurally and biologically this genus is related to *Sarcocystis*.

B. History

Frenkelia microti (Findlay and Middleton, 1934) Biocca, 1968 was discovered by Findlay and Middleton (1934) in voles from England. They named it *Toxoplasma microti*. Frenkel (1956) found a similar organism in *Microtus modestus* from Hamilton, Montana and separated it from *Toxoplasma* because he could not transmit it by passage into mice. He called it the M-organism. Biocca (1968) transferred it to a new genus, Frenkelia. Erhardova (1955) discovered *Frenkelia* cysts from the bank vole (*Clethrionomys glareolus*) from Czechoslovakia. This parasite is structurally different from *F. microti*. Tadros *et al.* (1972) named it *Frenkelia glareoli*. Rommel and Krampitz (1975) first described the life cycle of *F. glareoli* from the bank vole but they called it *Frenkelia clethrionomyobuteonis*, following the new classification of Heydorn *et al.* (1975c). Rommel and Krampitz (1975) found sporocysts in the feces of buzzards (*Buteo buteo*) after feeding infected bank voles. These sporocysts were identical to the sporocysts of *Isospora buteonis* Henry, 1932 described from hawks and an owl from California (Rommel and Krampitz, 1975; Tadros and Laarman 1976). Tadros and Laarman (1976) transferred *F. glareoli* to their new genus *Endorimospora* and called it *Endorimospora buteonis*. For reasons discussed in Section VIII of this chapter, the genus *Frenkelia* should be retained. Therefore, the proper name for this organism should be *Frenkelia buteonis* (Henry, 1932) nov. comb.

C. Structure and Life Cycle

Frenkelia microti cysts are up to one mm in diameter, lobulated, thin walled (Fig. 63) and occur in the brains of field voles (*Microtus agrestis*), meadow mice (*M. modestus*), and muskrats (*Ondatra zibethica*). *Frenkelia buteonis* cysts are spherical and occur in the brain of bank voles (*Clethrionomys glareolus*). Cysts of both species of *Frenkelia* can be seen through the skull (Bell *et al.*, 1964; Tadros and Laarman, 1976). Ultrastructurally, the cyst contains three types of cells, metrocytes at the periphery, intermediary cells, and endodyocytes (bradyzoites) in

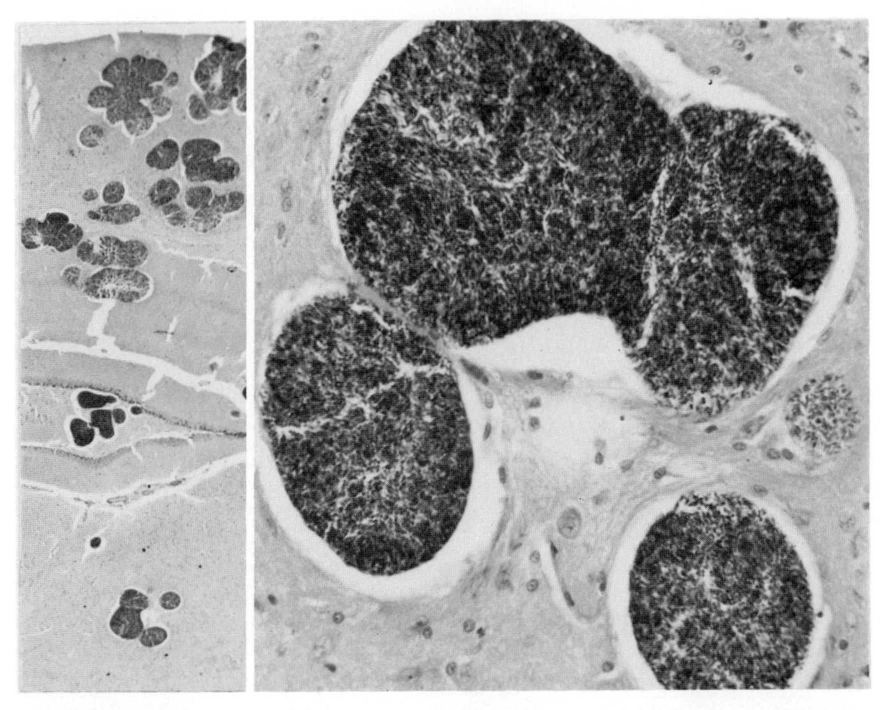

Fig. 63. *Frenkelia microti* cysts in the brain of a naturally infected vole (*Microtus modestus*). The cysts are lobulated, thin walled, and can even be seen through the skull. Mature cysts contain bradyzoites that stain bright red with PAS. Here they appear black. *Frenkelia microti* cysts structurally resemble *Sarcocystis*. × 32 (left) and × 315 (right). Periodic acid Schiff-hematoxylin. (Courtesy of Dr. J. K. Frenkel.)

the center. The metrocytes and bradyzoites structurally resemble those of *Sarcocystis* (Tadros *et al.*, 1972; Scholtyseck, 1973). Other stages of the life cycle of *F. microti* are not known.

The buzzard (*Buteo buteo*) is the definitive host for *F. buteonis* (Rommel and Krampitz, 1975; Tadros and Laarman, 1976). Gametogony occurs in the lamina propria of small intestine of the buzzard after the ingestion of cysts from the brain of mice. Macro- and microgamonts can be found 21 hours after infection. The microgamonts form about 12 comma-shaped microgametes. Unsporulated oocysts are found three–four days after infection. The oocysts sporulate within the lamina propria and sporulated sporocysts are excreted in feces 7–9 days after infection. The sporocysts are 11.0–13.8 × 7.8–10.0 (12.5 × 8.8) μm in size. They contain four sporozoites. The patent period is 7–57 days (Rommel and Krampitz, 1975; Rommel *et al.*, 1976). The intermediate host becomes infected by ingesting sporocysts. Schizonts occur in the liver of mice as early as 7 days after ingesting sporocysts. Schizonts contain 20–30 merozoites. Cysts are formed in the brain, beginning 18 days after infection. They measure 326 μm in diameter 120 days after infection (Rommel *et al.*, 1976).

Frenkelia is not transmissible by inoculation of infected tissues from one intermediate host to another but congenital infection can occur in *F. buteonis* (Tadros and Laarman, 1976).

V. *Besnoitia* Henry, 1913

A. Introduction

Besnoitia parasitizes mainly connective tissue cells and produces characteristic cysts with an extremely thick wall (Figs. 64 and 65) in the intermediate host (Chobotar *et al.*, 1970; Basson *et al.*, 1970; McCully *et al.*, 1966; Ernst *et al.*, 1968; Webster, 1976). Schizogony and gametogony occur in the definitive host.

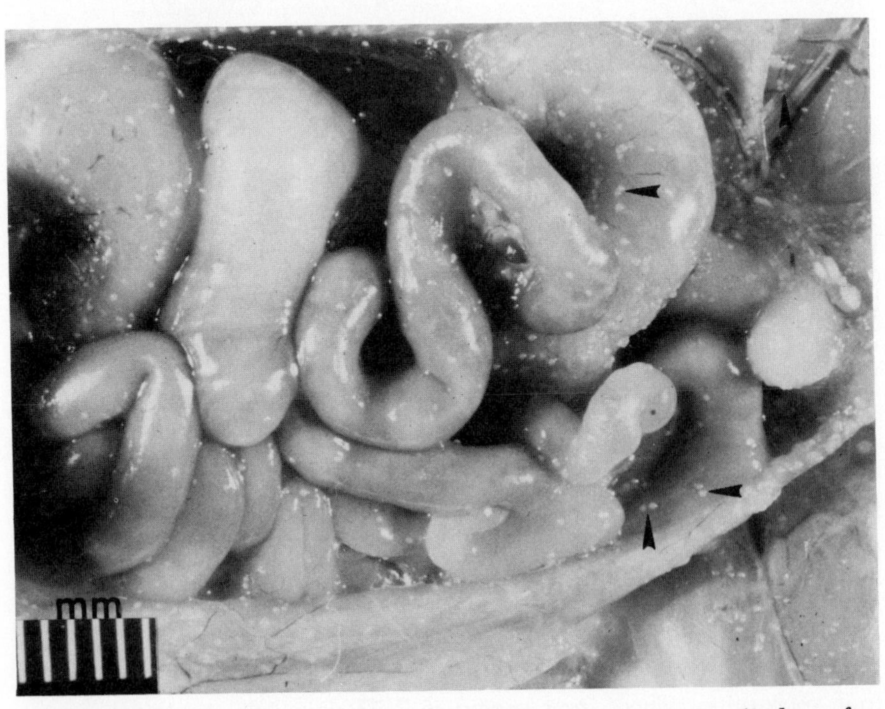

Fig. 64. *Besnoitia jellisoni* cysts in the peritoneum of a mouse 78 days after inoculation with *Besnoitia* cysts. There are numerous (arrows) white, shining cysts on the surface of visceral organs. Cysts naturally occur in these locations.

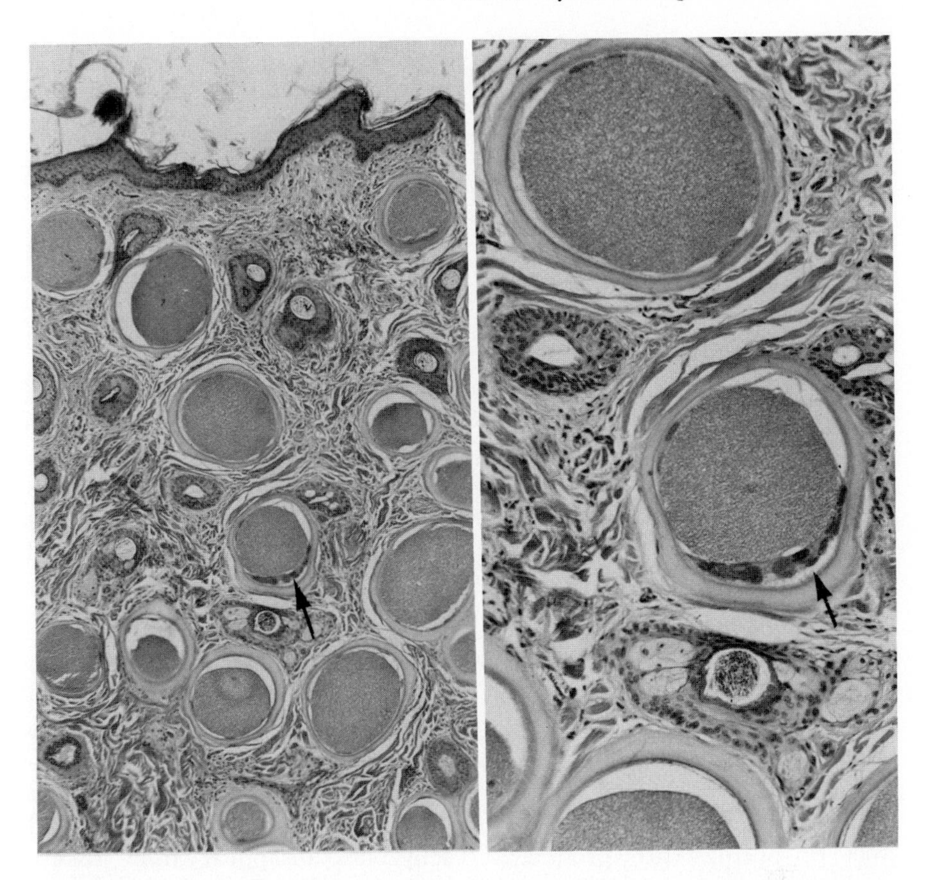

Fig. 65. *Besnoitia besnoiti* cysts in the skin of a naturally infected cow from South Africa. *Besnoitia* cysts have thick, PAS-positive walls enclosing the entire host cell. The host cell nuclei (arrows) are hypertrophic and hyperplastic. × 50 on left and × 125 on right. Hematoxylin and eosin. (Courtesy of Dr. R. Bigalke.)

B. History

Besnoitia besnoiti (Marotel, 1912) Henry, 1913, the type species, was first described from cattle. Its life history was discovered only recently. Peteshev *et al.* (1974) discovered the sexual phase of *B. besnoiti*, and Wallace and Frenkel (1975) independently discovered the life cycle of *Besnoitia* species of a murine.

C. Structure and Life Cycle

Besnoitia besnoiti and *B. jellisoni* are facultatively heteroxenous. Tachyzoites and bradyzoites multiply by endodyogeny in intermediate hosts (Fig. 66). Tachyzoites are crescent or banana-shaped and measure

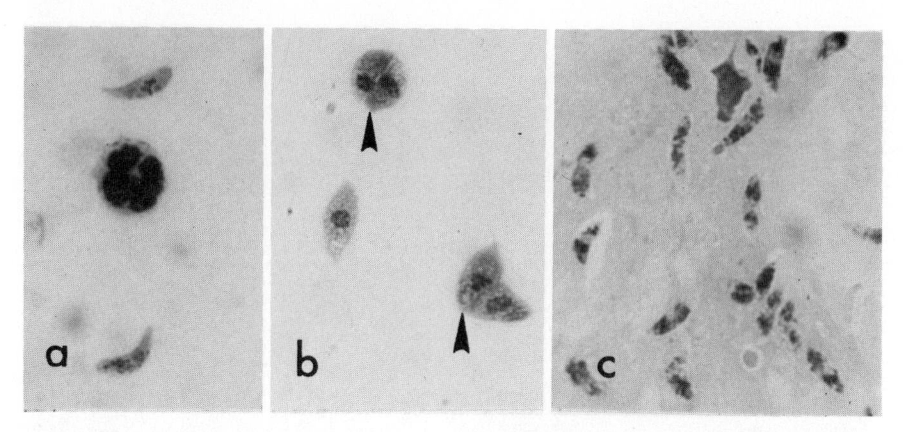

Fig. 66. Tachyzoites and bradyzoites of *Besnoitia jellisoni,* all × 1250. (a) Two tachyzoites from the peritoneal exudate of a mouse. They structurally resemble *Toxoplasma* tachyzoites. Giemsa. (b) Bradyzoites released from a cyst. Note division of mother cell into two daughter cells (arrows). Giemsa. (c) Bradyzoites contain amylopectin granules that stain bright red with PAS. Periodic acid Schiff-Hematoxylin. Here they appear black.

5–9 ×2–5 μm (Fig. 78). They are structurally similar to *Toxoplasma* (Frenkel, 1953b; Sheffield, 1966). However, unlike *T. gondii, B. jellisoni* tachyzoites are motile at 22°–26°C whereas *T. gondii* tachyzoites are motile between 30° to 37°C. After entry into the host cell tachyzoites multiply by endodyogeny and occasionally by endopolygeny (Sheffield, 1966; Senaud *et al.,* 1974). The host cell ruptures and foci of necrosis are produced. Tachyzoites disappear from tissues and cysts are formed in connective tissue cells. Cysts have a thick (1–30 μm), PAS-positive wall enclosing the entire host cell. The host cell nuclei are hypertrophic and hyperplastic. Ultrastructurally, the cyst consists of three areas: a cyst wall, a host cell, and a vacuole containing parasites (Sheffield, 1968). The outer wall is an extracellular layer consisting of fine fibrils and small dense granules embedded in an electron-lucid matrix. The plasma membrane of the host cell has pseudopodia with vesicles. The parasitophorous vacuole is limited by a membrane which has many blebs (Sheffield, 1968). Bradyzoites are crescent-shaped to piriform (Figs. 66–68) and structurally resemble *T. gondii* except for the presence of characteristic membrane-bound, spindle shaped bodies with an electron dense core called enigmatic bodies (Scholtyseck, *et al.,* 1973).

In the species studied by Peteshev *et al.* (1974) and Wallace and Frenkel (1975) cats became infected by ingesting cysts from intermediate hosts (Fig. 69). The bradyzoites form schizonts in the lamina

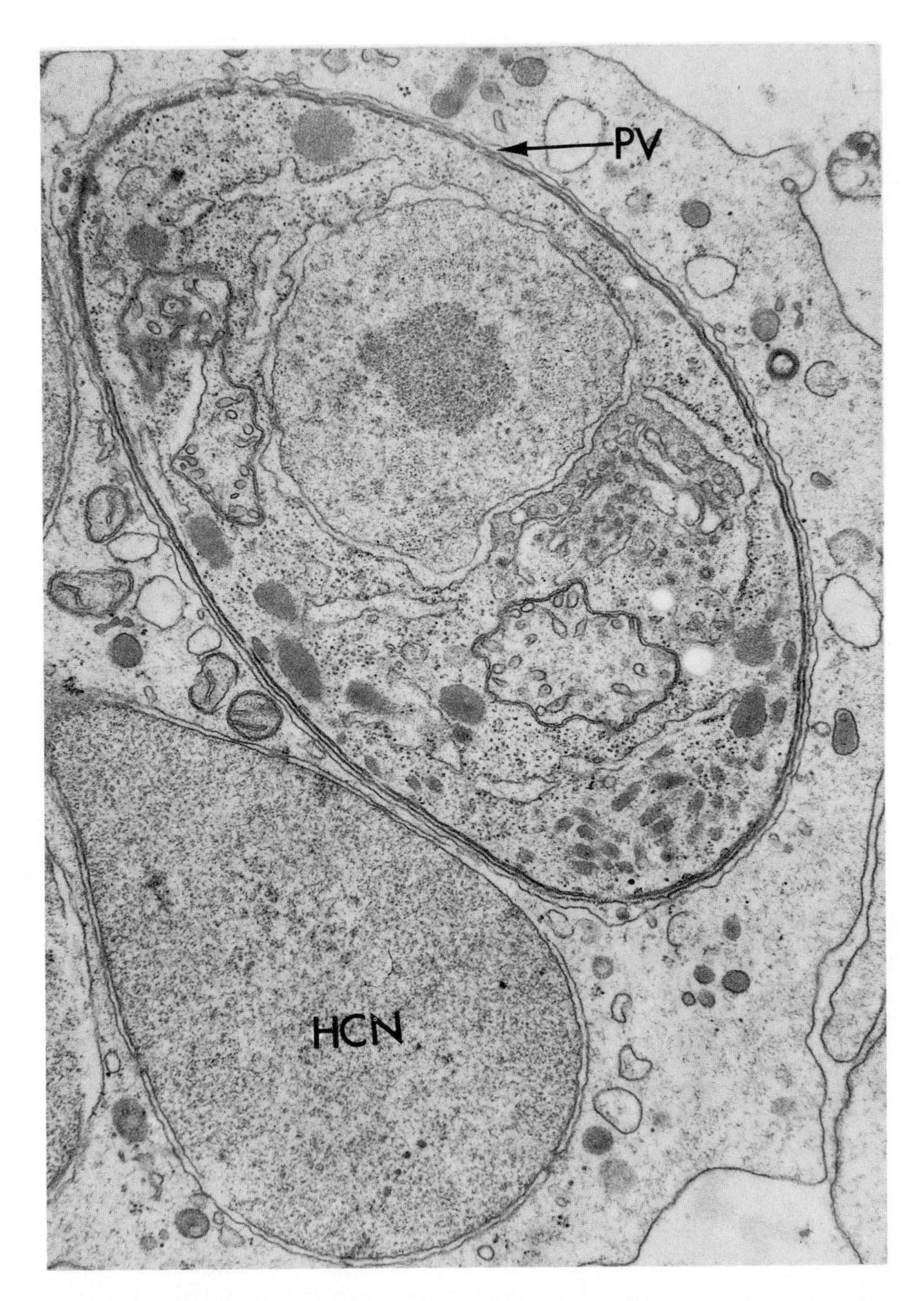

Fig. 67. Electron micrograph of a tachyzoite of *B. jellisoni* in a leukocyte of a mouse. This tachyzoite is ovoid and lies within a parasitophorous vacuole (PV). It contains most of the organelles described in *Toxoplasma* tachyzoite. HCN, host cell nucleus. × 30,000. (Courtesy of Dr. H. G. Sheffield.)

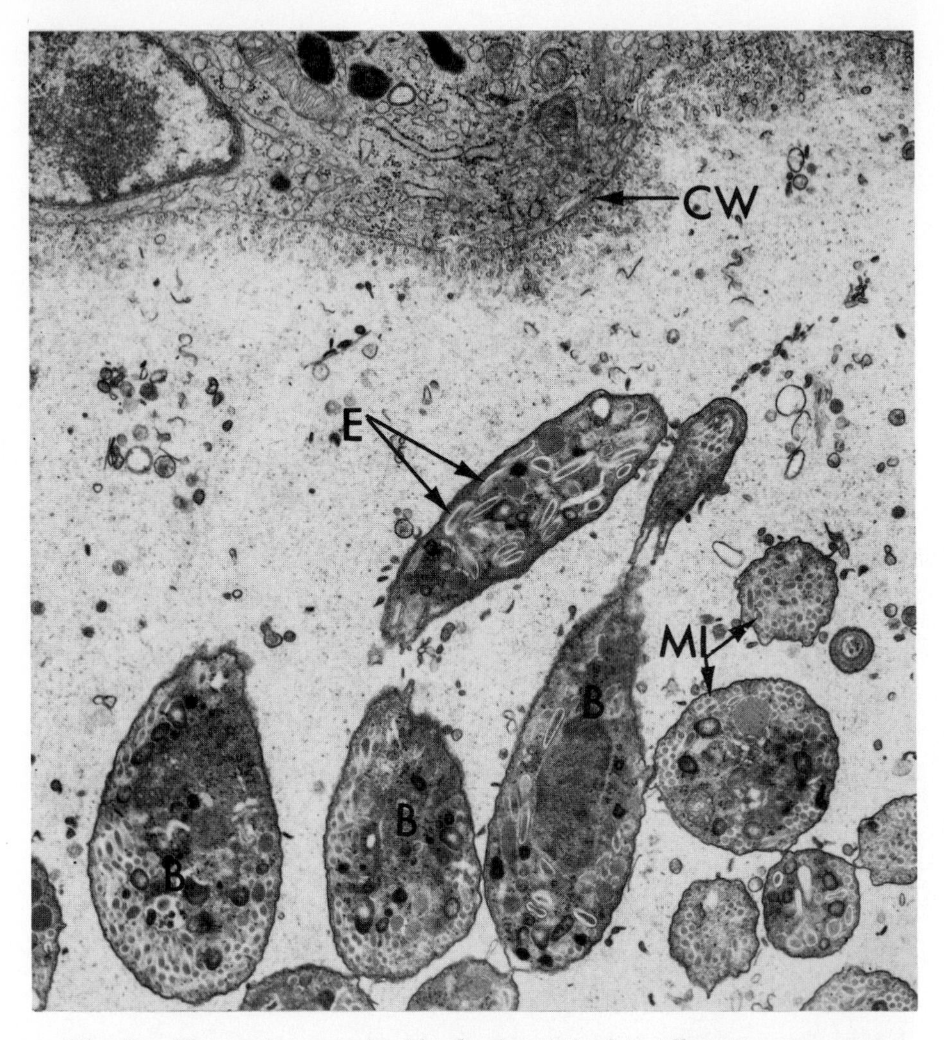

Fig. 68. Electron micrograph of bradyzoites (B) of *B. jellisoni* in a cyst. There are numerous micronemes (MI) and enigmatic (E) bodies characteristic of *Besnoitia*. CW, cyst wall. × 14,400. (Courtesy of Dr. H. G. Sheffield.)

propria and intestinal epithelial cells of the small intestine (Wallace and Frenkel, 1975). The number of generations of schizogony has not been determined. Macrogametocytes measuring 10 × 13 μm were seen in goblet cells (Fig. 70); microgametocytes were not reported (Wallace and Frenkel, 1975). Oocysts, measuring 12 × 17 μm are shed unsporulated in cat feces. Sporulation is completed within 2–4 days (Fig. 70). Sporulated oocysts are 13 × 16 μm, sporocysts are 8 × 11 μm, and sporo-

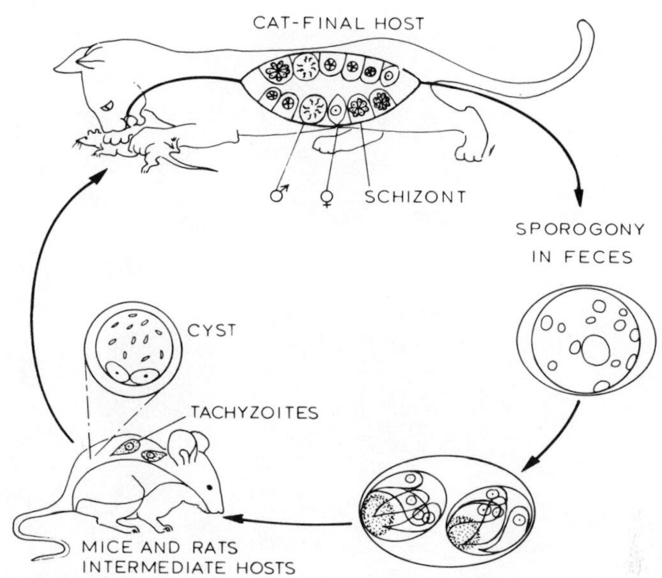

Fig. 69. Postulated life cycle of *Besnoitia wallacei* (Wallace and Frenkel, 1975). Cats, the definitive host, shed unsporulated oocysts in feces. After sporulation the oocysts are infectious to laboratory rats and mice. Tachyzoites develop in several organs of intermediate hosts. With the development of immunity, cysts are formed in connective tissue. The definitive host becomes infected by ingesting the cysts from the intermediate hosts. Schizonts, gametocytes, and oocysts are formed in the intestine. Extraintestinal organs are apparently not invaded in cats. The life cycle of *Besnoitia wallacei* is essentially similar to *Hammondia hammondi.*

zoites are 2×10 μm (Wallace and Frenkel, 1975) in size. No extraintestinal development of *Besnoitia* in cats has been reported. Oocysts are infectious to intermediate hosts but not to cats. Some species of *Besnoitia* can be maintained in mice indefinitely and methods of cultivation are essentially similar to those for *T. gondii*. *Besnoitia jellisoni* grows well in mice, hamsters, chick embryos, and in cell cultures (Frenkel, 1965; Fayer *et al.,* 1969).

D. Pathogenesis and Pathology

Besnoitia grow intracellularly destroying parasitized cells. Depending upon the virulence of the strain, generalized infection can occur and the intermediate host may die within a week. More often the host recovers from acute infection and cysts are formed primarily in connective tissue. Cysts may rupture and produce a granulomatous reaction depending upon the immune status of the host (Frenkel, 1956).

Besnoitia besnoiti can cause severe economic losses in cattle (Pols,

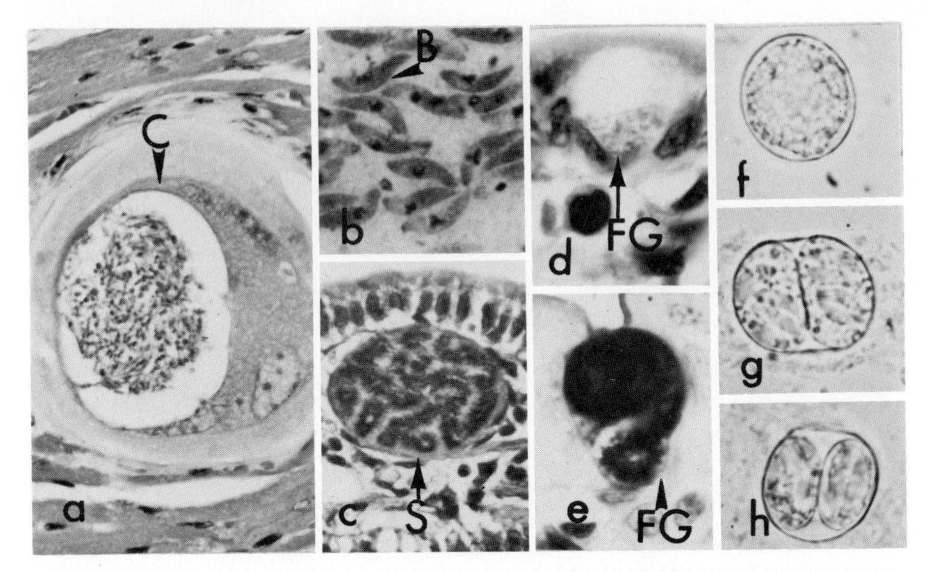

Fig. 70. Stages of *Besnoitia wallacei.* (From Wallace and Frenkel, 1975.) (a) A section of a cyst (C) in the heart of a mouse. Note hypertropic host cell nuclei. × 280. Hematoxylin and eosin. (b) Bradyzoites (B) from a cyst. Impression smear. × 700. Giemsa. (c) A section of a schizont (S) from the intestinal lamina propria of a cat infected for 11 days. × 280. Hematoxylin and eosin. (d and e) Macrogametocytes in goblet cells from the small intestine of a cat, 16 days after infection, FG, ♀ gametocyte. Each parasite (arrows) is at the base of the cell. The mucin remained unstained with hematoxylin and eosin (d) but stained deeply with PAS (e). This is the only feline coccidium known to develop in the goblet cell. (f) Unsporulated oocyst. × 700. Unstained. (g and h) Sporulated oocysts. × 700. Unstained.

1960; Bigalke, 1968). The disease in cattle can be clinically divided into acute febrile and chronic seborrheic stages. During the febrile stage, cattle may develop a fever of up to 107°F, photophobia, anasarca, diarrhea, and swelling of the lymph nodes. This may be followed by thickening and wrinkling of the skin and hair may fall out. During the seborrheic stages, denuded parts are covered by a thick scurfy layer. Death may occur in severe cases.

Besnoitia bennetti causes a similar but milder disease than *B. besnoiti* in which the skin of affected horses is thickened and mangelike.

The economic significance of other *Besnoitia* species (Table VII) is not well known.

E. Diagnosis

Definitive diagnosis can be made by finding pathognomic cysts in biopsied material or in scleral conjuctiva (Bigalke and Naudé, 1962).

Table VII

Characteristics of *Besnoitia* Species

	B. besnoiti	*B. bennetti*	*B. tarandi*	*B. jellisoni*	*B. darlingi*	*B. wallacei*
Authors	(Marotel, 1912) Henry, 1913	Babudieri, 1932	(Hadweir, 1922) Levine, 1961	Frenkel, 1953	(Brumpt, 1913) Schneider, 1967	(Tadros and Laarman, 1976) nov. comb
Hosts						
intermediate						
natural	Cattle, wildebeest, impala, kudu	Horse, ass, burrows	Reindeer, caribou	Deer mice, kangaroo, rats, opossum	Lizards, opossums	?
experimental	Mice, rabbits, hamsters, gerbils, sheep, goat	Unknown	Unknown	Mice, rats, hamsters, guinea pigs, rabbits	Mice, marmoset, free-tailed bats	Mice, rats
definitive	Cats	Unknown	Unknown	Unknown	Unknown	Cats
Geographic distribution	South Africa, Mediterranean countries, China, and the U.S.S.R.	Africa, Europe, Mexico, and United States	Alaska	United States, Peru	Panama	Hawaii
Location	Cutis, subcutis, connective tissue, fascia, serosas, mucosae	Cutis, subcutis	Fibrous connective tissue	Connective tissue and serosas	Connective tissue	Connective tissue

F. Treatment

Sulfadiazine and pyrimethamine are effective against murine besnoitiosis.

VI. *Isospora* Schneider, 1881

A. Introduction

Until 1970, *Isospora* contained all coccidians with two sporocysts and each sporocyst with four sporozoites. The life cycle of *Isospora* species, although not investigated, was considered direct, monoxenous, and similar to that of *Eimeria*. During the past 6 years, life cycles of several feline and canine *Isospora* species have been shown to be unlike that of *Eimeria*. At present, *Isospora* contains two structurally and biologically distinct types of organisms. One group (example: *Isospora canaria*) contains organisms with a direct fecal–oral life cycle without intermediate hosts. Infection is confined mainly to intestinal or other epithelial cells. The second group (examples: *Isospora felis*) contains organisms with a facultative or an obligatory two-host life cycle. Structurally, this group "encysts" in the tissues of intermediate hosts. The cystlike stages in *I. felis*, *I. rivolta*, *I. ohioensis*, and *I. canis* are unizoite and are formed in several organs of mice. Although intermediate hosts (optional or obligatory) are known for *Isospora arctopitheci* (Hendricks and Walton, 1974), and *Isospora vulpina* (Bledsoe, 1976), the structure of asexual stages in the intermediate host is not known. Currently, the isosporan genera *Toxoplasma*, *Hammondia*, *Besnoitia*, *Sarcocystis*, and *Frenkelia* are differentiated mainly on the life cycle pattern and the structure of cysts. Therefore, a new genus, *Levineia* is proposed to designate *Isospora* species that have unizoite stages in the intermediate host. Other *Isospora* species with a direct life cycle and without cysts are retained in the genus *Isospora* Schneider, 1881. *Isospora belli* of man and *Isospora canaria* of birds (Box, 1975) are retained in the genus *Isospora* Schneider, 1881. Isosporan infections of only man, dogs and cats are discussed in this chapter.

B. *Isospora* Infections of Man

1. *Isospora belli* Wenyon, 1923

A. INTRODUCTION. *Isospora belli* is uncommon in man. Most of the reported cases occurred in the tropics rather than in the temperate zone.

B. STRUCTURE AND LIFE CYCLE. *Isospora belli* oocysts are elongate, ellipsoidal, and measure 20–33 × 10–19 μm. Sporulated oocysts contain two ellipsoidal sporocysts without a Stieda body. Each sporocyst measures 9–14 × 7–12 μm and contains four crescent-shaped sporozoites and a residual body. No oocyst residuum is present. Sporulation occurs within five days, both within the host and in the external environment (Brandborg *et al.*, 1970; Trier *et al.*, 1974). Thus, both unsporulated and sporulated oocysts may be shed in feces.

Infection occurs by the ingestion of food contaminated by oocysts. Schizogony and gametogony occur in the upper small intestinal epithelial cells, from the level of the crypts to the tips of the villi (Figs. 71–73). Occasionally schizonts occur in the lamina propria. The number of generations of schizogony is unknown. Most information about the life cycle of *I. belli* is based on biopsy of infected patients (Brandborg *et al.*, 1970; Trier *et al.*, 1974). Intestinal stages can be found in the absence of oocysts in feces. It may be that oocysts sporulate and excyst in the duodenum thereby causing continued schizogony and gametogony. If this type of life cycle occurs it might explain the chronic nature of disease.

C. PATHOGENICITY. *Isospora belli* can cause severe clinical symptoms with an acute onset. Infection has been reported to cause fever, malaise,

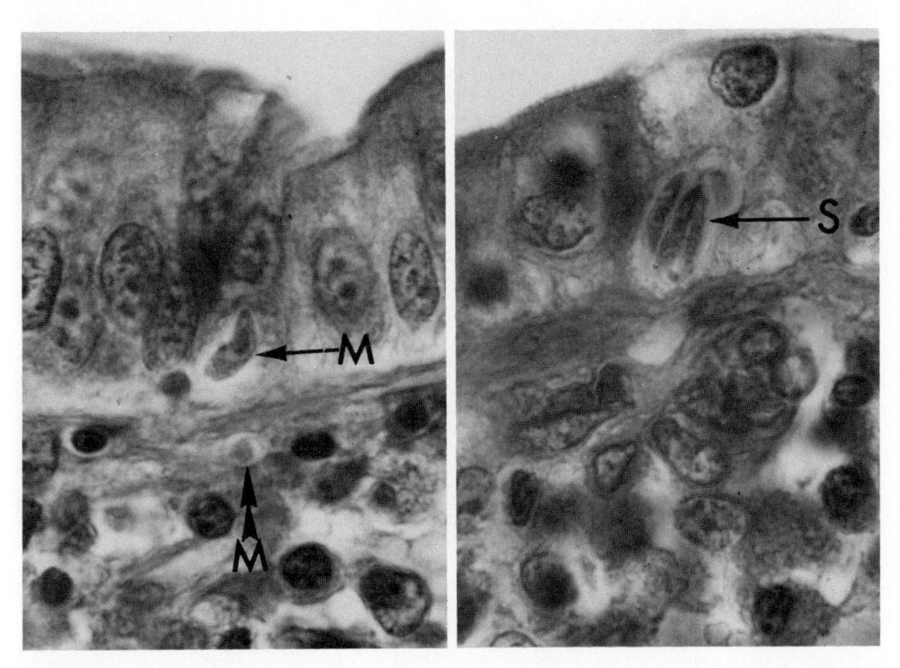

Fig. 71. A section containing schizonts of *Isospora belli* from a biopsy of the duodenojejunal junction of a human patient. A merozoite (M) (left) and a schizont (S) (right) in an epithelial cell are seen. One merozoite (double arrow) appears to be in the lamina propria. × 1250. Hematoxylin and eosin. (Courtesy of Dr. J. S. Trier.)

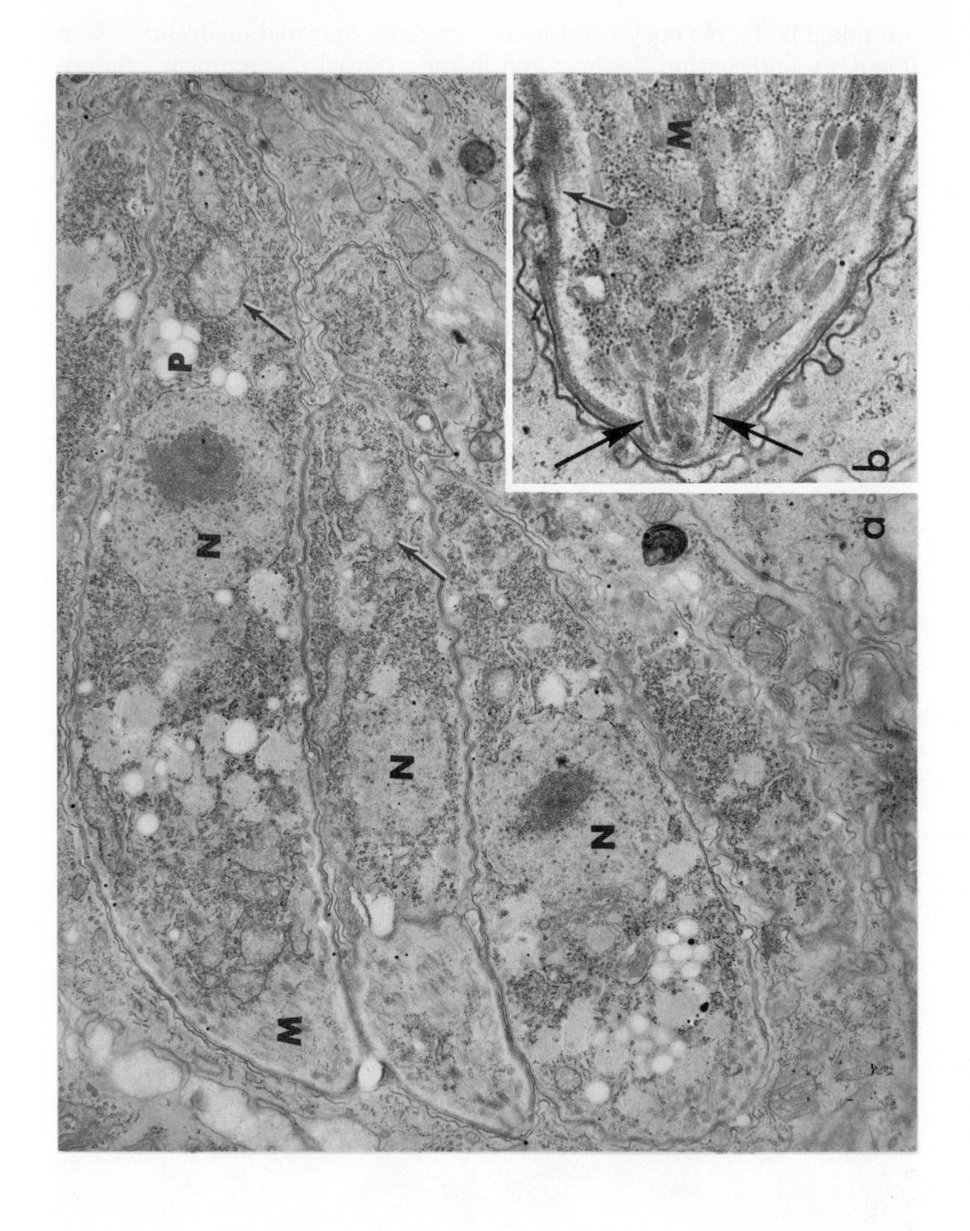
M
b
a
P
N
N
N
M

persistent diarrhea, weight loss, steatorrhea, and even death (Jarpa, 1966; Brandborg *et al.*, 1970).

D. DIAGNOSIS. Diagnosis can be established by finding oocysts in the feces (Fig. 74) or coccidian stages in a duodenal biopsy. Affected intestinal portions have a flat mucosa similar to that which occurs in sprue. The stools are fatty and at times very watery during infection.

E. TREATMENT. Trier *et al.* (1974) successfully treated a patient with pyrimethamine and sulfadiazine in doses similar to those used for treating toxoplasmosis (75 mg pyrimethamine and 4 gm of sulfadiazine in divided doses).

2. *Isospora natalensis* Elson-Dew, 1953

Elson-Dew (1953) found *I. natalensis* in two persons in South Africa. The oocysts are subspherical, 25–30 × 21–24 μm and contain two ellipsoidal sporocysts (17 × 12 μm) without a Stieda body. The sporulation time is 1 day. Its life cycle is unknown.

VII. *Levineia* nov. gen.

Definition: Oocysts with two sporocysts each with four sporozoites; sporogony outside the host; facultatively or obligatory heteroxenous, asexual cycle not restricted to intestine; cysts in many organs of intermediate hosts, typically unizoite; asexual stages preceding sexual development in carnivores; tachyzoites and bradyzoites nontransmissible to intermediate hosts, oocysts infectious to definitive and intermediate hosts.

A. *Levineia* Infections of Cats

Characteristics of *Levineia* infections are compared with those of other feline coccidia in Table VIII.

Fig. 72. Electron micrograph of a section from the duodenojejunal junction obtained from biopsy material of a human patient. (From Trier *et al.*, 1974.) (a) Several merozoites can be seen within a parasitophorous vacuole of an epithelial cell from a pretreatment biopsy. The nuclei (N) of three of the merozoites are evident. The anterior third of each merozoite contains many micronemes (M) and cytoplasmic granules presumed to contain polysaccharide (P). The posterior third of the merozoites is filled with many free ribosomes, a few strands of granular endoplasmic reticulum, and a few large mitochondria with tubular cristae (arrows). × 9310. (b) Higher magnification micrograph of the anterior end of a merozoite showing more detail of its complex structure. Note the presence of the conoid (large arrows), the micronemes (M), and subpellicular microtubules (small arrow). Both the inner and the outer membrane of the merozoite are apparent; the inner membrane is discontinuous at the extreme anterior pole of the organism. × 28,500.

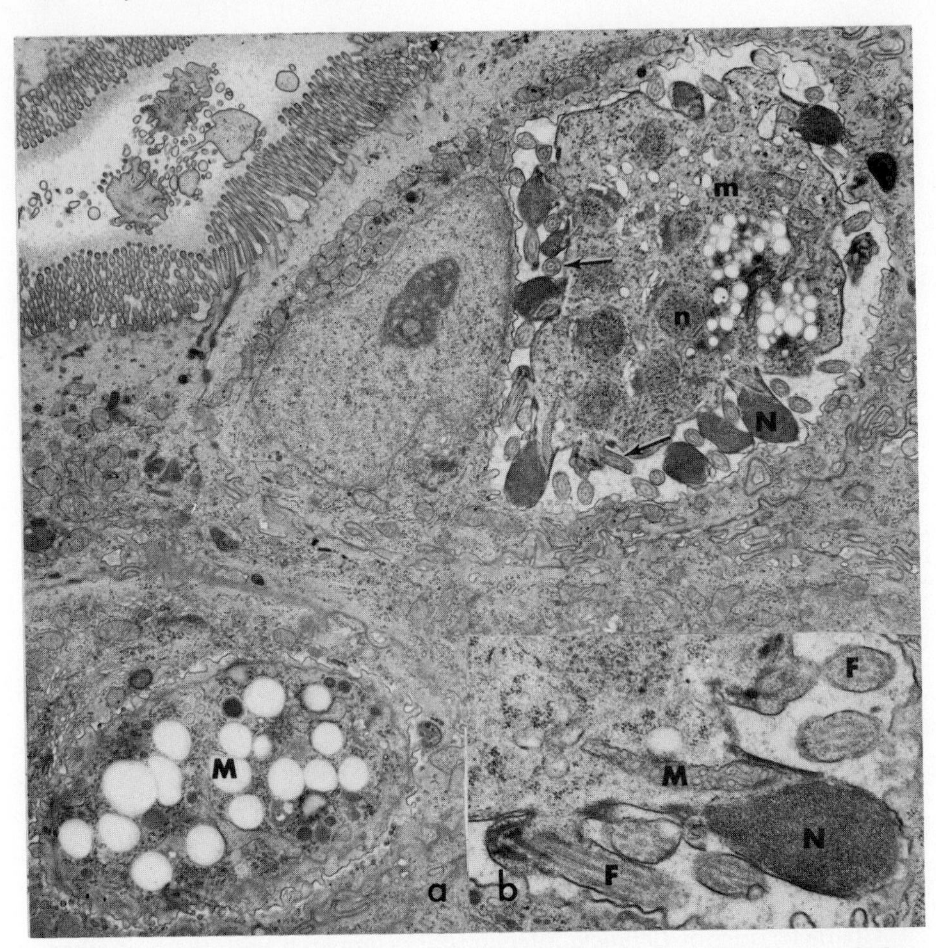

Fig. 73. Electron micrograph of the duodenojejunal junctional biopsy of the patient in Fig. 72. (From Trier *et al.,* 1974.) (a) A microgamont (m) and macrogametocyte (M), although in close proximity, are within separate epithelial cells and vacuoles in this pretreatment biopsy. Both light (wall-forming bodies) and dark (dark bodies) granules are seen in the tangentially sectioned macrogametocyte. The microgamont has many centrally located granular nuclei (n) and more peripherally located condensed nuclei (N). Many cross-sectioned and tangentially sectioned flagella (arrows) can be seen around the periphery of the microgamont. × 6600. (b) Higher magnification of the periphery of the microgamont shown in (a). The condensed nucleus (N) is closely associated with the long mitochondrion with tubular cristae (M) and with several flagella (F). × 19,000.

1. *Levineia felis* (Wenyon, 1923) nov. comb.

A. HISTORY. *Levineia felis* oocysts were described by Wenyon (1923). Hitchcock (1955), Lickfeld (1959), and Shah (1971) studied its life cycle. These authors described one-host life cycle of *L. felis.* Frenkel and

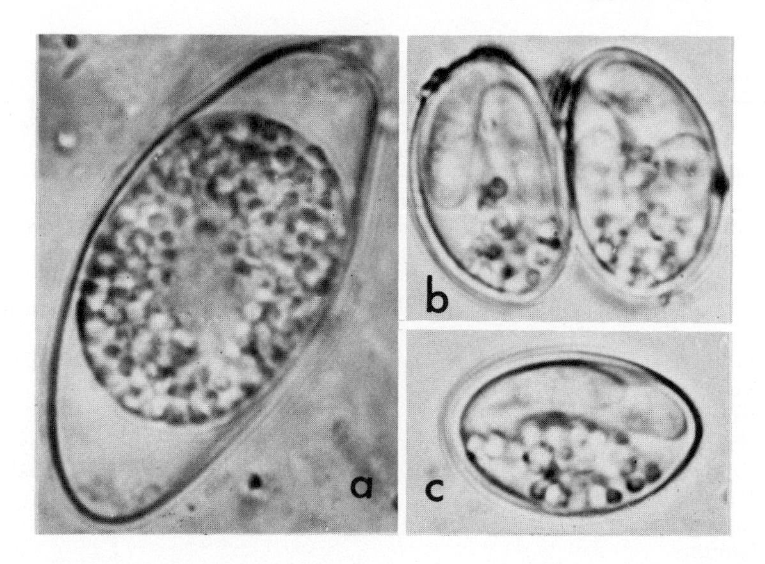

Fig. 74. *Isospora belli* and *Sarcocystis* sp. in the feces of human patients. All × 2200. Unstained. (Courtesy of Institute of Tropical Hygiene, Amsterdam.) (a) *Isospora belli* oocysts are shed unsporulated. (b and c) *Sarcocystis* is shed as a sporulated oocyst (b) or as a sporocyst (c).

Dubey (1972a) found that mice, rats and hamsters can act as intermediate hosts. *L. felis* invaded extraintestinal organs not only of intermediate hosts but also definitive hosts, cats (Frenkel and Dubey, 1972a; Dubey and Frenkel, 1972b).

B. STRUCTURE AND LIFE CYCLE. *Levineia felis* oocysts are the largest of all feline coccidia (Table IX; Fig. 75). Cats are the definitive hosts. Mice, rats, hamsters (Frenkel and Dubey, 1972a), dogs (Dubey, 1975), and chickens (J. P. Dubey, unpublished) may act as intermediate hosts (Fig. 76). Unsporulated oocysts of *L. felis* are shed by cats, after which they sporulate within 12 hours under optimal conditions. Cats and nonfeline hosts become infected by ingesting sporulated oocysts. Although generalized infections occur in the mouse, there is only a limited multiplication. *Levineia felis* "encysts" in the mouse, chiefly in the mesenteric lymph nodes. These cystlike stages have a thick PAS-positive cyst wall and contain a single zoite. *L. felis* can remain viable in mice for at least 23 months after inoculation.

After the ingestion of oocysts by the cat, *L. felis* multiplies extensively in the epithelium of small intestine, producing three generations of schizonts and gametocytes (Shah, 1971). The prepatent period is 7–8 days after the ingestion of oocysts (Shah, 1971). Although extraintestinal organs of the cat are parasitized, the rate of multiplication is slow and

Table VIII
Characteristics of Different Genera of Feline Coccidia

	Toxoplasma	*Hammondia*	*Levineia*	*Besnoitia*	*Sarcocystis*
Intermediate host	Wide [a] range	Probably wide range?	Probably wide range?	Wide range	Sheep, cattle, pig
Cyst					
Location	Many organs	Muscle	Many organs	Many organs	Muscle
Type	Non-septate	Non-septate	Non-septate	Non-septate	Septate
Number of organisms in cyst	Many	Many	One	Many	Many
Type of wall	Thin	Thin	Thick	Thick	Variable
Development in definitive host					
Intestine					
Schizogony	Yes	Yes	Yes	Yes	No
Gametogony	Yes	Yes	Yes	Yes	Yes
Oocyst sporulation	Outside of host	Outside of host	Outside of host	Outside of host	In host
Extraintestinal cycle	Yes	No	Yes	No	No
Mode of transmission					
Definitive hosts					
Carnivorism	+	+	+	+	+
Fecal	+	−	+	−	−
Congenital	−	−	−	?	−
Intermediate hosts					
Carnivorism	+	−	−	+ or −	−
Fecal	+	+	+	+	+
Congenital	+	−	−	?	−

[a] Several orders or families.

stages are difficult to demonstrate histologically in extraintestinal organs. The prepatent period after ingesting infected mice is 5–8 days (Frenkel and Dubey, 1972a).

C. PATHOGENICITY. Although *L. felis* infections are asymptomatic in cats and intermediate hosts (Dubey and Streitel, 1976c), superinfection with this coccidium can cause reshedding of *Toxoplasma gondii* oocysts from cats chronically infected with *T. gondii* (Chessum, 1972; Dubey, 1976a).

Table IX
Species of Coccidia in Cats [a]

Coccidium	Oocyst or sporocyst dimensions (μm)	Natural intermediate hosts	Mode of transmission to cats
	Oocysts		
Levineia felis (Wenyon, 1923) nov. comb.	40 × 30 (38–51 × 27–39)	Not known	Fecal
Levineia rivolta (Grassi, 1879) nov. comb.	25 × 20 (21–28 × 18–23)	Not known	Fecal
Besnoitia besnoiti (Marotel, 1912) Henry, 1913	15 × 13 (14–16 × 12–14)	Cattle	Carnivorism
Besnoitia wallacei (Tadros and Laarman, 1976) nov. comb.	17 × 12 (16–19 × 10–13)	Mouse	Carnivorism
Toxoplasma gondii (Nicolle and Manceaux, 1908) Nicolle and Manceaux, 1909	12 × 10 (11–13 × 9–11)	Mammals, birds	Carnivorism
Hammondia hammondi Frenkel and Dubey, 1975	12 × 11 (11–13 × 10–12)	Not known	Carnivorism
	Sporocysts		
Sarcocystis hirsuta Moulé, 1886	12 × 8 (11–14 × 7–9)	Ox	Carnivorism
Sarcocystis tenella Raílliét, 1886	12 × 8 (11–14 × 8–9)	Sheep	Carnivorism
Sarcocystis porcifelis Dubey, 1976	13 × 8 (13–14 × 7–8)	Pig	Carnivorism
Sarcocystis muris (Blanchard, 1885) Labbé, 1889	10 × 8 (9–12 × 7–9)	Mouse	Carnivorism
Sarcocystis sp., Janitschke, Protz and Werner, 1976	13 × 9 (11–15 × 8–12)	Gazelle	Carnivorism
Sarcocystis leporum Crawley, 1914	14 × 9 (9–11 × 13–17)	Cottontail rabbits	Carnivorism

[a] Modified from Dubey (1967b).

2. *Levineia rivolta* (Grassi, 1879) nov. comb.

A. HISTORY. *Levineia rivolta* was named *Coccidium rivolta* by Grassi (1879). Wenyon (1923) reviewed the early history of this coccidium and assigned it to the genus *Isospora.* Frenkel and Dubey (1972a) and Dubey and Frenkel (1972b) described its partial life cycle in cats and mice.

B. STRUCTURE AND LIFE CYCLE. Cats are the definitive hosts; mice, rats, hamsters, dogs, and chickens can act as experimental intermediate hosts.

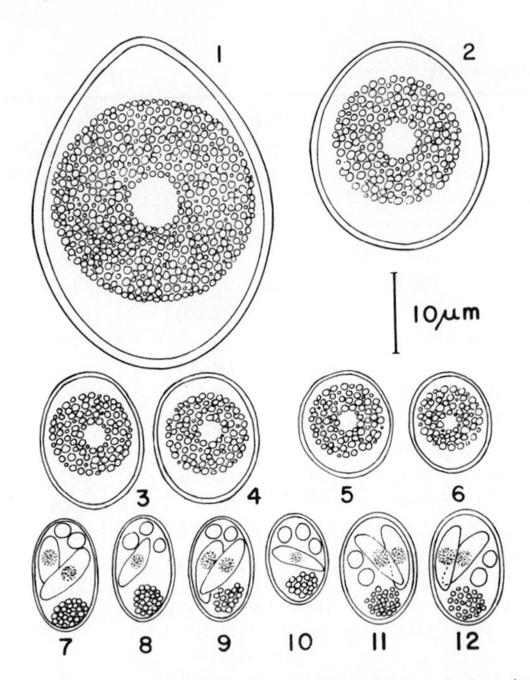

Fig. 75. Line drawings of coccidia from fresh feline feces. (1) *Levineia felis.* (2) *L. rivolta.* (3) *Besnoita besnoiti.* (4) *Besnoitia wallacei* (see Table III). (5) *Hammondia hammondi.* (6) *Toxoplasma gondii.* (7) *Sarcocystis porcifelis.* (8) *S. hirsuta.* (9) *S. tenella.* (10) *S. muris.* (11) *Sarcocystis* sp. from Grant's gazelle. (12) *Sarcocystis leporum.* (Modified from Dubey, 1976b.)

Although the stages that occur in the intestine of cats are not described in detail, its life cycle appears similar to that of *Levineia felis.*

C. PATHOGENICITY. *Levineia rivolta* infections are asymptomatic in cats and mice (Dubey and Streitel, 1976c).

B. *Levineia* Infections of Dogs

1. *Levineia canis* (Nemeséri, 1959) nov. comb.

A. HISTORY. Nemeséri (1959) named this coccidium from dogs because he could not transmit it to cats. Earlier to his report its oocysts were considered *Isospora felis* because they are structurally similar. Lepp and Todd (1974) described its direct life cycle. Dubey (1975d) found that mice and cats can act as intermediate hosts.

B. STRUCTURE AND LIFE CYCLE. *Levineia canis* oocysts are the largest of all canine coccidia (Table X, Fig. 77). Its life cycle is essentially similar to that of *L. felis* of cats. Dogs can acquire infection either by ingesting oocysts (Lepp and Todd, 1974) or infected mice (Dubey, 1975d).

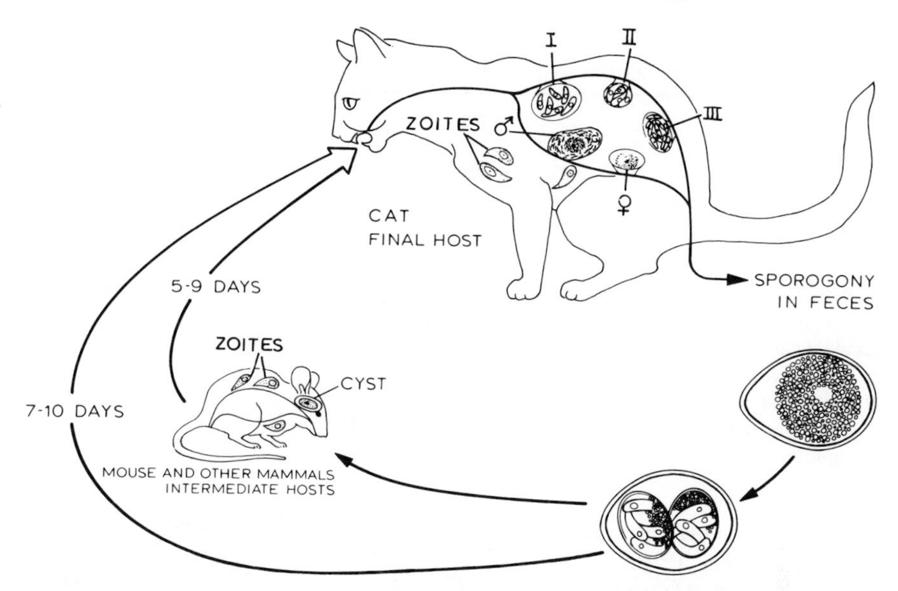

Fig. 76. Life cycle of *Levineia felis.* Cats are the definitive hosts and nonfeline hosts may act as intermediate or transport hosts. Unsporulated oocysts are shed in feline feces. *Levineia felis* can sporulate within 12 hours under optimal conditions. Three generations of schizonts (I, II, III) develop in the epithelium of the small intestine before the formation of gametes and oocysts. The life cycle of *L. rivolta* is essentially similar to that of *L. felis.* In nature *L. felis* and *L. rivolta* are probably transmitted to cats mainly through oocysts. (From Dubey, 1976b.)

Three generations of schizonts, and gametocytes occur just beneath the epithelium of the small intestine (Lepp and Todd, 1974). The prepatent period is 8–11 days.

C. PATHOGENICITY. *Levineia canis* infections are asymptomatic in mice and dogs (Dubey, 1976b).

2. *Levineia ohioensis* (Dubey, 1975) nov. comb.

A. HISTORY. This coccidium was considered the same as *Isospora rivolta* of cats because their oocysts are structurally similar. Dubey (1975d) proposed a new name, *I. ohioensis* for *I. rivolta* of the dog because it could not be transmitted to cats. Mahrt (1967) studied its life cycle in dogs after feeding oocysts. Dubey (1975d) showed that mice can act as experimental intermediate hosts.

B. STRUCTURE AND LIFE CYCLE. The life cycle of *L. ohioensis* is essentially similar to that of *L. canis.* Schizogony and gametogony occur in the epithelium and lamina propria of the small and large intestine (Mahrt, 1967). The prepatent period is 4–6 days (Mahrt, 1967; Dubey, 1975d).

Table X
Species of Coccidia in Dogs [a]

Coccidium	Oocyst or sporocyst dimensions (μm)	Natural intermediate hosts	Mode of transmission to dogs
	Oocysts		
Levineia canis (Nemeséri, 1959) nov. comb.	38 × 30 (32–42 × 27–33)	Not known	Fecal
Levineia ohioensis (Dubey, 1975) nov. comb.	24 × 20 (19–27 × 18–23)	Not known	Fecal
Hammondia heydorni (Tadros and Laarman, 1976) nov. comb.	12 × 11 (10–13 × 10–13)	Ox	Carnivorism
	Sporocysts		
Sarcocystis cruzi (Hasselmann, 1926) Wenyon, 1926	16 × 11 (14–17 × 9–13)	Ox	Carnivorism
Sarcocystis ovicanis Heydorn, Gestrich, Mehlhorn and Rommel, 1975	15 × 10 (13–16 × 8–11)	Sheep	Carnivorism
Sarcocystis miescheriana (Kühn, 1865) Lankester, 1882	13 × 10 (Not reported)	Pig	Carnivorism
Sarcocystis bertrami Doflein, 1901	15 × 10) (15–16 × 9–11)	Horse	Carnivorism
Sarcocystis fayeri Dubey, Streitel, Stromberg, Toussant, 1977	12 × 8 (11–13 × 7–9)	Horse	Carnivorism
Sarcocystis sp. Janitschke; Protz and Werner, 1976	16 × 10 (13–18 × 8–12)	Gazelle	Carnivorism
Sarcocystis hemionilatrantis Hudkins and Kistner, 1977	15 × 9 (14–16 × 9)	Mule deer	Carnivorism

[a] Modified from Dubey (1976b).

C. PATHOGENICITY. *Levineia ohioensis* infections are asymptomatic in experimentally inoculated mice or dogs (Dubey, 1976b).

VIII. "Sporozoan Orphans"

A. Sporozoan Encephalomyelitis in Horses

Toxoplasma can cause encephalomyelitis in all domestic animals. However, very little is known about clinical toxoplasmosis in horses. A fatal encephalomyelitis in association with an organism, somewhat structurally similar to *Toxoplasma gondii*, was reported from Illinois (Cusick *et al.*, 1974), Ohio (Dubey *et al.*, 1974), and Pennsylvania (Beech and Dodd,

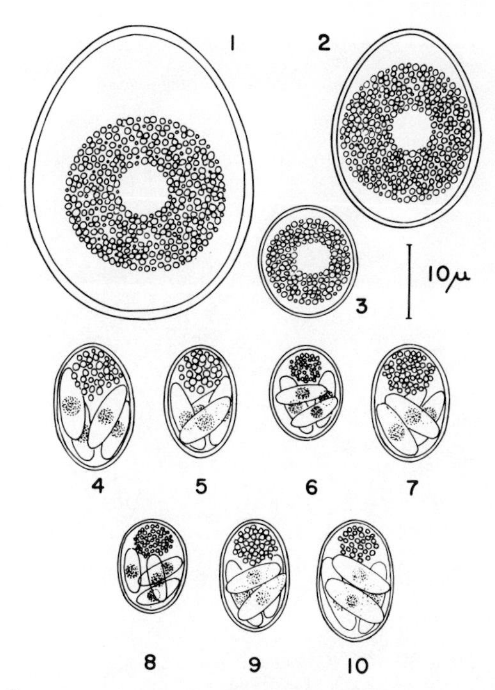

Fig. 77. Line drawings of coccidia from fresh canine feces. (1) *Levineia canis.* (2) *L. ohioensis.* (3) *Hammondia heydoeni.* (4) *Sarcocystis cruzi.* (5) *S. ovicanis.* (6) *S. miescheriana.* (7) *S. bertrami.* (8) *S. fayeri.* (9) *S. hemionilatrantis.* (10) *Sarcocystis* sp. (Grant's gazelle). (From Dubey 1976.)

1974). The affected horses had ataxia, muscle weakness, and difficulty in backing. Focal areas of malacia and hemorrhage were found more commonly in the spinal cord than in the brain. Neurons were severely parasitized by *Toxoplasma*-like organisms (Fig. 78). This unnamed organism is undoubtedly not *Toxoplasma gondii* because "tachyzoites" are arranged in rosettes, "cystlike" structures have indistinct walls that enclose PAS-negative organisms, and the organism has not been cultivated in mice by passage of equine tissues. The rosettes are similar to schizonts of *Sarcocystis cruzi* in bovines. However, unlike *S. cruzi*, the organisms in horses are within neurons and not in endothelial cells.

B. Sporozoan Encephalomyelitis in Sheep

Fatal encephalomyelitis in association with a *Toxoplasma*-like organism has been reported in sheep from Australia (Hartley and Blackmore, 1974). Affected sheep had severe malacia in the spinal cord (Fig. 79). The organisms were arranged in rosettes in astrocytes, and the cystlike structures were PAS-negative. A similar disease was reported from the

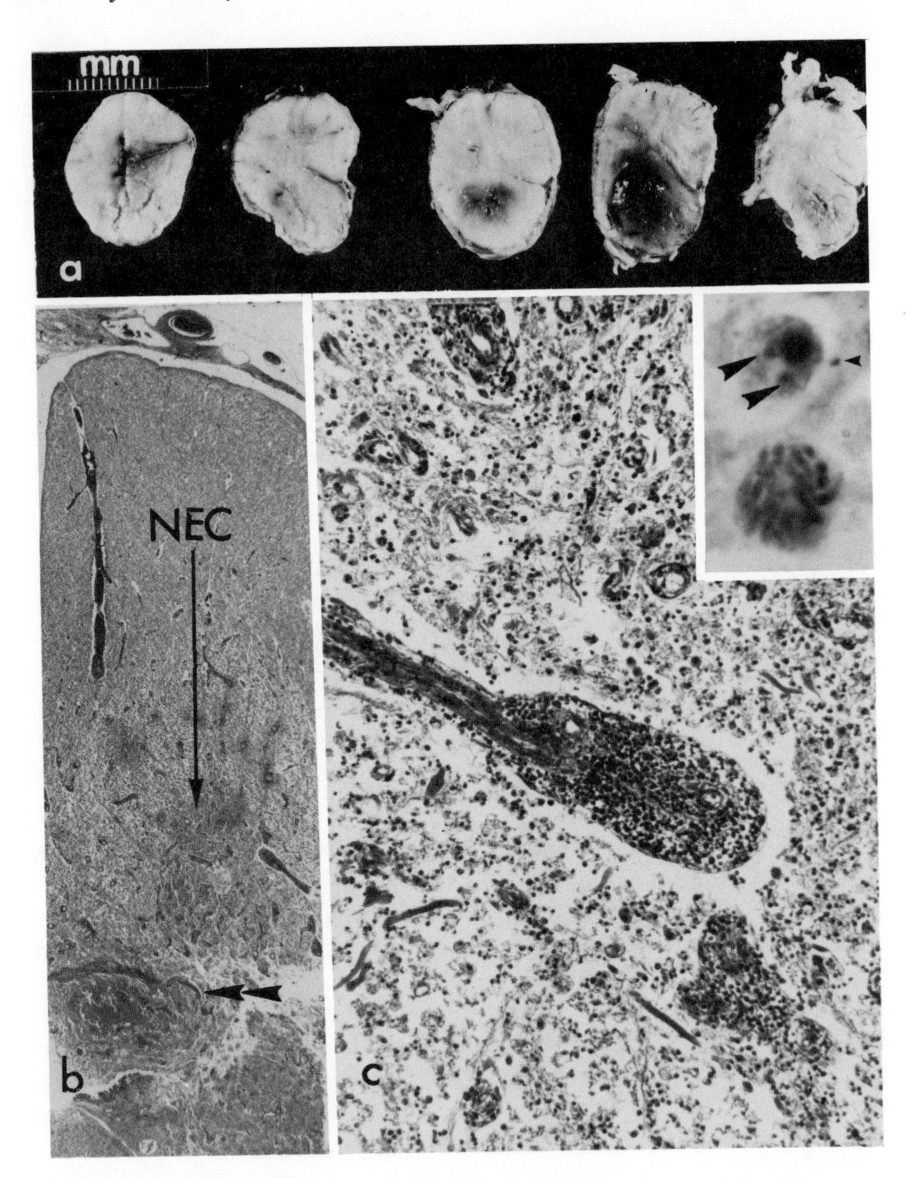

Fig. 78. Focal malacia and hemorrhage in the lumbar area of the spinal cord of a naturally infected horse. (From Dubey *et al.*, 1974.) (a) Gross lesions. The segment on the exreme left is L_1 followed by L_2, L_3, L_4, and L_5. (b) Photomicrograph illustrating an extensive area of necrosis (NEC) in dorsal horn of L_2 segment. Area of necrosis extends to ependyma (double arrow) and necrotic debris fills the central canal. Notice also congestion and cuffings of spinal and meningeal blood vessels. $\times$ 10. Hematoxylin and eosin. (c) A higher magnification of (b) depicting the marked

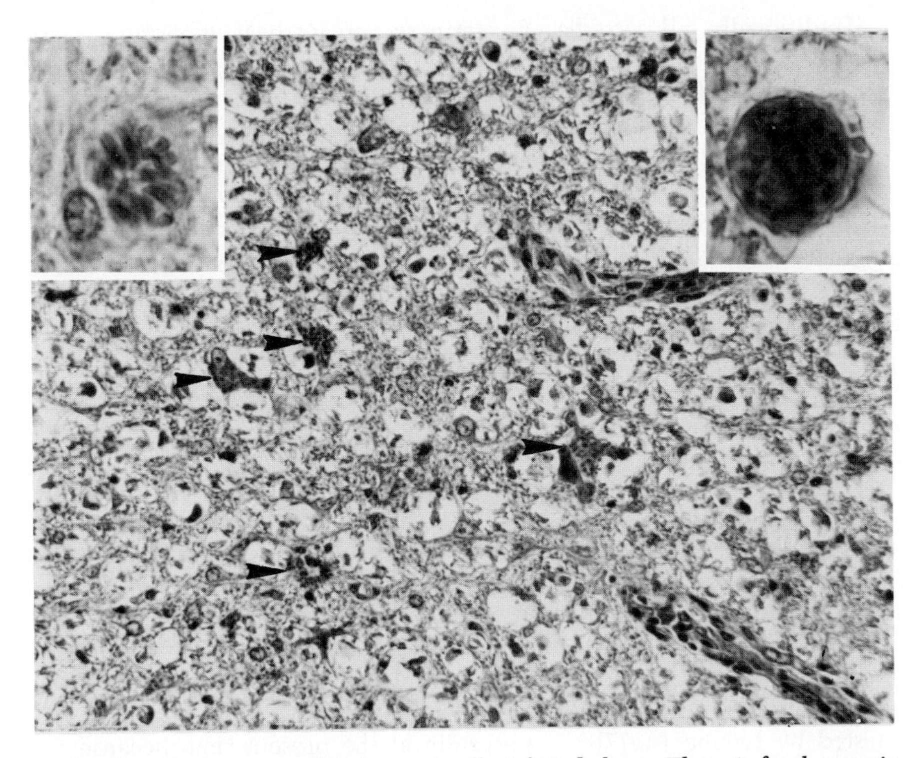

Fig. 79. Sporozoan myelitis in a naturally infected sheep. There is focal necrosis with several groups of organisms (arrows). × 315. Hematoxylin and eosin. Inset (left) shows a schizontlike structure with merozoites. × 1250. Inset (right) shows a cystlike structure. This protozoan is structurally similar to *Toxoplasma gondii* and may have been confused with it in the past. Unlike *T. gondii,* the cystlike structures are PAS negative. (Courtesy of Dr. W. J. Hartley.)

United States (Olafson and Monlux, 1942) and Ireland (McErlean, 1974). It appears that these reports were due to *Sarcocystis*-like organisms, not *Toxoplasma gondii.*

IX. Taxonomy of *Toxoplasma* and Related Genera

All seven genera discussed in this chapter are believed to be phylogenetically related and they are discussed together.

---

perivascular cuffing by lymphocytes and microglia within the area of necrosis. Numerous macrophages are discernible. × 125. Hematoxylin and eosin. Inset shows a cystlike extracellular structure filled with many crescent-shaped organisms, two intracellular organisms (large arrows), and one extracellular organism (small arrow). × 1300. Hematoxylin and eosin.

Traditionally, all coccidia of veterinary and medical importance were classified under the family Eimeriidae (Minchin, 1903), on the basis of the structure of their oocysts. Oocysts with four sporocysts each with two sporozoites are classified as *Eimeria*. *Eimeria* have one type of life cycle. Oocysts with two sporocysts, each with four sporozoites were classified as *Isospora* (Levine, 1961). *Toxoplasma*, *Hammondia*, *Isospora*, *Levineia*, *Besnoitia*, *Frenkelia*, and *Sarcocystis* all have an isosporan type of oocyst (oocysts with two sporocysts, each with four sporozoites). These genera belong to the Subphylum Apicomplexa (Levine, 1970), class Sporozoasida (Leukart, 1879), order Eucoccidiorida (Léger and Duboscq, 1970), suborder Eimeriorina (Léger, 1911). Opinions differ regarding further classification of these coccidia into families and subfamilies. In part, this is due to the fact that the life cycles of some of these coccidia are not fully known.

Frenkel (1974) placed the isosporan genera in three families, depending upon their life cycle patterns: (1) Toxoplasmatidae, Biocca, 1956, containing *Toxoplasma* and *Besnoitia*, both being facultatively heteroxenous; (2) Sarcocystidae, Poche, 1913, containing *Sarcocystis*, *Hammondia*, and *Frenkelia*, all three being obligatorily heteroxenous; and (3) Eimeridae, Minchin, 1903, containing *Isospora*, being principally monoxenous. Whether these genera should be placed in three separate families as suggested by Frenkel (1974) or in one family, Eimeriidae, as suggested by Levine (1977b) is uncertain at the present time because the life cycles of many species in these genera are unknown.

Tadros and Laarman (1976) proposed a new system of nomenclature because they considered that neither the morphology of tissue stages nor the type of life cycle pattern were suitable criteria for classifying atypical eimeriid coccidia. They classified isosporan coccidia into two main genera, *Isospora* and a new genus, *Endorimospora*, based on the sexual development of the parasite. *Toxoplasma*, *Besnoitia*, *Hammondia* were considered synonyms of *Isospora* because sporogony occurred outside the host. *Sarcocystis* and *Frenkelia* were synonymized with their new genus *Endorimospora* because sporogony occurred within the host. In my opinion, these seven genera should be retained to avoid confusion in elucidating the life cycles of poorly known species. The rules of zoological nomenclature permit the retention of well established names like *Toxoplasma gondii* to avoid confusion. At the present time, the structure of cysts and life cycle patterns appear to be convenient in separating the isosporan genera. This view might have to be modified as life cycles of other *Isospora* species are investigated. These genera are redefined here following the descriptions given by Frenkel (1974) and Levine (1977b).

A. Genus *Toxoplasma* Nicolle and Manceaux, 1909

Diagnosis: Oocysts with two sporocysts each with four sporozoites; sporogony outside the host; tachyzoites in almost any organ; cysts containing bradyzoites in many organs, principally in the brain; asexual reproduction not self-limited; facultatively heteroxenous; definitive host cats; asexual cycle preceding sexual cycle in feline intestine; transmissible from one host to another; tachyzoites, bradyzoites, and oocysts all infectious to both intermediate and definitive hosts.

Type species: *Toxoplasma gondii* (Nicolle and Manceaux, 1908) Nicolle and Manceaux, 1909. (Synonym: *Isospora gondii* Tadros and Laarman, 1976).

Diagnosis: Same as genus. For additional details see Section I of this chapter.

So far, only one valid species, *T. gondii* has been recognized in this genus. Whether this genus contains one or more species is debatable. Prior to the recognition of fatal congenital toxoplasmosis in man in 1939, several new species of *Toxoplasma* (for details of these species see Levine, 1977b) were described from several hosts. With the discovery of the Sabin-Feldman (dye) test for the diagnosis of toxoplasmosis in 1948, it became a common knowledge that only one species, *T. gondii*, parasitized many warm-blooded mammals and birds. Levine (1977b) has now officially synonymized these species with *T. gondii*.

The status of five species of *Toxoplasma*: *T. alencari*, *T. brumpti*, *T. colubri*, *T. renae*, and *T. serpai* (for details see Levine, 1977b) described from cold-blooded animals is not clear to me. Levine (1977b) considers them possibly valid. The life cycle of all five species is unknown. As a matter of fact, very little is known of their structure. For example, the only stage known in *T. brumpti* and *T. colubri* is the intravascular stage. For the remaining three species only "pseudocysts" are known. Therefore, until their structure and life cycles are described, I prefer to call them sporozoans of unknown taxonomic position.

B. Genus *Hammondia* Frenkel and Dubey, 1975

Diagnosis: Oocysts with two sporocysts, each with four sporozoites; sporogony outside the host; obligatorily heteroxenous; tachyzoites and cysts formed in intermediate hosts, schizonts and gametocytes in carnivores; cysts aseptate, like *Toxoplasma*; metrocytes not formed; oocysts noninfectious to definitive hosts by the oral route; tachyzoites not infectious to any host; transmission by oocysts to intermediate hosts, and by cysts to definitive host.

Type species: *Hammondia hammondi* Frenkel and Dubey, 1975 (Synonyms: *Isospora hammondi* Tadros and Laarman, 1976; *Toxoplasma hammondi* Levine, 1977)

Diagnosis: Same as genus. For additional details see Section II of this chapter.

There is structural similarity between *Toxoplasma gondii* and *Hammondia hammondi*. However, their life cycles are different. *H. hammondi* is obligatorily heteroxenous whereas *T. gondii* is not. Levine (1977b) considered that it was not a good idea to separate genera on the basis of cell or tissue parasitized. Therefore, he considered *Hammondia* to be a synonym of *Toxoplasma*. I do not agree with Levine (1977b) because *Hammondia* was separated from *Toxoplasma* not because it was dissimilar structurally with *T. gondii* but because it had an obligatory two-host life cycle. Therefore, *Hammondia* should be retained as a separate genus.

C. Genus *Sarcocystis* Lankester, 1882 (Synonym: *Endorimospora* Tadros and Laarman, 1976)

Obligatorily heteroxenous. Oocysts with two sporocysts, each with four sporozoites. Sexual cycle in carnivores; gametogony without prior schizogony in carnivores; sporogony within host, sporulated oocysts or sporocysts shed in feces, oocysts or sporocysts not infectious to definitive hosts. Schizogony in intermediate host; cysts formed principally in muscle; cysts often with septa and some species with radial "spines"; metrocytes preceding the formation of bradyzoites in cysts; schizonts and metrocytes noninfectious to definitive host.

Type species: *Sarcocystis miescheriana* (Kuhn, 1865) Lankester, 1882 (Synonym: *Endorimospora miescheriana* Tadros and Laarman, 1976)

D. Genus *Frenkelia* Biocca, 1968 (Synonym: *Endorimospora* Tadros and Laarman, 1976)

Diagnosis: Same as *Sarcocystis* except cysts formation principally in brain.

Type species: *Frenkelia microti* (Findlay and Middleton, 1932) Biocca, 1966 (Synonym: *Endorimospora microti* Tadros and Laarman, 1976).

E. Genus *Isospora* Schneider, 1881 (Synonym: *Diplospora* Labbé, 1893)

Monoxenous: Oocysts with two sporocysts, each sporocyst with four sporozoites; sporogony outside the host. Infection confined to mainly intestinal epithelium or other epithelial cells.

Type species: *Isospora rara* Schneider, 1881

F. Genus *Levineia* nov. gen.

Definition: Oocysts with two sporocysts each with four sporozoites; sporogony outside the host; facultatively or obligatorily heteroxenous, asexual cycle not restricted to intestine; cystlike stages in many organs of intermediate hosts, typically unizoite; asexual stages preceding sexual development in carnivores; tachyzoites and bradyzoites nontransmissible to intermediate hosts, oocysts infectious to definitive and intermediate hosts.

Type species: *Levineia felis* (Wenyon, 1923) nov. comb.

ACKNOWLEDGMENTS

I would like to thank Drs. C. P. Beattie, J. K. A. Beverley, R. Fayer, J. K. Frenkel, N. D. Levine, H. G. Sheffield, and G. D. Wallace for their help in the preparation of this manuscript.

REFERENCES

Abbas, A. M. A. (1967). Comparative study of methods used for the isolation of *Toxoplasma gondii*. *Bull. W.H.O.* **36**, 344–346.

Alford, C. A., Foft, J. W., Blankenship, W. J., Cassady, G., and Benton, J. W. (1969). Subclinical central nervous system disease of neonates: A prospective study of infants born with increased levels of IgM. *J. Pediat.* **75**, 1167–1178.

Alford, C. A., Jr., Stagno, S., and Reynolds, D. W. (1974). Congenital toxoplasmosis: Clinical, laboratory, and therapeutic considerations, with special reference to subclinical disease. *Bull. N.Y. Acad. Med.* **50** (2nd Ser.), 160–181.

Araujo, F. G., and Remington, J. S. (1974). Effect of clindamycin on acute and chronic toxoplasmosis in mice. *Antimicrobial Agents Chemother.* **5**, 647–651.

Araujo, F. G., Barnett, E. V., Gentry, L. O., and Remington, J. S. (1971). False-positive anti-*Toxoplasma* fluorescent-antibody tests in patients with antinuclear antibodies. *Appl. Microbiol.* **22**, 270–275.

Aryeetey, M. E., and Piekarski, G. (1976). Serologische *Sarcocystis*-Studien an Menschen und Ratten. *Z. Parasitenk.* **50**, 109–124.

Azab, M., and Beverley, J. K. A. (1974). Schizogony of *Toxoplasma gondii* in tissue cultures. *Z. Parasitenk.* **44**, 33–41.

Babudieri, B. (1932). I sarcosporidi e le sarcosporidiosi. *Arch. Protistenk.* **76**, 421–580 (cited by Schneider, 1967).

Barksdale, W. L., and Routh, C. F. (1948). *Isospora hominis* infections among American personnel in the southwest Pacific. *Am. J. Trop. Med.* **28**, 639–644.

Basson, P. A., McCully, R. M., and Bigalke, R. D. (1970). Observations on the pathogenesis of bovine and antelope strains of *Besnoitia besnoiti* (Marotel, 1912) infection in cattle and rabbits. *Onderstepoort J. Vet. Res.* **37**, 105–126.

Beech, J., and Dodd, D. C. (1974). Toxoplasma-like encephalomyelitis in the horse. *Vet. Pathol.* **11**, 87–96.

Bell, J. F., Jellison, W. L., and Glesne, L. (1964). Detection of *Toxoplasma microti* in living voles. *Exp. Parasitol.* **15**, 335–339.

Ben Rachid, M. S. (1970). Contribution ã l'étude de la toxoplasmose du gondi. II.

Comportement de *Ctenodactylus gundi* vis-à-vis de *Isospora bigemina. Arch. Inst. Pasteur Tunis* **47,** 33–35.

Beverley, J. K. A. (1958). A rational approach to the treatment of toxoplasmic uveitis. *Trans. Ophthalmol. Soc. United Kingdom* **78,** 109–121.

Beverley, J. K. A. (1959). Congenital transmission of toxoplasmosis through successive generations of mice. *Nature (London)* **183,** 1348–1349.

Beverley, J. K. A. (1961). Toxoplasmosis symposium. *Surv. Ophthalmol.* **6,** 871.

Beverley, J. K. A. (1974). Some aspects of toxoplasmosis, a world wide zoonosis. *In Parasitic Zoonoses:* Clinical and Experimental Studies (E. J. L. Soulsby, ed.), pp. 1–25. Academic Press, New York.

Beverley, J. K. A., and Watson, W. A. (1961). Ovine abortion and toxoplasmosis in Yorkshire. *Vet. Rec.* **73,** 6–11.

Beverley, J. K. A., and Watson, W. A. (1971). Prevention of experimental and of naturally occurring ovine abortion due to toxoplasmosis. *Vet. Rec.* **88,** 39–41.

Beverley, J. K. A., Watson, W. A., and Payne, J. M. (1971a). The pathology of the placenta in ovine abortion due to toxoplasmosis. *Vet. Rec.* **88,** 124–128.

Beverley, J. K. A., Watson, W. A., and Spence, J. B. (1971b). The pathology of the foetus in ovine abortion due to toxoplasmosis. *Vet. Rec.* **88,** 174–177.

Beverley, J. K. A., Archer, J. F., Watson, W. A., and Fawcett, A. R. (1971c). Trial of a killed vaccine in the prevention of ovine abortion due to toxoplasmosis. *Brit. Vet. J.* **127,** 529–535.

Beverley, J. K. A., Freeman, A. P., Henry, L., and Whelan, J. P. F. (1973). Prevention of pathological changes in experimental congenital Toxoplasma infections. *Lyon Méd.* **230,** 491–498.

Bigalke, R. D. (1968). New concepts on the epidemiological features of bovine besnoitiosis as determined by laboratory and field observations. *Onderstepoort J. Vet. Res.* **35,** 3–138.

Bigalke, R. D., and Naude, T. W. (1962). The diagnostic value of cysts in the scleral conjunctiva in bovine besnoitiosis. *J. S. Afr. Med. Asso.* **33,** 1–7.

Biocca, E. (1968). Class Toxoplasmatae: Critical review and proposal of the new name *Frenkelia* gen. n. for M-organism. *Parasitologia* **10,** 89–98.

Blanchard, R. (1885). Note sur les sarcosporidies et sur un essai de classification de ces sporozoaires. *Bull. Soc. Zool. Fr.* **10,** 244–277.

Bledsoe, B. (1976). Transmission of *Isospora vulpina* from silver fox to the dog; Establishment of the mouse as an intermediate host of *I. vulpina.* Proc. 51st Annual meeting Am. Soc. Parasitol. Abst. no. 96.

Bommer, W., Heunert, H. H., and Milthaler, B. (1969). Kinematographische Studien über die Eigenbewegung von *Toxoplasma gondii. Z. Tropenmed. Parasitol.* **20,** 450–458.

Box, E. D. (1975). Exogenous stages of *Isospora serini* (Aragão) and *Isospora canaria* sp. n. in the canary (*Serinus canarius* Linnaeus) *J. Protozool.* **22,** 165–169.

Brandborg, L. L., Goldberg, S. B., and Breidenbach, W. C. (1970). Human coccidiosis—a possible cause of malabsorption. *N. Engl. J. Med.* **283,** 1306–1313.

Brumpt, E. (1913). *Précis de Parasitologie.* 2nd ed. Masson, Paris (cited by Schneider, 1967).

Burtscher, H. and Meingassner, J. G. (1976). *Frenkelia*-Infection bei chinchillas. *Z. Parasitenk.* **50,** 220.

Campana-Rouget, Y., Dorchies, P., and Gourdon, L. (1974). Phénoménes intercurrents dans L'évolution de *Toxoplasma* et *Isospora* chez le Chat. *C. R. Séances Soc. Biol.* **168**, 514–516.

Campbell, R. S. F., Martin, W. B., and Gordon, E. D. (1955). Toxoplasmosis as a complication of canine distemper. *Vet. Rec.* **67**, 708–716.

Capen, C. C., and Cole, C. R. (1966). Pulmonary lesions in dogs with experimental and naturally occurring toxoplasmosis. *Pathol. Vet.* **3**, 40–63.

Carey, R. M., Kimball, A. C., Armstrong, D., and Lieberman, P. H. (1973). Toxoplasmosis. *Am. J. Med.* **54**, 30–38.

Čatár, G., Bergendi, L., and Holková, R. (1969). Isolation of *Toxoplasma gondii* from swine and cattle. *J. Parasitol.* **55**, 952–955.

Cathie, I. A. B. (1954). *Toxoplasma* adenopathy in a child with isolation of the parasite. *Lancet* **ii**, 115–116.

Chamberlain, D. M., Docton, F. L., and Cole, C. R. (1953). Toxoplasmosis, II. Intra-uterine infection in dogs, premature birth and presence of organisms in milk. *Proc. Soc. Exp. Biol. Med.* **82**, 198–200.

Chatton, E., and Blanc, G. (1917). Notes et reflexions sur le toxoplasmose et la toxoplasmose du gondi (*Toxoplasma gondii*, C. L. Nicolle et Manceaux, 1909). *Arch. Inst. Pasteur Tunis* **10**, 1–41.

Cheever, A. W., Valsamis, M. P., and Rabson, A. S. (1965). Necrotizing toxoplasmic encephalitis and herpetic pneumonia complicating treated Hodgkin's disease. *N. Engl. J. Med.* **272**, 26–29.

Chessum, B. S. (1972). Reactivation of *Toxoplasma* oocyst production in the cat by infection with *Isospora felis*. *Brit. Vet. J.* **128**, xxxiii–xxxvi.

Chinchilla, M. and Ruiz, A. (1976). Cockroaches as possible transport hosts of *Toxoplasma gondii* in Costa Rica. *J. Parasitol.* **62**, 140–142.

Chobotar, B., Anderson, L. C., Ernst, J. V., and Hammond, D. M. (1970). Pathogenicity of *Besnoitia jellisoni* in naturally infected kangaroo rats (*Didodomys ordii*) in northwestern Utah. *J. Parasitol.* **56**, 192–193.

Christie, E., and Dubey, J. P. (1977a). Cross immunity between *Hammondia* and *Toxoplasma* infections in mice and hamsters. Unpublished.

Christie, E., and Dubey, J. P. (1977b). Prevalence of *Hammondia hammondi* in Columbus Cats. *J. Parasitol.*, (in press).

Christie, E., Dubey, J. P., and Pappas, P. W. (1976). Prevalence of *Sarcocystis* infection and other intestinal parasitisms in cats from a humane shelter in Ohio. *J. Am. Vet. Med. Assoc.* **168**, 421–422.

Cole, C. R., Sanger, V. L., Farrell, R. L., and Kirnder, J. D. (1954). The present status of toxoplasmosis in veterinary medicine. *N. Am. Vet.* **53**, 265–270.

Colley, F. C., and Zaman, V. (1970). Observations on the endogenous stages of *Toxoplasma gondii* in the cat ileum. II. Electron microscope study. *Southeast Asian J. Trop. Med. Publ. Health* **1**, 465–480.

Cook, M. K., and Jacobs, L. (1958). *In vitro* investigations on the action of pyrimethamine against *Toxoplasma gondii*. *J. Parasitol.* **44**, 280–288.

Corner, A. H., Mitchell, D., Meads, E. B., and Taylor, P. A. (1963). Dalmeny disease. An infection of cattle presumed to be caused by an unidentified protozoon. *Canad. Vet. J.* **4**, 252–264.

Crawley, H. (1914). Two new sarcosporidia. *Proc. Acad. Nat. Sci. Philadelphia* **66**, 214–218.

Cusick, P. K., Sells, D. M., Hamilton, D. P., and Hardenbrook, H. J. (1974). Toxoplasmosis in two horses. *J. Am. Vet. Med. Assoc.* **164**, 77–80.

Cutchins, E., and Warren, J. (1956). Immunity patterns in the guinea pig following *Toxoplasma* infection and vaccination with killed *Toxoplasma*. *Am. J. Trop. Med. Hyg.* **5**, 197–209.

de Roever-Bonnett, H. (1961). Congenital toxoplasmosis. *Trop. Geogr. Med.* **13**, 27–41.

de Roever-Bonnet, H. (1964). Toxoplasma parasites in different organs of mice and hamsters infected with avirulent and virulent strains. *Trop. Geogr. Med.* **16**, 337–345.

de Roever-Bonnet, H. (1969). Congenital Toxoplasma infections in mice and hamsters infected with avirulent and virulent strains. *Trop. Geogr. Med.* **21**, 443–450.

de Souza, W. (1974). Fine structure of the conoid of *Toxoplasma gondii*. *Rev. Inst. Med. Trop. São Paulo* **16**, 32–38.

Desmonts, G. (1966). Definitive serological diagnosis of ocular toxoplasmosis. *Arch. Ophthalmol.* **76**, 839–851.

Desmonts, G., and Couvreur, J. (1974a). Congenital toxoplasmosis. A prospective study of 378 pregnancies. *N. Engl. J. Med.* **290**, 1110–1116.

Desmonts, G., and Couvreur, J. (1974b). Toxoplasmosis in pregnancy and its transmission to the fetus. *Bull. N.Y. Acad. Med.* (2nd Ser.) **50**, 146–159.

Desmonts, G., Couvreur, J., Alison, F., Baudelot, J., Gerbeaux, J., and Lelong, M. (1965). Étude épidémiologique, sur la toxoplasmose: de l'influence de la cuisson des viandes de boucherie sur la fréquence de l'infection humaine. *Rev. Fr. Etud. Clin. Biol.* **10**, 952–958.

Desmonts, G., Baufine-Ducrocq, H., Couzinea, P., and Peloux, Y. (1974). Natural antibodies against Toxoplasma. *Proc. 3rd Inf. Congr. Parasitol. Muenich,* **1**, 302–303.

Doflein, F. J. T. (1901). Die Protozoen als Parasiten und Krankheitserreger nach biologischen Gesichtspunkten dargestellt, xiii + 274 pp. G. Fisher, Jena.

Dubey, J. P. (1966). Transmission of toxoplasmosis in cats with special reference to associated *Toxocara cati* infestations. Ph.D. Thesis. pp. 1–169. University of Sheffield, Sheffield.

Dubey, J. P. (1967a). Distribution of *Toxoplasma gondii* in the tissues of infected cats. I. Isolation in mice. *Trop. Geogr. Med.* **19**, 199–205.

Dubey, J. P. (1967b). Studies on *Toxocara cati* larvae infected with *Toxoplasma gondii*. *J. Protozool.* **14**, 42.

Dubey, J. P. (1968a). *Toxoplasma* infections in English cats. *Vet. Rec.* **82**, 377–379.

Dubey, J. P. (1968b). Isolation of *Toxoplasma gondii* from the feces of a helminth free cat. *J. Protozool.* **15**, 773–775.

Dubey, J. P. (1968c). Feline toxoplasmosis and its nematode transmission. *Vet. Bull.* **38**, 495–499.

Dubey, J. P. (1973). Feline toxoplasmosis and coccidiosis: A survey of domiciled and stray cats. *J. Am. Vet. Med. Assoc.* **162**, 873–877.

Dubey, J. P. (1974). Effect of freezing on the infectivity of *Toxoplasma* cysts to cats. *J. Am. Vet. Med. Assoc.* **165**, 534–536.

Dubey, J. P., (1975a). Immunity to *Hammondia hammondi* infection in cats. *J. Am. Vet. Med. Assoc.* **167**, 373–377.

Dubey, J. P. (1975b). Experimental *Hammondia hammondi* infection in dogs. *Brit. Vet. J.* **131**, 742–743.

Dubey, J. P. (1975c). *Isospora ohioensis* sp. n. proposed for *I. rivolta* of the dog. *J. Parasitol.* **61**, 462–465.

Dubey, J. P. (1975d). Experimental *Isospora canis* and *Isospora felis* infection in mice, cats and dogs. *J. Protozool.* **22**, 416–417.

Dubey, J. P. (1976a). Reshedding of *Toxoplasma gondii* oocysts by chronically infected cats. *Nature* (*London*) **262,** 213–214.

Dubey, J. P. (1976b). A review of *Sarcocystis* of domestic animals and other coccidia of cats and dogs. *J. Am. Vet. Med. Assoc.* **169,** 1061–1078.

Dubey, J. P. (1977a). Attempted transmission of feline coccidia from chronically infected mother cats to their kittens. *J. Am. Vet. Med. Assoc.* **170,** 541–543.

Dubey, J. P. (1977b). Persistence of *Toxoplasma gondii* in the tissues of chronically infected cats. *J. Parasitol.* **63,** 156–157.

Dubey, J. P., and Beverley, J. K. A. (1967). Distribution of *Toxoplasma gondii* in the tissues of infected cats. II. Histopathology. *Trop. Geogr. Med.* **19,** 206–212.

Dubey, J. P., and Fayer, R. (1976). Development of canine *Isospora bigemina* in dogs and other mammals. *Parasitology* **73,** 371–380.

Dubey, J. P., and Frenkel, J. K. (1972a). Cyst-induced toxoplasmosis in cats. *J. Protozool.* **19,** 155–177.

Dubey, J. P., and Frenkel, J. K. (1972b). Extra-intestinal stages of *Isospora felis* and *I. rivolta* (Protozoa:Eimeriidae) in cats. *J. Protozool.* **19,** 89–92.

Dubey, J. P., and Frenkel, J. K. (1973). Experimental Toxoplasma infection in mice with strains producing oocysts. *J. Parasitol.* **59,** 505–512.

Dubey, J. P., and Frenkel, J. K. (1974). Immunity to feline toxoplasmosis: Modification by administration of corticosteroids. *Vet. Pathol.* **11,** 350–379.

Dubey, J. P., and Frenkel, J. K. (1976). Feline toxoplasmosis from acutely infected mice and the development of *Toxoplasma* cysts. *J. Protozool.* **23,** 537–546.

Dubey, J. P., and Hoover, E. A. (1977). Attempted transmission of *Toxoplasma gondii* from pregnant cats to kittens. *J. Am. Vet. Med. Assoc.* **170,** 538–540.

Dubey, J. P., and Streitel, R. H. (1976a). Further studies on the transmission of *Hammondia hammondi* in cats. *J. Parasitol.* **62,** 548–551.

Dubey, J. P., and Streitel, R. H. (1976b). Shedding of *Sarcocystis* in feces of dogs and cats fed muscles of naturally infected food animals in the midwestern United States. *J. Parasitol.* **62,** 828–830.

Dubey, J. P., and Streitel, R. H. (1976c). *Isospora felis* and *I. rivolta* infections in cats induced by mice or oocysts. *Brit. Vet. J.* **132,** 649–651.

Dubey, J. P., and Streitel, R. H. (1976d). Prevalence of *Toxoplasma gondii* infection in cattle. *J. Am. Vet. Med. Assoc.* **169,** 1197–1199.

Dubey, J. P., and Yeary, R. A. (1977). Anticoccidial activity of 2-sulfamoyl-4, 4-diaminodiphenylsulfane (SDDS), sulfadiazine, pyrimethamine, and clindamycin in cats infected with *Toxoplasma gondii*. *Canad. Vet. J.* **18,** 51–57.

Dubey, J. P., Miller, N. L., and Frenkel, J. K. (1970a). Characterization of the new fecal form of *Toxoplasma gondii*. *J. Parasitol.* **56,** 447–456.

Dubey, J. P., Miller, N. L., and Frenkel, J. K. (1970b). The *Toxoplasma gondii* oocyst from cat feces. *J. Exp. Med.* **132,** 636–662.

Dubey, J. P., Swan, G. V., and Frenkel, J. K. (1972). A simplified method of isolation of *Toxoplasma gondii* from cat feces. *J. Parasitol.* **58,** 1005–1006.

Dubey, J. P., Davis, G. W., Koestner, A., and Kiryu, K. (1974). Equine encephalomyelitis due to a protozoan parasite resembling *Toxoplasma gondii*. *J. Am. Vet. Med. Assoc.* **165,** 249–255.

Dubey, J. P., Hoover, E. A., and Walls, K. W. (1977a). Effect of age and sex on the acquisition of immunity to toxoplasmosis in cats. *J. Protozool.*, **24,** 184–186.

Dubey, J. P. Streitel, R. H., Stromberg, P. C., and Toussant, M. J. (1977b.) *Sarcocystis fayeri*, sp. n. from the horse. *J. Parasitol.* **63** (in press).

Dubey, J. P., Christie, E., and Pappas, P. W. (1977c). Characteristics of *Toxoplasma gondii* from the feces of naturally infected cats. *J. Infect. Dis.* (in press).

Durfee, P. T., Ma, C. H., Wang, C. F., and Cross, J. H. (1974). Infectivity and pathogenicity of *Toxoplasma* oocysts for swine. *J. Parasitol.* **60,** 886–887.

Elson-Dew, R. (1953). *Isospora natalensis* (sp. nov.) in man. *J. Trop. Med. Hyg.* **56,** 149–150.

Erhardova, B. (1955). Nález, cizopasníků podohných toxoplasmě v mozku norníka rudého *Clethrionomys glareolus. Schr. Cech. Biol.* **4,** 251–252.

Ernst, J. V., Chobotar, B., Oaks, E. C., and Hammond, D. M. (1968). *Besnoitia jellisoni* (Sporozoa:Toxoplasmea) in rodents from Utah and California. *J. Parasitol.* **54,** 545–549.

Eyles, D. E., and Coleman, N. (1953). Synergistic effect of Sulfadiazine and Daraprim against experimental toxoplasmosis in the mouse. *Antibiot. Chemother.* **3,** 483–490.

Farrell, R. L., Docton, F. L., Chamberlain, D. M., and Cole, C. R. (1952). Toxoplasmosis I. Toxoplasma isolated from swine. *Am. J. Vet. Res.* **13,** 181–185.

Fayer, R. (1972). Gametogony of *Sarcocystis* sp. in cell culture. *Science* **175,** 65–67.

Fayer, R. (1974). Development of *Sarcocystis fusiformis* in the small intestine of the dog. *J. Parasitol.* **60,** 660–665.

Fayer, R. (1975). Effect of refrigeration, cooking, and freezing on *Sarcocystis* in beef from retail food stores. *Proc. Helminthol. Soc. Wash.* **42,** 138–140.

Fayer, R., and Johnson, A. J. (1973). Development of *Sarcocystis fusiformis* in calves infected with sporocysts from dogs. *J. Parasitol.* **59,** 1135–1137.

Fayer, R., and Johnson, A. J. (1974). *Sarcocystis fusiformis:* development of cysts in calves infected with sporocysts from dogs. *Proc. Helminthol. Soc. Wash.* **41,** 105–108.

Fayer, R., and Johnson, A. J. (1975a). Effect of amprolium on acute sarcocystosis in experimentally infected calves. *J. Parasitol.* **61,** 932–936.

Fayer, R., and Johnson, A. J. (1975b). *Sarcocystis fusiformis* infection in the coyote (*Canis latrans*). *J. Infect. Dis.* **131,** 189–192.

Fayer, R., and Kradel, D. (1977). *Sarcocystis leporum* in cottontail rabbits and its transmission to carnivores. *J. Wild. Dis.* **13,** 170–173.

Fayer, R., and Leek, R. G. (1973). Excystation of *Sarcocystis fusiformis* sporocysts from dogs. *Proc. Helminthol. Soc. Wash.* **40,** 294–296.

Fayer, R., Hammond, D. M., Chobotar, B., and Elsner, Y. Y. (1969). Cultivation of *Besnoitia jellisoni* in bovine cell cultures. *J. Parasitol.* **55,** 645–653.

Fayer, R., Johnson, A. J., and Hildebrandt, P. K. (1976a). Oral infection of mammals with *Sarcocystis fusiformis* bradyzoites from cattle and sporocysts from dogs and coyotes. *J. Parasitol.* **62,** 10–14.

Fayer, R., Johnson, A. J., and Lunde, M. N. (1976b). Abortion and other signs of disease in cows experimentally infected with *Sarcocystis fusiformis* from dogs. *J. Infect. Dis.* **134,** 624–628.

Feldman, H. A. (1973). Effect of trimethoprim and sulfisoxazole alone and in combination on murine toxoplasmosis. *J. Infect. Dis.* **128,** 774–776.

Feldman, H. A. (1965). A nationwide serum survey of United States military recruits, 1962. VI. *Toxoplasma* antibodies. *Am. J. Epidemiol.* **81,** 385–391.

Feldman, H. A., and Miller, L. T. (1956). Serological study of toxoplasmosis prevalence. *Am. J. Hyg.* **64,** 320–335.

Ferguson, D. J. P., Hutchison, W. M., Dunachie, J. F., and Siim, J. C. (1974). Ultrastructural study of early stages of asexual multiplication and microgametogony of *Toxoplasma gondii* in the small intestine of the cat. *Acta Pathol. Microbiol. Scand.* **82B,** 167–181.

Ferguson, D. J. P., Hutchison, W. M., and Siim, J. C. (1975). The ultrastructural development of the macrogamete and formation of the oocyst wall of *Toxoplasma gondii*. *Acta Pathol. Microbiol. Scand.* **83B**, 491–505.

Findlay, G. M., and Middleton, A. D. (1934). Epidemic disease among voles (*Microtus*) with special reference to *Toxoplasma*. *J. Anim. Ecol.* **3**, 150–160.

Fischle, B. (1973). Untersuchungen über die Möglichkeit einer Sporozystenausscheidung bei Katze und Hund nach Verfütterung von *Sarcocystis tenella*-und *Sarcocystis fusiformis*-Sporozysten. Inaugural Dissertation, pp. 27. Freie, Universität, Berlin.

Folkers, C. (1962). Studies on toxoplasmosis in pigs with special reference to pathogenicity and immunity. Thesis, pp. 151, Utrecht.

Frelier, P., Mayhew, I. G., Fayer, R., and Lunde, M. N. (1976). *Sarcocystis:* A clinical outbreak in dairy calves. *Science* **195**, 1341–1342.

Frenkel, J. K. (1953). Infections with organisms resembling Toxoplasma, together with the description of a new organism: *Besnoitia jellisoni*. *Atti. VI Congr. Int. Microbiol.* **5**, 426–434.

Frenkel, J. K. (1956). Pathogenesis of toxoplasmosis and of infections with organisms resembling *Toxoplasma*. *Ann. N.Y. Acad. Sci.* **64**, 215–251.

Frenkel, J. K. (1965). Attempted transmission of *Toxoplasma* with arthropods. *Proc. 2nd Int. Conf. Protozool., London, 1965. Excerpta Med. Found. Int. Congr.*, Ser. **91**, 190.

Frenkel, J. K. (1967). Adoptive immunity to intracellular infection. *J. Immunol.* **98**, 1309–1319.

Frenkel, J. K. (1971a). Toxoplasmosis. *In* "Pathology of Protozoal and Helminthic Diseases" (R. Marcial-Rojas, ed.), pp. 254–290. Williams and Wilkins, Baltimore, Maryland.

Frenkel, J. K. (1971b). Toxoplasmosis. Mechanisms of infection, laboratory diagnosis and management. *Curr. Top. Pathol.* **54**, 29–75.

Frenkel, J. K. (1973a). Toxoplasmosis: Parasite life cycle, pathology and immunology. *In* "The Coccidia" (D. M. Hammond and P. L. Long, eds.), pp. 343–410. University Park Press, Baltimore, Maryland.

Frenkel, J. K. (1973b). *Toxoplasma* in and around us. *BioScience* **23**, 343–352.

Frenkel, J. K. (1974). Advances in the biology of sporozoa. *Z. Parasitenk.* **45**, 125–162.

Frenkel, J. K. (1975). Toxoplasmosis in cats and man. *Feline Pract.* **5**, 28–41.

Frenkel, J. K., and Dubey, J. P. (1972a). Rodents as vectors for feline coccidia, *Isospora felis* and *Isospora rivolta*. *J. Infect. Dis.* **125**, 69–72.

Frenkel, J. K., and Dubey, J. P. (1972b). Toxoplasmosis and its prevention in cats and man. *J. Infect. Dis.* **126**, 664–673.

Frenkel, J. K., and Dubey, J. P. (1972c). Effect of freezing on the viability of Toxoplasma oocysts. *J. Parasitol.* **59**, 587–588.

Frenkel, J. K., and Dubey, J. P. (1975). *Hammondia hammondi*, gen. nov., sp. nov., from domestic cats, a new coccidian related to *Toxoplasma* and *Sarcocystis*. *Z. Parasitenk.* **46**, 3–12.

Frenkel, J. K., and Friedlander, S. (1951). Toxoplasmosis. Pathogenesis of neonatal disease. Pathogenesis, diagnosis, and treatment. *Publ. Health Serv. Publ.* **141**, 1–105. U.S. Govt. Printing Office, Washington, D.C.

Frenkel, J. K., and Hitchings, G. H. (1957). Relative reversal by vitamins (p-aminobenzoic, folic, and folinic acids) of the effects of sulfadiazine and pyrimethamine on Toxoplasma, mouse and man. *Antibiot. Chemother.* **7**, 630–638.

Frenkel, J. K., Dubey, J. P., and Miller, N. Y. (1969). *Toxoplasma gondii:* fecal forms separated from eggs of the nematode *Toxocara cati*. *Science* **164**, 432–433.

Frenkel, J. K., Dubey, J. P., and Miller, N. L. (1970). *Toxoplasma gondii:* fecal stages identified as coccidian oocysts. *Science* **167**, 893–896.

Frenkel, J. K., Nelson, B., and Arias-Stella, J. (1975a). Immunosuppression and toxoplasmic encephalitis: Clinical and experimental aspects. *Human Pathol.* **6**, 97–112.

Frenkel, J. K., Ruiz, A., and Chinchilla, M. (1975b). Soil survival of *Toxoplasma* oocysts in Kansas and in Costa Rica. *Am. J. Trop. Med. Hyg.* **24**, 439–443.

Frenkel, J. K., Dubey, J. P., and Hoff, R. L. (1976). Loss of stages after continuous passage of *Toxoplasma gondii* and *Besnoitia jellisoni. J. Protozool.* **23**, 421–424.

Friedmann, C. T., and Knox, D. L. (1969). Variations in recurrent active toxoplasmic retinochoroiditis. *Arch. Ophthalmol.* **81**, 481–493.

Fulton, J. D., and Turk, J. L. (1959). Direct agglutination test for *Toxoplasma gondii. Lancet* **11**, 1068–1069.

Garin, J. P., and Eyles, D. E. (1958). Le traitement de la toxoplasmose expérimentale de la souris par la spiramycin. *Presse Med.* **66**, 957–958.

Garnham, P. C. C. (1971). The parasitic life. *In Progr. Parasitol.*, pp. 116–124. Athlone Press, London.

Gavin, M. A., Wanko, T., and Jacobs, L. (1962). Electron microscope studies of reproducing and interkinetic *Toxoplasma. J. Protozool.* **9**, 222–234.

Gestrich, R., Schmitt, M., and Heydorn, A. O. (1974). Pathogenität von *Sarcocystis tenella*—Sporozysten aus den Fäzes von Hunden für Lämmer. *Berlin Muenchen. Tieraerztl. Wochenschr.* **87**, 362–363.

Gibson, C. L., and Coleman, N. (1958). The prevalence of *Toxoplasma* antibodies in Guatemala and Costa Rica. *Am. J. Trop. Med. Hyg.* **7**, 334–338.

Gleason, T. H., and Hamlin, W. B. (1974). Disseminated toxoplasmosis in the compromised host. *Arch. Int. Med.* **134**, 1059–1062.

Golubkovan, D. I., and Kisliakova, Z. I. (1974). The sources of infection for swine *Sarcocystis* (In Russian). *Veterinariya* **11**, 85–86.

Grassi, B. (1879). Die protozoi parassiti e specialmente di quelli che sono nell'uomo. *Gaz. Med. Ital. Lombard* **39**, 445.

Gray, G. F., Kimball, A. C., and Kean, B. H. (1972). The posterior cervical lymph node in toxoplasmosis. *Am. J. Pathol.* **69**, 349–356.

Hadwen, S. Cyst-forming protozoa in reindeer and caribou, and a sarcosporidian parasite of the seal (*Phoca richardi*). *J. Am. Vet. Med. Assoc.* **61**, 374–382.

Hammond, D. M. (1973). Ultrastructure and development of coccidia. *Proc. Symp. Coccidia Related Organ.*, pp. 11–43. University of Guelph, Canada.

Harding, J. D. J., Beverley, J. K. A., Shaw, I. G., Edwards, B. L., and Bennett, G. H. (1961). Toxoplasma in English pigs. *Vet. Rec.* **73**, 3–6.

Hartley, W. J., and Blakemore, W. F. (1974). An unidentified sporozoan encephalomyelitis in sheep. *Vet. Pathol.* **11**, 1–12.

Hartley, W. J., and Kater, J. C. (1963). The pathology of *Toxoplasma* infection in the pregnant ewe. *Res. Vet. Sci.* **4**, 326–332.

Hartley, W. J., and Marshall, S. C. (1957). Toxoplasmosis as a cause of ovine perinatal mortality. *N.Z. Vet. J.* **5**, 119–124.

Hartley, W. J., and Moyle, G. (1968). Observations on an outbreak of ovine congenital toxoplasmosis. *Aust. Vet. J.* **44**, 105–107.

Hartley, W. J., and Moyle, G. G. (1974). Further observations on the epidemiology of ovine *Toxoplasma* infection. *Aust. J. Exp. Biol. Med. Sci.* **52**, 647–653.

Hartley, W. J., Lindsay, A. B., and MacKinnon, M. M. (1958). *Toxoplasma* meningoencephalomyelitis and myositis in a dog. *N.Z. Vet. J.* **6**, 124–127.

Hasselman, G. E. (1926). Alteracoes pathologicas do myocardio na Sarcoporideose. *Biol. Inst. Brasil. Sci.* **2**, 319–326.

Hendricks, L. D. and Walton, B. C. (1974). Vertebrate intermediate hosts in the life cycle of an isosporan from a non-human primate *Proc. 3rd Int. Congr. Parasitol.* Muenich. Facta Publications **1**, 96–97.

Henry, D. P. (1932). *Isospora buteonis* sp. nov. from the hawk and owl, and notes on *Isospora lacazii* (Labbé) in birds. *Univ. of California Publications in Zoology* **37**, 291–300.

Heydorn, A. O. (1973). Zum Lebenszyklus der kleinen Form von *Isospora bigemina* des Hundes. I. Rind und Hund als mögliche Zwischenwirte. *Berlin Muenchen. Tieraerztl. Wochenschr.* **86**, 323–329.

Heydorn, A. O., and Gestrich, R. (1976). Beiträge zum Lebenszyklus der Sarkosporidien VII. Entwicklungsstadien von *Sarcocystis ovicanis* im Schaf. *Berlin Muenchen. Tieraerztl. Wochenschr.* **89**, 1–5.

Heydorn, A. O., and Rommel, M. (1972a). Beiträge zum Lebenszyklus der Sarkosporidien. II. Hund und Katze als Überträger der Sarkosporidien des Rindes. *Berlin Muenchen. Tieraerztl. Wochenschr.* **85**, 121–123.

Heydorn, A. O., and Rommel, M. (1972b). Beiträge zum Lebenszyklus der Sarkosporidien. IV. Entwicklungsstadien von *S. fusiformis* in der Dünndarmschleimhaut der Katze. *Berlin Muenchen. Tieraerztl. Wochenschr.* **85**, 333–336.

Heydorn, A. O., Mehlhorn, H., and Gestrich, R. (1975a). Licht-und elektronenmikroskopische Untersuchungen an Cysten von *Sarcocystis fusiformis* in der Muskulatur von Kälbern nach experimenteller Infektion mit Oocysten und Sporocysten der grossen Form von *Isospora bigemina* des Hundes. 2. Die Feinstruktur der Cystenstadien. *Zentralbl. Bakteriol. Parasitenk. Infektionskr. Hyg.* 1. Abt. Orig. A **233**, 123–137.

Heydorn, A. O., Gestrich, R., and Ipczynski, V. (1975b). Zum Lebenszyklus der kleinen Form von *Isospora bigemina* des Hundes. II. Entwicklungsstadien im Darm des Hundes. *Berlin Muenchen. Tieraerztl. Wochenschr.* **88**, 449–453.

Heydorn, A. O., Gestrich, R., Mehlhorn, H., and Rommel, M. (1975c). Proposal for a new nomenclature of the *Sarcosporidia. Z. Parasitenk.* **48**, 73–82.

Hilgenfeld, M., and Punke, G. (1974). Zur Sarkosporidieninfektion des Zentralnervensystems des Schafes, ein Beitrag zur Differentialdiagnose von Protozoeninfektionen. *Arch. Exp. Veterinaermed.* **28**, 621–626.

Hitchcock, D. J. (1955). The life cycle of *Isospora felis* in the kitten. *J. Parasitol.* **41**, 383–397.

Hoare, C. A. (1972). The developmental stages of *Toxoplasma. J. Trop. Med. Hyg.* **75**, 56–58.

Hoff, R.L ., Dubey, J. P., Behbehani, A., and Frenkel, J. K. (1977). Biologic evidence of *Toxoplasma* cyst formation in cell culture. Unpublished.

Hogan, M. J., Zweigart, P. A., and Lewis, A. (1958). Recovery of *Toxoplasma* from a human eye. *Arch. Ophthalmol.* **60**, 548–554.

Hudkins-Vivion, G., Kistner, T. P., and Fayer, R. (1976). Possible species differences between Sarcocystis from mule deer and cattle. *J. Wild. Dis.* **12**, 86–87.

Hudkins, G. G., and Kistner, T. P. (1977). *Sarcocystis hemionilatrantis* (sp. nov.); life cycle in mule deer and coyote. *J. Wild. Dis.* **13**, 80–84.

Huldt, G. (1960). Experimental toxoplasmosis. Transplacental transmission in guinea pigs. *Acta Pathol. Microbiol.* **49**, 176–188.

Huldt, G., Gard, S., and Olovson, S. G. (1973). Effect of *Toxoplasma gondii* on the thymus. *Nature (London)* **244**, 301–303.

Hutchison, W. M. (1965). Experimental transmission of *Toxoplasma gondii. Nature (London)* **206**, 961–962.

Hutchison, W. M. (1967). The nematode transmission of *Toxoplasma gondii. Trans. Roy. Soc. Trop. Med. Hyg.* **61**, 80–89.

Hutchison, W. M., Dunachie, J. F., and Work, K. (1968). The fecal transmission of *Toxoplasma gondii. Acta Pathol. Microbiol. Scand.* **74,** 462–464.

Hutchison, W. M., Dunachie, J. F., Siim, J. C., and Work, K. (1969). Life cycle of *Toxoplasma gondii. Brit. Med. J.* **4,** 806.

Hutchison, W. M., Dunachie, J. F., Siim, J. C., and Work, K. (1970). Coccidian-like nature of *Toxoplasma gondii. Brit. Med. J.* **1,** 142–144.

Hutchison, W. M., Dunachie, J. F., Work, K., and Siim, J. C. (1971). The life cycle of the coccidian parasite, *Toxoplasma gondii,* in the domestic cat. *Trans. Roy. Soc. Trop. Med. Hyg.* **65,** 380–399.

Ito, S., Tsunoda, K., Nishikawa, H., and Matsui, T. (1974a). Small type of *Isospora bigemina:* isolation from naturally infected cats and relations with Toxoplasma oocysts. *Nat. Inst. Anim. Health Quart.* **14,** 137–144.

Ito, S., Tsunoda, K., Nishikawa, H., and Matsui, T. (1974b). Pathogenicity for piglets of Toxoplasma oocysts originated from naturally infected cat. *Nat. Inst. Anim. Health Quart.* **14,** 182–187.

Jacobs, L. (1964). The occurrence of *Toxoplasma* infection in the absence of demonstrable antibodies. *Proc. 1st Int. Congr. Parasitol.* **1,** 176–177.

Jacobs, L. (1967). *Toxoplasma* and toxoplasmosis. *In Advances in Parasitology* (B. Dawes, ed.), pp. 1–45. Academic Press, New York and London.

Jacobs, L. (1973). New knowledge of toxoplasmosis. *In Advances in Parasitology* (B. Dawes, ed.), pp. 631–669. Academic Press, New York and London.

Jacobs, L., and Lunde, M. N. (1957). A hemagglutination test for toxoplasmosis. *J. Parasitol.* **43,** 308–314.

Jacobs, L., and Melton, M. L. (1954). Modification in virulence of a strain of *Toxoplasma gondii* by passage in various hosts. *Am. J. Trop. Med. Hyg.* **3,** 447–457.

Jacobs, L., and Melton, M. L. (1966). Toxoplasmosis in chickens. *J. Parasitol.* **52,** 1158–1162.

Jacobs, L., and Hartley, W. J. (1964). Ovine toxoplasmosis: Studies on parasitemia, tissue infection, and congenital transmission in ewes infected by various routes. *Brit. Vet. J.* **120,** 347–364.

Jacobs, L., Remington, J. S., and Melton, M. L. (1960a). The resistance of the encysted form of *Toxoplasma gondii. J. Parasitol.* **46,** 11–21.

Jacobs, L., Remington, J. S., and Melton, M. L. (1960b). A survey of meat samples from swine, cattle, and sheep for the presence of encysted *Toxoplasma. J. Parasitol.* **46,** 23–28.

Jacobs, L., Moyle, G. G., and Ris, R. R. (1963). The prevalence of toxoplasmosis in New Zealand sheep and cattle. *Am. J. Vet. Res.* **24,** 673–675.

Janitschke, K., and Kühn, D. (1972). *Toxoplasma*—Oozysten im Kot natürlich infizierter Katzen. *Berlin Muenchen. Tieraerztl. Wochenschr.* **85,** 46–47.

Janitschke, K., and Nürnberger, F. (1975). Untersuchungen über die Bedentung der Ubertragnung von Toxoplasmen auf dem Geschlechtsweg. *Zentralbl. Bakteriol. Parasitenk. Infektionsk. Hyg.* Abst. 1 **231A,** 323–332.

Janitschke, K., and Werner, H. (1972). Untersuchungen über die Wirtsspezifität des geschlechtlichen Entwicklungszyklus von *Toxoplasma gondii. Z Parasitenk.* **39,** 247–254.

Janitschke, K., Protz, D., and Werner, H. (1976). Beitrag zum Entwicklungszyklus von Sarkosporidien der Grantgazelle (*Gazella granti*). *Z. Parasitenk.* **48,** 215–219.

Janitschke, K., Weiland, G., and Rommel, M. (1967). Untersuchungen über den Befall von Schlachtkälbern und-Schafen mit *Toxoplasma gondii. Fleischwirtschaft* **47,** 135–136.

Janitschke, K., Werner, H., and Hasse, W. (1974). Untersuchungen über die Moglichkeit der Übertragung von Toxoplasmen durch Bluttransfusionen. *Z. Gesamte Blutforschung* **29**, 407–415.

Janku, J. (1959). Die Pathogenese und pathologische Anatomie des Sogenannten angeborenen Koloboms des gelben fleckes in normal grossen sowie im mikropathalmischen auge mit Parasitenbefund in der Netzaut. *Cesk. Parazitol.* **6**, 9–57 (cited by Frenkel, 1973b).

Jarpa, G. A. (1966). Coccidiosis humana. *Biologica* **39**, 3–26.

Jeffrey, H. C. (1974). Sarcosporidiosis in man. *Trans. Roy. Soc. Trop. Med. Hyg.* **68**, 17–29.

Jewell, M. L., Frenkel, J. K., Johnson, K. M., Reed, V., and Ruiz, A. (1972). Development of *Toxoplasma* oocysts in neotropical felidae. *Am. J. Trop. Med. Hyg.* **21**, 512–517.

Johnson, A. J., Hildebrandt, P. K., and Fayer, R. (1975). Experimentally induced Sarcocystis infection in calves: Pathology. *Am. J. Vet. Res.* **36**, 995–999.

Jones, T. C. (1973). Multiplication of toxoplasmas in enucleate fibroblasts. *Proc. Soc. Exp. Biol. Med.* **142**, 1268–1271.

Jones, T. C. (1974). Macrophages and intracellular parasitism. *Res. J. Reticuloendothelial Soc.* **15**, 439–450.

Jones, T. C., Yeh, S., and Hirsch, J. G. (1972). The interaction between *Toxoplasma gondii* and mammalian cells. I. Mechanism of entry and intracellular fate of the parasite. *J. Exp. Med.* **136**, 1157–1172.

Jones, T. C., Len, L., and Hirsch, J. G. (1975). Assessment *in vitro* of immunity against *Toxoplasma gondii*. *J. Exp. Med.* **141**, 466–482.

Karstad, L. (1963). *Toxoplasma microti* (the M-organism) in the muskrat (*Ondatra zibethica*). *Can. Vet. J.* **4**, 249–251.

Kean, B. H. (1972). Clinical toxoplasmosis—50 years. *Trans. Roy. Soc. Trop. Med. Hyg.* **66**, 549–571.

Kean, B. H., Kimball, A. C., and Christenson, W. N. (1969). An epidemic of acute toxoplasmosis. *J. Am. Med Assoc.* **208**, 1002–1004.

Kimball, A. C., Kean, B. H., and Fuchs, F. (1971). Congenital toxoplasmosis: A prospective study of 4,048 obstetric patients. *Am. J. Obstet. Gynecol.* **111**, 211–218.

Knoch, W., Jungmann, R., and Hiepe, T. (1974). Zum koprologischen Nachweis von *Toxoplasma gondii*—Oozysten bei der Hauskatze. *Monatshefte für Veterinarmedizin* **29**, 247–250.

Koestner, A., and Cole, C. R. (1960a). Neuropathology of canine toxoplasmosis. *Am. J. Vet. Res.* **21**, 831–844.

Koestner, A., and Cole, C. R. (1960b). Neuropathology of porcine toxoplasmosis. *Cornell Vet.* **50**, 362–384.

Koppe, J. G., Kloosterman, G. J., de Roever-Bonnet, H., Eckert-Stroink, J. A., Loewer-Sieger, D. H., and de Bruijne, J. I. (1974). Toxoplasmosis and pregnancy, with a long-term follow-up of the children. *Europ. J. Obstet. Gynecol. Reprod. Biol.* **4**, 101–110.

Krahenbuhl, J. L., Ruskin, J., and Remington, J. S. (1972). The use of killed vaccines in immunization against an intracellular parasite: *Toxoplasma gondii*. *J. Immunol.* **108**, 425–431.

Krakowka, S., Cockerell, G., and Koestner, A. (1975). Effects of canine distemper virus infection on lymphoid function *in vitro* and *in vivo*. *Infect. Immunity* **11**, 1069–1078.

Kühn, D., and Weiland, G. (1969). Experimentelle *Toxoplasma*-infektionen bei der

Katze. I. Wiederholte Übertragung von *Toxoplasma gondii* durch Kot von mit Nematoden infizierten Katzen. *Berlin Muench. Tieraerztl. Wochenschr.* **82,** 401–404.

Labbé, A. (1893). Sur les coccidies des oiseux. *C. R. Acad. Sci.* CXVII, 407.

Labbé, A. (1899). Sporozoen. *Das Tierreich.* **5,** xx + 180 pp.

Lainson, R. (1972). A note on sporozoa of undetermined taxonomic position in an armadillo and a heifer calf. *J. Protozool.* **19,** 582–586.

Leek, R. G., and Fayer, R. (1976). Studies on ovine abortion and intrauterine transmission following experimental infection with *Sarcocystis* from dogs. *Fourth Congreso Latinamericano de Parasitologia, Resume de trabahas leiberes* 7–11 December. Page 33. San Jose, Costa Rica.

Lepp, D. L., and Todd, K. S. (1974). Life cycle of *Isospora canis* Nemeséri, 1959 in the dog. *J. Protozool.* **21,** 199–206.

Levine, N. D. (1961). *Protozoan Parasites of Domestic Animals and of Man,* 1st ed., p. 412. Burgess, Minneapolis, Minnesota.

Levine, N. D. (1973). *Protozoan Parasites of Domestic Animals and of Man,* 2nd ed., p. 406. Burgess, Minneapolis, Minnesota.

Levine, N. D. (1977a). Nomenclature of *Sarcocystis* in the ox and sheep and of fecal coccidia of the dog and cat. *J. Parasitol.* **63,** (in press).

Levine, N. D. (1977b). Taxonomy of *Toxoplasma. J. Protozool.* **24,** in press.

Levine, N. D., and Ivens, V. (1965). *Isospora* species in the dog. *J. Parasitol.* **51,** 859–864.

Lickefeld, K. G. (1959). Untersuchungen über das Katzencoccid *Isospora felis* (Wenyon, 1923). *Arch. Protistenk.* **103,** 427–456.

Lotze, J. C., Shalkop, W. T., Leek, R. G., and Behin, R. (1964). Coccidial schizonts in mesenteric lymph nodes of sheep and goat. *J. Parasitol.* **50,** 205–208.

Lunde, M. N., and Fayer, R. (1977). Serologic tests for the detection of antibody to *Sarcocystis* in cattle. *J. Parasitol.* **63,** in press.

McCully, R. M., Basson, P. A., van Niekerk, J. W., and Bigalke, R. D. (1966). Observations on *Besnoitia* cysts in the cardiovascular system of some wild antelopes and domestic cattle. *Onderstepoort J. Vet. Res.* 33, 245–275.

McErlean, B. A. (1974). Ovine paralysis associated with spinal lesions of toxoplasmosis. *Vet. Rec.* **94,** 264–266.

McMaster, P. R. B., Powers, K. G., Finerty, J. F., and Lunde, M. N. (1973). The effect of two chlorinated lincomycin analogues against acute toxoplasmosis in mice. *Am. J. Trop. Med. Hyg.* **22,** 14–17.

Mahmoud, A. A., Warren, K. S., and Strickland, G. T. (1976). Acquired resistance to infection with *Schistosoma mansoni* induced by *Toxoplasma gondii. Nature (London)* **263,** 56–57.

Mahrt, J. L. (1967). Endogenous stages of the life cycle of *Isospora rivolta* in the dog. *J. Protozool.* **14,** 754–759.

Mahrt, J. L., and Fayer, R. (1975). Hematologic and serologic changes in calves experimentally infected with *Sarcocystis fusiformis. J. Parasitol.* **61,** 967–969.

Maloney, E. D., and Kaufman, H. E. (1964). Multiplication and therapy of *Toxoplasma gondii* in tissue culture. *J. Bacteriol.* **88,** 319–321.

Marotel, G. (1921). Sur une nouvelle coccidie du chat. *Bull. Soc. Sci. Vet. Lyon.* **1921,** 3–4 (cited by Levine, 1977a).

Meads, E. B. Dalmeny disease—another outbreak—probably sarcocystosis. *Canad. Vet. J.* **17,** 271.

Mehlhorn, H., and Scholtyseck, E. (1973). Elektronenmikroskopische Untersuchungen an Cystenstadien von *Sarcocystis tenella* aus der Oesophagus-Muskulatur des Schafes. *Z. Parasitenk.* **41,** 291–310.

Mehlhorn, H., and Scholtyseck, E. (1974). Licht-und elektronenmikroskopische Untersuchungen an Entwicklungsstadien von *Sarcocystis tenella* aus der Darmwand der Hauskatze. I. Die Oocysten und Sporocysten Z. *Parasitenk.* **43**; 251–270.

Mehlhorn, H., Sénaud, J., Chobotar, B., and Scholtyseck, E. (1975). Electron microscope studies of cyst stages of *Sarcocystis tenella:* The origin of micronemes and rhoptries. *Z. Parasitenk.* **45**, 227–236.

Mehlhorn, H., Hartley, W. J., and Heydorn, A. O. (1976). A comparative ultrastructural study of the cyst wall of 13 Sarcocystis species. *Protistologica* **12**, 451–467.

Meier, H., Holzworth, J., and Griffiths, R. C. (1957). Toxoplasmosis in the cat—fourteen cases. *J. Am. Vet. Med. Assoc.* **131**, 395–414.

Mello, U. (1910). Un cas de toxoplasmose du chien observé á Turin. *Bull. Soc. Pathol. Exot.* **3**, 359–363.

Melton, M. L., and Sheffield, H. G. (1975). Activity of anticoccidial compound, lasalocid against *Toxoplasma gondii* in cultured cells. *J. Parasitol.* **61**, 713–717.

Meyer, H. F. (1963). Primeros aislamientos de *Toxoplasma gondii* de retina de bovinos. *Revista de Medicina Veterinaria* (Argentina) **44**, 423–430.

Miller, N. L., Frenkel, J. K., and Dubey, J. P. (1972). Oral infections with Toxoplasma cysts and oocysts in felines, other mammals, and in birds. *J. Parasitol.* **58**, 928–937.

Møller, T., Fennestad, K. L., Eriksen, L., Work, K., and Siim, J. C. (1970). Experimental toxoplasmosis in pregnant sows. II. Pathological findings. *Acta Pathol. Microbioli. Scand.* **A78**, 241–255.

Moriwaki, M., Hayashi, S., Minami, T., and Ishitani, R. (1976). Detection of congenital toxoplasmosis in piglet. *Jap. J. Vet. Sci.* **38**, 377–381.

Moulé, L. (1888). Des sarcosporidies et de leur fréquence, principalement chez les animaux de boucherie. *Mem. Soc. Arts Vitry-le-Fr.* **14**, 3–42.

Munday, B. L. (1972a). Serological evidence of *Toxoplasma* infection in isolated groups of sheep. *Res. Vet. Sci.* **13**, 100–102.

Munday, B. L. (1972b). Transmission of *Toxoplasma* infection from chronically infected ewes to their lambs. *Brit. Vet. J.* **128**, lxxi.

Munday, B. L., Barker, I. K., and Rickard, M. D. (1975). The developmental cycle of a species of *Sarcocystis* occurring in dogs and sheep, with observations on pathogenicity in the intermediate host. *Z. Parasitenk.* **46**, 111–123.

Nemeséri, L. (1959). Adatok a kutya coccidiosisahoz I. *Isospora canis. Magyar Allatorvosok Lapja* **14**, 91–92.

Neufeld, J. L., and Brandt, R. W. (1974). Cholangiohepatitis in a cat associated with a coccidia-like organism. *Canad. Vet. J.* **15**, 156–159.

Nicolle, C., and Manceaux, L. (1908). Sur une infection á corps de Leishman (ou organisms voisins) du gondi. *C. R. Hebd. Seances Acad. Sci.* **147**, 763–766.

Nicolle, C., and Manceaux, L. (1909). Sur un Protozoaire nouveau du Gondi. *C. R. Hebd. Seances Acad. Sci.* **148**, 369–372.

Nobuto, K., Suzuki, K., Omuro, M., and Ishii, S. (1960). *Studies on Toxoplasmosis in domestic animals, Bull. No. 40.* Nat. Inst. Animal Health, Tokyo.

Norrby, R. (1970). Host cell penetration of *Toxoplasma gondii. Infect. Immunity* **2**, 250–255.

Norrby, R., Eilard, T., Svedhem, Å., and Lycke, E. (1975). Treatment of toxoplasmosis with trimethoprim-sulphamethoxazole. *Scand. J. Infect. Dis.* **7**, 72–75.

Nozik, R. A., and O'Connor, G. R. (1969). The so-called toxin of *Toxoplasma. Am. J. Trop. Med. Hyg.* **18**, 511–515.

O'Connor, G. R. (1974). Manifestations and management of ocular toxoplasmosis. *Bull. N.Y. Acad. Med.* **50** (2nd Ser.), 192–210.

O'Connor, G. R. (1975). Ocular toxoplasmosis. *Japan. J. Ophthalmol.* **19,** 1–24.

Ohshima, S., Tanaka, H., and Inami, Y. (1967). The chemotherapeutic effect of 2-sulfamoyl-4,4′-diaminodiphenylsulfone (SDDS) on acute experimental toxoplasmosis in mice. *Japan. J. Parasitol.* **16,** 331–338.

Olafson, P., and Monlux, W. S. (1942). Toxoplasma infection in animals. *Cornell Vet.* **32,** 176–190.

Overdulve, J. P. (1970). The identity of *Toxoplasma* Nicolle and Manceaux, 1909 with *Isospora* Schneider, 1881. I. *Proc. Kon. Ned. Akad. Wetensch. Ser. C* **73,** 129–151.

Overdulve, J. P. (1974). Immunity to *Toxoplasma* in cats. *Proc. 3rd. Int. Congr. Parasitol. Muenich,* Facta Publications **1,** 302–303.

Pampiglione, S., Poglayen, G., Arnone, B., and de Lalia, F. (1973). *Toxoplasma gondii* oocysts in the faeces of naturally infected cat. *Brit. Med. J.* **2,** 306.

Pelster, B., and Piekarski, G. (1971). Electronenmikroskopische Analyse der Mikrogametenentwicklung bei *Toxoplasma gondii. Z. Parasitenk.* **37,** 267–277.

Pelster, B., and Piekarski, G. (1972). Untersuchungen zur Feinstruktur des Makrogameten von *Toxoplasma gondii. Z. Parasitenk.* **39,** 225–232.

Perkins, E. S. (1973). Ocular toxoplasmosis. *Brit. J. Ophthalmol.* **57,** 1–17.

Perrotto, J., Keister, D. B., and Gelderman, A. H. (1971). Incorporation of precursors into *Toxoplasma* DNA. *J. Protozool.* **18,** 470–473.

Peteshev, V. M., Galuzo, I. G., and Polomoshnov, A. P. (1974). Cats—definitive hosts of Besnoitia (*Besnoitia besnoiti*) (In Russian). *Izv. Akad. Nauk Kaz. SSR, B* **1,** 33–38.

Petrak, M., and Carpenter, J. (1965). Feline toxoplasmosis. *J. Am. Vet. Med. Assoc.* **146,** 728–734.

Pettersen, K. (1971). An explanation of the biological action of toxotoxin based on some *in vitro* experiments. *Acta Pathol. Microbiol. Scand.* **B79,** 33–36.

Piekarski, G., and Witte, H. M. (1971). Experimentelle und histologische Studien zur *Toxoplasma*–Infektion der Hauskatze. *Z. Parasitenk.* **36,** 95–121.

Piekarski, G., Pelster, B., and Witte, H. M. (1971). Endopolygenie bei *Toxoplasma gondii. Z. Parasitenk.* **36,** 122–130.

Pinkerton, H., and Weinman, D. (1940). *Toxoplasma* infection in man. *Arch. Pathol.* **30,** 374–392.

Pols, J. W. (1960). Studies on bovine besnoitiosis with special reference to the aetiology. *Onderspoort J. Vet. Res.* **28,** 265–356.

Raíllíet, A. (1886). (*Miescheria tenella*). *Bull. Mem. Soc. Centr. Med. Vet.* **40,** 130.

Raíllíet, A. (1897). La douve pancréatique. *Bull. Soc. Centr. Med. Vet.* **51** (N.S.), 371–377.

Raíllíet, A., and Lucet, A. (1891). Note sur quelques espéces de coccidies encore peu étudiées. *Bull. Soc. Zool. Paris* **16,** 246–250.

Rawal, B. D. (1959). Toxoplasmosis. A dye-test survey on sera from vegetarians and meat eaters in Bombay. *Trans. Roy. Soc. Trop. Med. Hyg.* **53,** 61–63.

Remington, J. S. (1968). Toxoplasmosis and congenital infection. *Birth Defects* **4,** 49–56.

Remington, J. S. (1969). The present satus of the IgM fluorescent antibody technique in the diagnosis of congenital toxoplasmosis. *J. Pediat.* **75,** 1116–1124.

Remington, J. S. (1970). Toxoplasmosis: Recent developments. *Annu. Rev. Med.* **21,** 201–218.

Remington, J. S. (1976). Trimethoprim-sulfamethoxazole in murine toxoplasmosis. *Antimicrob. Agents Chemother.* **9**, 222–223.

Remington, J. S., and Merigan, T. C. (1968). Resistance to virus challenge in mice infected with protozoa or bacteria. *Proc. Soc. Exp. Biol. Med.* **131**, 1184–1188.

Remington, J. S., Efron, B., Cavanaugh, E., Simon, H. J., and Trejos, A. (1970a). Studies on toxoplasmosis in El Salvador. Prevalence and incidence of toxoplasmosis as measured by the Sabin-Feldman dye test. *Trans. Roy. Soc. Trop. Med. Hyg.* **64**, 252–267.

Remington, J. S., Earle, P., and Yagura, T. (1970b). *Toxoplasma* in nucleus. *J. Parasitol.* **56**, 390–391.

Remington, J. S., Bloomfield, M. M., Russel, E., Jr., and Robinson, W. S. (1970c). The RNA of *Toxoplasma gondii*. *Proc. Soc. Exp. Biol. Med.* **133**, 623–626.

Rifaat, M. A., Arafa, M. S., Sadek, M. S. M., Nasr, N. T., Azab, M. E., Mahmoud, W., and Khalil, M. S. (1976). Toxoplasma infection of stray cats in Egypt. *J. Trop. Med. Hyg.* **79**, 67–70.

Rivolta, S. (1878). Della gregarinosi dei polli e dell' ordinarmento delle gregarine e dei psorospermi degli animali domestici. *Gior. Anat. Fisiol. Patol. Anim.* **10**, 220–235.

Rommel, M., and Geisel, O. (1975). Untersuchungen über die Verbreitung und den Lebenszyklus einer Sarkosporidienart des Pferdes (*Sarcocystis equicanis* n. spec.). *Berlin Muench. Tieraerztl. Wochenschr.* **88**, 468–471.

Rommel, M., and Heydorn, A. O. (1972). Beiträge zum Lebenszyklus der Sarkosporidien. III. *Isospora hominis* (Railliet und Lucet, 1891) Wenyon, 1923, eine Dauerform der Sarkosporidien des Rindes und des Schweins. *Berlin Muench. Tieraerztl. Wochenschr.* **85**, 143–145.

Rommel, M., and Krampitz, H. E. (1975). Beiträge zum Lebenszyklus der Frenkelien. I. Die Identität von *Isospora buteonis* aus dem Mäusebussard mit einer Frenkelienart (*F. clethrionomyobuteonis* spec. n.) aus der Rötelmaus. *Berlin Muench. Tieraerztl. Wochenschr.* **88**, 338–340.

Rommel, M., Heydorn, A. O., and Gruber, F. (1972). Beiträge zum Lebenszyklus der Sarkosporidien. I. Die Sporozyste von *S. tenella* in den Fäzes der Katze. *Berlin Muench. Tieraerztl. Wochenschr.* **85**, 101–105.

Rommel, M., Heydorn, A. O., Fischle, B., and Gestrich, R. (1974). Beiträge zum Lebenszyklus der Sarkosporidian. V. Weitere Endwirte der Sarkosporidien von Rind, Schaf und Schwein und die Bedeutung des Zwischenwirtes für die Verbreitung dieser Parasitose. *Berlin Muench. Tieraerztl. Wochenschr.* **87**, 392–396.

Rommel, M., Krampitz, H. E., Göbel, E., Geisel, O., and Kaiser, E. (1976). Untersuchungen Über den Lebenszyklus von *Frenkelia clethrionomyobuteonis*. *Z. Parasitenk.* **50**, 204–205.

Roth, J. A., Siegel, S. E., Levine, A. S., and Berard, C. W. (1971). Fatal recurrent toxoplasmosis in a patient initially infected *via* a leukocyte transfusion. *Am. J. Clin. Pathol.* **56**, 601–605.

Ruiz, A., and Frenkel, J. K. (1976). Recognition of cyclic transmission of *Sarcocystis muris* by cats. *J. Infect. Dis.* **133**, 409–418.

Ruiz, A., Frenkel, J. K., and Cerdas, L. (1973). Isolation of *Toxoplasma* from soil. *J. Parasitol.* **59**, 204–206.

Ruskin, J., and Remington, J. S. (1968a). Role for the macrophage in acquired immunity to phylogenetically unrelated intracellular organisms. *Antimicrob. Agents Chemother.* pp. 474–477.

Ruskin, J., and Remington, J. S. (1968b). Immunity and intracellular infection: resistance to bacteria in mice infected with a protozoan. *Science* **160**, 72–74.

Ruskin, J., and Remington, J. S. (1971). Resistance to intracellular infection in mice immunized with *Toxoplasma* vaccine and adjuvant. *Res. J. Reticuloendothel. Soc.* **9**, 465–479.

Ruskin, J., McIntosh, J., and Remington, J. S. (1969). Studies on the mechanisms of resistance to phylogenetically diverse intracellular organisms. *J. Immunol.* **103**, 252–259.

Sabin, A. B. (1941). Toxoplasmic encephalitis in children. *J. Am. Med. Assoc.* **116**, 801–807.

Sabin, A. B. (1942). Toxoplasmosis, a recently recognized disease of human beings. *Adv. Pediat.* **1**, 1–60.

Sabin, A. B., and Feldman, H. A. (1948). Dyes as microchemical indicators of a new immunity phenomenon affecting a protozoon parasite (Toxoplasma). *Science* **108**, 660–663.

Sanger, V. L., and Cole, C. R. (1955). Toxoplasmosis. VI. Isolation of Toxoplasma from milk, placentas, and newborn pigs of asymptomatic carrier sows. *Am. J. Vet. Res.* **16**, 536–539.

Sanger, V. L., Chamberlain, D. M., Chamberlain, K. W., Cole, C. R., and Farrell, R. L. (1953). Toxoplasmosis. V. Isolation of *Toxoplasma* from cattle. *J. Am. Vet. Med. Assoc.* **123**, 87–91.

Sasaki, Y., Iida, T., Oomura, K., Tsutsumi, Y., Tsunoda, K., Ito, S., and Nishikawa, H. (1974). Experimental Toxoplasma infection in pigs with oocysts of *Isospora bigemina* of feline origin. *Japan. J. Vet. Sci.* **36**, 459–465.

Schneider, A. (1881). Sur les psorospermies oviformes ou coccidies, espeĉes nouvelles ou peu connues. *Arch. Zool. Exp.* **9**, 387–404.

Schneider, C. R. (1967). *Besnoitia darlingi* (Brumpt, 1913) in Panama. *J. Protozool.* **14**, 78–82.

Scholtyseck, E. (1973). Ultrastructure. *In The Coccidia* (D. M. Hammond and P. L. Long, eds.), pp. 81–144. University Park Press, Baltimore, Maryland.

Scholtyseck, E., Mehlhorn, H., and Müller, E. G. (1973). Identifikation von Merozoiten der vier cystenbildenden Coccidien (*Sarcocystis, Toxoplasma, Besnoitia, Frenkelia*) auf Grund feinstruktureller Kriterien. *Z. Parasitenk.* **42**, 185–206.

Sénaud, J., Mehlhorn, H., and Scholtyseck, E. (1974). *Besnoitia jellisoni* in macrophages and cysts from experimentally infected laboratory mice. *J. Protozool.* **21**, 715–720.

Seneviratana, P., Edward, A. G., and DeGiusti, D. L. (1975). Frequency of *Sarcocystis* spp. in Detroit, metropolitan area, Michigan. *Am. J. Vet. Res.* **36**, 337–339.

Sethi, K. K., Pelster, B., Piekarski, G., and Brandis, H. (1973). Multiplication of *Toxoplasma gondii* in enucleated L cells. *Nature* (*London*), New Biol. **243**, 255–256.

Shah, H. L. (1971). The life cycle of *Isospora felis,* Wenyon, 1923, a coccidium of the cat. *J. Protozool.* **18**, 3–17.

Sheffield, H. G. (1966). Electron microscope study of the proliferative form of *Besnoitia jellisoni. J. Parasitol.* **52**, 583–594.

Sheffield, H. G. (1968). Observations on the fine structure of the "cyst stage" of *Besnoitia jellisoni. J. Protozool.* **15**, 685–693.

Sheffield, H. G. (1970). Schizogony in *Toxoplasma gondii:* An electron microscopic study. *Proc. Helminthol. Soc. Wash.* **37**, 237–242.

Sheffield, H. G., and Melton, M. L. (1968). The fine structure and reproduction of *Toxoplasma gondii*. *J. Parasitol.* **54,** 209–226.

Sheffield, H. G., and Melton, M. L. (1969). *Toxoplasma gondii:* Transmission through feces in absence of *Toxocara cati* eggs. *Science* **164,** 431–432.

Sheffield, H. G., and Melton, M. L. (1970). *Toxoplasma gondii:* The oocyst, sporozoite, and infection of cultured cells. *Science* **167,** 892–893.

Sheffield, H. G., and Melton, M. L. (1974). Immunity to *Toxoplasma gondii* in cats. *Proc. 3rd. Int. Congr. Parasitol. Muenich.,* Facta Publications **1,** 106–107.

Sheffield, H. G., and Melton, M. L. (1975). Effect of pyrimethamine and sulfadiazine on the fine structure and multiplication of *Toxoplasma gondii* in cell cultures. *J. Parasitol.* **61,** 704–712.

Sheffield, H. G., and Melton, M. L. (1976). Effect of pyrimethamine and sulfadiazine on the intestinal development of *Toxoplasma gondii* in cats. *Am. J. Trop. Med. Hyg.* **25,** 379–383.

Sheffield, H. G., Melton, M. L. and Neva, F. A. (1976). Development of *Hammondia hammondi* in cell cultures. *Proc. Helminthol. Soc. Wash.* **43,** 217–225.

Shimada, K., O'Connor, G., and Yoneda, C. (1974). Cyst formation by *Toxoplasma gondii* (RH strain) *in vitro.* The role of immunologic mechanisms. *Arch. Ophthalmol.* **92,** 496–500.

Shimizu, K., Shirahata, T., and Inami, Y. (1968). Therapeutic effect of SDDS (2-sulfamoyl-4,4′-diaminodiphenyl sulfone) on experimental toxoplasmosis. *Japan. J. Vet. Sci.* **30,** 183–195.

Siegel, S. E., Lunde, M. N., Gelderman, A. H., Halterman, R. A., Brown, J. A., Levine, A. S., and Graw, R. G., Jr. (1971). Transmission of toxoplasmosis by leukocyte transfusion. *Blood* **37,** 388–394.

Siim, J. C. (1952). Studies on acquired toxoplasmosis. II. Report of a case with pathological changes in a lymph node removed at biopsy. *Acta Pathol. Microbiol. Scand.* **30,** 104–108.

Siim, J. C., Biering-Sørensen, U., and Møller, T. (1963). Toxoplasmosis in domestic animals. *Adv. Vet. Sci.* **8,** 335–429.

Smart, M. E., Downey, R. S., and Stockdale, P. H. G. (1973). Toxoplasmosis in a cat associated with cholangitis and progressive pancreatitis. *Can. Vet. J.* **14,** 313–316.

Splendore, A. (1908). Un nuovo protoaoz parassita de conigli incontrato nelle lesioni anatomiche d'une malattiache ricorda in moltopunti il Kalaazar dell' uomo. Nota preliminaire pel. *Rev. Soc. Sci. São Paulo* **3,** 109–112.

Stiles, C. W. (1891). Note préliminaire sur quelques parasites. *Bull. Soc. Zool. Fr.* **16,** 163–165.

Stiles, C. W. (1892). Notes on parasites, No. II *J. Comp. Med. Vet. Arch.* **13,** 517–526.

Streitel, R. H., and Dubey, J. P. (1976). Prevalence of *Sarcocystis* infection and other intestinal parasitisms in dogs from a humane shelter in Ohio. *J. Am. Vet. Med. Assoc.* **168,** 423–424.

Strickland, G. T., Ahmed, A., and Sells, K. W. (1975). Blastogenic response of *Toxoplasma*-infected mouse spleen cells to T- and B-cell mitogens. *Clin. exp. Immunol.* **22,** 167–176.

Sulzer, A. J., Strobel, P. L., Springer, E. L., Roth, I. L., and Callaway, C. S. (1974). A comparative electron microscopic study of the morphology of *Toxoplasma gondii* by freeze-etch replication and thin sectioning technic. *J. Protozool.* **21,** 710–714.

Swartzberg, J. E., Krahenbuhl, J. L., and Remington, J. S. (1975). Dichotomy between macrophage activation and degree of protection against *Listeria monocytogenes* and *Toxoplasma gondii* in mice stimulated with *Corynebacterium parvum*. *Infect. Immunt.* **12**, 1037–1043.

Tabbara, K. F., Nozik, R. A., and O'Connor, G. R. (1974). Clindamycin effects on experimental ocular toxoplasmosis in the rabbit. *Arch. Ophthalmol.* **92**, 244–247.

Tabbara, K. F., O'Connor, G. R., and Nozik, R. A. (1975). Effect of immunization with attenuated *Mycobacterium bovis* on experimental toxoplasmic retinochorioditis. *Am. J. Ophthol.* **79**, 641–647.

Tadros, W., and Laarman, J. J. (1976). *Sarcocystis* and related coccidian parasites: a brief general review, together with a discussion on some biological aspects of their life cycles and a new proposal for their classification. *Acta Leidensia.* **44**, 1–107.

Tadros, W. A., Bird, R. G., and Ellis, D. S. (1972). The fine structure of cysts of *Frenkelia* (the M-organism). *Fol. Parasitol.* **19**, 203–209.

Trier, J. S., Moxey, P. C., Schimmel, E. M., and Robles, E. (1974). Chronic intestinal coccidiosis in man: intestinal morphology and response to treatment. *Gastroenterology* **66**, 923–935.

Uhlíková, M., and Hübner, J. (1973). Congenital transmission of toxoplasmosis in domestic rabbits. *Fol. Parasitol.* **20**, 285–291.

Vietzke, W. M., Gelderman, A. H., Grimley, P. M., and Valsamis, M. P. (1968). Toxoplasmosis complicating malignancy. Experience at National Cancer Institute. *Cancer.* **31**, 816–827.

Virshinin, I. I. (1975). Sarkotsisty krupnogo rogatogo skota. *Dokl. Veesoiuz. Akad. Sel'skoph. Nauk, Selskokh. Inst. Sverdlovsk* **1**, 30–32 (cited by Levine, 1977a).

Vivier, E. (1970). Observations nouvelle sur la reproduction asexueé de *Toxoplasma gondii* et considérations sur la notion d'endogènese. *C. R. Hebd. Seances Acad. Sci.* **271D**, 2123–2126.

Vivier, E., and Petitprez, A. (1969). Le complexe membranaire superficiel et son evolution lors de l'elaboration des individus-fils chez *Toxoplasma gondii*. *J. Cell. Biol.* **43**, 329–342.

Vogelsang, E. G. (1938). Contribucion al estudio de la parasitologia animal en Venezuela. VIII *Sarcocystis iturbei* sp. n. del bovino (*Bos taurus* L.). *Bol. Soc. Venez. Cienc. Natur.* **4**, 279–280.

Wallace, G. D. (1969). Serologic and epidemiologic observations on toxoplasmosis on three pacific atolls. *Am. J. Epidemiol.* **90**, 103–111.

Wallace, G. D. (1971a). Isolation of *Toxoplasma gondii* from the feces of naturally infected cats. *J. Infect. Dis.* **124**, 227–228.

Wallace, G. D. (1971b). Experimental transmission of *Toxoplasma gondii* by filthflies. *Am. J. Trop. Med. Hyg.* **20**, 411–413.

Wallace, G. D. (1972). Experimental transmission of *Toxoplasma gondii* by cockroaches. *J. Infect. Dis.* **126**, 545–547.

Wallace, G. D. (1973a). Sarcocystis in mice inoculated with Toxoplasma-like oocysts from cat feces. *Science* **180**, 1375–1377.

Wallace, G. D. (1973b). The role of the cat in the natural history of *Toxoplasma gondii*. *Am. J. Trop. Med. Hyg.* **22**, 313–322.

Wallace, G. D. (1973c). Intermediate and transport hosts in the natural history of *Toxoplasma gondii*. *Am. J. Trop. Med. Hyg.* **22**, 456–464.

Wallace, G. D. (1975). Observations on a feline coccidium with some characteristics of *Toxoplasma* and *Sarcocystis*. *Z. Parasitenk.* **46**, 167–178.

Wallace, G. D., and Frenkel, J. K. (1975). *Besnoitia* species (protozoa, sporozoa, toxoplasmatidae): recognition of cyclic transmission by cats. *Science* **188**, 369–371.

Wallace, G. D., Marshall, L., and Marshall, M. (1972). Cats, rats, and toxoplasmosis on a small Pacific Island. *Am. J. Epidemiol.* **95**, 475–482.

Wallace, G. D., Zigas, V., and Gajdusek, D. C. (1974). Toxoplasmosis and cats in New Guinea. *Am. J. Trop. Med. Hyg.* **23**, 8–14.

Walton, B. D., de Arjona, I., and Benchoff, B. M. (1966). Relationship of Toxoplasma antibodies to altitude. *Am. J. Trop. Med. Hyg.* **15**, 492–495.

Wanko, T., Jacobs, L., and Gavin, M. A. (1962). Electron microscope study of *Toxoplasma* cysts in mouse brain. *J. Protozool.* **9**, 235–242.

Weiland, G., and Kühn, D. (1970). Experimentelle Toxoplasma-infektionen bei der Katze. II. Entwicklungstadien des Parasiten im Darm. *Berlin Muench. Tieraerztl. Wochenschr.* **83**, 128–132.

Weinman, D., and Chandler, A. H. (1954). Toxoplasmosis in swine and rodents. Reciprocal oral infection and potential human hazard. *Proc. Soc. Exp. Biol. Med.* **87**, 211–216.

Wenyon, C. M. (1923). Coccidiosis of cats and dogs and the status of the *Isospora* of man. *Ann. Trop. Med. Parasitol.* **17**, 231–288.

Wenyon, C. M. (1926a). Coccidia of the genus *Isospora* in cats, dogs and man. *Parasitology* **18**, 253–266.

Wenyon, C. M. (1926b). *Protozoology*. Vol. I. pp. 778. William Wood & Co., New York.

Wenyon, C. M., and Sheather, L. (1925). Exhibition of specimens illustrating *Isospora* infections of dogs. *Trans. Roy. Soc. Trop. Med. Hyg.* **19**, 10.

Wilder, H. C. (1952). *Toxoplasma* chorioretinitis in adults. *Arch. Ophthalmol.* **48**, 127–137.

Witte, H. M., and Piekarski, G. (1970). Die oocysten-Ausscheidung bei experimentel infizierten Katzen in Abhängigkeit vom *Toxoplasma*-Stamm. *Z. Parasitenk.* **33**, 358–360.

Wobster, G. (1976). Besnoitiosis in a woodland caribou. *J. Wild. Dis.* **12**, 566–571.

Woke, P. A., Jacobs, L., Jones, F. E., and Melton, M. L. (1953). Experimental results on possible arthropod transmission of toxoplasmosis. *J. Parasitol.* **39**, 523–532.

Wolf, A., and Cowen, D. (1937). Granulomatous encephalomyelitis due to an encephalitozoon (encephalitozic encephalomyelitis). A new protozoan disease of man. *Bull. Neurol. Inst. New York* **6**, 306–335.

Work, K. (1967). Isolation of *Toxoplasma gondii* from the flesh of sheep, swine and cattle. *Acta Pathol. Microbiol. Scand.* **71**, 296–306.

Work, K. (1968). Resistance of *Toxoplasma gondii* encysted in pork. *Acta Pathol. Microbiol. Scand.* **73**, 85–92.

Work, K., Eriksen, L., Fennestad, K. L., Møller, T., and Siim, J. C. (1970). Experimental toxoplasmosis in pregnant sows. I. Clinical, parasitological and serologic observations. *Acta Pathol. Microbiol. Scand.* **B78**, 129–139.

Wright, I. (1972). Transmission of *Toxoplasma gondii* across the guinea-pig placenta. *Lab. Anim.* **6**, 169–180.

Yilmaz, S. M., and Hopkins, S. H. (1972). Effects of different conditions on duration of infectivity of *Toxoplasma gondii* oocysts. *J. Parasitol.* **58**, 938–939.

Zaman, V., and Colley, F. C. (1970). Observations on the endogenous stages of *Toxoplasma gondii* in the cat ileum. I. Light microscopic study. *Southeast Asian J. Trop. Med. Publ. Health* **1**, 457–464.

5

On Species of *Leucocytozoon*, *Haemoproteus*, and *Hepatocystis*

A. Murray Fallis and Sherwin S. Desser

I. Introduction

Many species of *Leucocytozoon* and *Haemoproteus* have been described since Danilewsky (1884) saw these parasites in the blood of birds almost 100 years ago. The first of these parasites to be described were species of *Leucocytozoon* in owls and of *Haemoproteus* in pigeons and crows. Only gametocytes of these parasites are seen in the peripheral blood. The male, or microgametocyte, is distinguishable from the female, or macrogametocyte, by its larger and more diffuse nucleus. Although gametocytes of both parasites occur in erythrocytes, those of *Leucocytozoon* also develop in white blood cells. There are differences in metabolism of the two types of parasites as pigment is produced by the erythrocytic stages of species of *Haemoproteus* but not by those of *Leucocytozoon.* The life cycles of these two groups of parasites and the related species of *Plasmodium* possess many common features.

Extensive studies would be required to determine the validity of the more than seventy named species of *Leucocytozoon* and over eighty of *Haemoproteus.* Surveys indicate that the incidence of infection differs in different kinds of birds and in different localities (Levine, 1972; Fallis *et al.,* 1974). Because many birds migrate, the occurrence of parasites in birds in a particular locality is not necessarily indicative of local transmission. Incidence and levels of parasitemia may indicate, among other things, the susceptibility of the bird, the availability of suitable vectors of the parasite, preference of vectors for certain birds, and the relative abundance of different kinds of birds and vectors.

The majority of mammalian haemoproteids are included in the genus *Hepatocystis. Hepatocystis* parasites like plasmodia of mammals, undergo preerythrocytic development in the liver and form gametocytes in erythrocytes. Schizogony in species of *Hepatocystis* like schizogony in species of *Leucocytozoon* and *Haemoproteus,* however, is restricted to the tissues.

II. *Leucocytozoon* *

A. Morphology and Life Cycles

The morphology of the gametocytes as seen in peripheral blood and the appearance of the cells in which they live are the criteria most commonly used for distinguishing species of *Leucocytozoon,* although names are often assigned simply because of the occurrence of a parasite in a bird in which it had not been reported previously. Most species of *Leucocytozoon* produce gametocytes of one of two types: (1) those which appear round in round cells whose nucleus has been pushed aside to form a narrow or broad band, or cap, and (2) those which appear round, oval, or elliptical in a cell which becomes elongated as the parasite grows. The host cell nucleus in these elongated cells is pushed aside and may undergo little distortion, or may be altered to a pointed ellipse, or into a narrow band that extends the length of the gametocyte, or, in some species, well beyond (Figs. 1–9). In some species, such as *L. smithi* of turkeys, the flattened, elongate nucleus of the host cell is separated into two parts by the gametocyte. A type of parasitized cell

* Much of the information in this section is taken from a review article "On Species of *Leucocytozoon*" by Fallis *et al.* (1974).

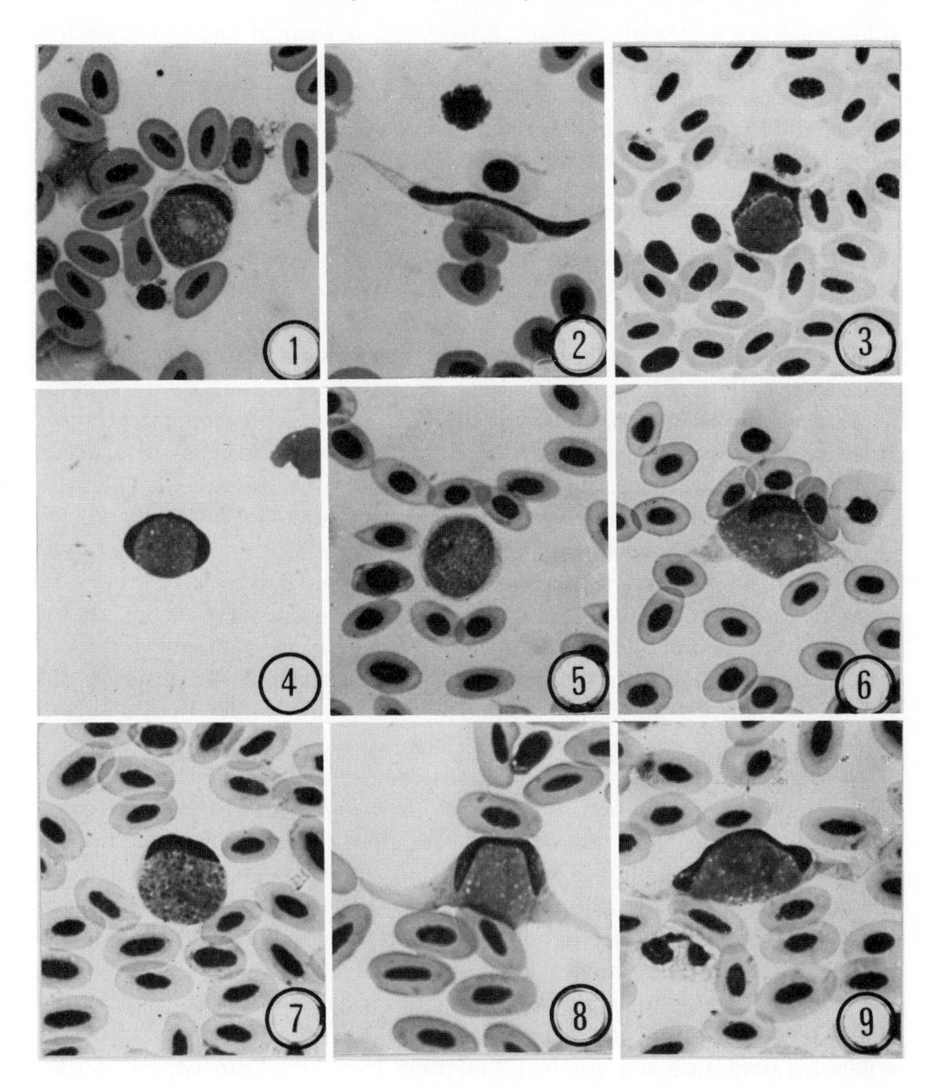

Figs. 1–9. Methanol fixed, Giemsa-stained gametocytes of species of *Leucocytozoon.* × 800. **Fig. 1.** Round gametocyte of *L. simondi* from a domestic duck. **Fig. 2.** Elongate gametocyte of *L. simondi.* **Fig. 3.** *Leucocytozoon fringillinarum* from purple grackle (*Quiscalus quiscula*). **Fig. 4.** *Leucocytozoon dubreuli* from robin (*Turdus migratorius*). **Fig. 5.** *Leucocytozoon caulleryi* from domestic chicken. **Fig. 6.** *Leucocytozoon lovati* from grouse (*Bonasa umbellus*). **Fig. 7–9.** Three morphological types of gametocytes all characteristic of *L. danilewskyi* from owl (*Athena noctua*). Courtesy of Dr. A. Corradetti.

differing from either of the two above types is seen in chickens infected with *L. caulleryi*. Gametocytes of this species are round. The nucleus of the cell is pushed aside with little change in its shape and is eventually extruded (Fig. 5). Similar loss of the nuclei of cells harboring gametocytes has been noted, although rarely, in blood of animals infected with other species of *Leucocytozoon*.

Gametocytes of some species, e.g., *L. fringillinarum*, are exclusively round (Fig. 3). Gametocytes of other species are round in round cells at some point in their life cycles and elongate in elongated cells at other times, e.g., *L. simondi, L. smithi, L. danilewskyi*, and *L. lovati* (Figs. 1, 2, and 6–9). Limited evidence indicates that species with gametocytes in elongated cells have, earlier in their asexual cycles, round gametocytes in round cells. Round gametocytes develop in erythroblasts and erythrocytes, although possibly round gametocytes of some species also occur in lymphocytic cells. Elongation of the cell develops as the parasite grows, not after the parasite has attained its full size.

A species of *Leucocytozoon* in owls has round gametocytes in round cells in early patency. Later, most of the gametocytes are found in elongated cells and may be round, broadly, or narrowly oval, or almost elliptical (Figs. 7–9). The variation in the shape of the gametocytes could lead to the belief that they belong to different species if the observer did not know that all the forms resulted from a single rather than a mixed infection.

Merozoites which develop into gametocytes are 1–2 μm in diameter when they enter cells and, because of their small size, are easily overlooked. Soon after their entry they locate close to the nucleus of the cell which becomes indented. The indentation increases and the nucleus of the cell is pushed aside as the parasite grows. The degree of distortion of the invaded cell probably depends on its type as well as on the genetic characteristics of the gametocyte.

The gametocytes mature and fertilization occurs in the midgut of the simuliid vector after a blood meal. The resulting motile zygotes or ookinetes penetrate the gut wall and then transform into oocysts (Figs. 10–13). The process of sporogony, whereby sporozoites are produced, occurs in the oocysts in the wall of the midgut. In most species examined, sporogony requires approximately 1 week at 20°C. Sporogony of *L. caulleryi* and possibly other species occurs in biting midges, *Culicoides* spp. When the insect feeds on a suitable avian host the sporozoites which enter the wound find and penetrate appropriate cells in which they develop into asexual stages called schizonts. Schizogony culminates in the production of thousands of tiny, uninucleate merozoites. These merozoites enter blood cells and form the sexual stages, the gametocytes, which await ingestion by another suitable vector.

Schizogony occurs in hepatic parenchymal cells, renal epithelial cells, and in reticuloendothelial cells throughout the body, especially in the spleen and lymph glands (Figs. 14–17). Noticeable differences in schizogony occur among species. Primary schizogony occurs in parenchymal cells of the liver and in several species is completed in less than 2 weeks. It is not known if a similar primary schizogony occurs in *L. caulleryi*. The life cycle of *L. simondi* is understood more fully than that of other leucocytozoa. Consequently, unless otherwise indicated, the following descriptions will refer to this species.

Sporogony of *L. simondi* occurs in *Simulium rugglesi, S. anatinum, S. innocens,* and *Eusimulium rendalense* and *E. fallisi* from Norway. Other species of *Simulium* will undoubtedly be found capable of supporting sporogony although their feeding behavior and their ecology may be such that they are unlikely to be important vectors of the parasite. For example, *Simulium venustum* is a suitable host for development of *L. simondi,* although this fly feeds normally on mammals rather than on birds. Similarly, although *Simulium adersi* is a natural vector of *L. schoutedeni, Simulium vorax* and *S. nyasalandicum* will also support sporogony but these flies feed on cattle and are unlikely vectors of this parasite of chickens. Vectors of *L. simondi* are usually found close to the ground at the waters' edge and prefer to feed on anseriform birds. Flies that transmit *L. fringillinarum, L. schoutedeni,* and *L. dubreuili,* although taken more often in some habitats than others, are less specific in their feeding behavior than those which transmit *L. simondi.*

In the midgut of the vector the rapid formation of micro- and macrogametes from the gametocytes is stimulated by a change in the O_2 and CO_2 tension rather than by the change of temperature that occurs when the gametocytes are transferred from the warm blood of the bird to the lower environmental temperature of the fly. The maturation of the gametocytes begins soon after removal of blood from the vertebrate host. Maturation is initiated by the appearance of an opening in the membrane of the host cell through which the parasite escapes. The nucleus of the host cell adheres temporarily to the parasite. The nuclear chromatin in the microgametocyte, which is at first arranged in a spireme, condenses into a solid mass. The mass of chromatin then divides into eight separate parts which move to the peripheral cytoplasm. Long filamentous projections of the plasma membrane are formed rapidly and into each moves an axoneme and one of the eight nuclear masses (Fig. 10). The gametes during this process of exflagellation break free from a residual

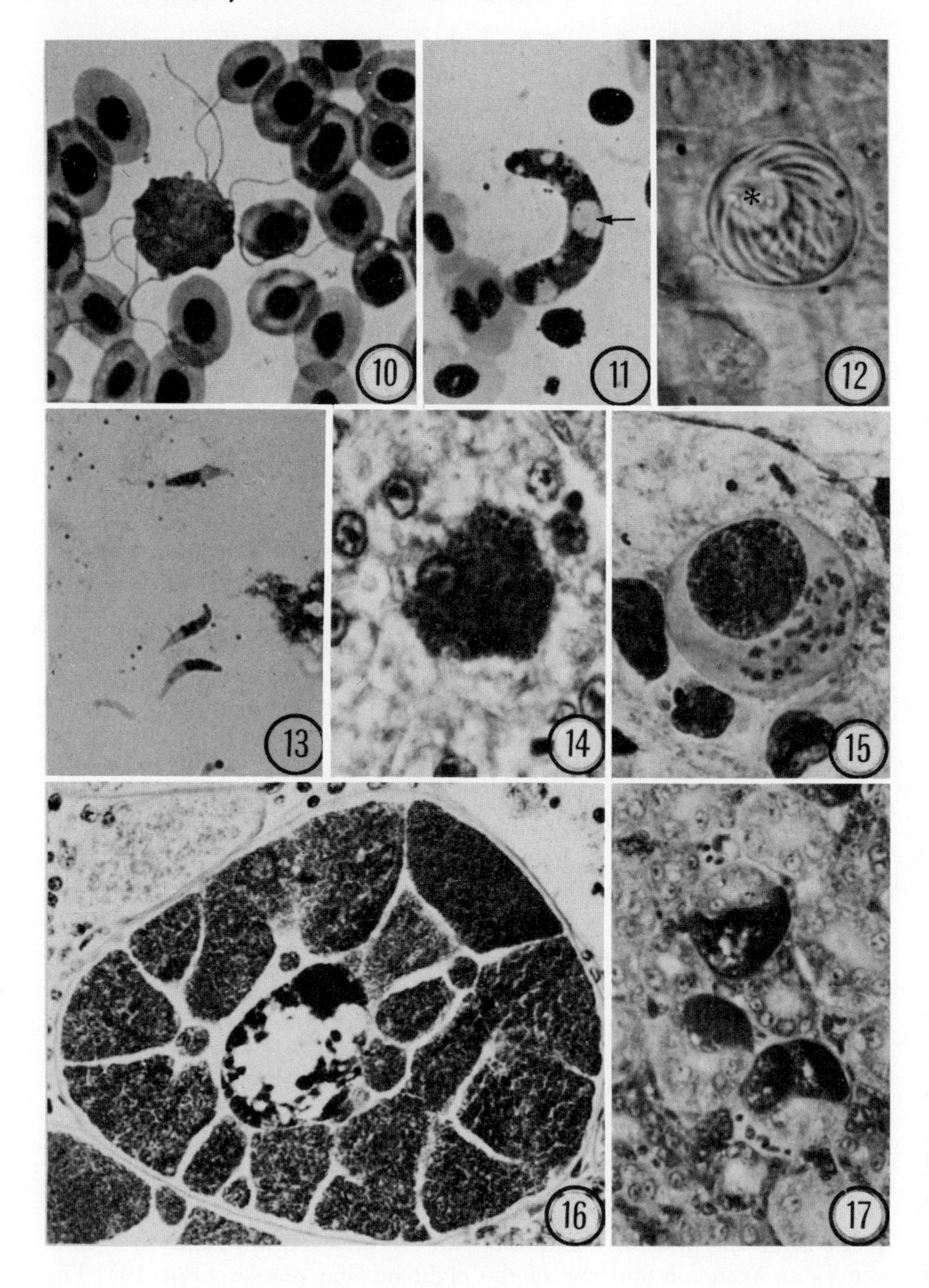
*
10
11
12
13
14
15
16
17

mass of the gametocyte and move rapidly away either to contact and fertilize a macrogamete or to die.

The zygote is formed after fertilization of the female gamete and within a few hours it transforms into an elongate, motile ookinete. The ookinete is about 30 μm in length and 4 μm in maximum breadth, although its size and shape can vary considerably. One end is more pointed than the other. The nucleus lies slightly nearer the blunt end. Several clear areas are noticeable in the cytoplasm of the ookinete (Fig. 11). The clear areas resemble vacuoles and are actually lipoprotein inclusions (see Fig. 22). The slow bending movements and gregarine-like forward motion result presumably from folds that form on the outer surface and move as a wave from one end of the ookinete to the other in the manner described by Desser and Weller (1973) for a haemogregarine. The ookinete has a heavily reinforced "apical cap" region with a specialized pellicular structure which can be demonstrated by electron microscopy.

Some ookinetes penetrate the gut wall of the fly soon after their formation. Penetration by others may be retarded or prevented by the peritrophic membrane that forms around the ingested blood. Those ookinetes that do not succeed in moving into the wall of the gut before the membrane forms may have to wait until it ruptures 3 or 4 days later. The ookinetes penetrate the midgut epithelium and lodge beneath the basal lamina, where they round up and become enclosed by a clear cyst wall. Oocysts do not always protrude into the hemocoel. They may even cause an indentation in the wall of the gut. For this reason and because of their small size, oocysts of leucocytozoa are seen less readily in dissected simuliids than are oocysts of plasmodia in dissected mosquitoes.

Viewed in a wet preparation in saline, early oocysts have a clear, somewhat granular appearance and a more or less central translucent core of lipoprotein. Oocysts of *L. simondi* are about 10 to 12 μm in diameter. Those of other species, such as *L. dubreuili,* may be up to 25 μm in diameter.

The process of sporozoite formation in oocysts of *L. dubreuili* is clearly revealed by electron microscopy (Fig. 18, A-G) (for details see

---

Fig. 10. Exflaggelation of microgametocyte of *L. simondi.* Note eight microgametes. × 1200. **Fig. 11.** Ookinete of *L. simondi* with large crystalloid inclusions (arrow). × 1200. **Fig. 12.** Maturing oocyst of *L. dubreuili* with sporozoites attached to residual body (*). × 1000. **Fig. 13.** Sporozoites of *L. simondi.* × 1200. **Fig. 14.** Mature hepatic schizont of *L. simondi.* × 1200. **Fig. 15.** Young megaloschizont in phagocyte in brain impression smear of duckling infected experimentally 7 days previously. × 1200. **Fig. 16.** Maturing megaloschizont in section of spleen of duckling infected experimentally 9 days previously. × 350. **Fig. 17.** Renal schizonts of *L. fringillinarum.* × 350.

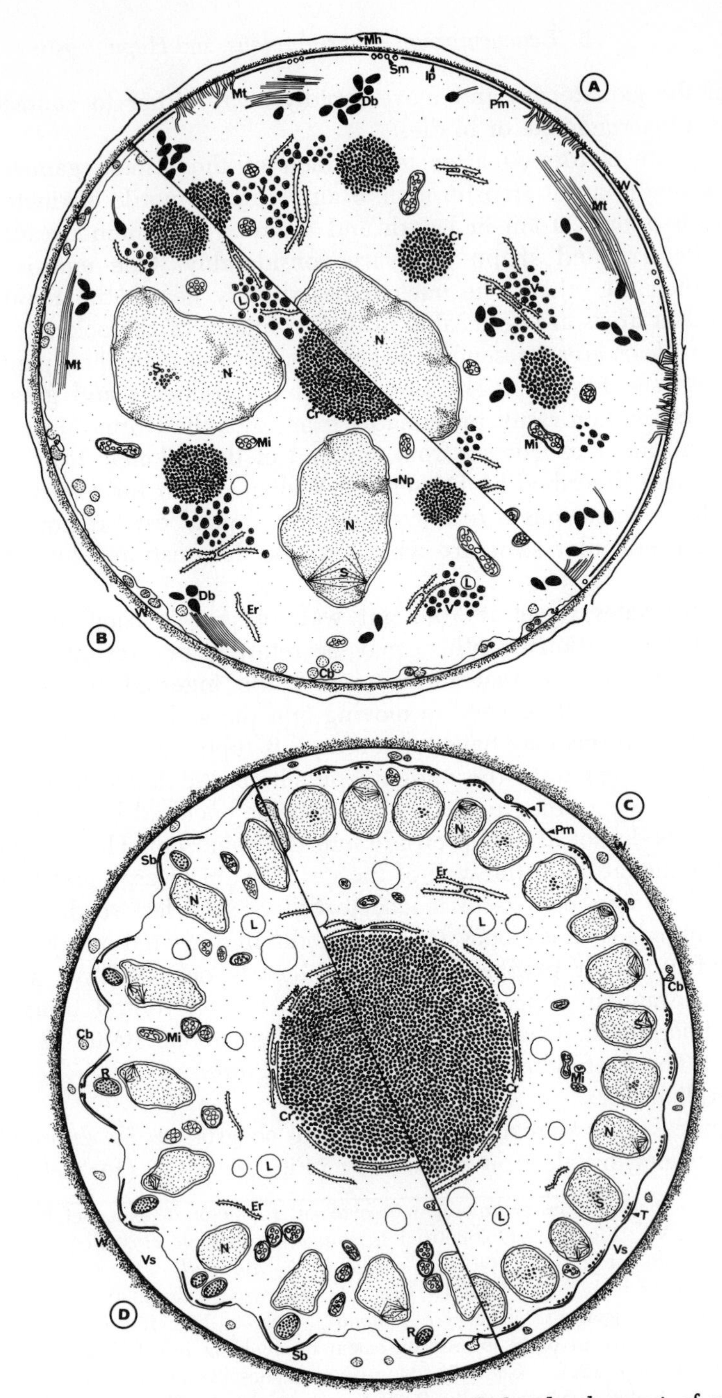

Fig. 18(A–G). Diagrammatic representation of the development of sporozoites within oocysts of *L. dubreuili* as seen in thin sections. Ac, apical complex; Bl, basal lamina of host cell; Cb, membrane bounded bodies; Cr, crystalloid; Cy, cytosome; Db, electron dense bodies; Er, granular endoplasmic reticulum; G, Golgi apparatus; Ip, inner interrupted membrane layer; L, lipid; Mh, host cell membrane; Mi, mito-

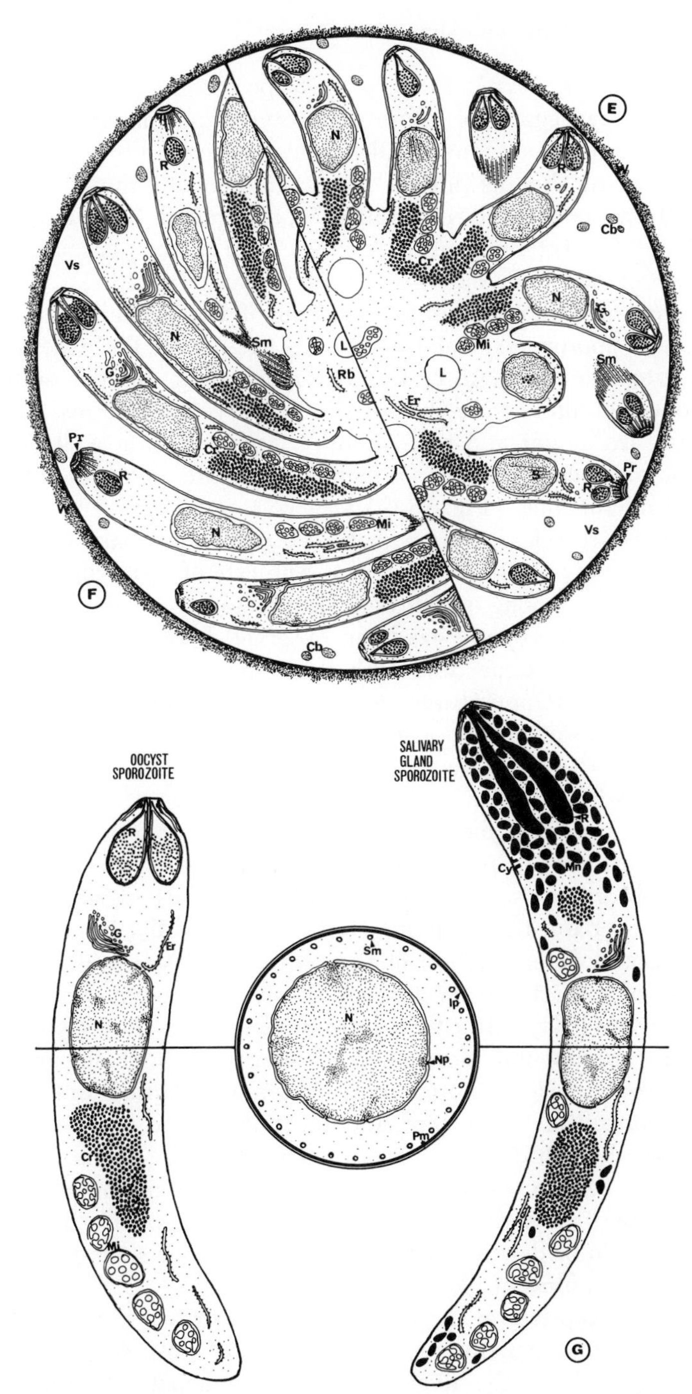

chondria; Mn, microneme; N, nucleus; Ne, nuclear envelope; Np, nuclear pore; Pm, outer membrane of pellicle; Pr, polar rings of sporozoite; R, rhoptry; Rb, residual body of oocyst; S, nuclear spindles; Sb, sporozoite bud; Sm, subpellicular microtubule; T, membrane thickening; Vs, subcapsular space; W, oocyst wall. Reproduced with permission from *J. Protozool.*

Wong and Desser, 1976). Early oocysts possess large dividing nuclei and scattered aggregates of lipoprotein, crystalloid material (Fig. 18-A, B). Subsequently, numerous smaller nuclei become aligned around the circumference of the peripheral cytoplasm beneath the plasma membrane and the crystalloid material is concentrated in a large central core (Fig. 18-C). Intermittent thickenings form on the plasma membrane and the sporozoite buds emerge in these areas (Fig. 18-D). A nucleus, several small mitochondria and a portion of the crystalloid move into each forming sporozoite, which continues to grow at the expense of the residual cytoplasm (Fig. 18-E, F). The morphology of sporozoites from mature oocysts differs from that of the salivary gland forms. The latter sporozoites are longer and more slender, and possess a well developed apical complex, consisting of a pair of elongate dense rhoptries and associated micronemes (Fig. 18-G).

Sporozoites are released from mature oocysts and make their way to the salivary glands of the fly (Fig. 13). The rate of sporogony may differ within, as well as among, species and is affected by temperature as well as other factors. Sporogony has been observed between 7 to 18 days in flies infected at the same time with the same species of parasite. Low rather than high temperatures may be optimal for some varieties of parasites as it was noted that a strain of *L. simondi* in Norway developed as rapidly at 15°C in its vector as did another strain at 20°C in other species of flies in Canada.

Schizogony occurs in the tissues of the avian host following introduction of sporozoites by the insect vector. Knowledge of schizogony is based on observations on tissues of infected birds of unknown prior history and on tissues of birds infected experimentally. Interpretation of the process and determination of the sequence of stages is difficult, for sporozoites of at least two species may survive with no apparent development for up to 11 days after their introduction into a host. Consequently, even after experimental inoculation of sporozoites the age of any observed asexual schizont is uncertain. Minimum elapsed time from the entry of sporozoites into a bird and the appearance of gametocytes in the blood is about 6 days for birds infected with *L. simondi* and somewhat less for certain other species. The prepatent period may be considerably longer in some individuals than in others. Certainly it will appear longer in birds with relatively light infections since parasites may be overlooked in films of blood.

With the exception of *L. caulleryi,* and further studies of it are desirable, the first generation of schizonts of most species occurs in parenchymal cells of the liver. Thus, in the hepatic localization, the exoerythrocytic development of species of *Leucocytozoon* resembles that

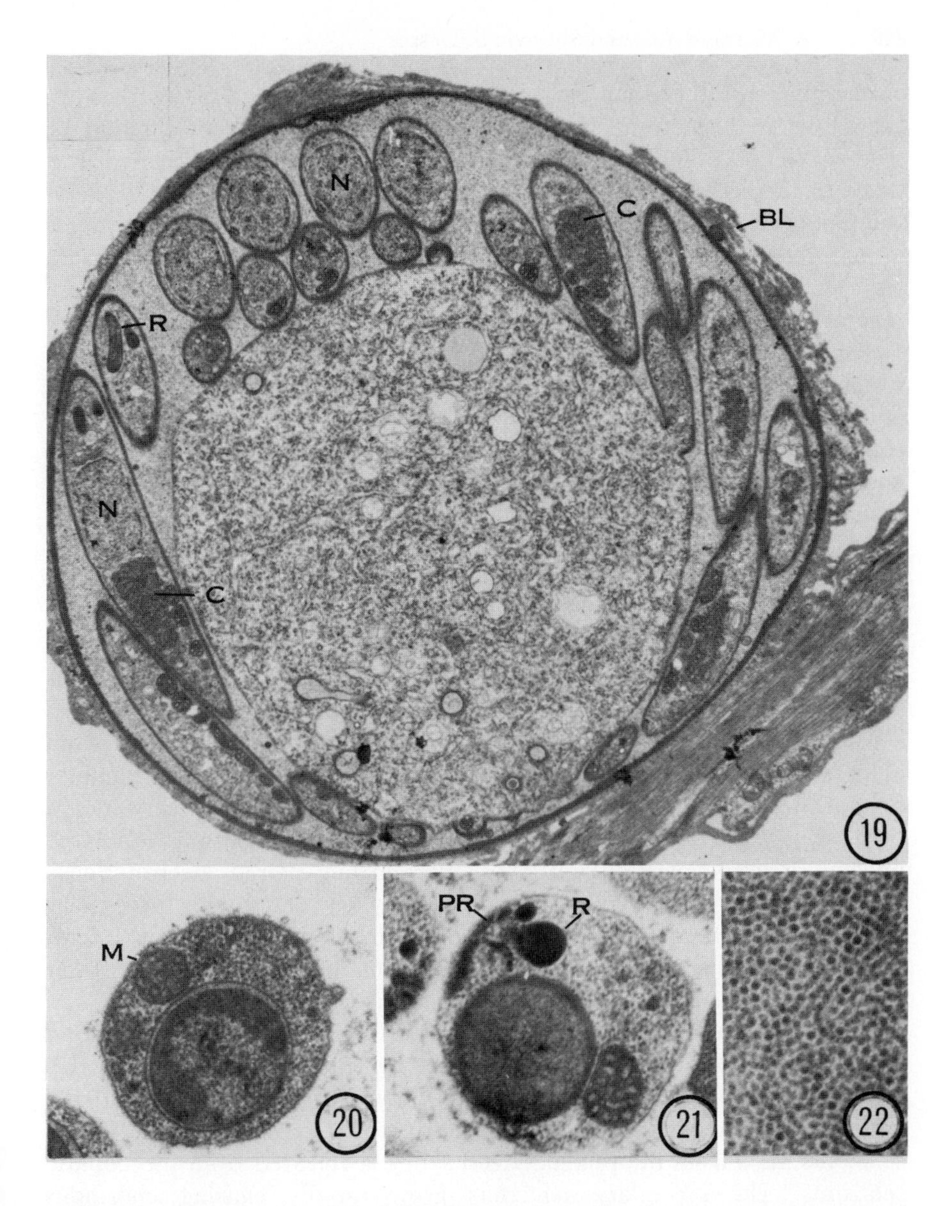

Fig. 19. Young oocyst of *L. dubreuili* with sporozoites forming around peripheral cytoplasm. Note sporozoite nuclei (N), crystalloid inclusions (C), and rhoptries (R). BL, basal lamina of midgut. × 8200. Courtesy of Mr. S. Wong. **Figs. 20 and 21.** Two types of merozoites observed in megaloschizonts of *L. simondi*. Each is bounded by a trilaminar plasma membrane and contains a large central nucleus, a single mitochondrion (M), electron-dense rhoptries (R), and dense polar rings (Pr). × 28,500. Reproduced with permission from *Can. J. Zool.* **Fig. 22.** Portion of crystalloid inclusion in ookinete of *L. simondi*. × 43,700. Reproduced with permission from *Can. J. Zool.*

of mammalian rather than avian species of *Plasmodium*. Relatively rapid growth of at least some hepatic schizonts leads to the production of merozoites in 4 to 5 days after infection with *L. simondi* and in even less time after infection with *L. fringillinarum* and *L. dubreuili* (Fig. 19). Development is not completely synchronous and schizonts in different stages of maturation can be found in the same liver. These hepatic schizonts are variable in size ranging from 20 to 40 μm and, when mature, release several thousand merozoites. Development of the schizont to produce merozoites begins with repeated nuclear division within the cytoplasm which becomes highly invaginated and eventually divides into multinucleate cytomeres. Nuclear division and the process of invagination continues in the cytomeres. Ultimately, small uninucleate merozoites about 1 μm in diameter are produced. In *L. simondi*, and possibly in other species, hepatic schizonts may rupture before the merozoites are completely differentiated with the result that small fragments of the cytomeres with two or more nuclei, as well as merozoites, are released from the schizont. These fragments have been referred to as syncytia. It is not known if these syncytia are abnormal forms arising when hepatic schizonts rupture prematurely in well-adapted parasites or whether their presence indicates a lack of adaptation by the parasite for its host.

Merozoites escaping from hepatic schizonts flood into the blood stream, enter erythrocytes and erythroblasts, and develop into round gametocytes, some of which mature in 48 hours. The merozoites are surrounded by a trilaminar plasma membrane. They possess, in addition to a large nucleus, a mitochondrion with vesicular cristae, paired electron-dense rhoptries, and smaller dense inclusions closely associated with three polar rings (Figs. 20 and 21). More macrogametocytes develop than microgametocytes. The gametocytes may circulate for weeks although their viability decreases with time and they disappear gradually.

Some merozoites from hepatic schizonts probably initiate a second asexual generation in hepatic parenchymal cells, although no proof of this is available. Second-generation schizonts, however, occur throughout the body within phagocytic cells, particularly vascular endothelium. These schizonts arise from the phagocytized syncytia released from the hepatic schizonts. The secondary schizonts grow rapidly causing enormous hypertrophy of the parasitized cells whose nuclei are hypertrophied and altered almost beyond recognition. These large schizonts, 100–200 μm in diameter, called megaloschizonts by Huff (1942), produce a million or more merozoites each 1 μm in diameter in about 5 days (Fig. 16). Electron microscopy reveals that there are two types of merozoites, one of which is more electron dense. It is believed the darker merozoites develop into macrogametocytes and the lighter into microgametocytes

(Figs. 20 and 21). *Leucocytozoon simondi* may develop asynchronously in a given cell if several syncytia enter the same cell at different times. Probably megaloschizonts of *L. caulleryi* arise in a similar fashion to those of *L. simondi,* as Pan (1963) reported fragments of cytomeres in the peripheral circulation of chickens infected with *L. caulleryi.* Merozoites from megaloschizonts of *L. simondi* enter white blood cells and grow into oval gametocytes, or possibly enter liver cells and initiate another cycle of hepatic schizogony. At present, two basic types of life cycle are known for species of *Leucocytozoon* (Scheme 1).

In contrast to *L. simondi,* megaloschizonts do not occur in the second asexual generations of *L. fringillinarum, L. berestneffi,* and *L. dubreuili* where schizogony occurs in liver and kidney cells whose nuclei may become enlarged. Asexual schizogony of *L. simondi* may continue for long periods as infections in ducks have been followed for more than 2 years. *Leucocytozoon dubreuili* has been seen in a robin 5 years after it was infected. This would seem possible only if schizogony had occurred repeatedly during this time. Parasitemia during periods of chronicity is low and, at times, gametocytes are too few to be detected in the peripheral blood. Irregular resurgences of parasitemia are believed to be associated with fluctuations of the host's resistance. Release of merozoites from megaloschizonts is believed to initiate the spring relapse reported for ducks with chronic infections of *L. simondi.* Probably there is also a relationship between occurrence of megaloschizonts and pathogenicity,

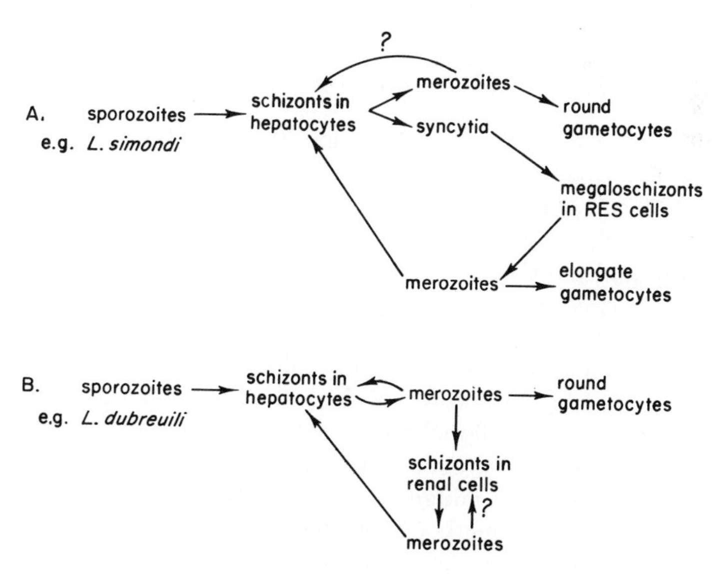

Scheme 1. Basic types of life cycles of species of *Leucocytozoon.*

since particularly pathogenic species such as *L. simondi, L. caulleryi,* and *L. sakharoffi* all have megaloschizonts.

B. Pathology

Anemia is the most marked pathological feature associated with infection. Anemia appears in ducks about 1 week after infection and is more severe in domestic than in wild species. Destruction of parasitized cells alone would not account for the severity of the anemia. Kocan (1968) showed that intravascular hemolysis of nonparasitized erythrocytes occurs. Anemia is reported also in chickens infected with *L. caulleryi.* Severely ill birds are listless, lose their appetite, and may pass diarrheic green feces. Inflammatory and necrotic foci occur in the liver. Mortality is associated commonly, although not exclusively, with infections of young birds. Pathogenicity of several species is in doubt. Khan and Fallis (1970) demonstrated no serious pathology in robins and fringillids with high parasitemias. Impressions of pathogenicity in birds infected with species of *Leucocytozoon* may be the result, in some instances, of concurrent infections with other organisms.

Antimalarial drugs have been used with only limited success to prevent and cure infections by *L. simondi.* Japanese workers reported successful treatment of infections caused by *L. caulleryi* by administration of sulfamethoxine and pyrimethamine alone or in combination (Akiba, 1970).

III. *Haemoproteus* and *Parahaemoproteus*

Species of *Haemoproteus* occur predominantly in birds although some have been described from lizards and turtles. Many species are morphologically indistinguishable and identification is based extensively on the host in which the parasite occurs. Not until we acquire knowledge of life histories and the asexual stages of the parasites will it be possible to determine if many of the named species are valid. Bennett *et al.* (1965) proposed the creation of the genus, *Parahaemoproteus,* for species of *Haemoproteus* transmitted by biting midges. Only those species of *Haemoproteus* of pigeons, doves, and quail which undergo sexual development, sporogony, in hippoboscids remain in the old genus in this system. Differences in sporogony exist between members of the Haemoproteidae transmitted by midges and those transmitted by hippoboscids, although some authorities consider that these differences only justify division at the subgeneric level. DeGuisti *et al.* (1973) challenged the validity of separation into two genera on the basis of vector because *H.*

metchnikovi of turtles develops in a third type of insect, namely, *Chrysops* sp.

A. Morphology and Life Cycles

Only the gametocytes of Haemoproteidae are seen in peripheral blood. When mature, gametocytes of some species may completely, and of others partially, encircle the nucleus of the red blood cell (Figs. 23–28). The gametocytes may assume a spherical shape soon after blood is withdrawn from an infected animal with the result that round gametocytes are occasionally seen in blood films. Bennett *et al.* (1973) reported that the gametocyte of *H. enucleator* from a kingfisher, fills the erythrocyte. The host cell nucleus is absent from the infected cells as it is from erythrocytes which contain the gametocytes of *L. caulleryi.*

The sexual cycle of species of the Haemoproteidae begins in the vector's stomach after the insect ingests blood which contains macro- and microgametocytes (Fig. 32). The first step in the formation of the gametes of members of the Haemoproteidae is similar to the first step in formation of gametes of *Leucocytozoon* (Fig. 34). Exflagellation of the microgametocyte of *H. columbae* of pigeons in the hippoboscid, *Lynchia maura,* was described in 1906 by Sergent and Sergent. They also noted the transformation of the fertilized macrogamete to the zygote and ookinete in the fly, but it remained for Adie (1915, 1924) to trace development beyond this point. Adie demonstrated oocyst formation beneath the basement membrane in the midgut of the fly. The ookinetes were about 25 μm in length and when stained showed an eccentric nucleus and clear areas, which contained lipoprotein (crystalloid). Ookinetes of most species of *Haemoproteus* and *Parahaemoproteus* are generally similar in appearance (Fig. 30). As in *Leucocytozoon,* the apical end of ookinetes of *Parahaemoproteus* is reinforced (Fig. 33). Oocysts of *H. columbae* may be up to 36 μm in diameter, and contain pigment carried over from the fertilized gamete. Sporozoites develop from multiple germinal centers. Baker (1957, 1963) working with *H. palumbae* of wood pigeons in England, observed oocysts up to 28 μm in diameter in the gut of hippoboscids and sporozoites 8.5 μm in length. O'Roke (1932) reported up to 5000 sporozoites in one oocyst of *H. lophortyx* of California Valley quail. The pattern of sporozoite development in those parasites which develop in *Culicoides* is different from that in hippoboscids as the oocysts undergo little growth and the sporozoites form around a single germinal center (Fig. 31).

Sporozoites of *H. columbae* are about 10 μm long and one end is more blunt than the other. They are somewhat like those of *Leucocytozoon.* The ends of sporozoites of *P. nettionis* in *Culicoides* sp. are more uni-

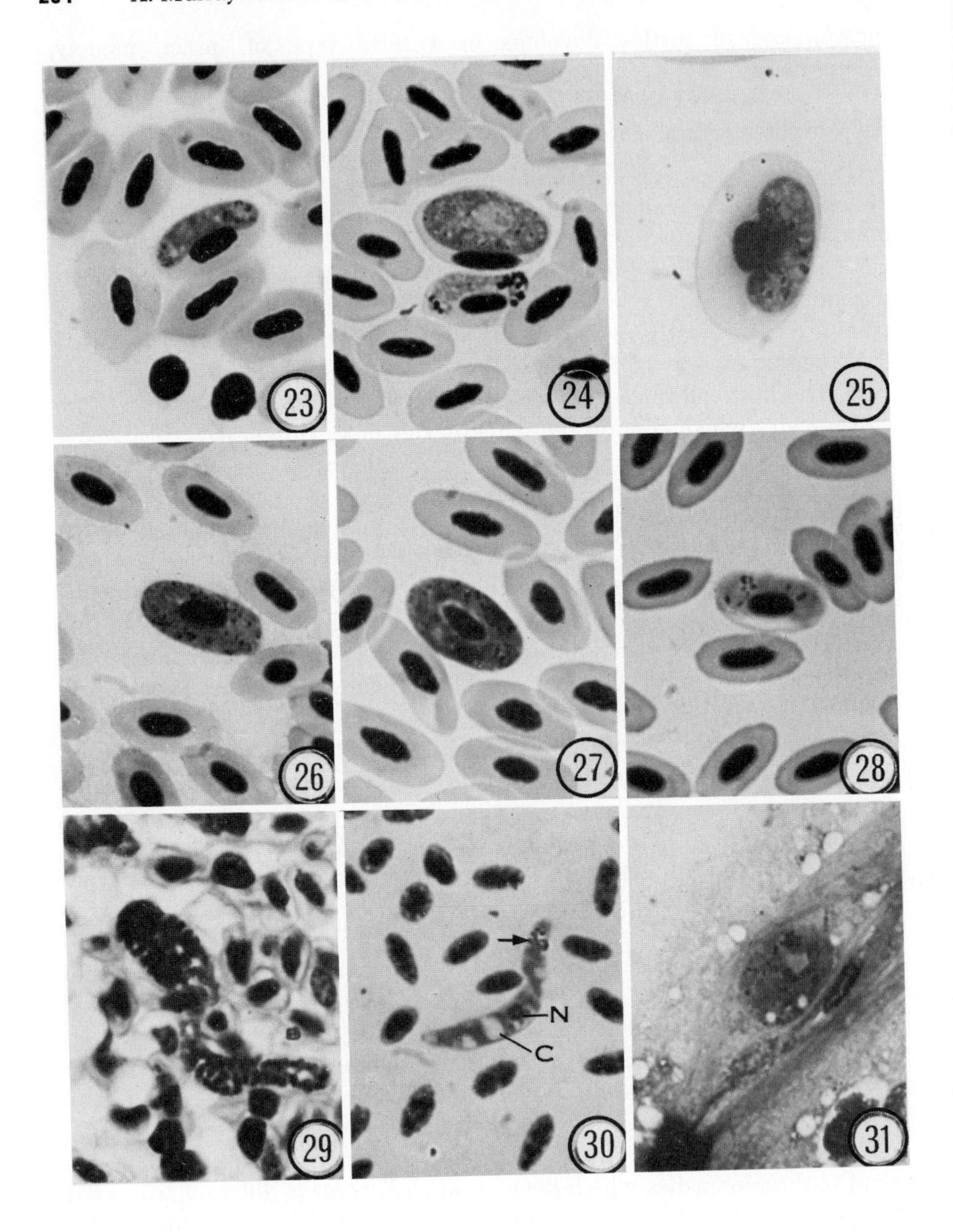
23
24
25
26
27
28
29
30
N
C
31

formly pointed than those of *H. columbae* and thus are similar to sporozoites of *Plasmodium* spp. Adie demonstrated that sporozoites of *H. columbae* are released by rupture of the oocyst wall. Sporozoites of *P. nettionis,* in contrast, escape gradually from oocysts in *Culicoides downsi.* The minimum time for the sporogonic cycle of *H. columbae* is 10–12 days while that of *P. nettionis* is 7–10 days. The cycles of *P. velans* and *P. fringillae* in *Culicoides sphagnumensis* like that of *P. nettionis* are 7–12 days in duration (Khan and Fallis, 1970). Sporozoites of all species accumulate in the salivary glands of their vectors.

Knowledge of the asexual development of *H. columbae* stems mostly from observations made several years ago by Aragão (1908), Adie (1915, 1924), and Wenyon (1926). Isolated reports on other species, whether transmitted by hippoboscids or biting midges, provide only fragmentary information on the asexual cycle. Aragão noted elongated, twisted schizonts in the vascular endothelium of the lung of pigeons infected with *H. columbae.* Schizonts of the Haemoproteidae have been observed more often in the lung than elsewhere, although Wenyon and others have also seen them in blood vessels of the liver, kidney, and spleen (Fig. 29). Bradbury and Gallucci (1971, 1972) described the fine structure of schizonts of *H. columbae.* The parasites were segregated from the host–cell cytoplasm by a trilaminar plasma membrane which contained several micropores. Numerous nuclei and mitochondria were seen in the ribosome-rich cytoplasm of immature schizonts. Merozoite formation begins with the appearance of periodic dense areas beneath the plasma membrane of the schizont, which becomes deeply invaginated, dividing the parasites into pseudocytomeres. Rhoptries and polar rings develop beneath the dense areas which begin to elongate, forming the merozoite buds. With further development, a cytostome, dense bodies, and a spherical structure closely associated with a mitochondrion were seen in the budding merozoites.

Figs. 23–28. Methanol fixed, Giemsa-stained gametocytes of *Haemoproteus* and *Parahaemoproteus* species. × 1200. **Fig. 23.** *Haemoproteus columbae* from pigeon (*Columbia livia*). **Fig. 24.** *Haemoproteus saccharovi* (above) and *H. maccallumi* (below) from mourning dove (*Zenaidura macroura*). **Fig. 25.** *Haemoproteus metchnikovi* from turtle (*Chrysomys picta*). **Fig. 26.** *Parahaemoproteus velans* from yellow-bellied sapsucker (*Sphyrapicus varius*). **Fig. 27.** *Parahaemoproteus nettionis* from the domestic duck. **Fig. 28.** *Parahaemoproteus fringillae* from white-throated sparrow (*Zonotrichia albicollis*). **Fig. 29.** Elongate schizont of *P. fringillae* in vascular endothelium of lung of sparrow. × 1200. **Fig. 30.** Ookinete of *P. nettionis.* Note central nucleus (N), crystalloid inclusions (C), and pigment at posterior end (arrow). × 1200. **Fig. 31.** Young oocyst of *P. velans* in gut smear of *Culicoides sphagnumensis.* Note fine pigment granules and central crystalloid inclusion. × 1200.

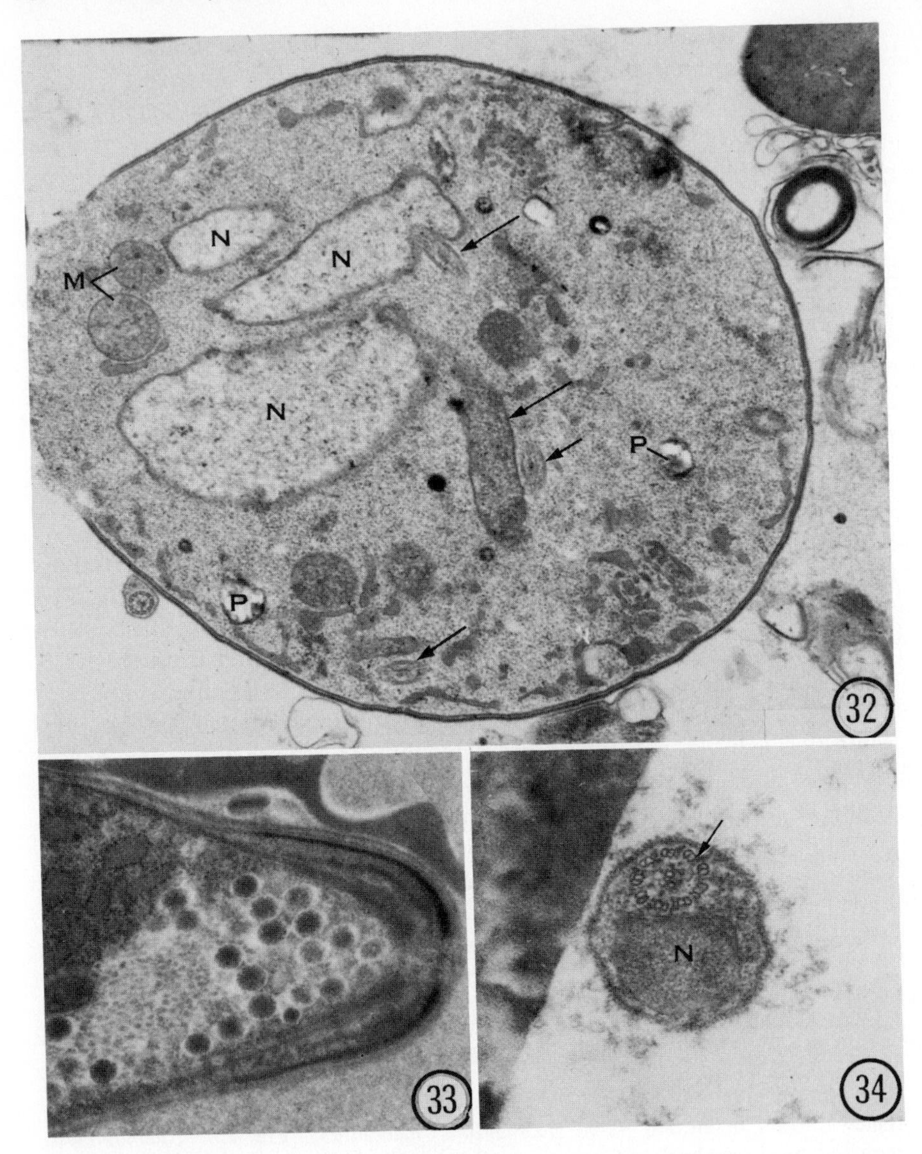

Fig. 32. Early zygote of *P. velans*. The nucleus and axoneme of the microgamete (arrows) can be seen in the cytoplasm. M, mitochondria; N, nucleus of macrogamete; P, pigment. × 12,600. Reproduced with permission from *J. Protozool.* **Fig. 33.** Reinforced apical end of ookinete of *P. velans*. × 28,500. Reproduced with permission from *Can. J. Zool.* **Fig. 34.** Transection through microgamete of *P. fringillae* illustrating single axoneme (arrow) and nucleus (N). × 47,500. Reproduced with permission from *J. Protozool.*

Schizonts of *H. metchnikovi* occur primarily in the spleen of infected turtles. In a study of the ultrastructure of these schizonts, Sterling and DeGiusti (1972) described microvillus-like projections around the periphery of immature stages which, they suggested, might function to increase the surface area for absorption of nutrients. During the primary growth phase, cytoplasmic cleavage and nuclear division result in the formation of multinucleate syncytia or pseudocytomeres. Merozoites are produced by budding and the process resembles that described above for *H. columbae.* Mature merozoites of *H. metchnikovi* and *H. columbae* are ultrastructurally similar to those of *Plasmodium* species.

It has generally been accepted that there is a single schizogonic cycle before gametocytes appear in the blood. This assumption should be reexamined in view of the fact that there is a prepatent period of 3 weeks for *H. columbae,* 17 days for *H. palumbae,* and 14 days or less for *P. nettionis.*

Merozoites of the Haemoproteidae are 1–2 μm in diameter when seen in the blood and are similar to those of species of *Plasmodium.* More than one merozoite may penetrate the same cell. Merozoites require 4–6 days to mature to gametocytes. Some gametocytes persist in the blood for weeks after maturation although their capacity to exflagellate undoubtedly decreases with time. As stated earlier, the mature gametocyte partially or completely surrounds the nucleus of the host cell. In some species the nucleus is pushed to one edge of the cell and the cell may be enlarged. Gametocytes may "round up" quickly in a drop of blood exposed to air. Pigment granules of various sizes, numbers, and distribution appear in the gametocytes as a result of the incomplete utilization of hemoglobin. Usually, although there are exceptions, the granules are more numerous in the extremities of the gametocyte than in the central portion (Figs. 23–28).

Bradbury and Trager (1968a,b) described microgametogenesis and the mature gametes of *H. columbae.* The macrogametes are spherical and contain a prominent central nucleus whose outer membrane is drawn out into extensive anastomosing evaginations. The cytoplasm contains numerous ribosomes and glycogenlike granules. Several mitochondria and vesicles containing hemozoin pigment were observed. Soon after blood containing gametocytes was removed from the pigeon, axonemes were observed in the cytoplasm of microgametocytes. The axonemes appeared to be conneceted to bundles of intranuclear spindle fibers. Following their escape from the host's erythrocytes, the microgametocytes assumed a "dumbell" shape. Half of the parasite was surrounded by a single membrane and contained most of the nucleus. The other half was surrounded by the remains of the multiple membranes of the gametocyte and contained hemozoin granules, mitochondria, axonemes, and nuclear extensions. The axonemes and nuclear extensions were situated at the periphery of the host cell where they were assembled into microgametes, each with a nucleus and two axonemes. A recent high-voltage electron microscope study by Aikawa and Sterling (1974) revealed that microgametes of *H. columbae* contain a single axoneme.

Gallucci (1974a,b) described fertilization and ookinete formation in *H. columbae.* Penetration of the macrogamete by the microgamete occurred rapidly through a fusion of their plasma membranes. The cytoplasm, nucleus, and axoneme of the microgamete enter the macrogamete and nuclear fusion occurs. Transformation of the zygote to ookinete begins with the accumulation of dense material beneath the plasma membrane. The conoid and associated rings form in this area which gradually extends to form the apical end of the developing ookinete. The elaborate and specialized anterior end of the ookinete becomes differentiated with further outgrowth. Crystalloid granules are formed in association with granular endoplasmic reticulum and are later concentrated into two or more large inclusions in the cytoplasm. The ultrastructure of ookinetes of *H. columbae* is similar to that described for other Haemosporina.

Sterling (1972) reported on the ultrastructure of gametocytes and gametogenesis of *H. metchnikovi.* The gametocytes are strikingly similar to those of other Haemosporina and the process leading to the release of the microgametes resembles that reported for *H. columbae* and *Leucocytozoon simondi.* Atypical centrioles were observed in both micro- and macrogametocytes. Axoneme development begins before the microgametocytes escape from the host's erythrocytes. The microgametes are typically filiform and contain a nucleus and an axoneme.

The fine structure of sporogonic stages of *H. metchnikovi* in *Chrysops callidus* was described by Sterling and DeGiusti (1974). The oocysts lie beneath the basal lamina of the midgut epithelial cells and are surrounded by a distinct capsule. Subcapsular vacuolization precedes the appearance of sporozoite buds around the peripheral cytoplasm in areas of membrane thickening. One to two hundred sporozoites are formed around a single sporoblastoid center. Sporozoites of *H. metchnikovi,* like those of *L. simondi,* contain crystalloid inclusions.

B. Pathology

The parasitemias in birds infected with species of *Haemoproteus* are often higher than those in birds infected with species of *Leucocytozoon.* Woodpeckers in some localities may have haemoproteids in 50–75% of their erythrocytes without gross signs of disease. Markus and Oosthuizen (1972) consider *H. columbae* a pathogen, although adequate proof of pathogenicity is lacking. The parasites stimulate a reaction in the host as there is evidence of resistance to infection. An increased parasitemia in late winter and in the spring indicates that relapses or recrudescences in the infection may occur. The recrudescences appear in nonmigratory as well as migratory birds prior to the beginning of nesting. This suggests that the recrudescence may be associated with an increase in gonadotropins as indicated by Haberkörn's studies (1968).

IV. *Hepatocystis* *

Most haemoproteid parasites of mammals except the plasmodia are included in the genus *Hepatocystis.* These parasites occur mainly in arboreal tropical mammals of the Old World. Most species of *Hepatocystis* occur in lower monkeys, fruit bats, and squirrels. A species has been described from mouse deer and another from a hippopotamus.

Two additional genera of bat haemoproteids, *Nycteria* and *Polychromophilus,* have recently been established. These genera contain five species and little is known of their biology. *Nycteria* and *Polychromophilus* species differ from *Hepatocystis* of bats in the morphology of their gametocytes and preerythrocytic stages, and presumably in their vectors.

Thirteen species of *Hepatocystis,* their hosts, and distribution are shown in Table I.

A. Morphology and Life Cycles

Sporozoites of *Hepatocystis,* when introduced by the vector, give rise to schizonts primarily in hepatic parenchymal cells. Young schizonts resemble those of mammalian *Plasmodium.* Unlike the latter which mature in about a week, schizonts of *Hepatocystis* require 1–2 months to reach maturity. When mature they may be 2 or more mm in diameter and are referred to as merocysts. The size and morphology of the merocysts of *Hepatocystis* are variable. Merocysts range in morphology from the relatively simple type produced by *H. kochi* to the bizarre forms of *H. fieldi* and *H. rayi.* Merozoites are released from mature merocysts and enter erythrocytes where they develop into gametocytes. Some specialized merozoites apparently reinvade liver cells to initiate secondary tissue schizogony. Occasionally, in animals with heavy infections, multinucleate bodies are released from the merocysts and may be observed in the peripheral circulation. These multinucleate bodies are referred to as "corps plasmatiques" and may function in the initiation of secondary

* Much of the information in this section is taken from Garnham (1966).

Table I
Some Hosts and Distribution of Species of *Hepatocystis* [a]

Species	Host		Distribution
H. kochi	African monkeys:	*Cercopithecus aethiops*	Tropical Africa
		Papio cynocephalus	Tropical Africa
H. bouillezi		*Cercopithecus mona*	Tropical Africa
H. cercopitheci		*Cercopithecus nictitans*	Tropical Africa
H. simiae		*Papio papio*	Tropical Africa
H. semnopitheci	Oriental monkeys:	*Macaca irus*	Southern Asia
H. taiwanensis		*Macaca cyclopis*	Taiwan
H. epomophori	African fruit bats:	*Epomops fangueti*, *Epomophorus gambianus*, *Epomophorus wahlberghi*	Tropical Africa
H. peronnae			
		Myonycteris torquata	Central African Republic
		Lissonycteris angolensis	Congo-Brazzaville
H. pteropti [b]	Oriental fruit bats:	*Pteropus gouldii*, *P. conspicillatus*, *P. scapulatus*	Asia and Australia
H. vassali [c]	Oriental squirrels:	*Sciurus griseimanus*, *Ratufa bicolor*, *Callosciurus finlaysoni*, *Tamiops macclellandi*	India, Ceylon, Thailand, Malaya, Taiwan
H. rayi	Oriental squirrels:	*Petaurista inornata*, *Petaurista grandis*	Formosa, Taiwan
H. fieldi	Mouse deer:	*Tragulus meminna*, *Tragulus javanicus*	Malaya, Ceylon
H. hippopotami	Hippopotamus:	*Hippopotamus amphibius*	Northern Rhodesia

[a] Modified from Garnham (1966).
[b] With one subspecies.
[c] With three subspecies.

tissue cycles. Landau and Adam (1971) in their study of *H. peronnae* in African bats discovered, in addition to typical merocysts in hepatic parenchymal cells, smaller schizonts apparently within endothelial cells in hepatic vessels. Possibly the multinucleate bodies released from merocysts of *Hepatocystis* are phagocytized by vascular endothelial cells and continue their growth there. If this occurs, this part of the life cycle of species of *Hepatocystis* would be similar to the portion of the cycle of *Leucocytozoon simondi* which gives rise to megaloschizonts. Large vacuoles form in the maturing merocysts of *H. kochi* and these vacuoles eventually coalesce to form a central core within the merocyst which is filled with a colloid (Fig. 38). Mature merocysts of *H. kochi* are large ovoid bodies with central fluid-filled cores and a peripheral band of merozoites (Figs. 39 and 40). After release from the merocyst, the merozoites, which initially resemble the small ring forms characteristic of *Plasmodium* species, enter erythrocytes (Fig. 35). A vacuole is present in the erythrocytic parasite during its early development and the vacuole persists in the amoeboid gametocytes. Like those of the plasmodia, intra-erythrocytic stages of *Hepatocystis* ingest hemoglobin and residual hemozoin pigment may be seen in young gametocytes. As with other species of Haemosporina, the macrogametocytes of *Hepatocystis* are more basophilic than the microgametocytes when stained with Giemsa's stain and their nuclei are smaller and contain denser chromatin. The chromatin of the microgametocytes is surrounded by a pink halo (Fig. 36). The microgametocytes of *Hepatocystis* when stained with Giemsa's have a peculiar nucleus which consists of a large, oval, pink area occupying one-third to one-half of the parasite's body. Scattered in the nucleus are numerous red granules or chromatin threads (Fig. 37). Mature macrogametocytes are 9.5 μm in diameter; the microgametocytes are slightly smaller.

Exflagellation occurs rapidly. About eight slender microgametes are formed by the process from each microgametocyte. Garnham and his colleagues (1967) found that the microgametes of *H. kochi* were similar to those of species of *Plasmodium* as they contained a single axoneme and a centrally positioned nucleus. Other organelles were not observed. Ookinetes of *H. kochi* unlike those of other species of Haemosporina penetrate the basal lamina of the vector's midgut and enter the hemocoel. Many ookinetes of *H. kochi* accumulate in the anterior part of the insect's body, particularly in the region between the dorsal ganglion and the eyes.

The oocysts develop rapidly and contain several germinal centers around which the sporozoites are formed. The majority of the oocysts reach maturity in approximately 5 days at 27°C. They are about 40 μm in diameter and contain hundreds of sporozoites (Fig. 41). The sporozoites of *H. kochi* are 11–13 μm long and are rarely seen in the salivary glands of infected *Culicoides.* Garnham (1966) stated that the salivary glands may not be an important site of sporozoite localization and that sporozoites of *H. kochi* may invade the insect's mouthparts.

B. Vectors and Ecology of *Hepatocystis kochi*

After many years of unsuccessful attempts to discover the vector of *H. kochi,* Garnham discovered that the oocysts developed in *Culicoides fulvithorax* and *C. adersi* which had fed on infected monkeys. Sporogony

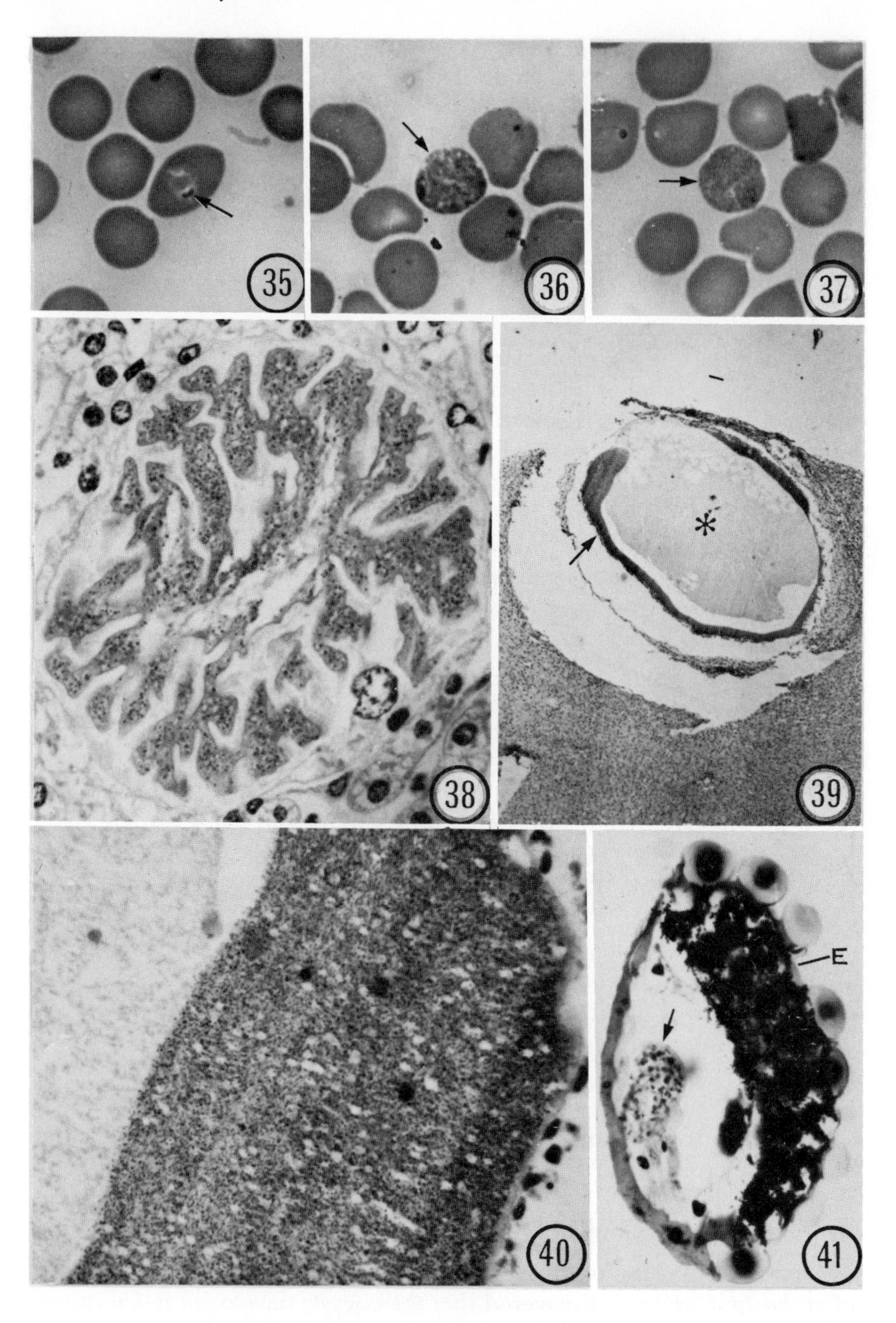
35
36
37
38
39
40
41
E

of the other species of *Hepatocystis* has not been demonstrated. Garnham feels that, in all cases, *Culicoides* spp. are the probable vectors. *Hepatocystis kochi* has been reported from almost every country in tropical Africa. It is most prevalent in second growth forest with moderate rainfall and is less common in dry scrub or tropical rain forest. The parasite has not been reported from areas with altitudes above 2000 m in the equatorial regions. In some areas, 100% of certain species of monkeys are infected, while fewer than one-half harbor the parasite in other regions. The variation in infection rate may be related in part to the differences in the susceptibility of the host, and prevalence of vectors and their feeding preferences. Garnham noted that the higher primates (chimpanzee, gorilla, orangutan, and man) are not susceptible to *H. kochi.*

C. Pathology

It is doubtful that liver function is impaired in naturally infected animals since they usually harbor relatively few merocysts. Inflammatory cells are often seen around developing merocysts. As a rule, the inflammatory cells are unable to enter cells which contain intact parasites. Following rupture of the cell containing the merocyst, however, the phagocytes may destroy some of the freed parasites. That immunity to *H. kochi* is slight or absent is indicated by the fact that there is a progressive increase in the parasitemia in animals as they become older. This situation is the reverse of that which occurs in animals infected with *Plasmodium* spp. where, through a gradual acquisition of immunity, the parasite shows a decline in numbers with aging from a peak in childhood.

V. Summary

There is a lack of precise knowledge about the parasites in the genera *Leucocytozoon, Haemoproteus,* and *Hepatocystis.* Taxonomic uncertainties could be clarified if we obtained information on the biology of these

---

Figs. 35–37. Methanol fixed, Giemsa-stained erythrocytic stages of *Hepatocystis kochi* from *Cercocebus aterriums.* × 1200. Slides for illustrations of *Hepatocystis* courtesy of Prof. P. C. C. Garnham. **Fig. 35.** Ring-shaped trophozoite (arrow). **Fig. 36.** Mature macrogametocyte (arrow). **Fig. 37.** Mature microgametocyte (arrow). **Fig. 38.** Immature schizont (merocyst) of *H. kochi* from liver of *Cercocebus aterrimus.* × 480. **Fig. 39.** Mature merocyst with large central colloid-filled cavity (*) and peripheral rim of merozoites (arrow). × 32. **Fig. 40.** Enlargement of previous figure illustrating colloid and merozoites. × 480. **Fig. 41.** Immature oocyst of *H. kochi* (arrow) in section of head of *Culicoides adersi.* E, eye. × 480.

parasites. Species of *Haemoproteus* and *Parahaemoproteus* are considerably less pathogenic than are those of some species of *Leucocytozoon.* This may indicate a lack of adaptation between the species of *Leucocytozoon* and the host. Diverse patterns of schizogony have been reported among species of *Leucocytozoon.* We do not know if the pattern of schizogony of *Parahaemoproteus* species is the same as that of *H. columbae* or if sporogony of the latter will occur in biting midges as well as in hippoboscids. There is a striking similarity between oocysts of *H. columbae* and of many *Plasmodium* species, while those of species of *Parahaemoproteus* resemble *Leucocytozoon.* These similarities may indicate phylogenetic relationships. The development of oocysts of *H. kochi* in the head of *Culicoides* spp. is an unusual phenomenon among haemosporidia. We do not know if the more typical haemosporidian pattern of sporogony will be found when the vectors of other species of *Hepatocystis* are discovered. Additional studies are needed on the life history of most species of *Hepatocystis.* Almost nothing is known of their fine structure, physiology, or of the immunology and pathology of the infected animals. In general, few data are available on host specificity and little is known concerning recrudescence and relapse of infection in the host. Because of the many unanswered questions, there is still much work for parasitologists interested in these parasites.

VI. Addendum

Miltgen and colleagues have recently observed the development of oocysts of *Hepatocystis brayi* between the epithelium and basal membrane of the stomach of *Culicoides variipenis* and *C. nubeculosus.* This development is similar to the pattern of sporogony among other species of Haemosporina.

REFERENCES

Adie, H. (1915). The sporogony of *Haemoproteus columbae. Indian J. Med. Res.* **2,** 671–680.

Adie, H. (1924). The sporogony of *Haemoproteus columbae. Bull. Soc. Pathol. Exot.* **17,** 605–613.

Aikawa, M., and Sterling, C. R. (1974). Microgametogenesis in *Haemoproteus columbae. Z. Mikrosk. Anat.* **147,** 353–362.

Akiba, K. (1970). Leucocytozoonosis of chickens. *Nat. Inst. Anim. Health Q.* **10,** Suppl., 131–147.

Aragão, H. de B. (1908). Über den Enwicklungsgang und die Übertragung von *Haemoproteus columbae Arch. Protistenkd.* **12,** 154–167.

Baker, J. R. (1957). A new vector of *Haemoproteus columbae* in England. *J. Protozool.* **4,** 204–208.

Baker, J. R. (1963). The transmission of *Haemoproteus* sp. of English wood-pigeons by *Ornithomyia avicularia. J. Protozool.* **10**, 461–465.

Bennett, G. F., Garnham, P. C. C., and Fallis, A. M. (1965). On the status of the genera *Leucocytozoon* Ziemann, 1898 and *Haemoproteus* Kruse, 1890 (Haemosporidiida: Leucocytozoidae and Haemoproteidae). *Can. J. Zool.* **43**, 927–932.

Bennett, G. F., Okia, N. A., Ashford, R. G., and Campbell, A. G. (1973). Avian Haemoproteidae: *Haemoproteus enucleator* sp. n. from the Kingfisher, *Ispidina picta* (Boddaert). *J. Parasitol.* **58**, 1143–1147.

Bradbury, P. C., and Gallucci, B. B. (1971). The fine structure of differentiating merozoites of *Haemoproteus columbae* Kruse *J. Protozool.* **18**, 679–686.

Bradbury, P. C., and Gallucci, B. B. (1972). Observations on the fine structure of the schizonts of *Haemoproteus columbae* Kruse. *J. Protozool.* **19**, 43–49.

Bradbury, P. C., and Trager, W. (1968a). The fine structure of the mature gametes of *Haemoproteus columbae* Kruse *J. Protozool.* **15**, 89–102.

Bradbury, P. C., and Trager, W. (1968b). The fine structure of microgametocytogenesis in *Haemoproteus columbae* Kruse. *J. Protozool.* **15**, 700–712.

Danilewsky, B. (1884). On the parasites of blood (Haematozoa). *Russ. Med.* pp. 44 and 48 (from Garnham, 1966).

DeGiusti, D. L., Sterling, C. R., and Dobrzechowski, D. (1973). Transmission of the chelonian Haemoproteid *Haemoproteus metchnikovi* by a tabanid fly, *Chrysops callidus. Nature (London)* **242**, 50–51.

Desser, S. S., and Weller, I. (1973). Structure, cytochemistry & locomotion of *Haemogregarina* sp. from *Rana berlandieri. J. Protozool.* **20**, 65–73.

Fallis, A. M., Desser, S. S., and Khan, R. A. (1974). On species of *Leucocytozoon. Adv. Parasitol.* **12**, 1–67.

Gallucci, B. B. (1974a). Fine structure of *Haemoproteus columbae* Kruse during macrogametogenesis and fertilization. *J. Protozool.* **21**, 254–263.

Gallucci, B. B. (1974b). Fine structure of *Haemoproteus columbae* Kruse during differentiation of the ookinete. *J. Protozool.* **21**, 264–275.

Garnham, P. C. C. (1966). "Malaria Parasites and Other Haemosporidia." Blackwell, Oxford.

Garnham, P. C. C., Bird, R. G., and Baker, J. R. (1967). Electron microscope studies of motile stages of malaria parasites. V. Exflagellation of *Plasmodium, Hepatocystis* and *Leucocytozoon. Trans. R. Soc. Trop. Med. Hyg.* **61**, 58–68.

Haberkörn, A. (1968). Zur hormonellen Beeinflussung von *Haemoproteus* infektionen. *Z. Parasitenkd.* **31**, 108–112.

Huff, C. G. (1942). Schizogony and gametocyte development in *Leucocytozoon simondi* and comparisons with *Plasmodium* and *Haemoproteus. J. Infect. Dis.* **71**, 18–32.

Khan, R. A., and Fallis, A. M. (1970). Life cycles of *Leucocytozoon dubreuili* Mathis & Léger, 1911 and *L. fringillinarum* Woodcock, 1910 (Haemosporidia: Leucocytozoidae). *J. Protozool.* **17**, 642–658.

Khan, R. A., and Fallis, A. M. (1971). A note on the sporogony of *Parahaemoproteus velans* (*Haemoproteus velans* Coatney & Roudabush) (Haemosporidia: Haemoproteidae) in species of *Culicoides. Can. J. Zool.* **49**, 420–421.

Kocan, R. M. (1968). Anemia and mechanism of erythrocyte destruction in ducks with acute *Leucocytozoon* infections. *J. Protozool.* **15**, 445–462.

Landau, I., and Adam, J. P. (1971). Description de schizontes de rechute chez un nouvel Haemoproteidae, *Hepatocystis peronnae* n. sp. parasite de Megachiroptères africains. *Cah. ORSTOM, Ser. Entomol. Med. Parasitol.* **4**, 373–378.

Levine, N. D., and Campbell, G. R. (1971). A check-list of the species of the genus *Haemoproteus* (Apicomplexa, Plasmodiidae). *J. Protozool.* **18,** 475–484.

Markus, M. B., and Oosthuizen, J. H. (1972). Pathogenicity of *Haemoproteus. Trans. R. Soc. Trop. Med. Hyg.* **68,** 186–187.

Miltgen, F., Landau, I., Canning, E. U., Boorman, J., and Kremer, M. (1976). *Ann. Parasitol. Hum. Comp.* **51,** 299–307.

O'Roke, E. C. (1932). Parasitism of the California Valley Quail by *Haemoproteus lophortyx,* a blood protozoan parasite. *Calif. Fish Game* **18,** 223–238.

Pan, I. C. (1963). A new interpretation of the gametogony of *Leucocytozoon caulleryi* in chickens. *Avian Dis.* **7,** 361–368.

Sergent, Ed., and Sergent, Et. (1906). Sur le second hôte de l'*Haemoproteus* (Halteridium) du pigeon. *C. R. Seances Soc. Biol. Ses Fil.* **61,** 494–496.

Sterling, C. R. (1972). Ultrastructural study of gametocytes and gametogenesis of *Haemoproteus metchnikovi. J. Protozool.* **19,** 69–76.

Sterling, C. R., and DeGiusti, D. L. (1972). Ultrastructural aspects of schizogony, mature schizonts, and merozoites of *Haemoproteus metchnikovi. J. Parasitol.* **58,** 641–652.

Sterling, C. R., and DeGiusti, D. L. (1974). Fine structure of differentiating oocysts and mature sporozoites of *Haemoproteus metchnikovi* in its intermediate host *Chrysops callidus. J. Protozool.* **21,** 276–283.

Wenyon, C. M. (1926). "Protozoology," Vol. II. Baillière, London.

Wong, T. C., and Desser, S. S. (1976). Fine structure of oocyst transformation and the sporozoites of *Leucocytozoon dubreuili. J. Protozool.* **23,** 115–126.

6

Plasmodia of Reptiles

Stephen C. Ayala

I. Introduction

Malaria has been found in reptiles on five continents. It occurs in lizards of all of the major families and occasionally in snakes, but not in crocodiles, turtles, or tuataras. It is most common in wet, tropical forests, but some species occur as far north as Wyoming and Japan, and as far south as New Zealand.

Reptilian malaria is especially well suited for studies on the evolution and epidemiology of natural infections. Like their ancestors for thousands of generations, lizards normally live their entire lifetime within a small home range. Marked individuals can be studied, released, and often recaptured throughout a season or over several years. Studies on infections of reptiles are made easier because of our growing knowledge of the physiology, ecology, and historical zoogeography of many reptile groups.

Small, inexpensive, and easily maintained lizards are good research animals. The total blood volume of many species is less than 1 cm^3. Their body temperature can be precisely controlled by varying the environmental temperature, and their malaria infections develop more slowly than those in endothermic vertebrates. Nevertheless, a 1- or 2-gm lizard shows much the same responses to infections as a 10-kg monkey, or man himself.

Plasmodia of reptiles have received comparatively little attention. Garnham (1966) reviewed the information on 24 reptilian *Plasmodium* species in only 88 pages, compared to 267 pages for the 25 species from birds and 507 pages for 37 species from mammals. You seldom get a feeling from the literature for the unity of *Plasmodium* in its different hosts. Only a few of the 80 authors of the 153 papers on reptile malaria have also made significant original contributions on the biology of plasmodia in birds or mammals.

The tabulation below summarizes the number of publications and new species designations (in parentheses) in the world literature on reptile plasmodia (Ayala, 1977). The outstanding surge of interest in the New World during the last decade is especially evident.

There is an increasing number of regional studies on malaria in lizard populations, such as those of David Peláez and Rodolfo Pérez Reyes in Central Mexico, Stephen Ayala in California, Helen Jordan in Georgia, Sam Telford in Panama and the Caribbean islands, and Ralph Lainson, Irene Landau, and Jeffrey Shaw in northeastern Brazil.

Serious studies are available on the biology of six saurian *Plasmodium* species: *P. colombiense*: population epidemiology; *P. floridense*: ultrastructure, course of infection, epidemiology and historical zoogeography, infection dynamics, and morphological variation in different host species;

	To 1965	1966–1975	Total
Africa	24 (9)	6 0	30 (9)
Australasia	5 (4)	6 (2)	11 (6)
Americas	43 (16)	69 (25)	112 (41)
	72 (29)	81 (27)	153 (56)

P. mexicanum: blood and paraerythrocytic cycles, host range, epidemiology, regional and historical zoogeography, and sporogony; *P. sasai*: course of infection; *P. tropiduri*: ultrastructure and blood pathology during infection; *P. zonuriae*: hematological parameters.

Some recent significant advances in the study of reptile plasmodia include: a realization of the surprising frequency of infections in lizards of savanna and forest habitats; discovery of the wide variety of host blood cell types; clarification of the ecological, evolutionary, and biogeographical patterns of some species; and the finding of sporogony of a *Plasmodium* species in hematophagous Diptera other than culicid mosquitoes. For most recently designated species, attention has been given to thorough, quantified descriptions emphasizing intrapopulation variance.

II. Morphology and Life Cycles

Most reports include descriptions of the parasites in erythrocytes and sometimes in leukocytes. The ultrastructure of the blood forms of described species is similar to that of avian forms (Aikawa and Jordan, 1968; Scorza, 1970), and the appearance of reptile, bird, and mammal plasmodia is also similar in the insect stages. Knowledge of preerythrocytic development awaits identification of the transmitting hosts.

A. Course of Infection

The course of natural or induced infections has been reported for the species *P. azurophilum*, *P. carinii* (=*rhadinurum*), *P. chiricahuae*, *P. floridense*, *P. mexicanum*, *P. multiformis*, *P. sasai*, and *P. tropiduri*. Typical infections in reptiles follow the same general course as those in birds or mammals: a period of initial rise, a peak, an abrupt or gradual decline, and a period of chronicity that may be interrupted by relapse. In reptiles the infections develop and subside more slowly than in birds or mammals with peak parasitemias usually occurring 1 to 2 months or more after initial infection.

Prepatent development following the inoculation of sporozoites is unknown. Ralph Lainson *et al.* (1975) suggested that a preerythrocytic cycle in hemopoietic sinuses might explain the high frequency of infection in immature erythrocytes that occurs with some species.

As the parasitemia increases, the plasmodia show exuberant growth, highly amoeboid asexual forms and segmenters which develop their maximum numbers of merozoites. The parasites may continue reproducing "out of control" and kill the host before it can mount any effective response. The opposite extreme is seen in benign infections, like those of *P. carinii* in iguanas, where the parasite reproduces slowly, maintaining a persistent, low-grade parasitemia with no clear peak.

The height of peak parasitemia depends upon the host and parasite species and on other physiological or epidemiological factors. *Plasmodium carinii* rarely surpasses 1–5% parasitemia. *Plasmodium floridense* in *Anolis carolinensis* may show peak parasitemias of 10 to 30%, but the same parasite in *Sceloporus undulatus* reaches 40–60% or more (Jordan, 1975). During the period of spring relapse, *P. mexicanum* parasitemia in yearling *Sceloporus occidentalis* often reaches 60–80%, but in adults it seldom exceeds 30–40%. In fatal infections with several *Plasmodium* species, the number of parasites may exceed the number of erythrocytes in circulation.

Crises follow the peak of the growth curve in most of the infections which reach high parasitemias. Within a few days, the number of parasites drops to a much lower level. They may be hard to find after the crisis, but some usually remain in circulation for many months. In contrast, crises are often not seen in infections producing only low-level parasitemias.

"Crisis forms," showing much less vigorous growth, occur during the crisis in birds, mammals, and reptiles, but in reptiles "crisis forms" continue to be found throughout the subsequent chronic phase. Vacuolated, dead or dying parasites with ragged, pale-staining cytoplasm are the predominant forms during the crisis. Most trophozoites, schizonts, segmenters, and gametocytes which occur during and after the crisis seem stunted and confined or perhaps even crushed within a tight parasitophorous membrane. Segmenters that manage to complete division yield only a low number of merozoites. With *P. colombiense*, for example, segmenters developing during the early acute infection produce eight to fourteen, with an average of nine or ten merozoites. In postcrisis, chronic infections, they produce an average of six merozoites, and segmenters with more than eight nuclei are almost never seen (Ayala and Spain, 1976). Similar observations are available for *P. floridense* (Jordan, 1975) and for *P. tropiduri* (Scorza, 1970).

Surviving lizards may remain infected for life. When parasites are found in immature erythrocytes, it is clear that the infection is still active. Although they may eventually disappear altogether, there are several reports of infections persisting for over a year. We followed one low-grade *P. carinii* infection in a Colombian iguana for 3 years, and parasites could always be found in smears taken every 2 months. Reactivation after a period of latency has been reported for several species.

B. Schizogony in Erythrocytes

Reproduction in the peripheral blood has been considered typical of the genus *Plasmodium* for more than half a century. The segmenters assume various shapes: cruciform, fan-shaped or rosette in the smaller plasmodia, and round, elongate, and amorphous in the larger species.

Synchronous multiplication may occur on 4- to 5-day intervals during the initial generations in the blood. Synchrony has rarely been reported in more advanced infections, although *P. sasai* may show peak schizogony between 10 PM and 2 AM (Telford, 1972a). It seems unlikely that rapid, synchronized multiplication occurs with reptile malaria, but it would be surprising if the normal daily body temperature fluctuation of most lizards had no effect on the reproduction of their plasmodia.

Infection of immature erythrocytes probably depends on the stage of infection, and segmenters in young cells usually produce a larger number of merozoites. In some cases, a predilection for immature host cells is reported to aid identification of the species. Perhaps there is a "selective invasion," or perhaps the final stages of schizogony and reinvasion occur in hemopoietic sinuses where immature cells naturally predominate (Ayala and Spain, 1976). This mechanism could explain the seeming predilection for reticulocytes, as well as why a parasite like *P. mexicanum* invades mainly erythrocytes in its normal *Sceloporus* hosts, and leukocytes in experimentally infected *Phrynosoma* or *Crotaphytus* species. *Plasmodium* species described as occurring predominantly in thrombocytes or other leukocytes may undergo the final stages of schizogony and invasion of new cells in sinuses of the spleen or other organs where these cells are produced. Granulocyte formation occurs outside the vascular system (Pienaar, 1962), and this may explain why these cells are so rarely infected; *P. azurophilum* is an outstanding exception (Telford, 1975).

Precise identification of hemocytoblasts and their lymphoid precursors in lizards is difficult. Erythroblasts and even early polychromatic erythrocytes may be confused with leukocytes. Pienaar's "Haematology of Some South African Reptiles" (1962) is the most thorough reference, but workers often rely on the better known avian blood cell types (e.g., Lucas and Jamroz, 1961).

C. Paraerythrocytic Schizogony

Schizonts inhabiting nonerythrocytes in the peripheral circulation may be part of either the normal erythrocytic or a different paraerythrocyte cycle. The only serious study of exoerythrocytic schizogony of reptile plasmodia was made by Paul Thompson and Clay Huff in the early 1940's when similar forms of avian plasmodia were first being studied. A "*P. mexicanum*-type" of exoerythrocytic schizogony was proposed to include parasites capable of invading both endothelial and other "fixed" tissue cells, and also circulating leukocytes, corresponding to the two distinct forms seen in avian plasmodia. The significance of these two forms in reptilian plasmodia is not yet understood (see also Huff, 1969; Lainson *et al.*, 1975; Telford, 1975). Massive invasion of endothelial cells has been noted only for *P. mexicanum.*

Occasional schizonts in circulating leukocytes or fixed cells have been reported for the following species: *P. agamae, P. azurophilum, P. basilisci, P. chiricahuae, P. cnemidophori, P. diploglossi, P. giganteum, P. gonatodi, P. lacertiliae, P. lygosomae, P. mexicanum, P. morulum, P. multiformis, P. pitmani, P. telfordi, P. tropiduri, P. tupinambi, P. uranoscodoni, P. utingensis,* and *P. vautieri.* Furthermore, the species *P. audaciosa, P. effusa, P. mabuyi, P. modesta,* and *P. simplex* are reported to undergo schizogony mainly or perhaps exclusively in cells other than erythrocytes. With *P. mexicanum* and *P. azurophilum* and probably many other species, the host cells utilized depend largely on the host lizard species. Natural or acquired immunity may also favor the development of paraerythrocytic forms.

In fulminating *P. mexicanum* and *P. multiformis* infections, large hypertrophied leukocytes are found containing schizonts segmented into "islands" or cytomeres. Upon rupture, these may give rise to the numerous circulating "syncytia" containing two to twenty nuclei, similar to those found in severe *Leucocytozoon* infections after the hepatic schizonts rupture. Large, branching schizonts have been seen with both *P. mexicanum* and *P. multiformis.* Mature merozoites in intact leukocytes are usually retained in the parasitophorous vacuole instead of occurring free in the cytoplasm as in erythrocytes.

D. Gametocytes

Gametocytes of saurian and avian plasmodia are similar at both light microscope and ultrastructure levels. Macrogametocytes may sometimes exceed by five to ten times the numbers of microgametocytes. Gametocyte shape is one of the most useful characters for identification. Nevertheless, it may vary during the course of infection. *Plasmodium mexicanum* gametocyes are larger in the late fall of the year than during the summer (Ayala, 1970). Lainson *et al.* (1975) recorded a surprising change in the appearance of *P. multiformis* gametocytes from round to irregular-shaped over the course of 3 weeks.

Many *Plasmodium* species persist with schizogony limited to certain seasons or epizootic periods. In advanced or chronic infections, gametocytes are common; they may predominate or they may be the only forms seen. Since they cause minimal damage to the host cell, they may remain in circulation throughout the long life of the reptile's erythrocyte. It may be a short step from the persistent gametocyte to the development of a haemoproteid-like life cycle where the peripheral schizogony can be abandoned completely. This step may have occurred in the evolution of leucocytozoid-like *P. tupinambi* and *P. mabuyi*. The continued viability of such gametocytes over several months has not yet been assessed.

E. The Invertebrate Phase

The skin of lizards and snakes is not resistant to insect bites. Between and under the scales there is an extensive vascular bed, easily accessible to capillary-feeding insects like mosquitoes and hemiptera, as well as to pool-feeding ceratopogonids, simulids, phlebotomids, and tabanids.

The transmitting insect hosts of reptile plasmodia remain an enigma. Acarine ectoparasites do not seem able to support sporogonic development, although normal appearing plasmodia may be retained for many days in their stomachs. Triatomid bugs are sometimes used to take blood samples, but no development of the parasites has been reported in Hemiptera.

Hematophagous, nocturnal Diptera are the most likely suspects, including ceratopogonids, phlebotomids, and culicid mosquitoes. Haemosporidians are known to develop in representatives of each of these groups, and the complete extrinsic cycle of *P. mexicanum* has been followed in two phlebotomine species (Figs. 1–11).

Host feeding patterns of hematophagous Diptera are receiving increasing attention. Blood meal identification has revealed several groups of lizard-feeding species. In Panama, where malaria is a widespread infection of reptiles, some of these species include *Culex egcymon, C. tecmarsis, C. dunni,* and *C. elevator; Deinocerites dyari, D. epitedeus, D. melanophyllum,* and *D. pseudes*; and the phlebotomines *Lutzomyia micropyga, L. rorotaensis,* and *L. trinidadensis*. Further identification of reptile-feeding Diptera should simplify the search for lizard malaria vectors.

Microgametocyte exflagellation has been reported for a few reptile malaria species. It begins a few minutes after the blood meal has been taken and yields up to eight microgametes within the following hour. Fertilization and formation of an ookinete may occur in nonvector insects, but subsequent development is infrequent and abnormal.

In some Neotropical forest lizards, simultaneous infection with two or more species of *Plasmodium* is quite common. For example, in 1975 about

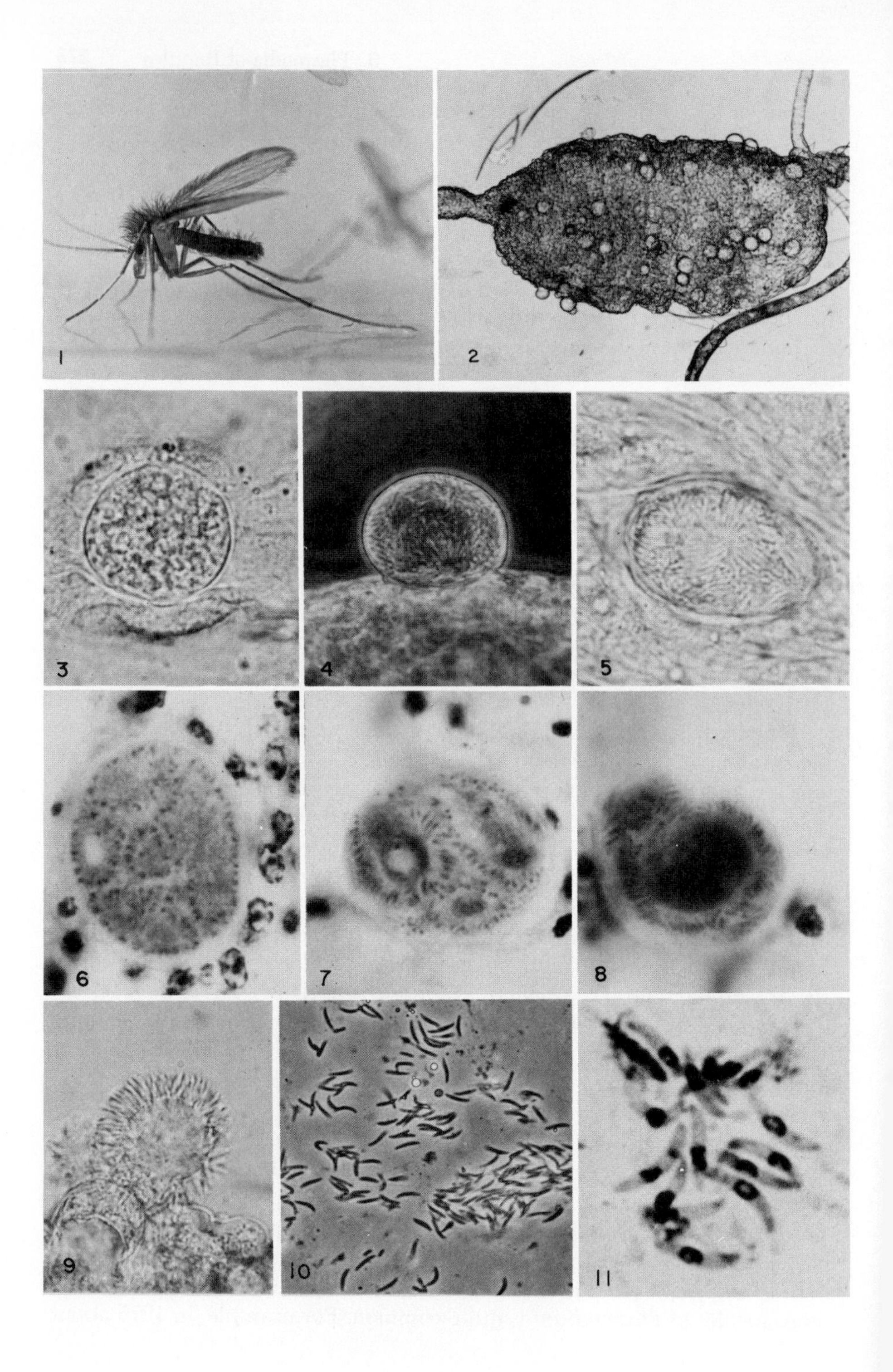
1
2
3
4
5
6
7
8
9
10
11

40% of the *Anolis limifrons* we sampled (Fig. 16, and Guerrero *et al.*, 1977) on Barro Colorado Island in Panama harbored malaria: 41% of these had mixed infections of *P. balli*, *P. floridense*, and *P. "tropiduri"* in the sense of Telford (1974). The potential for cross-fertilization must be considerable in the stomachs of Diptera biting these animals.

Ookinetes sometimes occur in such large numbers that the insect dies following massive penetration of the stomach wall. Their appearance is similar to ookinetes from most other haemosporidian species.

Oocysts have been found on the stomach wall of some Diptera after feeding on infected lizards: *Lutzomyia stewarti* and *L. vexatrix* in California (*P. mexicanum*), one *Aedes aegypti*, four *Culex territans*, two *C. quinquefasciatus*, and one *Culex* sp. in Georgia (*P. floridense*), one *Aedes aegypti* in Liberia (*P. agamae*), and three *Culex pipens* (*P. tupinambi* from Brazil). They appear large on the small stomach wall of a phlebotomine sandfly (Fig. 2), but their diameter is well within the range of *Plasmodium* oocysts of avian or mammalian types. Sporogony (Figs. 3–9) of *P. mexicanum* involves the development of a single large sporoblast and migration of the individual nuclei to its periphery, followed by extensive infolding and continued nuclear division. One or several sporoblast islands are thus formed, surrounded by maturing sporozoites (Figs. 6–8).

Mature oocysts of *P. mexicanum* in sandflies measure about 30–40 μm across and contain approximately a thousand or more sporozoites. David Young at the University of Florida, Gainesville (personal communication) has obtained similar mature oocysts in *Lutzomyia vexatrix* fed on *Sceloporus undulatus* harboring *P. floridense*. The rare oocysts of *P. floridense*, *P. agamae*, and *P. tupinambi* in culicine mosquitoes develop to a similar size and, where known, contain a like number of sporozoites, typical of the genus *Plasmodium*.

Development beyond the oocyst stage is known only for *P. mexicanum*. Released sporozoites invade the hemocoel (Figs. 10 and 11). Some enter the salivary glands, but others are seen in the foregut of the sandfly.

Figs. 1–11. Sporogony of *Plasmodium mexicanum*. **Fig. 1.** Female *Lutzomyia vexatrix occidentis* (Diptera:Psychodidae). **Fig. 2.** *Plasmodium mexicanum* oocysts developing on the stomach wall of *L. vexatrix*. **Fig. 3.** Unsporulated oocyst. **Figs. 4 and 5.** Mature oocysts in fresh preparation, sporozoites visible through oocyst wall. **Figs. 6–8.** Cross section through mature oocyst showing sporoblast "islands," sporozoites, and residual material in the oocysts. **Fig. 9.** Ruptured oocyst with sporozoites attached to residual body (see also *Science:* cover photograph, February 6, 1970). **Fig. 10.** Sporozoites released from ruptured oocysts. **Fig. 11.** Air-dried, Giemsa stained sporozoites from hemocoel of infected *L. vexatrix* (from Ayala, 1971).

Transmission to new lizards may take place by inoculation of salivary fluid or regurgitation during the bite, or it may possibly occur when the insect is eaten. Sporozoites from heavily infected sandflies were capable of producing infections in new lizards following inoculation (Ayala, 1971; Ayala, 1973).

F. The Described Species

Drawings of the accepted *Plasmodium* species of reptiles described through 1975 are shown in Text Fig. 1. For each, what I interpret to be a mature segmenter and a gametocyte of the species is shown. The reader should be aware that although these drawings are usually taken directly from the figures accompanying the original descriptions, they may, if necessary, be based on more precise descriptions when available by later workers, or my own judgment about the normal range of morphological variation in a population, the effect of postcrisis immunity, and structural changes accompanying the different stages of infection. For example, a species said to have segmenters with four nuclei may be judged to typically have six or eight merozoites if the author says that his description is based on rare schizonts found in a low-grade (=chronic) infection. This was the case for *P. minasense* Carini and Rudolph 1912, which I feel was probably an old *P. tropiduri* infection. Some well described plasmodia have unfortunately been labeled with names (like *P. tropiduri* or *P. minasense*) which may properly belong to unrelated parasite populations.

The drawings are organized into five groups. Group I (*Carinamoeba*) species have tiny schizonts, usually with four merozoites, and round gametocytes slightly larger than the host cell nucleus. Group II (*telfordi*) species have segmenters averaging six to fifteen merozoites and round, oval, elliptical, or elongate gametocytes about the same size as the host cell nucleus. Group III (*tropiduri*) species form a large complex having slightly larger segmenters averaging eight to twenty-five merozoites, and gametocytes almost twice the size of the host cell nucleus; host cells may include leukocytes; Group IV (*mexicanum*) species have large schizonts averaging eight to thirty merozoites and large round or elongate gametocytes; host cells include both erythrocytes and leukocytes; Group V (*Sauramoeba*) species have large round or elongate schizonts averaging forty to over 100 merozoites, and very large round or elongate gametocytes.

III. Taxonomic Considerations

". . . It could be assumed that lizard malaria is the earliest form of malaria on the evolutionary scale" (Garnham, 1966). Several authors suggest that reptiles could have had malaria long before the divergence of avian or mammalian lines. However, as there is little available information on reptile plasmodia, most accounts concentrate on the better known avian or mammalian forms and their three clearly defined families: Plasmodiidae, Haemoproteidae, and Leucocytozoidae.

A summary of the taxonomy of reptilian plasmodia is not difficult. It concerns historical controversies now largely resolved, and developing controversies centered on recently discovered parasites that do not seem to fit the previously defined categories.

A. Historical Controversies

Plasmodia were originally described under several different names: *Haemamoeba, Haemocystidium, Halteria, Halteridium, Haemoproteus, Hepatocystis, Plasmodium,* and *Proteosoma,* for example. These names still appear in synonym and host lists, but as the differences between *Plasmodium* and the haemoproteids became more clearly defined, these nomenclature problems were largely settled.

Although their life cycles are largely unknown, haemoproteids (*Haemocystidium* or *Haemoproteus*) may occur in three lizard groups—geckoes from Australia: *Phyllurus platurus, Heteronota binoei, Oedura tyroni,* and *Gehyra variagata*; Ceylon: *Hemidactylus leschenaulti*; India: *Hemidactylus brookei*; Palestine: *Hemidactylus turcicus*; Iran: *Phyllodactylus elisae*; Algeria: *Tarentola mauritanica*; and Sudan: *Tarentola annularis*; lacertids from Italy: *Lacerta sicula*; and possibly also in the agamids *Agama nupta* of Iran, *A. cyanogaster* of Uganda, and *A. agama* of Liberia and other parts of tropical Africa. Bray suggested that his difficulty in finding schizonts in the peripheral blood of some *A. agama* from Liberia thought to have *P. giganteum* infections might imply the presence of a haemoproteid with gametocytes indistinguishable from those of the *Plasmodium* species. With these possible exceptions, all the remaining plasmodia of lizards are thought to be species of the family Plasmodiidae, genus *Plasmodium* (but see below, Table I).

Haemoproteids have never been reported from New World lizards, although in the early name shuffling some vaguely described species, like *P. gonzalesi,* wound up briefly in the genus *Haemoproteus.* For the same reason, turtles and snakes are listed as hosts for some "*Plasmodium*" species. *Plasmodium metchnikovski, P. simondi,* and *P. roumie* are haemoproteids of New and Old World turtles. *Plasmodium passerita* of the snake *Passerita mycterizans* of Goa seems to be a piroplasm, and *P. mesnili* of the West African snakes *Naja naja, N. nigricollis,* and *Sepedon haemachotes* is another haemoproteid, *H. mesnili.*

There are only two good published descriptions of *Plasmodium* infections in snakes: *P. tomodoni* from one *Tomodon dorsatus* (Pessoa and Fleury, 1968) and *P. wenyoni* from one *Thamnodynastes pallidus* from Brazil (Garnham, 1965). Undescribed *Plasmodium* species have been

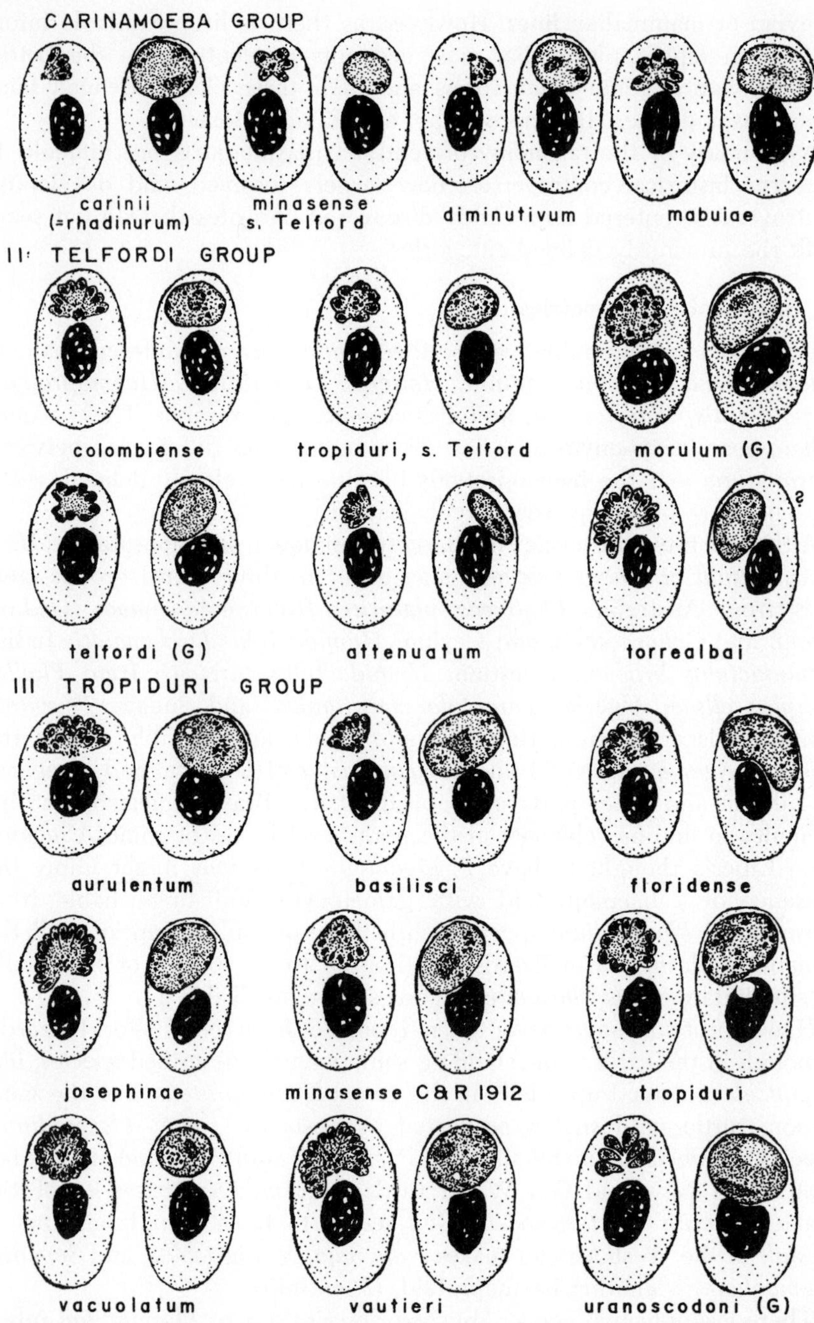

Text Fig. 1. Representative mature schizonts and gametocytes of the *Plasmodium* species described from reptiles. Typical merozoite numbers and relative segmenter

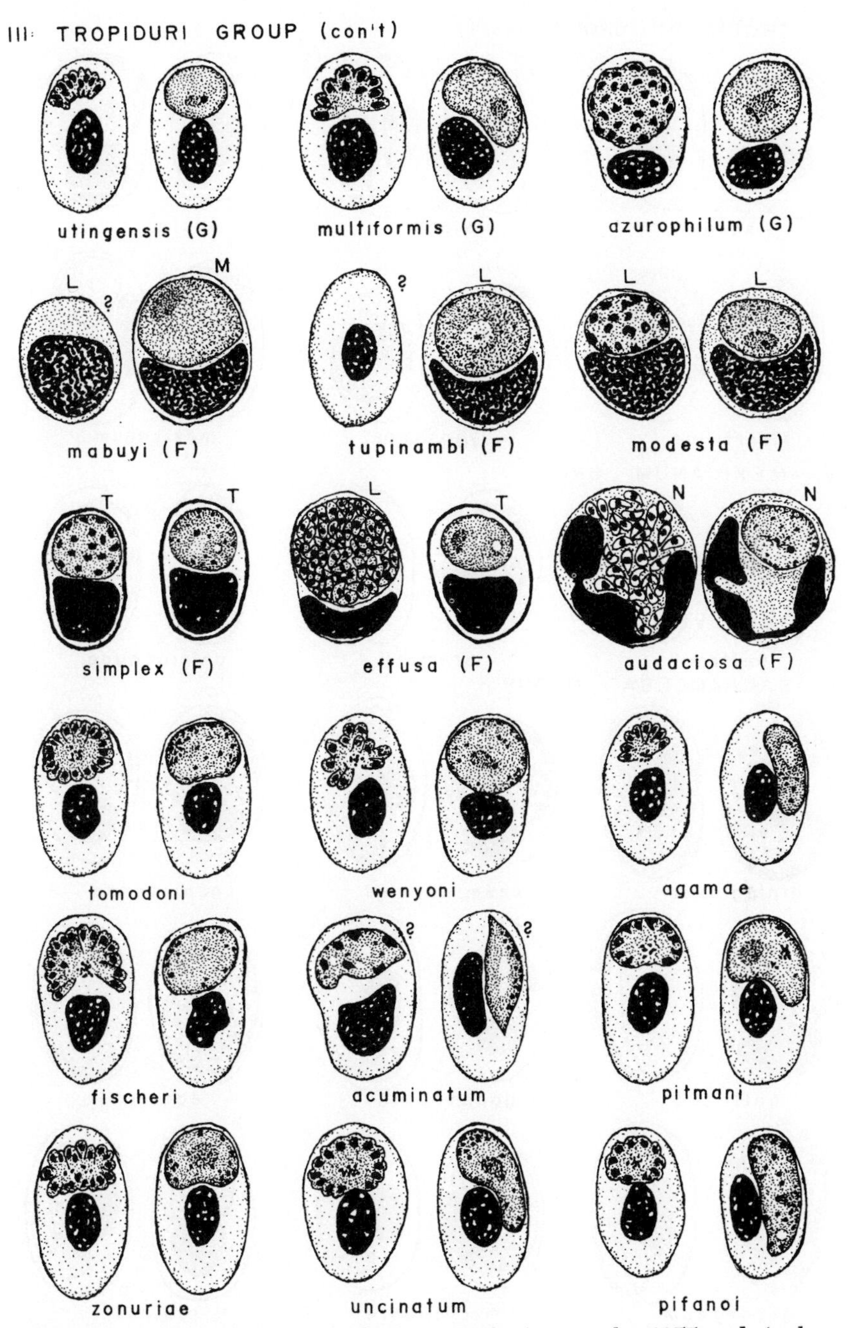

and gametocyte sizes are shown, drawn to scale (see Ayala, 1977 and Ayala and Spain, 1976). Hosts and localities of each species are listed in Table II. L, lympho-

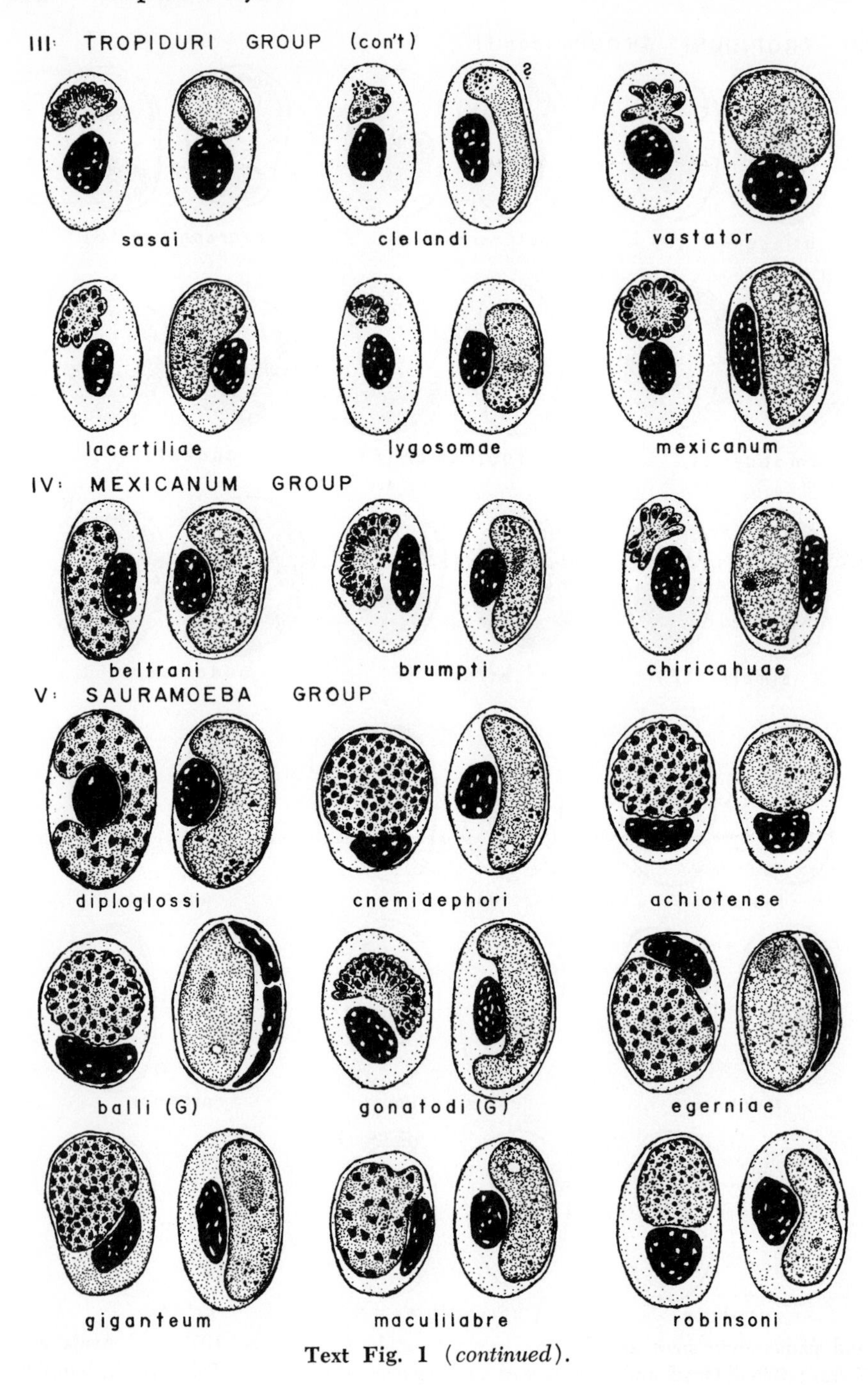
III: TROPIDURI GROUP (con't)
?
sasai
clelandi
vastator
lacertiliae
lygosomae
mexicanum
IV: MEXICANUM GROUP
beltrani
brumpti
chiricahuae
V: SAURAMOEBA GROUP
diploglossi
cnemidephori
achiotense
balli (G)
gonatodi (G)
egerniae
giganteum
maculilabre
robinsoni

Text Fig. 1 (*continued*).

reported from several other snakes in Brazil: *Bothrops moojeni, Corallus enydris, Chironius bicarinatus, Thamnodynastes strigatus,* and *Tomodon dorsatus* (Pessôa *et al.*, 1974); and eastern Panama: *Leptophis ahaetulla, Oxybelis aeneus, Psuestes poecilonotus,* and *Clelia* sp. by workers at the Gorgas Memorial Laboratory. We are currently studying a *Plasmodium* from *Lachesis muta* and *Spilotes pullatus* from Costa Rica. Many of these parasites are probably saurian plasmodia capable of occasionally infecting snakes.

B. Developing Controversies

Garnham placed the saurian plasmodia species known up until 1965 into two subgenera: *Carinamoeba,* with small segmenters, and *Sauramoeba,* with large segmenters. Although his type species: *P. carinii* (="*minasense*") and *P. diploglossi* certainly fit these criteria, the largest group of species with intermediate size segmenters can not be easily separated into either category (see Ayala and Spain, 1976).

Ralph Lainson and co-workers (Lainson and Shaw, 1969; Lainson *et al.*, 1971, 1974) at the Wellcome Laboratories in Brazil suggested reorganizing the taxonomy of reptile plasmodia to account for several unusual parasites from the New World tropics. Their classification is shown in Table I.

Recently, Sam Telford (1973) placed both *Garnia* and Garniidae in synonomy as *Plasmodium* and Plasmodiidae, and Hsu *et al.* (1973) placed *Saurocytozoon* in synonomy with *Leucocytozoon.* Telford thought, and it seems to me too, that these leucocytozoid-like forms are saurian *Plasmodium* with the special characteristic of reduced or possibly absent peripheral schizogony, perhaps in an abnormal host.

It is understandable that some aspects of the older classification of *Plasmodium* and Plasmodiidae do not strictly apply to reptile plasmodia since the knowledge of malaria in its avian and mammal host has until recently been the only available source of information for constructing taxonomies of the plasmodia (Fig. 12). Rather than construct whole new taxonomic categories, it seems to me preferable to admit that the earlier classifications were based on partial information and that the recent discoveries will permit a more refined inclusive definition (e.g., Telford, 1973).

Some reasons for not considering lack of pigment, occurrence in leuko-

---

cyte; M, monocyte; N, neutrophil; T, thrombocyte; (G), *Garnia* group, showing little or no pigment in host cells; (F), *Fallisia* group, developing predominantly or perhaps exclusively in leukocytes; ?, unknown, or no clear description is available.

Table I
An Alternative Classification of Plasmodia from Reptiles [a]

Family	Genus	Definition
1. Plasmodiidae	*Plasmodium* (Included: most described malaria parasites)	Pigmented haemosporidia with both gametocytes and schizonts in circulating erythrocytes
2. Garniidae	*Garnia* (Included: *azurophilum* (?), *balli, gonatodi, morula, multiformis, telfordi, uranoscodoni, utingensis*)	Nonpigmented haemosporidia with gametocytes and schizonts in erythrocytes and/or leukocytes
3. Garniidae	*Fallisia* (Included: *audaciosa, effusa, modesta, simplex*)	Nonpigmented haemosporidia with both gametocytes and schizonts exclusively in leukocytes
4. Leucocytozoidae	*Saurocytozoon* (Included: *mabuyi, tupinambi*)	Nonpigmented haemosporidia with only gametocytes in circulating erythrocytes or leucocytes

[a] Proposed by Lainson and Shaw (1969) and Lainson *et al.* (1971, 1974).

cytes, or apparent absence of schizonts as necessary criteria for separation from the genus *Plasmodium* include the following:

1. *Lack of pigment.* Pigmentless strains of well known mammal and avian *Plasmodium* species may occur naturally or under experimental conditions, such as with the chloroquine-resistant strains of *P. berghei* or *P. gallinaceum* (Peters *et al.*, 1965; Garnham, 1966). Occasionally, infections in particular animals (with *P. mexicanum* in my experience) may consist mostly or entirely of otherwise normal parasites lacking pigment. Furthermore, unless they have been phagocytized, parasites in host cells

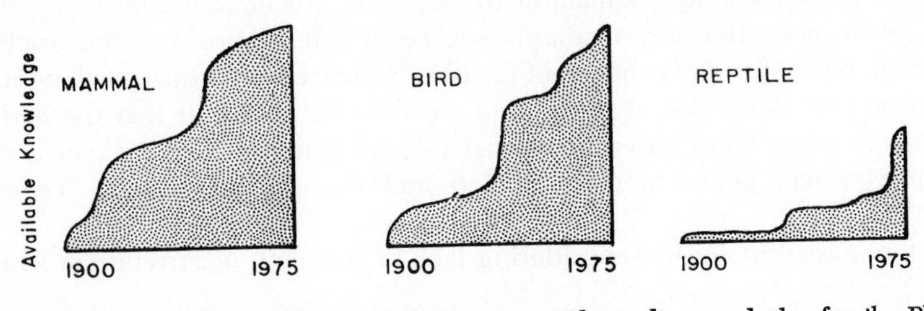

Fig. 12. Taxonomic definitions of the genus *Plasmodium* and the family Plasmodiidae have, until recently, been based almost exclusively on the available knowledge of plasmodia in mammals and birds.

lacking hemoglobin (immature erythrocytes, leukocytes, tissue cells, etc.) could not normally produce pigment, so a nonpigment criterion can not be applied.

2. *Host cells.* Many *Plasmodium* species of birds or mammals have para-erythrocytic phases proceeding concurrently with the erythrocyte cycle. In some lizards this phase may occur in circulating leukocytes, and may even predominate, depending on the host lizard species, the parasite, and the stage of infection. Several saurian plasmodia inhabit both leukocytes and erythrocytes. Gametocytes and asexual stages of *P. azurophilum* and another plasmodium we are studying in *Anolis frenatus* from Panama apparently inhabit leukocytes and erythrocytes with equal frequency. In some lizard species, the leukocytes seem to provide better host cells than the erythrocytes, although some of the infections appear abnormal. Eventually, the leukocyte-inhabiting species may prove to be more closely related to each other than to other plasmodia and thus justify inclusion in a separate subgenus *Fallisia*, and, similarly, with the nonpigmented species, *Garnia*. *Fallisia* and *Garnia* are certainly useful descriptive terms.

3. *Absence of schizonts.* Gametocytes persist in the circulation for many months in reptiles. They may greatly outnumber asexual forms in animals with chronic infections, or they may be the only forms present. Some species undergo most peripheral schizogony during a brief annual period, and it seems reasonable that some could abandon it altogether. If this happened with pigmented species like *P. giganteum* or *P. cnemidophori,* it might be identified as a haemoproteid; if it happened with a nonpigmented species, like *P. balli,* it might be identified as a leucocytozoid. In fact, schizonts are rare in advanced infections of *P. balli.* Knowledge about other factors, like sporogony and oocyst size, the invertebrate host, or the tissue phases, may help us decide in doubtful cases.

C. Biological Species Descriptions

Ecological criteria have broad application in distinguishing otherwise similar *Plasmodium* species. A thorough species account includes information on the habitat, population structure, seasonal dynamics, and course of infection, as well as the range of morphological variance. Fortunately, species descriptions are becoming increasingly quantified, making possible comparisons of plasmodia populations from different sites and in different host species, and analyses of geographical variation.

Parasites in individual infections can so widely differ that without additional information it may be hard to know they come from the same population. Criteria originally cited as distinguishing some *Plasmodium* species become insignificant as the normal variation of the type populations become more fully known. For *P. floridense,* average merozoite numbers differ considerably in different stages of the infection, in different parasite populations, in different lizards of the same species, and in lizards of different species (Ayala, 1975; Jordan, 1975). Redescription of the more poorly known type populations with emphasis on intraspecific variation would be of tremendous help in clarifying the taxonomy of saurian plasmodia.

Reliance on secondary descriptions has sometimes been misleading. For example, Wenyon's account of *P. carinii* (=*P. rhadinurum*) from a Trinidad iguana, which he erroneously called *P. minasense,* has been used as if it were the true description of *P. minasense.* The two parasites may not even be closely related, and the form originally named *P. minasense* by Carini and Roudolph was probably *P. tropiduri* (see Table II, under Scincidae).

IV. Host–Parasite Interactions

A. Pathogenicity

Some saurian *Plasmodium* species do much more damage than others. *Plasmodium mexicanum,* with its endothelial schizogony and severe spring recrudescences is especially pathogenic. The *Sauramoeba* group with their large segmenters and frequent invasion of immature erythrocytes may also cause significant morbidity or mortality. The *tropiduri* group parasites are probably of intermediate pathogenicity and the *Carinamoeba* group comparatively innocuous.

A parasite may be much more pathogenic in one host species that in another. This has been well documented by Helen Jordan (1975) for *P. floridense* in Georgia. Infections in *Sceloporus undulatus,* as compared with *Anolis carolinensis,* are less frequent (5 vs. 36%), of longer patency (150 vs. 71 days), and more intense (peak parasitemias 11,600 vs. 1,600 per 10,000 erythrocytes). Deaths are rare in infected anoles but not in *Sceloporus.* Such differences could explain the absence of one lizard species from an area where less susceptible hosts carried pathogenic plasmodia, as was proposed to account for the disappearance of native bird species from most lowland habitats on the Hawaiian Islands.

The pathogenesis of saurian malaria is essentially similar to the better described avian or mammal infections, modified by the ectothermy, small size, and small total blood volume of lizards of most species. It may also be altered by concurrent viral, rickettsial or bacterial infections in the blood of the lizard host (Ayala, 1973; Telford, 1972a; Johnston, 1975).

Plasmodium mexicanum is the only saurian species in which exoerythrocytic schizonts are known to cause significant tissue damage. The picture is similar to *P. gallinaceum* with capillary bed blockage in the brain and other organs. In natural infections, this condition occurs during the spring recrudescence of latent or chronic infections. Yearling lizards first infected in the late spring or early summer may also suffer severe pathology, but in the late summer or fall even hatchlings seem to have only mild infections (Ayala, 1970, 1973).

Erythrocyte destruction is minimal in benign infections like those produced by *P. carinii*. With acute infections the entire mature erythrocyte population is often replaced, but it is not yet known whether an autoimmune mechanism is involved. The blood picture is nearly normal during the period of developing parasitemia, except for the large number of infected erythrocytes. During the crisis, erythrocyte destruction may be rapid, their numbers reduced to less than one-half the normal range of around 1 million/mm^3, and the hematocrit and hemoglobin values reduced proportionately (Duguy, 1970; Scorza, 1971). The blood becomes thin and watery and the plasma yellow. Deaths may occur during the fulminating infection or later due to the postcrisis anemia.

For whatever reason, some species seem to selectively invade very immature blood cells. When this happens with certain bird or mammal plasmodia, the infections can be fulminating, and the result might be the same in reptiles, although the effects would not be seen so quickly.

Throughout the infection, but especially during and soon following the crisis, many of the parasites are phagocytized. Accumulated pigment is seen in the kidneys, liver, and spleen, but it is not yet known to what degree pigment deposition affects the normal functioning of these organs.

B. The Host Response

The infected lizard has two immediate problems: control the parasite growth before it causes excessive damage, and reestablish the normal blood cell concentration. The intensity of the parasite-destroying crisis, with its direct relation to the degree of parasitemia, and the inhibition of parasite growth resulting in small, stunted segmenters has already been mentioned.

The initial response to erythrocyte loss and developing anemia seems to be a release into the circulation of the available, nearly mature polychromatic erythrocytes. If this is insufficient to reestablish balance, the peripheral circulation is flooded with early and midpolychromatic erythrocytes and even blast cells. The blood picture following a severe crisis clearly reflects the compensating hyperplastic erythropoiesis underway.

Severely deteriorated blood pictures are probably more common in small lizards than in those with greater total blood volume and oxygen transport reserve. Age immunity has not yet been seriously studied.

Severe anemia in lizards may call into play a special mechanism permitting erythrocyte replacement directly from stem cell sources (Pienaar, 1962). Indeed, the hyperplastic postcrisis blood picture looks much like one from an early embryo. A blood system capable of falling back in such an emergency to a rapid embryolike erythropoietic mechanism would be valuable to animals with such small total blood volumes.

C. Temperature and Behavioral Effects

Temperature in ectotherms affects both the parasite growth rate and the capability of the animal itself to mount an effective immune response. In recent experiments with pathogenic bacteria, Mathew Kluger and co-workers (1975) found that infected lizards preferentially maintained higher body temperatures, and when allowed to do so showed increased survival. Apparently several components of the lizard defense mechanism are temperature-dependent, including phagocytic activity, leukocyte mobility, and humoral mediators of inflammation.

Only Paul Thompson and Winder (1947) have studied the effect of maintenance at different temperatures on saurian malaria infections. In their observations of *P. floridense* in *Anolis carolinensis*, they found that growth curves, levels of peak parasitemia, and manifestation of acquired immunity were similar at 30° and 20°C, but the infections developed much more slowly at the lower temperature. In my own experience, blood-passaged infections often fail to develop in lizards maintained at a constant and abnormally low (for the lizard) "room temperature."

In contrast to the experiments with pathogenic bacteria, José Scorza (1971) observed that lizards with heavy *P. tropiduri* infections tended to remain inactive in the coolest areas of their cages. Reptiles have the unique capacity to reduce their metabolic rate, including oxygen needs, by lowering their body temperature. This, added to the normal high tolerance of reptiles for anoxia, may permit them to survive hemolytic infections that would be fatal to birds or mammals.

The difficulty of a weakened, anemic lizard to engage in energetic activity could prevent it from gaining and defending a territory during the mating season (which in temperate zones coincides with the period of spring relapse). In another sense, whoever catches a wild lizard soon realizes that it is a terrifying experience for the animal: heartbeat and respiratory rates are greatly increased. An anemic lizard which might have survived in the wild may die under the stress of a collector's bag or captivity.

V. Epidemiology

A. Zoogeography

Plasmodium species occur in lizards from each of the world's six zoogeographical regions, and many more species await discovery. Even in the Neotropics, the majority of reports come from workers in only three countries: central Mexico, Panama, and eastern Brazil. Table II and Figs. 13 and 14 summarize the known distribution of named reptile *Plasmodium* species.

Reptile plasmodia occur in forested or sometimes grassland habitats, being especially abundant in wet tropical forests. They are conspicuously absent in desert reptiles, although haemococcidians and haemogregarines are often common. They are found from sea level to over 3000 m elevation in Arizona; as far north as 38° in Japan, 39° in California, and 41° in Wyoming; and as far south as 25° in South Africa, and 39° in New Zealand. Most species occur between 20° north and 20° south latitude, their frequency increasing progressively toward the equatorial tropical forests, just like the species diversity of their reptile hosts.

Substantial gaps between the haemosporidian and coccidia groups, as well as between haemoproteids, *Leucocytozoon,* and the plasmodia, make tracing the origin of the malarial parasites highly speculative. There are far too many missing links. These gaps speak for themselves as evidence for the haemosporidian's great antiquity. It could well be that the early evolution of the group took place in the blood of long extinct Mesozoic reptile species. Biting Diptera, including psychodids and even the subfamily Phlebotominae are believed to have been present in the early Mesozoic (Downes, 1971).

Lizards existed throughout most of the Mesozoic "Age of Reptiles," witnessing the rise and disappearance of the dinosaurs. All modern lizard superfamily groups were present by the end of the Jurassic, and fossils from the late Cretaceous onward can usually be placed without difficulty into current lizard families. Nevertheless, lizards and snakes are a modern and still evolving order and many groups have undergone considerable speciation in the late Tertiary and Quaternary.

Nearly all saurian plasmodia are found in, or can be derived directly from, species of Gondwanaland-derived regions, especially Africa and the American tropics. Except for the single report of malaria in a varanid from Ceylon, their hosts are all believed to have a Gondwanaland origin (Cracraft, 1973). The impoverishing, arid climatic changes undergone by continental Africa since the Miocene may in part account for the comparatively fewer species found there thus far. South America is so rich in

Table II

Known Distribution of *Plasmodium* Species in Reptiles, According to Zoogeographical Realm, Host Family and Species, Study Site, and a Recent Reference for Each Host from Which a Named Species Was Reported

Family	*Plasmodium*	Hosts	Location	Recent Reference [a]
Palearctic Realm (2 reports)		*Takydromus tachydromoides*	Japan	Telford (1972a)
Lacertidae	*sasai*	*Takydromus smaragdinus*	Amami Island	Telford (1972b)
Oriental Realm (6 reports)				
Agamidae	*"minasense"*	*Gonyocephalus borneensis*	Malaya	Laird (1960)
		Gonyocephalus grandis	Malasia	Yap *et al.* (1967)
	vastator	*Draco volans*	Malaya	Laird (1960)
Varanidae	*clelandi*	*Varanus cepedianus* (=*bengalensis*)	Ceylon	Manawadu (1972)
Austrialian Realm (4 reports)				
Agamidae	*giganteum-australis*	*Amphibolurus barbatus*	Australia	Mackerras (1961)
Scincidae	*egerniae*	*Egernia major*	Australia	Mackerras (1961)
	lacertiliae	*Carlia* (=*Leiolopisma*) *fusca*	Goodenough Island	Thompson and Hart (1946)
	lygosomae	*Leiolopisma* (=*Lygosoma*) *moco*	New Zealand	Laird (1951)
		Emoia (=*Lygosoma*) *cyanura*	Solomon Islands	Garnham (1966)
		Prasinohaema virens anolis	Solomon Islands	Garnham (1966)
		Lepidodactylus (?) *guppyi*	Solomon Islands	Garnham (1966)
Ethiopian Realm (30 reports)				
Agamidae	*agamae*	*Agama agama* (=*colonorum*)	Gambia, Liberia	Baker (1961)
			Sudan, Nigeria(?)	Macfie (1914)
			Sierra Leone(?), Congo	Schwetz (1931)
			Kenya	Ball (1967)

		Agama atricollis	South Africa	Pienaar (1962)
	giganteum	*Agama agama* (=*colonorum*)	Liberia, Nigeria, Sierra Leone, Mali	Bray (1959)
			Kenya, Tanzania	Ball (1967)
		Agama cyanogaster	Kenya	Southgate (1970)
		Agama sp	Ethiopia	R. W. Ashford [c]
Cordylidae	*zonuriae*	*Cordylus* (=*Zonurus*) *vittifer*	South Africa	Pienaar (1962)
Gerrhosauridae	*zonuriae*(?)	*Gerrhosaurus vallidus*	South Africa	Garnham (1966)
Chamaeleonidae	*acuminatum*	*Chamaeleo fischeri*	Tanganyika	Pringle (1960)
	fischeri	*Chamaeleo fischeri*	Tanganyika	Ball and Pringle (1965)
	robinsoni	*Chamaeleo brevicornis*	Madagascar	Brygoo (1962)
Scincidae	*mabuiae*	*Mabuya* (=*Mabuia*) *quinquetaeniata*	Sudan	Wenyon (1926)
	maculilabre	*Mabuya* (=*Mabuia*) *maculilabris*	Congo	Schwetz (1931)
	pitmani	*Mabuya maculilabris*	Kenya, Uganda	Garnham (1966)
		Mabuya striata	Kenya, Uganda	Krampitz (1970)
Nearctic Realm (46 reports)				
Anguidae	*mexicanum*	*Gerrhonotus multicarinatus*	California	Ayala (1973)
Iguanidae	*beltrani*	*Sceloporus variabilis*	Veracruz,[b] Oaxaca	Peláez (1967)
		Sceloporus teapensis	Veracruz [b]	Peláez (1967)
	brumpti	*Sceloporus horridus*	Morelos	Peláez and Pérez-Reyes (1952)
	chiricahuae	*Sceloporus clarkii*	Arizona	Telford (1970b)
		Sceloporus jarrovii	Arizona	Telford (1970b)
	floridense	*Sceloporus undulatus*	Florida, Georgia	Jordan (1975)
		Anolis carolinensis	Florida, Georgia	Jordan (1975)
	mexicanum	*Sceloporus graciosus*	California	Ayala (1973)
		Sceloporus grammicus (=*microlepidotus*)	Michoacán, México DF	Peláez and Pérez-Reyes (1952)

Table II (continued)

Family	*Plasmodium*	Hosts	Location	Recent Reference [a]
		Sceloporus horridus	Michoacán	Peláez and Pérez-Reyes (1952)
		Sceloporus pyrocephalus (?)	Michoacán	Thompson and Huff (1944b)
		Sceloporus occidentalis	California	Ayala (1973)
		Sceloporus torquatus (=*ferrariperezi*)	México, DF, Michoacán	Peláez *et al.* (1948)
		Sceloporus undulatus	Wyoming	Greiner and Daggett (1973)
Neotropical Realm (62 reports)				
Anguidae	*diploglossi*	*Diploglossus fasciatus*	Rio (Brazil)	Aragão and Neiva (1909)
Gekkonidae	*aurulentum*	*Thecadactylus rapicaudus*	Panama, Venezuela [c]	Telford (1971)
	gonatodi	*Gonatodes albogularis*	Panama	Telford (1970a)
		Gonatodes humeralis	Pará	Lainson *et al.* (1971)
Iguanidae	*achiotensis*	*Basiliscus basiliscus*	Panama	Telford (1972b)
	audaciosum	*Plica umbra*	Pará	Lainson *et al.* (1975)
	azurophilum	*Anolis cybotes*	Haiti	Telford (1975)
		Anolis grahami	Jamaica	Telford (1975)
		Anolis krugi	Puerto Rico	Telford (1975)
		Anolis lineatopus	Jamaica	Telford (1975)
	balli	*Anolis chloris*	W. Colombia	S. C. Ayala [d]
		Anolis fuscoauratus	Peru	Guerrero and Ayala (1976)
		Anolis limifrons	Panama	Guerrero *et al.* (1977)
		Anolis lionotus	Panama	Telford (1974)
		Anolis poecilopus	Panama	Telford (1974)
		Anolis vittigerus	W. Colombia	S. C. Ayala [d]
	basilisci	*Basiliscus basiliscus*	Panama	Telford (1972b)

	Basiliscus galeritus	W. Colombia	S. C. Ayala [d]
	Basiliscus plumifrons	Panama	Telford (1972b)
	Basiliscus vittatus	Beliz, Panama, Veracruz	Telford (1972b)
carinii (=*rhadinurum*)	*Ctenosaura similis*	Belize	Garnham (1966)
	Iguana iguana	Mexico: Colima, Guerrero, Nayarit, Oaxaca, Veracruz; Belize, Panama, W. Colombia [d], Venezuela, Trinidad, Brazil: Goias, Amazonas	Peláez (1967), Scorza (1970)
cnemidophori	*Plica plica*	Guyana	Telford (1973a)
colombiense	*Anolis auratus*	Colombia, Venezuela [c]	Ayala and Spain (1976)
floridense	*Anolis biporcatus*	Panama	Telford (1974)
	Anolis concolor	San Andrés Island	Ayala (1975)
	Anolis conspersus	G. Cayman Island	Telford (1975)
	Anolis cybotes	Haiti	Telford (1975)
	Anolis distichus	Haiti	Telford (1975)
	Anolis frenatus	Panama	Guerrero *et al.* (1977)
	Anolis garmani	Jamaica	Telford (1975)
	Anolis grahami	Jamaica	Telford (1975)
	Anolis limifrons	Panama	Guerrero *et al.* (1977)
	Anolis lineatopus	Jamaica	Telford (1975)
	Anolis opalinus	Jamaica	Telford (1975)
	Anolis pentaprion	Panama	Telford (1974)
	Anolis pulchellus	Puerto Rico	Telford (1975)
	Anolis sagrei	N. Bimini, Bahamas	Telford (1975)
	Sceloporus malachiticus	Panama	Huff and Marchbank (1953)

Table II (continued)

Family	*Plasmodium*	Hosts	Location	Recent Reference [a]
	gonzalezi(?)	*Anolis squamulatus* (?)	Venezuela	Iturbe and González (1921)
	"minasense"	*Anolis capito* (?)	Panama	Telford (1974)
		Anolis frenatus (?)	Panama	Telford (1974)
		Anolis limifrons (?)	Panama	Telford (1974)
		Plica plica (?)	Guyana	Telford (1973a)
		Plica umbra (?)	Guyana	Telford (1973a)
	modestum	*Tropidurus torquatus*	Pará	Lainson *et al.* (1974)
	multiformis	*Plica umbra*	Pará	Lainson *et al.* (1975)
	simplex	*Plica umbra*	Pará	Lainson *et al.* (1975)
	tropiduri	*Anolis biporcatus* (?)	Panama	Telford (1974)
		Anolis frenatus (?)	Panama	Telford (1974)
		Anolis limifrons (?)	Panama	Telford (1974)
		Anolis lionotus (?)	Panama	Telford (1974)
		Anolis pentaprion (?)	Panama	Telford (1974)
		Anolis poecilopus (?)	Panama	Telford (1974)
		Plica umbra	Guyana	Telford (1973a)
		Tropidurus hispidis	Venezuela, Guyana?	Scorza (1970, 1971) Telford (1973a)
		Tropidurus torquatus	Bahía, Goias, Minas, São Paulo	Rocha e Silva (1975)
	uncinatum	*Plica plica*	Guyana	Telford (1973)
	uranoscodoni	*Uranoscodon superciliosa*	Pará	Lainson *et al.* (1975)
	utingensis	*Anolis punctatus*	Pará	Lainson *et al.* (1971)
	vacuolatum	*Plica umbra*	Pará	Lainson *et al.* (1975)
	vautieri	*Urostrophus vautieri*	São Paulo	Pessoa and Biasi (1973)

Scincidae	*diploglossi*	*Mabuya mabouya*	Pará, Panama, W. Colombia [d]	Telford (1970)
	mabuyi	*Mabuya mabouya*	Pará	Lainson *et al.* (1974)
	"minasense"	*Mabuya mabouya*	Minas	Carini and Rudolph (1912)
	morulum	*Mabuya mabouya*	Panama, Pará	Lainson *et al.* (1974)
	tropiduri	*Mabuya mabouya*	Pará	Lainson and Shaw (1969)
Teiidae	*attenuatum*	*Ameiva ameiva*	Guyana, Venezuela [c]	Telford (1973)
	basilisci	*Ameiva ameiva*	Venezuela	Scorza (1970)
	cnemidophori	*Ameiva ameiva*	Amazonas (Brazil), Guyana, Venezuela, Panama, E. Colombia [d]	Telford (1973), Scorza (1970)
		Cnemidophorus lemniscatus	Brazil: Amazonas, Goias	Lainson and Shaw (1969)
	diminutivum	*Ameiva ameiva*	Panama	Telford (1973b)
	effusum	*Neusticurus bicarinatus*	Pará	Lainson *et al.* (1974)
	josephinae	*Ameiva undulata*	Veracruz [b]	Peláez (1967)
	"minasense"	*Ameiva ameiva*	Venezuela	S. R. Telford [c]
	pifanoi	*Ameiva ameiva*	Venezuela	Scorza and Dagert (1956)
	"Saurocytozoon" sp.	*Ameiva ameiva*	Venezuela	S. R. Telford [c]
		Tupinambis tequixin	Venezuela	S. R. Telford [c]
	telfordi	*Ameiva ameiva*	Mato Grosso, Guyana, Venezuela [c]	Lainson *et al.* (1971)
	tropiduri (?)	*Ameiva ameiva*	Amazonas, Brazil	Walliker (1966)
		Cnemidophorus lemniscatus	Amazonas, Brazil	Walliker (1966)
	tupinambi	*Crocodilurus lacertinus*	Pará (?)	Lainson *et al.* (1974)
		Tupinambus tequixin	Pará	Lainson *et al.* (1974)

[a] A complete host-parasite list and annotated bibliography are found in Ayala (1977).
[b] Borderline Nearctic–Neotropical.
[c] Personal communication, unpublished.
[d] Personal observation, unpublished.

species that it must have been a center of evolution during its long Tertiary isolation from the other world landmasses.

Some patterns are evident at the lizard family level. Agamids are historically an Asian group. If the agamid parasites *P. agamae* and *P. giganteum* were not included, the distribution map for Africa would be quite bare (Fig. 13). Plasmodia are widely distributed in the two "old" South American families Iguanidae and Teiidae. The Gekkonidae are mainly an Old World group, but *Plasmodium* species have been found only in a few New World representatives, so these parasites would seem to have been secondarily acquired. Skinks form a huge family with over 600 species, but outside the Australian region plasmodia have been found only in a few members of the genus *Mabuya*. In the New World, they are known from the single species *Mabuya mabouya*, and at least two of its four described *Plasmodium* species also occur in other lizard groups. Once again, these infections seem to have been acquired since the relatively recent arrival of their hosts in the New World.

An analysis of the recent distribution of saurian plasmodia proves especially fruitful. Unlike birds, lizards do not migrate, and individuals spend their entire lifetime within a sharply delimited area. This has been characteristic of lizards throughout their Quaternary history, with most species populations closely associated with specific climatic and geofloral environments. Supported by a growing knowledge of the evolution and zoogeography of their host groups, the saurian plasmodia provide excellent models for tracing the evolution and speciation of malaria parasites.

Two iguanid genera are especially well suited for these studies: *Anolis*, a tropical group with around 200 species, and *Sceloporus*, a temperate genus with nearly 60 species (Peters and Donoso-Barros, 1970). Both underwent extensive radiation during the Quaternary and there are numerous examples of closely related species showing geographical or ecological isolation—favoring the development of isolated hemoparasite races. Both genera are the object of a multidisciplinary effort to reconstruct the main patterns of their evolutionary history (morphology, chromosome analysis, biochemistry, behavioral comparisons, zoogeography, demography, and reproductive patterns, etc.).

Our understanding of the current distribution of the *P. mexicanum* and *P. floridense* groups of parasites seems to correlate well with the known facts of climatic-related events in North and Central America during the Quaternary. Parasites of the *P. mexicanum* complex infect spiny lizard species inhabiting grassland, chaparral, or oak-pine woodland "ecological island" remnants of the Madro-Tertiary geofloral assemblage. They are likely derived from a basal stock that once infected *Sceloporus* ancestors in the Central Mexican plateau (Ayala, 1970). Pleistocene drying trends

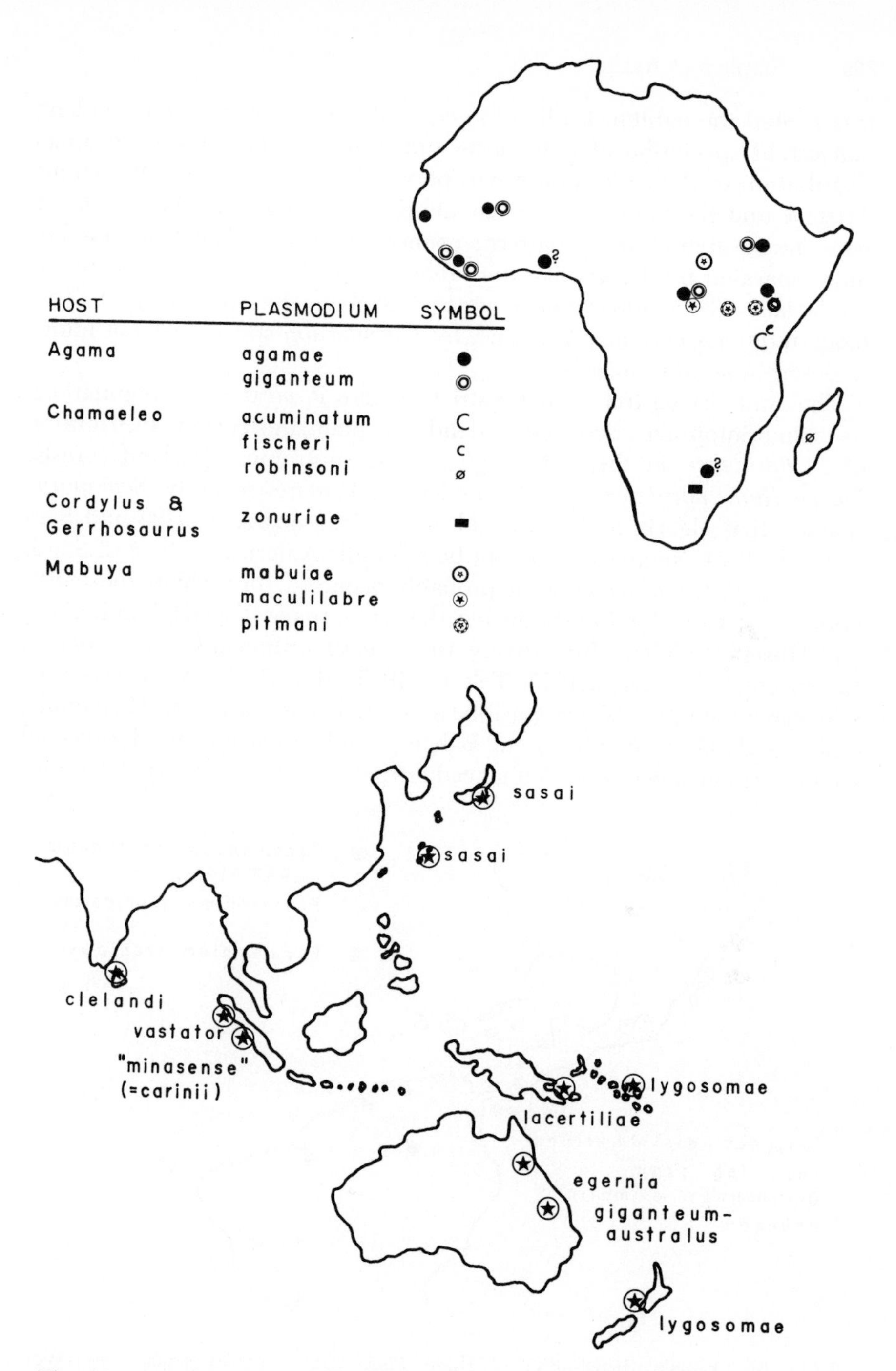

Fig. 13. Distribution of the described species of *Plasmodium* from Old World lizards.

fragmented the habitat, leading to geographic isolation and independent, concurrent speciation of both hosts and parasites. The present disjunct distribution of *P. mexicanum* group populations in California, Wyoming, Arizona, and the Sierra Madre Mountains of Mexico (Fig. 14) is similar to a "mountaintop" relict pattern: connected during glacial periods but now separated by deserts like the Mohave that developed during the interglacials. The comparative newness (geologically speaking) of most modern deserts may help explain why *Plasmodium* species are not found in desert-inhabiting lizards.

The same drying trends that restrict modern *P. mexicanum* populations to mountaintop-like "ecological islands," explain the current distribution of *P. floridense* in disjunct, tropical, and subtropical lowland forests. *Plasmodium floridense* infects anoles, and more recently *Sceloporus* species. It is clearly of Neotropical origin; José Scorza (1970) and Sam Telford (1974) suggested it might be a Middle America offshoot of South American *P. tropiduri*. If so, it probably occurred throughout the moist, Gulf Coast Corridor forests during the Quaternary glacials (Auffenberg and Milstead, 1965). Alternatively, the recent Caribbean findings confirm the speculation (Ayala, 1975; Telford, 1975) that *P. floridense* was once widespread on the larger Caribbean islands like Jamaica, Hispaniola, Cuba, and the now submerged Bahama bank. Invasion of Florida and Central America was surely enhanced during the 100-m sea level lowerings

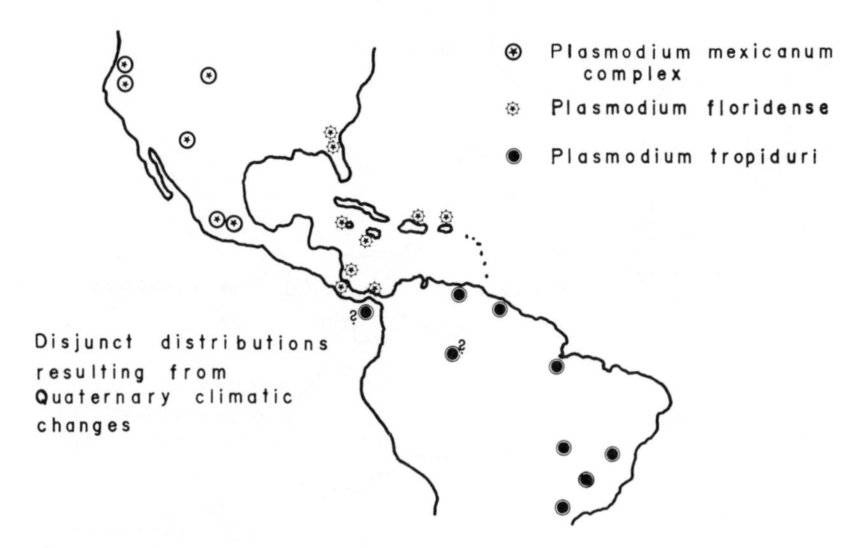

Fig. 14. Known distribution of three *Plasmodium* species from New World lizards, showing the effect of climate changes during the Quaternary in isolating host and parasite populations.

accompanying the Pleistocene glacials. Many of the hosts of *P. floridense,* such as, *Anolis carolinensis, A. sagrei, A. grahami,* and *A. distichus* are notorious for their versatility and ability to invade new areas across water barriers. Furthermore, parasite species like *P. floridense,* which reach such high prevalence in anole populations inhabiting coastal areas, have a high probability of "waifing" via overwater dispersal.

Although the situation in South America is still largely unknown, data is accumulating on the distribution patterns of several Neotropical saurian plasmodia. *Plasmodium tropiduri* is widespread in *Tropidurus* species from southeastern Brazil to coastal Venezuela* and apparently infects several other sympatric lizard populations. With growing attention to quantitative descriptions, it will soon be possible to clarify the degree of relationship between *P. tropiduri, P. floridense,* and the other *tropiduri*-complex populations (Ayala and Spain, 1976). *Plasmodium balli* occurs in anoles in Central America, western Colombia, eastern Peru, and Brazil. Several reptilian plasmodia occur in both Panamá and northeastern Brazil, including the species *P. cnemidophori, P. diploglossi, P. gonatodi, P. balli, P. tropiduri* (?), *P. carinii, P. morulum,* and *P. diminutivum.* Dry interglacial periods during the Pleistocene seem to account for the isolation and extensive speciation of bird and lizard groups in confined forest refugia (Müller, 1973; Raven and Axelrod, 1975; Vanzolini and Williams, 1970), and probably explain the large numbers of somewhat distinct but clearly related *Plasmodium* species now being described from lizards of Central and South America.

B. Host Specificity

Host specificity of saurian plasmodia is probably much less strict than is often assumed. *Plasmodium mexicanum* (Fig. 15) demonstrates specificity at several levels. As a temperate species, it has long occupied a place in a distinct lizard–arthropod–habitat assemblage. Phylogenetic specificity and concurrent evolution in isolated *Sceloporus* species is well established; known natural hosts include *Sceloporus torquatus* (=*ferrari-perezi*), *S. graciosus, S. horridus, S. grammicus* (=*microlepidotus*), *S. occidentalis, S. pyrocephalus,* and *S. undulatus.* An ecological specificity is also well documented: grassland, chaparral, and oak-pine woodlands derived from Madro-Tertiary origins, isolated into several "ecological islands" during interglacial periods like the present, but continuous during glacial times. A physiological specificity is also indicated by comparing the predominantly erythrocyte invasion in its natural *Sceloporus* hosts

* An excellent qualitative and quantitative description of *Plasmodium tropiduri* in *Tropidurus hispidis* from Venezuela is found in Scorza (1970).

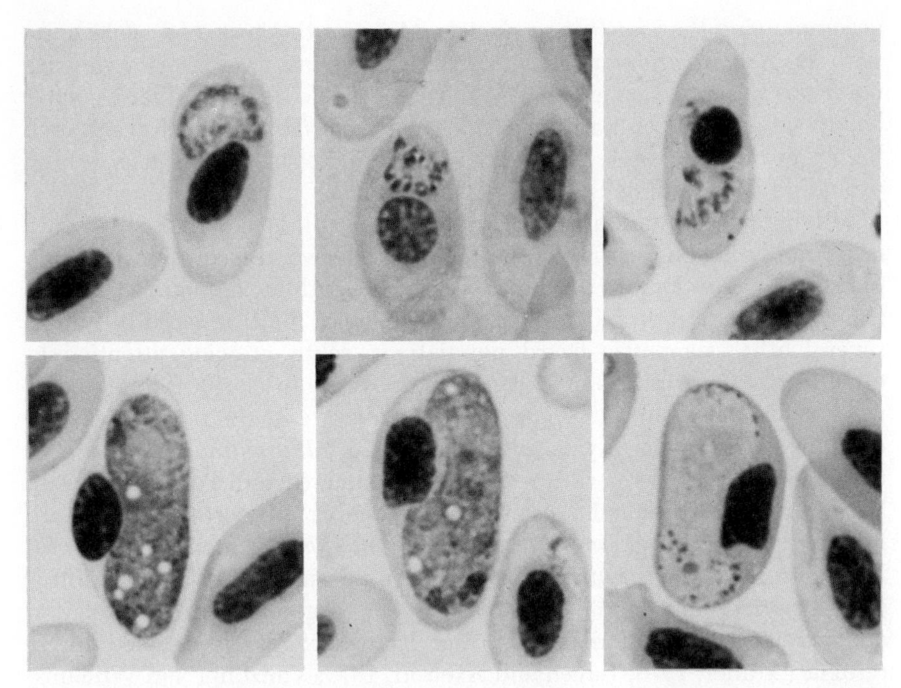

Fig. 15. *Plasmodium mexicanum* gametocytes and mature schizonts in *Sceloporus undulatus* from southeastern Wyoming. Slide prepared by Ellis Greiner and Pierre Daggett.

versus the largely leukocyte invasion in several experimentally infected *Phrynosoma* species (Ayala, 1970, 1973; Thompson and Huff, 1944a).

Outside the *Sceloporus* species, temperate *P. mexicanum* probably has few other natural hosts, except where ecological circumstances bring other species (like *Gerrhonotus multicarinatus* in California) into the natural cycle. In contrast, widespread tropical species like *P. balli* or *P. tropiduri* must have countless opportunities to explore new host–parasite–vector–microhabitat combinations. The epidemiology of these infections will become much clearer when the intermediate hosts are known.

Island populations are another promising area for research. Several saurian malaria parasites, like *P. clelandi* or *P. floridense,* inhabit isolated continental shelf islands, separated from the mainland only during interglacial periods. The present distribution of *P. sasai* may be historically related to the fact that Japan was spared the effects of the last Ice Age, and has since maintained a warm, moist climate. The high frequency of malaria in Caribbean anoles and skinks on the South Pacific islands reflects the facility which anoles and skinks, respectively, bridge overwater gaps. Land areas in both regions were much larger due to the lower sea levels of the Pleistocene and potential vectors must be widespread. Searches for saurian plasmodia on other islands in the South Pacific, the Philippines, and the Japan island chain will surely prove successful. One practical advantage in studying small island populations is the low frequency of simultaneous infections with two or more *Plasmodium* species, so common in tropical mainland populations.

C. Epidemiological Studies

The impact of malaria on lizard populations, as with most other wildlife diseases, has never been evaluated. If judged by human morbidity standards, it would have to be classified as a widespread, debilitating disease.

Most studies provide information on how many blood smears were positive, but they seldom include a sampling program deliberately planned to evaluate prevalence. Infection foci in lizard populations are often closely circumscribed, surrounded by areas of little or no infection. Most frequently infected host species live on or near the ground, but some, like *Iguana iguana* or *Draco volans*, live high in the trees.

For previously unstudied areas, samples of five to ten adult lizards from several different rural sites are preferable to much larger numbers from fewer areas. Some typical findings where prevalence estimates have been attempted include: *Agama agama* (Baker, 1961, in Liberia), 30% infection rate; *Anolis auratus* (Ayala and Spain, 1976, in western Colombia), 28%; *Anolis carolinensis* (Jordan and Friend, 1971, in southern Georgia), 10–50%; *Anolis limifrons* (Guerrero *et al.*, 1977, on Barro Colorado Island, Panama), 40%; *Anolis limifrons* (Ayala, 1973, unpublished, in Finca La Selva, Heredia, Costa Rica), 38%; *Basiliscus basiliscus* (Telford, 1972b, in Panama), 25–50%; *Sceloporus occidentalis* (Ayala, 1970, 1973, in central California), 25%; *Sceloporus undulatus* (Jordan and Friend, 1971, in southern Georgia), 2–10%; and *Tropidurus hispidus* (Scorza, 1970, in north-central Venezuela), 28%.

Malaria infections in reptiles normally remain patent for several months, at least. The continuous patency of a *P. carinii* infection during more than 3 years was already mentioned. As a result, natural infections are easier to detect in reptiles than in birds or mammals, and information obtained in a single blood film survey (point prevalence) provides a much closer estimate of the true prevalence of malaria in the population than can be expected with similar surveys of endothermic vertebrates.

Seasonal infection patterns have been studied with one temperate and one tropical zone species. *Plasmodium mexicanum* infections in California are essentially latent throughout the late summer, fall, and winter. Like many avian plasmodia from temperate regions, it undergoes a period of intense reactivation in the spring, a regional pattern visible in all the local populations (Ayala, 1970, 1973). *Plasmodium sasai* in Japan seems to have a similar annual pattern.

In contrast, *P. colombiense* in Colombia is active in every month of the year, even though the majority of infections seen in a particular survey will usually be chronic. There are local epidemics, but at the regional population level there is little overall seasonal synchronization.

Where periodic, synchronized relapses occur, they seem to coincide with host reproductive seasons. Thus, *P. mexicanum* in *Sceloporus* species shows a "spring relapse," while *P. colombiense* shows active infections year-round in *Anolis auratus*, which, like most tropical anoles, reproduces continuously (Ayala and Spain, 1976). Tropical lizard populations, with their wide variety of contrasting annual reproductive patterns (Fitch, 1970), provide numerous "experiments of nature" that could help clarify these different epidemiologic cycles.

Plasmodium species also undergo long-term fluctuation in population density. This was well documented by Helen Jordan and Margaret Friend (1971) for *P. floridense* in the Okefenokee Swamp near Fargo, Georgia. Infection prevalence in *Anolis carolinensis* decreased every year from over 50% in 1958 to about 10% in 1964, and then began a gradual but continuous rise back to over 45% by 1970. At the same time, prevalence in *Sceloporus undulatus* varied from 10 to 2%, with a corresponding low in 1963–1964.

Sampling recently hatched lizards is a reliable way of detecting newly acquired infections, or determining the season of transmission. In anoles, infections may be uncommon before the lizards reach maturity. Prevalence usually increases progressively with age, with most infections being chronic in the older lizards. However, most of the *Plasmodium* biomass occurs in "middle age" lizards because the older animals form only a small proportion of most populations (Fig. 16).

When deaths occur, they will be most likely seen in the initial weeks of a fulminating infection, in the period of anemic stress following the crisis, or again many months later during a relapse. In temperate zones where severe infections are seasonally synchronized, highest mortality will probably occur in yearlings in the spring following their birth. A selective die-off of infected lizards may also occur during the winter hibernation. The most practical way to evaluate the impact of malaria on the wild populations would be in a joint effort, by routinely preparing blood smears from each of the animals censused in the demographic studies currently underway on several lizard populations. The extremely high rate of population turnover in some tropical lizards, like *Anolis limifrons*, combined with a widespread, high prevalence of infections would seem to make them especially well-suited models for disease impact studies.

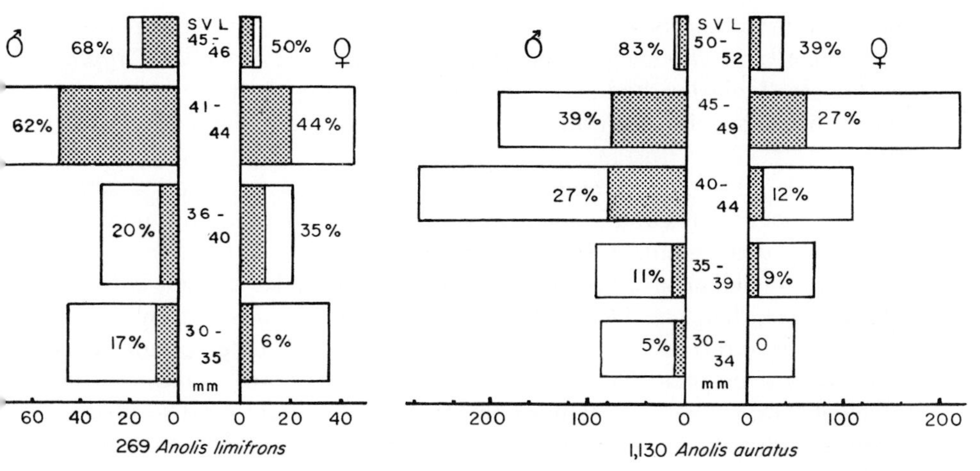

Fig. 16. Age/size-specific malaria prevalence (shaded areas and percentages) in two lizard population samples: 269 *Anolis limifrons* from Barro Colorado Island, Panama (August–October 1974; Guerrero *et al.*, 1977) and 1130 *Anolis auratus* from Cali, Colombia (August 1973–July 1974; Ayala and Spain, 1976). SVL—snout-vent length.

VI. Diagnosis and Identification

Identification of saurian plasmodia species is usually made from stained thin blood smears. Smears can be made from a single drop of blood obtained from a toe clip, with a micropipette from the retroorbital sinus, by allowing a blood-sucking insect to feed (for example, a mosquito or a kissing bug like *Rhodnius prolixus*) and then quickly removing the blood from the insect's stomach, or at autopsy. Joyce Spain in our laboratory was able to make rapid surveys by placing fresh blood drops from each of many lizards under cover slips and scanning for refractile pigment granules with a phase contrast microscope.

Many other poorly known organisms invade reptile blood cells and are sometimes confused with plasmodia. Haemogregarines (*Haemogregarina, Hepatozoon*) and haemococcidia (*Schellackia* and *Lankesterella*) inhabit both erythrocytes and leukocytes. The round, weakly staining gametocytes of *Hepatozoon iguanae*, for example, may easily be misidentified as a nonpigmented *Plasmodium* gametocyte. There are many unconfirmed reports of piroplasms from reptiles, especially turtles and snakes. Tissue coccidians (*Besnoitia* and *Sarcocystis*) sometimes flood impression smears or blood smears made during autopsy or from a toe clip. Bacterial, rickettsial, and viral organisms attach to the erythrocyte surface or form inclusion bodies within the erythrocytes. Although not governed by the rules of zoological nomenclature, these have been reported under a variety of names: *Pirhaemocyton, Toddia, Sauroplasma, Aegyptianella, Cytamoeba, Eperythrozoon, Haemabartonella, Cryptococcus,* etc. Inclusion bodies are reported from the erythrocytes of numerous lizard species, including several also carrying malaria (Johnston, 1975).

A reliable plasmodia identification must be based on a measured series of mature segmenters and gametocytes. Some estimate of variation within the sample is essential. A series of twenty examples of each form will usually be sufficient to show the variation in a particular animal, unless it harbors more than one *Plasmodium* species. Camera lucida drawings or photographs of the series are especially helpful. Except in temperate areas where only *one* species is present, a single character will usually not be sufficient. Unfortunately, many of the biological criteria used to characterize and identify avian plasmodia species (Garnham, 1966) are not yet applicable or practical for saurian plasmodia.

Telford (1974) summarized the criteria used to distinguish saurian malaria in fixed blood films. This is much more information than most species descriptions contain, especially the earlier ones, but with it the task of comparing your parasite with others previously reported becomes much easier. In this form, the data can be easily tabulated manually or by computer.

Direct characters: (1) segmenter length; (2) width; (3) number of nuclei; (4) shape: cruciform, fan, rosette, morulum, amorphous; (5) pigment form; (6) color; (7) distribution; (8) gametocyte length; (9) width; (10) size: L x W; (11) length/width ratio: L/W; (12) shape: round, oval, lenticular, elongate; (13) pigment form: individual granules or concentrated masses; (14) pigment distribution: scattered, localized; (15) sex ratio.

Relative characters: (16) host cell type; (17) position in host cell: polar, lateropolar, lateral; (18) size relative to host cell nucleus: gametocyte L x W/host cell nucleus L x W.

Indirect characters: (19) host cell nucleus displaced; (20) host cell nucleus distorted and/or hypertrophied; (21) host cell distorted or hypertrophied.

In each case, the mean, range, and standard deviation are calculated. If properly interpreted, the mean of the sample means gives a reliable estimate of the characteristic phenotype of the parasite in that particular host species. A typical species description is given below to illustrate how this information can be presented. It is adapted from Sam Telford's (1974) study of malaria in Panamanian anoles (Fig. 17).

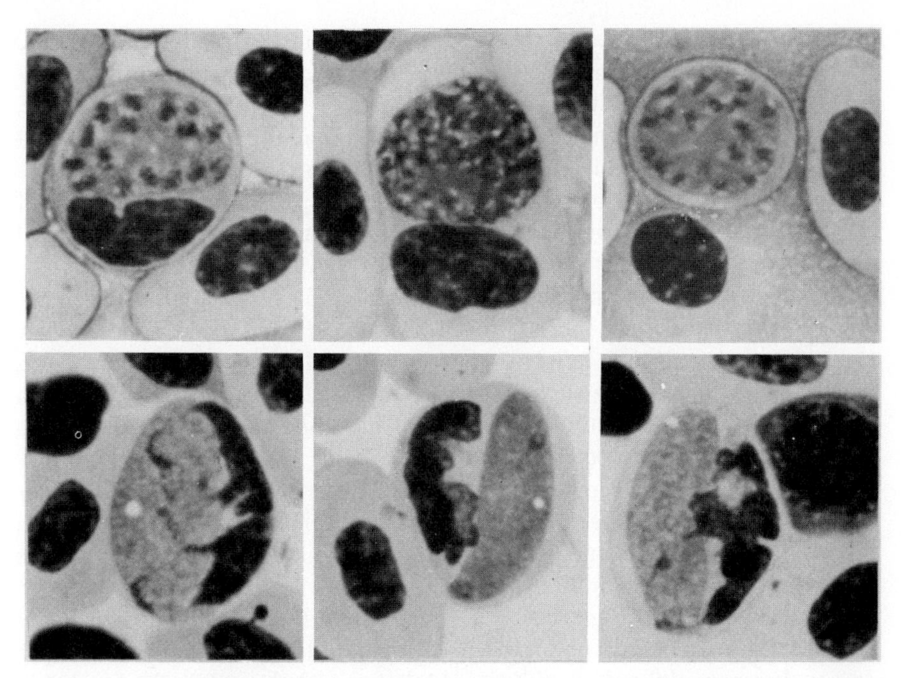

Fig. 17. *Plasmodium balli* gametocytes and mature schizonts in distorted erythrocytes of *Anolis limifrons* from Barro Colorado Island, Panama (see Guerrero *et al.*, 1977). This parasite is also known from *Anolis* species in amazonian Peru and Brazil, western Colombia, and Costa Rica.

Plasmodium balli Telford 1969

Description: This is a saurian malaria parasite which has 13 to 100 nuclei arranged amorphously in mature schizonts. The mean number of nuclei in individual infections ranges from 22.7 to 55.9 and may be influenced by the host species. Gametocytes are oval to bulky, with mean L/W ratios of 1.3 to 2.4. Gametocytes measure 8–25 × 4–14 μm; their mean size (LW) ranges from 77 to 140. Mean ratios of gametocyte LW to host cell nucleus LW range from 1.5 to 1.9 in erythrocytes. There is no visible pigment.

Trophozoites, schizonts, and gametocytes may parasitize erythroblasts, proerythrocytes, or erythrocytes, usually being found in the immature cells. The type of host cell parasitized may be influenced by the host species. Host cells and nuclei are almost always hypertrophied and distorted by both schizonts and gametocytes. Nuclei are displaced and are usually disorganized by a lytic effect from the parasites. No exoerythrocytic forms have been seen in circulating or fixed cells.

When different parasite populations are to be compared, the following additional information should be available for each animal: *ecological:* collection site, habitat type, date, season (rainfall, temperature), host size, and sex; *blood picture:* percentage normocytes and principal types of immature cells, evidence of erythropoiesis, absolute parasitemia per 1000 cells, relative parasitemia (percentage gametocytes, asexual forms); *infection state:* active (initial, acute, fulminating), recent postcrisis, chronic.

With this information, by using the accompanying host–parasite–location lists (Table II), and drawings of typical forms of each species (see Text Fig. 1), it should be possible to decide the probable identification of most species. For confirmation, it may be necessary to refer to the original descriptions.

Unfortunately as quantified information is not yet available for many species, the taxonomist faces a dual problem of judging how accurately the printed description reflects what was actually present on the smear, and to what degree the sample described reveals the range of variation in the population. Quantified descriptions are and should be applied even when only a single infected animal is available, but it pays to be highly suspicious of the degree to which such a description reflects the larger population. This is especially important if the host comes from an area where there are seasonal differences in infection dynamics, or where mixed infections are common.

Simultaneous infection with different *Plasmodium* species is extremely common in some lizard populations of mainland Neotropical rain forests. It can be very difficult deciding which forms belong to which species when the number of lizards examined is small, yet a few seemingly well-quantified descriptions are based precisely on samples of this type (e.g., Telford, 1973).

It may help to be aware of other possible pitfalls in describing or identifying saurian plasmodia. They could have affected the material used for someone else's original description as well as your own material. Some records can only be treated with caution until they have been more thoroughly investigated. When few lizards or only low-level infections are available, the identification may best be left tentative. The way of defining a "segmenter" can affect the number of nuclei counted; the criteria: division underway and cytoplasm beginning to concentrate around compact nuclei is probably the most reliable (Ayala and Spain, 1976; Thompson, 1944). Parasites on slides made from dead lizards may show postmortem deterioration or other structural changes. Malarial parasites show somewhat different structure in different host species. Immature erythrocytes, especially erythroblasts, are easily confused with medium-sized lymphocytes. Pigment particles from melanocytes, uric acid crystals, or formalin-hemoglobin artifacts in fixed tissues may be mistaken for malaria pigment. The methyl alcohol used for fixation may dissolve the pigment granules of some plasmodia (Garnham, 1966). Giemsa is not a stable stain; it may fade after many months and, although old smears can be restrained, often the changes cannot be corrected. The acidity of the staining solution can make a striking difference in the appearance of the parasites. In some mounting media the cytoplasm fades considerably, making the chromatin appear more prominent, and other media seem to dissolve the pigment. Parasites in such preparations may be described as having prominent vacuoles, where none were originally present. It may not be possible to determine how slides described by other workers were treated before they were examined. Erythrocytes of different lizard species vary considerably in size; normal for one species may be 7 x 13 μm, and for another 12 x 18 μm (Saint Girons, 1970). The size relationship between the cell and its nucleus is quite different in different lizard species, affecting the relative size and appearance of the parasites. Mechanical distortion of host cells in the preparation of the blood smear can alter the normal dimensions of the parasites; if possible, only parasites in undistorted cells should be measured. If erythrocytes and their nuclei appear abnormally hypertrophied, the smear may not be suitable for taking reliable measurements.

VII. Treatment

The early promise of saurian plasmodia as models for testing therapeutic agents was not followed up once mouse plasmodia became widely available. Few of the antimalarial drugs of current choice have been tried in the treatment of saurian plasmodia infections, although there is no reason to believe they would be any less effective than they are in avian or mammal infections. Quinine decreases the rate of growth and segmentation of *P. mexicanum* and *P. floridense* in erythrocytes, but is ineffective against *P. mexicanum* exoerythrocytic forms in circulating or fixed cells (Thompson, 1946a). Atabrine inhibits development of *P. floridense* in *Anolis carolinensis* (Thompson, 1946b). José Scorza (1971) successfully used an iron–dextran complex containing 50 mg/ml of iron to treat severe postcrisis anemia in *Tropidurus hispidis* infected with *P. tropiduri.*

The animal, the lizard with its malaria,
  character of this time, this space
    but more.
This year's program in eons of heritage.
  Shapes and processes of time;
    seen only in the moment now.
What can it tell us,
    if we know how to ask?

ACKNOWLEDGMENT

Research supported in part by Grant 1 RO1 AI 12511–01 TMP from the National Institutes of Health, United States Public Health Service.

REFERENCES

Aikawa, M., and Jordan, H. B. (1968). Fine structure of a reptilian malarial parasite. *J. Parasitol.* **54,** 1023–1033.

Auffenberg, W., and Milstead, W. W. (1965). Reptiles in the Quaternary of North America. *In* "The Quaternary of the United States" (H. E. Wright and D. G. Frey, eds.), pp. 557–568. Princeton Univ. Press, Princeton.

Aragão, H. de B., and Neiva, A. (1909). A contribution to the study of the intraglobular parasites of the lizards. Two new species of *Plasmodium, Pl. diploglossi* n. sp. and *Pl. tropiduri* n. sp. *Mem. Insto. Oswaldo Cruz.* **1,** 44–50.

Ayala, S. C. (1970). Lizard malaria in California; description of a strain of *Plasmodium mexicanuum,* and biogeography of lizard malaria in western North America. *J. Parasitol.* **56,** 417–425.

Ayala, S. C. (1971). Sporogony and experimental transmission of *Plasmodium mexicanum. J. Parasitol.* **57,** 598–602.

Ayala, S. C. (1973). The Phlebotomine sandfly—protozoan parasite community of central California grasslands. *Am. Midl. Nat.* **89,** 266–280.

Ayala, S. C. (1975). Malaria and hemogregarines from lizards of the western Caribbean islands of San Andrés and Providencia. *Rev. Inst. Med. Trop. Sao Paulo* **17,** 218–224.

Ayala, S. C. (1977). Check-list, host-index and annotated bibliography of the plasmodia of reptiles. *J. Protozool.* (in press).

Ayala, S. C., and Spain, J. (1976). A population of *Plasmodium colombiense* n. sp. in the Iguanid lizard, *Anolis auratus. J. Parasitol.* **62,** 177–189.

Baker, J. R. (1961). Attempts to find the vector of the Plasmodiidae of the lizard *Agama agama agama* in Liberia. *Ann. Rep. Res. Act. Liberian Inst. Am. Found. Trop. Med.* pp. 28–40.

Ball, G. H. (1967). Some blood sporozoans from East African reptiles. *J. Protozool.* **14,** 198–210.

Ball, G. H., and Pringle, G. (1965). *Plasmodium fischeri* n. sp. from *Chamaeleo fischeri. J. Protozool.* **12,** 479–482.

Bray, R. S. (1959). On the parasitic protozoa of Liberia. II. The malaria parasites of agamid lizards. *J. Protozool.* **6,** 13–18.

Brygoo, E. R. (1962). Un noveau *Plasmodium* de caméléon *Haemamoeba robinsoni* n. sp. *Arch. Inst. Pasteur de Madagascar* **30,** 161–169.

Carini, A., and Rudolph, M. (1912). Sur quelques hématozoaires de lézards au Brésil. *Bull. Soc. Path. Exot.* **5,** 592–595.

Cracraft, J. (1973). Vertebrate evolution and biogeography in the Old World tropics: Implications of continental drift and palaeoclimatology. *In* "Implications of Continental Drift to the Earth Sciences" (D. H. Tarling and S. K. Runcorn, eds.), pp. 373–393. Academic Press, New York.

Downes, J. A. (1971). The ecology of blood-sucking Diptera: an evolutionary perspective. *In* "Ecology and Physiology of Parasites" (A. M. Fallis, ed.), pp. 232–258. Univ. of Toronto Press, Toronto.

Duguy, R. (1970). Numbers of blood cells and their variation. *In* "Biology of the Reptilia" (C. Gans, ed.), Vol. 3, pp. 93–109. Academic Press, New York.

Fitch, H. S. (1970). Reproductive cycles in lizards and snakes. *Misc. Publ. Univ. Kansas Mus. Nat. Hist.* No. **52,** pp. 1–247.

Garnham, P. C. C. (1965). *Plasmodium wenyoni* sp. nov., a malaria parasite of a Brazilian snake. *Trans. R. Soc. Trop. Med. Hyg.* **59,** 277–279.

Garnham, P. C. C. (1966). "Malaria Parasites and Other Haemosporidia." Blackwell, Oxford.

Greiner, E. C., and Daggett, P. M. (1973). A saurian *Plasmodium* in a Wyoming population of *Sceloporus undulatus*. *J. Herpetol.* **7**, 303–304.

Guerrero, S., and Ayala, S. C. (1976). Hemoparásitos de algunos reptiles y anfibios de la selva amazónica del Perú. *Rev. Inst. Med. Trop. Sao Paulo* **18**, (in press).

Guerrero, S., Rodríguez, C., and Ayala, S. C. (1977). Prevalencia de hemoparásitos en lagartijas de la Isla Barro Colorado, Panamá. *Biotropica* **9** (1) (in press).

Hsu, C. K., Cambell, G. R., and Levine, N. D. (1973). A checklist of the species of the genus *Leucocytozoon* (Apicomplexa, Plasmodiidae). *J. Protozool.* **20**, 195–203.

Huff, C. G. (1969). Exoerythrocytic stages of avian and reptilian malarial parasites. *Exp. Parasitol.* **24**, 383–421.

Huff, C. G., and Marchbank, D. F. (1953). Saurian malaria in Panama. *Nav. Med. Res. Inst. Rept.* **11**, 509–516.

Iturbe, J., and González, E. (1921). Sobre algunos datos de protozoología y parasitología recogidos en San Juan de los Moros. I. *Plasmodium gonzalezi* nov. sp. *Gaceta Méd. Caracas (Venezuela)* **28**, 275–276, 283.

Johnston, M. R. L. (1975). Distribution of *Pirhemocyton* Chatton & Blanc and other, possibly related, infections of poikilotherms. *J. Protozool.* **22**, 529–535.

Jordan, H. B. (1975). The effect of host constitution on the development of *Plasmodium floridense*. *J. Protozool.* **22**, 241–244.

Jordan, H. B., and Friend, M. B. (1971). The occurrence of *Schellackia* and *Plasmodium* in two Georgia lizards. *J. Protozool.* **18**, 485–487.

Kluger, M. J., Ringler, D. H., and Anver, M. R. (1975). Fever and survival. *Science* **188**, 166–168.

Krampitz, H. E. (1970). Plasmodien im Reptilienblut. *Natur u. Museum* **100**, 85–90.

Lainson, R., and Shaw, J. J. (1969). A new haemosporidian of lizards, *Saurocytozoon tupinambi* gen. nov., sp. nov., in *Tupinambus nigropunctatus* (Teiidae). *Parasitology* **59**, 159–162.

Lainson, R., Landau, I., and Shaw, J. J. (1971). On a new family of non-pigmented parasites in the blood of reptiles: Garniidae Fam. Nov. (Coccidiida: Haemosporidiidae). Some species of the new genus *Garnia*. *Int. J. Parasitol.* **1**, 241–250.

Lainson, R., Landau, I., and Shaw, J. J. (1974). Further parasites of the family Garniidae (Coccidiida: Haemosporidiidae) in Brazilian lizards. *Fallisia effusa gen.* nov., sp. nov. and *Fallisia modesta* gen. nov., sp. nov. *Parasitology* **68**, 117–125.

Lainson, R., Shaw, J. J., and Landau, I. (1975). Some blood parasites of the Brazilian lizards *Plica umbra* and *Uranoscodon superciliosa* (Iguanidae). *Parasitology* **70**, 119–141.

Laird, M. (1951). *Plasmodium lygosomae* n. sp., a parasite of a New Zealand skink, *Lygosoma moco* (Gray, 1839). *J. Parasitol.* **37**, 183–189.

Laird, M. (1960). Malayan protozoa. 3. Saurian malaria parasites. *J. Protozool.* **7**, 245–250.

Lucas, A. M., and Jamroz, C. (1961). Atlas of Avian Hematology, *U.S. Dep. Agric., Agric. Monogr.* **25**, 1–271.

Macfie, J. W. (1914). Notes on some blood parasites collected in Nigeria. *Ann. Trop. Med. Parasitol.* **8**, 439–463.

Mackerras, M. J. (1961). The haemotozoa of Australian reptiles. *Aust. J. Zool.* **9**, 61–122.

Manawadu, B. R. (1972). A new saurian malaria parasite *Plasmodium clelandi* sp. n. from Ceylon. *J. Protozool.* **19,** 587–589.

Muller, P. (1973). "The Dispersal Centres of Terrestrial Vertebrates in the Neotropical Realm. A Study in the Evolution of the Neotropical Biota in its Native Landscapes." Junk, The Hague.

Peláez, D. (1967). Estudios sobre hematozoarios. XIII. Un nuevo *Plasmodium* de *Ameiva* en México. *Ciencia* (*Mexico City*) **25,** 121–130.

Peláez, D., and Pérez-Reyes, R. (1952). Estudios sobre hematozoarios. III. Las especies americanas del género *Plasmodium* en reptiles. *Rev. Palud. Med. Trop.* (*Mexico*) 3/4, 137–160.

Peláez, D., Pérez-Reyes, R., and Barrera, A. (1948). Estudios sobre hematozoarios. I. *Plasmodium mexicanum* Thompson y Huff 1944 en sus huéspedes naturales. *Anales Esc. Nac. Cienc. Biol.* **5,** 197–215.

Pessôa, S. B., and Biasi, P. (1973). Plasmódio de uma lagartixa, *Urostrophus vautieri* D. & B. (Sauria, Iguanidae). *Mem. Inst. Butantan* **37,** 309–316.

Pessôa, S. B., Biasi, P., and Puorto, G. (1974). Nota sobre a freqüência de hemoparasitas em serpentes do Brasil. *Mem. Inst. Butantan* **38,** 69–118.

Pessôa, S. B., and Fleury, G. C. (1968). *Plasmodium tomodoni* sp. n. parasita da serpente *Tomodon dorsatus* D. & B. *Rev. Bras. Biol.* **28,** 525–530.

Peters, J. A. and Donoso-Barros, R. (1970). Catalogue of the Neotropical Squamata. Part II. Lizards and Amphisbaenians. *U.S. Nat. Mus., Bull.* **297,** pp 1–293.

Peters, W., Fletcher, K. A., and Staubli, W. (1965). Phagotrophy and pigment formation in a chloroquine resistant strain of *Plasmodium berghei* Vincke and Lips 1948. *Ann. Trop. Med. Parasitol.* **59,** 126–134.

Pienaar, U. de V. (1962). "Haematology of Some South African Reptiles." Witwatersrand Univ. Press, Johannesburg.

Pringle, G. (1960). Two new malaria parasites from East African vertebrates. *Trans. R. Soc. Trop. Med. Hyg.* **54,** 411–414.

Raven, P. H., and Axelrod, D. I. (1975). History of the flora and fauna of Latin America. *Am. Sci.* **63,** 420–429.

Rocha e Silva, E. O. (1975). Ciclo evolutivo de *Hepatozoon triatomae* (Sporozoa, Haemogregarinidae) parasita de triatomíneos. *Rev. Saúde Públ.* (*Sao Paulo*) **9,** 383–391.

Saint Girons, M. (1970). Morphology of the circulating blood cells. *In* "Biology of the Reptilia" (C. Gans, ed.), Vol. 3, pp. 73–91. Academic Press, New York.

Schwetz, J. (1931). Sur quelques hématozoaires des lézards de Stanleyville et du lac Albert. *Ann. Parasit. Hum. et Comp.* **9,** 193–201.

Scorza, J. V. (1970). Lizard Malaria. Doctoral dissertation, Univ. of London, pp. 1–300. (avail. University Microfilms, Ann Arbor, No. 76–29–075).

Scorza, J. V. (1971). Anaemia in lizard malaria infections. *Parassitologia* **13,** 391–405.

Scorza, J. V., and Dagert, C. (1956). *Plasmodium pifanoi,* nov. sp. parásito de *Ameiva ameiva ameiva. Nov. Cient., Museo Hist. Nat. La Salle* (*Caracas*), *Ser. Zool.* **20,** 3–7.

Southgate, B. A. (1970). *Plasmodium* (*Sauramoeba*) *giganteum* in *Agama cyanogaster:* a new host record. *Trans. R. Soc. Trop. Med. Hyg.* **64,** 12–13.

Telford, S. R. (1969). A new saurian malarial parasite *Plasmodium balli* from Panama. *J. Protozool.* **16,** 431–437.

Telford, S. R. (1970a). Saurian malarial parasites in eastern Panama. *J. Protozool.* **17,** 566–574.

Telford, S. R. (1970b). *Plasmodium chiricahuae* sp. nov. from Arizona lizards. *J. Protozool.* **17**, 400–405.

Telford, S. R. (1971). A malaria parasite, *Plasmodium aurulentum* sp. nov. from the Neotropical forest gecko *Thecadactylus rapicaudus. J. Protozool.* **18**, 308–311.

Telford, S. R. (1972a). The course of infection of Japanese saurian malaria (*Plasmodium sasai,* Telford & Ball) in natural and experimental hosts. *Jap. J. Exp. Med.* **42**, 1–21.

Telford, S. R. (1972b). Malarial parasites of the "Jesu Cristo" lizard *Basiliscus basiliscus* (Iguanidae) in Panama. *J. Protozool.* **19**, 77–81.

Telford, S. R. (1973a). Saurian malarial parasites from Guyana: their effect upon the validity of the family Garniidae and the genus *Garnia,* with descriptions of two new species. *Int. J. Parasitol.* **3**, 829–842.

Telford, S. R. (1973b). Malaria parasites of the "borriguerro" lizard, *Ameiva ameiva* (Sauria: Teiidae) in Panama. *J. Protozool.* **20**, 203–207.

Telford, S. R. (1974). The malarial parasites of *Anolis* species (Sauria: Iguanidae) in Panama. *Int. J. Parasitol.* **4**, 91–102.

Telford, S. R. (1975). Saurian malaria in the Caribbean: *Plasmodium azurophilum* sp. nov., a malarial parasite with schizogony and gametogony in both red and white blood cells. *Int. J. Parasitol.* **5**, 383–394.

Thompson, P. E. (1944). Changes associated with acquired immunity during initial infections in saurian malaria. *J. Infect. Dis.* **75**, 138–150.

Thompson, P. E. (1946a). Effects of quinine on saurian malarial parasites. *J. Infect. Dis.* **78**, 160–166.

Thompson, P. E. (1946b). The effects of atabrine on the saurian malarial parasite, *Plasmodium floridense. J. Infect. Dis.* **79**, 282–288.

Thompson, P. E., and Hart, T. A. (1946). *Plasmodium lacertiliae* n. sp. and other saurian blood parasites from the New Guinea area. *J. Parasitol.* **32**, 79–82.

Thompson, P. E., and Huff, C. G. (1944a). A saurian malarial parasite, *Plasmodium mexicanum,* n. sp., with both elongatum- and gallinaceum-types of exoerythrocytic stages. *J. Infect. Dis.* **74**, 48–67.

Thompson, P. E., and Huff, C. G. (1944b). Saurian malarial parasites of the United States and Mexico. *J. Infect. Dis.* **74**, 68–79.

Thompson, P. E., and Winder, C. V. (1947). Analysis of saurian malarial infections as influenced by temperature. *J. Infect. Dis.* **81**, 84–95.

Vanzolini, P. E., and Williams, E. E. (1970). South American anoles: the geographic differentiation and evolution of the *Anolis chrysolepis* species group (Sauria, Iguanidae). *Arq. Zool.* (*Sao Paulo*) **19**, 1–124.

Walliker, D. (1966). Malaria parasites of some Brazilian lizards. *Parasitology* **56**, 39–44.

Wenyon, C. M. (1926). Plasmodia of lizards. *In,* Protozoology. A Manual for Medical Men, Veterinarians and Zoologists. Vol. **II**, pp. 982–983. Hafner Publ. Co., New York (1965 republ).

Yap, L. F., Fredericks, H. J., and Omar, I. (1967). A new host for *Plasmodium minasense* Carini & Rudolph, 1912. *Med. J. Malaya* **21**, 369.

7

Plasmodia of Birds

Thomas M. Seed and Reginald D. Manwell

I. Introduction

Malaria of birds has only slight veterinary interest but there are, in fact, many good reasons for knowing something about it. The major reason is that very much of what has been learned about avian malaria has proved applicable to malaria of man, a most important disease through most of his history.

Human malaria has decreased in importance in the last two decades largely because its vector, the anopheline mosquito, has been curbed. Control of malaria by control of mosquitoes was the natural outgrowth of the discovery that mosquitoes spread the disease. Sir Ronald Ross, the second Nobel Prize winner in medicine, proved that mosquitoes transmitted malaria of sparrows. Shortly thereafter, the Italians, Grassi, Bignami, and Bastianelli, demonstrated that transmission of human malaria also occurred by mosquitoes.

Avian malaria is not one disease, but many. It is caused by many quite distinct species of *Plasmodium.* Human malaria is also a disease complex, but there are only four species of plasmodia which infect man, not nearly as many as infect birds.

II. Historical Review

Avian malaria parasites were first seen almost as early as were those of man, and quite independently. Laveran in 1880 discovered the human plasmodia and realized their causal role in malaria. Danielewski first saw malaria parasites in the blood of birds in 1884. For several years, however, there was considerable confusion about the identity of the parasites he and others had seen and, in particular, there was confusion between the true malaria parasites and *Haemoproteus.* Such confusion is understandable because there are species of avian *Plasmodium* whose gametocytes are almost indistinguishable from those of *Haemoproteus.* Furthermore, mixed infections are common. To be sure, *Haemoproteus* multiplies only in the internal organs (usually the lungs) whereas *Plasmodium* multiplies in blood cells as well. This distinction, however, may be hard to make in practice since, in chronic avian malarial infections, the form most often seen in the peripheral blood is the gametocyte.

For some time there were those who thought avian and human malaria parasites were the same. Fortunately for all of us, they are not; for if they were, malaria-infected birds would threaten human health almost everywhere since mosquitoes infected from them could then infect man.

W. G. MacCallum, who later became one of the world's greatest pathologists, discovered the significance of "exflagellation" or microgametogenesis in pigeon malaria (really *Haemoproteus* infection), while he was a medical student at Johns Hopkins. This observation led to the demonstration of the sexual cycle of the parasites and resolved the question of the relation of the microgametes to the plasmodia. Some had thought the microgametes were a degeneration product of the parasites, and others, that they were another parasite to which they had given the name "*Polymitus*."

After the epic work of the late 1800's, little additional work of significance was done in the field until the mid-1920's, when malaria-infected canaries (first used by the great bacteriologist, Robert Koch) proved very useful for screening programs for the detection of antimalarial activity of synthetic compounds. As a result of the tests, the effectiveness of an antimalarial, Plasmochin, was established, although the drug never gained wide acceptance because of its relatively high toxicity for man. Shortly thereafter, however, the same screening programs led to the discovery of atebrin (now usually called mepacrine or quinacrine). It was not long before this drug almost completely replaced quinine, which had been the only effective medicinal agent in malaria therapy for three centuries.

At first, only canaries infected with *Plasmodium cathemerium* or *P. relictum* were available for experimental work in avian malaria. The need for a larger and less expensive host was satisfied with the discovery of *Plasmodium gallinaceum*, a species from the East Indies which infects chickens, discovered by E. Brumpt, an eminent French scientist, in 1935 and *P. lophurae* from the Philippines which infects ducks, discovered by L. T. Coggeshall in 1938.

James and Tate (1937) made a discovery of great importance soon after the introduction of *P. gallinaceum* as an experimental tool. It had long been supposed that erythrocytes were the only cells infected by malaria parasites in the vertebrate, but these two English scientists proved that in chicken malaria, at least, endothelial cells of capillaries in the brain were also parasitized. However, the complete life cycle for this species was not fully elucidated until 1944 by C. G. Huff and F. Coulston. The discovery of tissue stages in the plasmodial life cycle was of great significance for at least two reasons: (1) the existence of such stages, which are not susceptible to some antimalarial drugs, helped explain the failure of therapy to completely eradicate malarial infection, and (2) the persistence of such stages after cessation of the infection in the blood furnished a good explanation for relapse. It took prolonged and laborious study to locate these stages in lower mammals and man, but they were finally revealed to be in the liver. The demonstration of tissue stages in the malaria parasite life cycle furnished important new evidence of the close relationship of the three genera of haemosporidia, *Haemoproteus*, *Leucocytozoon*, and *Plasmodium*.

The amount of recent research in avian malaria has declined somewhat, largely because the discovery of species of *Plasmodium* which will infect rats and mice has made mammalian hosts inexpensive and easily obtainable. The use of the lower primates for malaria research has also become more practical than it previously was.

Much of the recent research in avian malaria has involved study of the physiology of the parasites and attempts at *in vitro* culture (for general reviews, see Hewitt, 1940; Huff, 1963, 1968).

III. Parasites and Life Cycles

When Laveran discovered malaria parasites in human blood in 1880, he did so without benefit of staining; quite certainly this was also the case when Danielewski first saw avian malaria parasites 5 years later. Danielewski did not have the advantage of having his attention drawn to the parasitized red cells by amoeboid movement as he might have had he been looking at vivax-infected erythrocytes of man. Even the so-called amoeboid stages of avian plasmodia exhibit little or no amoeboid movement (though they sometimes appear to have pseudopodial processes when observed in stained preparations). The living avian plasmodia usually appear only as small, clear spaces, or vacuoles, with perhaps a little pigment in the cytoplasm of the nucleated host erythrocyte. The larger stages of some species often distort the host cell and displace its nucleus. Malaria parasites of any kind are seldom studied today except in Romanowski-stained thick or thin blood films (Figs. 1–12). Giemsa or Wright stains are usually used in the United States. Leishman's stain is more often used elsewhere.

Ordinarily both sexual and asexual forms may be found in the same blood film, but in species with a well-defined periodicity, the relative proportions of parasites of various ages may differ greatly, depending on the time of day. The youngest asexual parasites ("rings") may be ring-like in form, and are often located at the ends of the erythrocyte. Pigment granules are usually only visible in the larger forms. Chromatin and cytoplasm become readily distinguishable as the parasite becomes older. Parasites in which the chromatin has commenced division are known as schizonts (Fig. 2) and are often called segmenters when schizogony is complete. Offspring are known as merozoites (Fig. 5).

Sexual stages (gametocytes) may be round or elongate (Figs. 1 and 9). The two sexes always show marked differences in staining. The cytoplasm of macrogametocytes (female) takes a slaty blue color and the nucleus, which is usually quite compact, a deeper reddish tint. The male cells (microgametocytes) exhibit a much more diffuse nucleus, which stains lightly, while the cytoplasm hardly stains at all. When mature, both sexes show widely scattered pigment granules which may vary slightly in color according to the species.

The type of host cell invaded varies to some extent with the species of *Plasmodium*. Some parasites prefer reticulocytes to older erythrocytes, and one (*Plasmodium elongatum*) may parasitize any type of blood or blood-forming cell, even thrombocytes. *Plasmodium elongatum* may also segment in the bone marrow as well as in the peripheral blood.

Effects on the host cells during development vary greatly with the species. *Plasmodium cathemerium* (Fig. 1) and *P. relictum* distort the host cell greatly as they increase in size and their gametocytes often push out the nucleus entirely (Fig. 1). By contrast, species of *Novyella* hardly affect it at all.

The life cycles of the avian malaria parasites in the blood are similar to those of the mammalian plasmodia, and are probably also like those of the reptilian malaria parasites from which both the avian and mammalian forms may have descended. Indeed, the morphology, at least in the blood stages, of the malaria parasites of reptiles and birds is very similar.

The life cycle of *Plasmodium relictum,* probably the first of the avian malaria parasites to be discovered, will serve as an example. Infection of the bird begins with the bite of an infected culicine mosquito. Sporozoites introduced with the salivary secretion are carried to appropriate tissue sites of the bird. Huff (1954) found these usually to be the reticular cells of the splenic Malpighian body, not only for *P. relictum* but for some other species as well. Later generations of the parasites are less selective in host cell preference. The mammalian species of *Plasmodium,* in contrast, undergo preerythrocytic schizogony only in the liver parenchyma.

After a minimum of three generations of exoerythrocytic development, the parasites (known as cryptozoites, metacryptozoites, and phanerozoites, respectively) may spill over in the circulation and invade erythrocytes, or they may continue asexual reproduction as phanerozoites. Such reproduction may occur in a variety of tissue sites; the lungs, brain and spleen are favorite ones. Human malaria parasites may be limited to a single exoerythrocytic generation, and this occurs in the liver.

These merozoites transform into rings, then trophozoites, schizonts, and finally presegmenters and segmenters after invasion of erythrocytes. The time required to complete a cycle in some species is quite definite, although this is not especially so for *P. relictum. Plasmodium matutinum,* in many ways like *P. relictum,* takes almost 24 hours to mature in the erythrocytes and the cycle is generally synchronized so that a brood

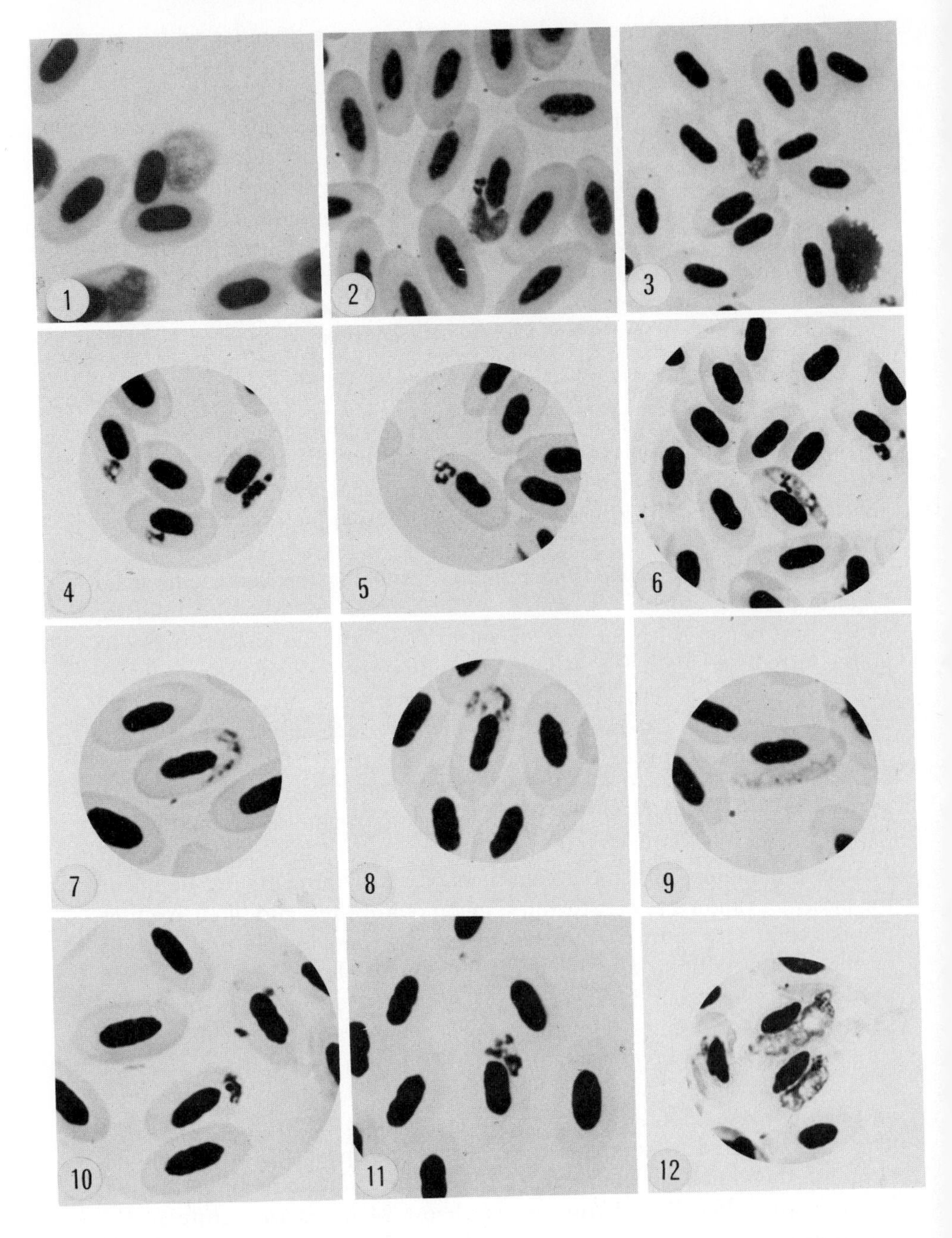
1
2
3
4
5
6
7
8
9
10
11
12

matures rather early in the morning. *Plasmodium cathemerium* has an equally sharply defined asexual cycle but segmentation occurs at about 6:00 P.M. Species of *Novyella* seem to have almost no periodicity.

The cause of periodicity is still unclear but it has been shown to be dependent both on the physiology of the host and the genetic constitution of the parasite. Parasites in which development has been arrested by refrigeration so that they are 12 hours out of phase will adjust their asexual cycle to normal when reintroduced into a noninfected bird. Similarly, if an infected bird is kept under conditions of artificially controlled light and darkness, the parasites will slow or accelerate their reproductive cycle in accordance with the changed environment.

Erythrocytic avian merozoites appear to be able to choose among four possible fates. Most of them continue schizogony in erythrocytes, but some may become gametocytes (either male or female) or they may reinitiate asexual reproduction in the tissues as phanerozoites. This last option is apparently not available to the malaria parasites of mammals. It is unknown what factors determine which course is taken. Since, when there is multiple infection of an erythrocyte each of the parasites may go its own way, becoming a gametocyte of either sex or proceeding to reproduction the determining factors must lie within each parasite.

Maturation of the gametocytes never occurs in the circulation, although it will on a slide. Normally, of course, maturation takes place only in the gut of a female mosquito (males do not suck blood). The factors triggering the process are not completely known, although a rise in carbon dioxide tension is thought to be one. If the microgametocyte is completely mature, the change will begin within a few minutes after

---

Fig. 1. Microgametocyte of *Plasmodium cathemerium* (center). Both sexual and asexual stages of this species displace (and often push out) the host cell nucleus as they increase in size. An advanced schizont is shown in the left-hand lower corner. 2000 ×. **Fig. 2.** A moderately advanced schizont of *Plasmodium anasum* is present in the center red cell. The U-shape is rather characteristic of the larger asexual stages of this species, so far known with certainty only from the Formosan shoveller duck (*Anas clypeata*). 2000 ×. **Fig. 3.** An advanced schizont of *Plasmodium nucleophilum* (center). A peculiarity of this species is its tendency to adhere to the host cell nucleus. 1500 ×. **Figs. 4–6.** *Plasmodium vaughani.* (4) Schizonts of various ages; (5) segmenter (usual number of merozoites is 4, but there may be up to 8); (6) macrogametocyte. 1500 ×. **Figs. 7–9.** *Plasmodium polare.* (7) Schizont of medium age; (8) segmenter (8 to 14 merozoites; usual about 10); (9) microgametocyte. 2000 ×. **Figs. 10 and 11.** *Plasmodium hexamerium.* (10) Segmenter (usual number of merozoites 6) (lower center); (11) more advanced segmenter, perhaps destined to produce eight merozoites. 2000 ×. **Fig. 12.** *Plasmodium fallax,* a species native to Africa and first seen in an owl (*Syrnium nuchale*). Stages shown are schizonts, which usually lie alongside the host cell nucleus. 2000 ×.

the blood is drawn and culminate in the production of about eight minute, active, filamentous microgametes, each of which awaits fertilization of a macrogamete. Fertilization results in the production of a motile zygote, or ookinete. The ookinete penetrates the gut wall and forms a tumorlike body, or oocyst, on the outside of the gut. Within a few days, the time depending on the temperature and species of parasite, schizogony within the cyst produces hundreds, or even thousands of sporozoites. These are liberated into the body cavity by rupture of the oocyst wall, and some find their way into the salivary glands.

IV. Taxonomy

The malaria parasites, no matter what the hosts, are Sporozoa. Beyond that they are classified as follows.

Subclass: Telosporidia. Elongate sporozoites, no polar capsules, usually with spore formation only at the end of trophic or growth period.

Order: Coccidiomorphida. Intracellular parasites at all stages; with unlike (anisogamous) gamonts.

Suborder: Haemosporidina. Parasites with two hosts, a vertebrate and invertebrate; sporozoites naked in latter.

Genus: *Plasmodium.* Blood parasites transmitted by mosquitos. Within the erythrocyte they metabolize hemoglobin, producing the pigment hematin as a by-product.

The actual number of species of *Plasmodium* infecting birds is uncertain, and it is important to remember that birds are not subject to infection with any of the species of malaria parasites to which mammals and reptiles are susceptible. Many more species have been described from birds than actually exist; many of such names are known to be synonyms. In the older literature, too, other generic names may occasionally be found. One such example is *Proteosoma,* usually with the species name *grassii.*

At the present time about twenty-five species are generally believed to be valid (Table I). Many of these species have received little or no laboratory study other than a determination of their morphology in blood films. Included among these are *P. anasum* (Fig. 2), *P. durae, P. formosanum, P. gundersi,* and *P. hegneri.*

There is also often the problem of whether a parasite should be re-

Table I
The Recognized Species of Avian Plasmodia

Plasmodium
anasum Manwell and Kuntz, 1965
cathemerium Hartman, 1927
circumfexum Kikuth, 1931
durae Herman, 1941
elongatum Huff, 1930 (syn. *P. praecox* Grassi & Feletti, 1890) [a]
fallax Schwetz, 1930
formosanum Manwell, 1962
gallinaceum Brumpt, 1935
garnhami Guindy, Hoogstraal and Mohammed, 1965
giovannolai Corradetti, Verolini, and Neri, 1963
gundersi Bray, 1962
hegneri Manwell, 1966
hexamerium Huff, 1935 (syn. *P. oti* Wolfson, 1936)
juxtanucleare Versiani and Gomes, 1941 (syn. *P. japonicum* Ishiguro, 1957)
lophurae Coggeshall, 1938
matutinum Huff, 1937
nucleophilum Manwell, 1935 (syn. *P. huffi* in part)
octamerium Manwell, 1968
paranucleophilum Manwell and Sessler, 1971
pinottii Muniz and Soares, 1953
polare Manwell, 1934
relictum (Grassi and Feletti, 1891) (syn. *P. inconstans* Hartman, 1927; *P. praecox* G. & F. in part)[a]
rouxi Sergent, Ed. & Et., and Catanei, 1928
tenue Laveran and Marullaz, 1914
vaughani Novy and MacNeal, 1904

[a] Hartman believed *P. elongatum* to be the same as Grassi and Feletti's *P. praecox*, and called what has since been known as *P. relictum*, *P. inconstans*, a new species. It is uncertain just what parasite Grassi and Feletti were describing under the species name *praecox;* some think it was one of the species of human *Plasmodium.*

garded as a species or subspecies on the basis of available information. *Plasmodium matutinum,* for example, was at first ranked as subspecies of *P. relictum,* which it much resembles, but it has sharp periodicity, with the height of segmentation in the morning, and apparently occurs more often in columbiform hosts than does *P. relictum.* There are also other less important differences. Some workers consider these differences sufficient to merit species rank, while others do not. Several other plasmodia are still ranked as subspecies, but are considered by some workers to be valid species. An example is *Plasmodium nucleophilum toucani* from Swainson's toucan.

It has recently been proposed to divide the genus *Plasmodium* into a number of subgenera. This has several advantages, one of them being convenience, since the number of recognized species in the genus has multiplied greatly in the last few years.

For malaria parasites of birds the following subgenera have been proposed:

Species having round gametocytes: *Haemamoeba*
Species having elongate gametocytes
and schizogony in primitive red cells: *Huffia*
without such schizogony but with relatively
large asexual stages: *Giovannolaia*
without such schizogony, but with relatively
small asexual stages: *Novyella*

This scheme may be as good as any that can be devised but it has disadvantages. Where, for example, does one draw the line between large and small asexual stages? How does one allow for the fact that a given species may exhibit a somewhat different morphology in different host species? On the whole, however, the groups listed as subgenera seem to be natural ones. Certainly, *Huffia* and *Novyella* qualify, although certain species in the last two groups overlap somewhat in their characteristics. More knowledge of the life histories of some species may also lead to changes in their taxonomic disposition.

V. Mosquito Vector

It was previously assumed that avian malaria is always transmitted by culicine mosquitoes and that anophelines are always the vectors of mammalian species. However, it is now known that anophelines may, at least experimentally, act as vectors of both. It still seems likely, however, that culicines are the usual, if not the exclusive, transmitters of the disease in birds under natural conditions.

Huff (1965), in a comprehensive review of the subject, listed twenty-five species of mosquitoes known to be susceptible to infection with *Plasmodium relictum*. Thirteen belonged to the genus *Culex*, three to *Aedes*, and two to *Culiseta*; the others were anophelines (although in two cases the experimental data were contradictory). Some 2700 species of mosquitoes are known, and it is very apparent that our present knowledge is extremely scanty, especially when it is recalled that there are numerous species of avian *Plasmodium*.

Even within a mosquito species, individuals differ in susceptibility to malaria infection. It was shown long ago by Huff (1931) that this was a matter of genetics, susceptibility of *Culex pipiens* to *Plasmodium cathe-*

merium being dependent on a single recessive gene. A mosquito receptive to one species of avian *Plasmodium* may not be receptive to infection with others.

A variety of environmental factors contribute to the development of plasmodia in mosquitoes, of which temperature is probably the most important. Ball and Chao (1963) found that 26°C is close to the optimum for the development of *Plasmodium relictum* in the mosquito. Temperatures much above this are deleterious and exposure to markedly lower temperatures (e.g., 4°C) for several days after biting prevents the maturation of sporozoites. Other species of malaria parasites would probably have their own temperature optima.

Once infected, a mosquito probably remains infected for life which, in nature, is usually not long. High temperatures, particularly if combined with low humidity, are lethal to mosquitoes, as is cold. Few survive a temperate zone winter. Birds are probably always reservoirs of malarial infection. There is no reason to think that malaria parasites harm the mosquito even when there are many oocysts or numerous sporozoites.

VI. Epizootiology

It is doubtful whether epizootics of malaria ever occur among birds. Instead, it is an enzootic disease just as human malaria is endemic in many parts of the world. Undoubtedly, the incidence of avian malaria varies with the season; it is certain that where it occurs there are always infections.

Factors governing the occurrence of avian malaria are probably similar to those governing the occurrence of human malaria but are less well known. The chief differences are in the number of host and parasite species involved and the greater pathogenicity of the human disease. The high parasitemias usually characterizing avian malaria make for great infectivity for mosquitoes, and birds are constantly outdoors and therefore exposed to infection.

We actually know little of the ecological importance of avian malaria as a cause of morbidity or mortality.

Birds, like man, are inveterate travelers. Most of them migrate to warmer climates in winter unless they already reside there. There is reason to think that many of them acquire malaria in the warmer climates if they did not have it before. Migrant species usually breed in the temperate latitudes and nestlings are especially susceptible to malarial infection just as are the very young of the human species.

Avian malaria is as common in many species as is human malaria where control measures are lacking. However, avian species differ greatly in this respect. The Psittacidae (parrots and parrotlike birds) are very seldom infected, and their relative immunity is certainly not explained by any absence of culicine mosquitoes in their environment. The incidence of malaria frequently differs greatly even in different species of the same genus. Of 156 tree sparrows (*Spizella s. arborea*) trapped in Fayetteville, New York, only one was found infected, whereas 18 of 329 chipping sparrows (*S. p. passerina*) caught in the same areas had malarial infections—0.01 vs. 5%. Examination of blood films indicated that in Fayetteville, song sparrows (*Melospiza m. melodia*) had a malaria incidence of 19% (188 infections in 1005 birds examined). When blood subinoculation was used as a test of infection, the incidence was tripled.

Just as certain species of birds are infected more commonly than others, so the incidence of a given species of *Plasmodium* may vary among different host species. *Plasmodium tenue* is extremely common in Pekin robins (*Liothrix luteus*) and is thus far known only from this and one other species of babbler (Pekin robins are neither from Pekin nor are they robins; they belong to the Timaliidae, or babblers). Similarly, *P. vaughani* is much more common in American robins (*Turdus m. migratorius*) than in other host species.

The incidence of malaria has been studied by one of us (RDM) in birds of many species, both native and foreign. A total of more than 4000 birds of sixty species trapped in central New York has been examined as well as 1234 individuals of 186 species imported from Africa, Asia, South America, and Australia. Other birds native to the High Rockies in Colorado, as well as some from Washington, D.C. and still others from Cape Cod, have also been studied. To these may be added blood films from many birds of numerous species from Taiwan. A variety of types of malaria was found in many of them. There is no doubt that avian malaria has a worldwide distribution. Some of the causal species of *Plasmodium* are evidently cosmopolitan, but others are limited to certain regions. The introduction of birds infected with exotic species may have disastrous effects on native bird populations. This occurred in Hawaii with resulting near extinction of some indigenous avian species (Warner, 1968).

VII. Ultrastructure

Avian malaria parasites develop a variety of morphological forms during their complex life cycle (Figs. 13–19). Organisms developing in the mosquito (Figs. 18 and 19) are distinct both morphologically and physiologically from those found in the blood and tissues of infected birds (Figs. 13–17).

A. Stages in the Vertebrate

The growth and reproductive cycles of plasmodia begin with the invasion of susceptible host cells by sporozoites from the mosquito. Intracellularly the sporozoites become trophozoites and grow. Merozoites are formed later intracellularly by the asexual budding of the trophozoites. When the host cell ruptures, the merozoites escape into the extracellular spaces and are either destroyed by the defense mechanisms of the bird or penetrate new susceptible host cells.

Merozoites are, in general, oval to elliptical in shape, ranging in size from 1 to 2 μm in length to 0.75 μm in width (Fig. 16). The ultrastructural features of merozoites have been studied by Aikawa (1966, 1967, 1971). They have large centrally positioned nuclei. Nucleoli are evident in exoerythrocytic merozoites but absent from erythrocytic ones. A single cresent-shaped mitochondrion with tubular cristae is present posterior to the nucleus. In close association to the mitochondrion is a "spherical body" whose function is unknown although it has been speculated that it serves as an energy source providing substrate for enzymes involved in mitochondrial phosphorylation. Apical organelles, called micronemes and rhoptries, are found only in the motile forms, the merozoites (Fig. 16), sporozoites (Fig. 19), and possibly in ookinetes. Micronemes are smaller and more numerous than are rhoptries. Rhoptries are flask-shaped and there are only two of them. Both micronemes and rhoptries are membrane-bound and have microducts extending to the tip of the truncated anterior end of the parasite. Because of their location and orientation within the cell and their dramatic change in density following the penetration of the host cell, these organelles are thought to play an active role in host cell invasion.

The merozoite, like its counterpart found in the insect, the sporozoite, has a complex surface architecture. These complex surface layers of the motile forms are probably designed to permit the parasite to survive extracellularly and to migrate to and penetrate new host cells. The apical region of the merozoite contains a series of polar rings which contribute to the "light bulb" shape of the organism. Microtubules radiate from the polar rings and extend posteriorly (Fig. 13). The middle layer of the pellicle of the merozoite is termed the "labyrinthine structure" and has a latticed appearance when viewed from the surface (Aikawa, 1967). This layer underlies the plasmalemma and covers the entire body of the merozoite with the exception of the cytostomal area. The outermost layer of the three-layered pellicle, the plasmalemma, is

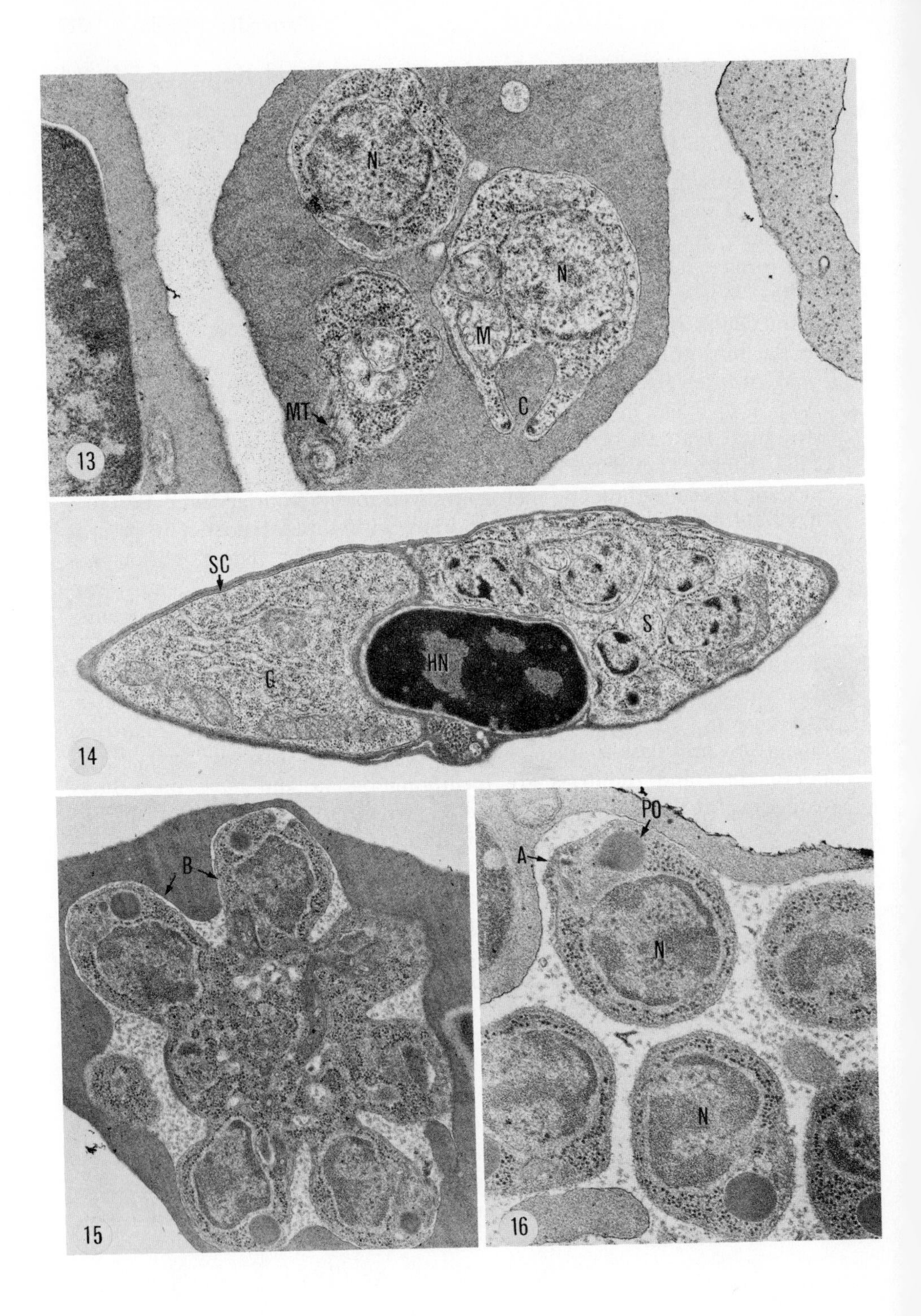

N
N
M
C
MT
13
SC
S
HN
G
14
B
15
PO
A
N
N
16

a thin trilaminar unit membrane. This plasma membrane functions to control metabolite transport, regulate osmotic balance, and physically limit the cell's cytoplasm. The surface of the extracellular merozoite is coated with a rather thick layer of mucopolysaccharides, termed the glycocalyx.

The mechanism of host cell penetration by merozoites has recently been elucidated by electron microscopy (Ladda *et al.*, 1969). Initially, the extracellular merozoite contacts the host cell and then orients itself so that its anterior end presses into the erythrocyte's plasma membrane. A depression is formed which enlarges as the merozoite advances into the cell. The host cell membrane, despite being greatly deformed, remains intact. During the last stages of penetration, the orifice of the parasitic vacuole reseals itself. Intracellularly, the highly specialized and differentiated merozoite begins to dedifferentiate (Fig. 13). The first change which the penetrating merozoite undergoes is the decrease in the electron density of the paired organelles. As the parasite completes the penetration process it rounds up and loses it light bulb shape. The specialized organelles such as the pellicular microtubules and the labyrinthine structure quickly degenerate.

The result of dedifferentiation of the intracellular merozoite is a small uninucleated trophozoite with a single surface membrane surrounded by a vacuolar membrane (Figs. 13 and 17). The trophozoite is a great deal more pleomorphic than the merozoite. The trophozoite has a prominent cristate mitochondrion. There is usually little endoplasmic reticulum (ER) within the trophozoite. What ER is present is generally

---

Fig. 13. Three young *Plasmodium gallinaceum* parasites (transitional merozoites) beginning to "dedifferentiate" into trophozoites. Note the centrally located nucleus (N) and the tubular mitochondrion (M). A prominent cytostome (C) is evident. The parasite was fixed in the process of ingesting hemoglobin. Remnants of the thick intermediate layer of the merozoite's pellicle are also shown along with pellicular microtubules (MT) radiating from the merozoite's anterior end. 26,520 ×. **Fig. 14.** An infected avian erythrocyte containing two *Plasmodium gallinaceum* parasites, separated by the host cell nucleus (HN). The parasite on the left is a gametocyte (G) characterized by a tripled membrane surface complex (SC) along with a high concentration of ribosomes and lipid inclusions in its cytoplasm. The parasite to the right of the nucleus is a multinucleated parasite (S) in the process of cytoplasmic differentiation. (Electron micrograph supplied through the courtesy of Dr. Charles Sterling.) 17,230 ×. **Fig. 15.** *Plasmodium gallinaceum* in the terminal stages of cell division. Each bud (B), segmenting from a central body, is destined to become a separate infective merozoite. 21,818 ×. **Fig. 16.** A crop of mature, intra-erythrocytic merozoites (*P. gallinaceum*). Note the typical truncated anterior (A) containing it's apical organelles [i.e., pair organelles PO)] and also the large centrally located nucleus (N). 33,300 ×.

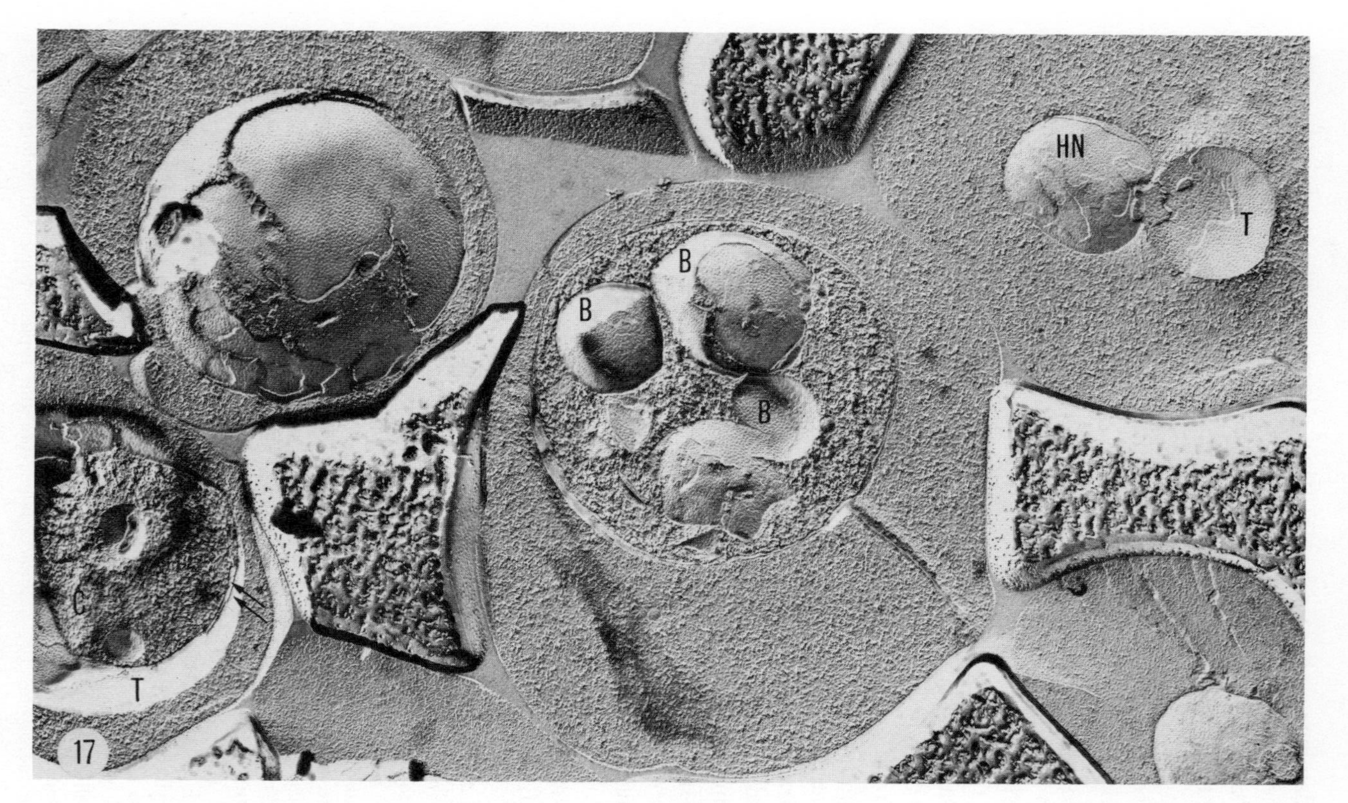

Fig. 17. Freeze-cleaved *P. gallinaceum*-infected erythrocytes. In upper right, a small trophozoite (T) lies next to the host cell nucleus (HN). In the center a schizont is shown with the maturing buds (B) lying within the coarsely granular cytoplasm of the parasitic vacuole. A surface view of a parasite in possibly a similar stage of division is shown in the upper left. Just below, lies a large trophozoite (T) whose granular cytoplasm (C) is limited by two membranes (arrow). 14,063 ×.

vesiculated and lined with ribosomes. Ribosomes are abundant and occur mainly in the free form.

One way the trophozoite obtains nutrients is through the engulfment and digestion of host cell hemoglobin. Hemoglobin is ingested through a contractile sacklike structure called a cytostome (Fig. 13) (Aikawa *et al.*, 1966). Other processes such as pinocytosis may be functional in the uptake of hemoglobin. That large quantities of hemoglobin are ingested is indicated by the large numbers of hemoglobin-containing vacuoles within the growing trophozoite. Digestion of hemoglobin occurs within these food vacuoles by lysosomal hydrolytic enzymes. A by-product of hemoglobin digestion is hematin or "malaria pigment." This is the iron-containing portion of hemoglobin which the growing parasite cannot metabolize. The parasite packages aggregates of this by-product in membrane envelops and discards them when schizogony occurs.

Following a period of active growth the uninucleated trophozoite undergoes nuclear division. The result of this mitotic activity is the formation of a schizont (Fig. 14). The earliest morphological sign of mitosis is the transformation of the nuclear chromatin from the form called heterochromatin, the inactive form, to euchromatin, the active form. Morphologically, this change is seen as the transformation of the coarse granular chromatin of the resting nucleus to the finely fibrous form of the active nucleus. Biochemically, the growth and schizogony period is one of active DNA synthesis, as indicated experimentally by the uptake and incorporation of exogenously supplied nucleic acid precursors. The mitotic apparatus has spindle fibers which radiate in a fanlike fashion from centriolar plaques. The two sets of spindle fibers meet approximately midway between the centrioles. On occasion, a pair of small electron-dense bars are seen in close association with the spindle fibers. Originally these bars were thought to be the chromosomes of the avian malarial parasites, but it is now known through cytochemical studies that these are kinetochores—structures which bind chromosomes to spindle fibers (Aikawa, 1971). In spite of the many studies of the processes of nuclear division, chromosomes have not yet been observed. Metaphase processes are followed by morphological changes in the dividing nucleus typical of telophase. The nucleus elongates and then constricts at its center; the mitotic apparatus then degenerates and disappears. As in most other protozoa, the nuclear membrane of the avian malarial parasite remains intact during mitosis. Concurrent with nuclear division, mitochondrial division occurs. This process which occurs through a budding process, produces many mitochondria, one for each of the subsequently formed daughter merozoites. Cytoplasmic division follows shortly after nuclear division. Cytoplasmic division begins with a

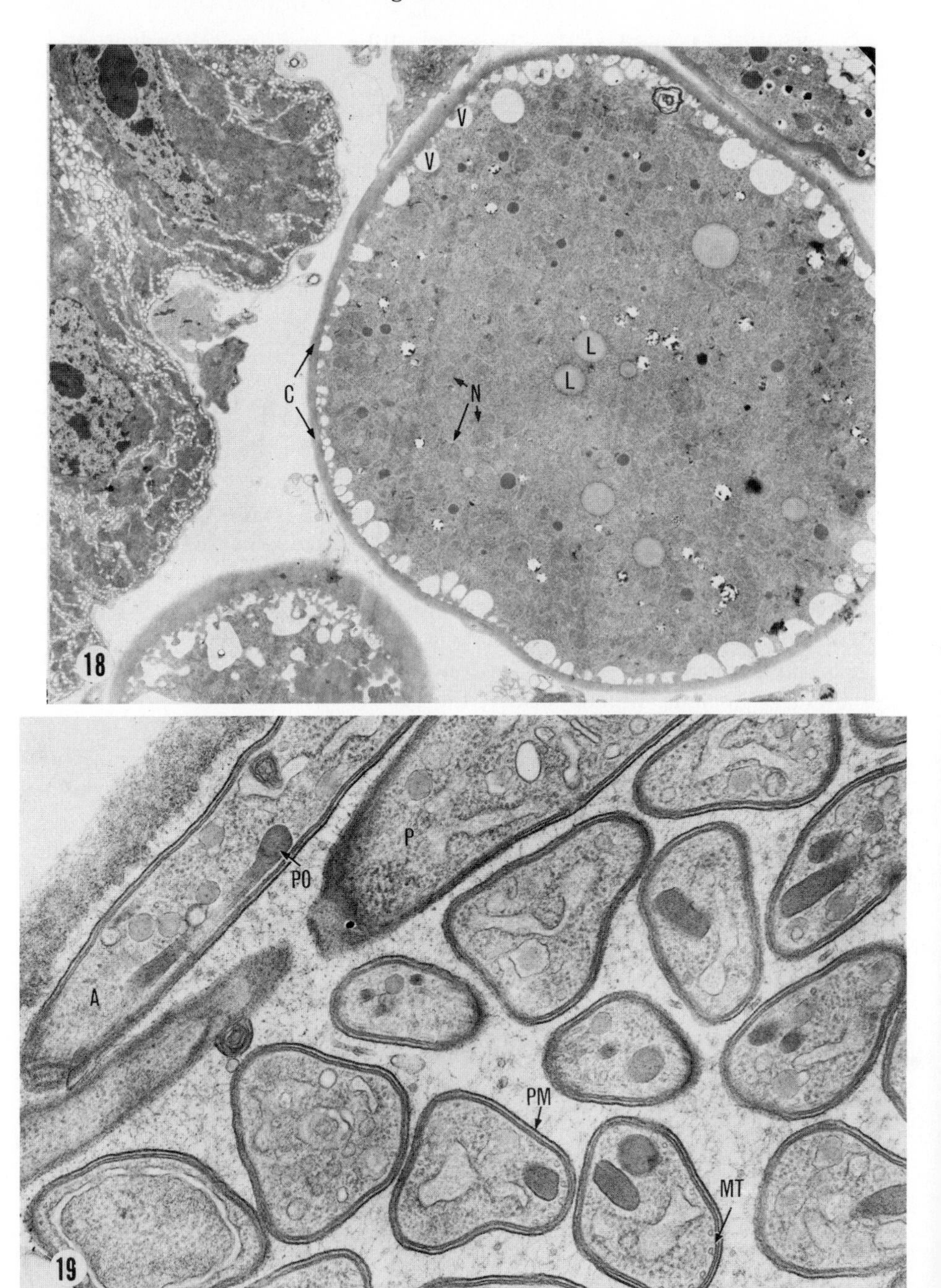
V
V
C
N
L
L
18
P
PO
A
PM
MT
19

bulging at various sites on the surface of the schizont. Portions of the thick intermediate layer of the merozoite pellicle are laid down beneath the bulging areas. Specialized apical organelles common to the fully differentiated merozoite appear in the bulge. A single nucleus, mitochondrion, and spherical body, along with endoplasmic reticulum and ribosomes, migrate into each of the buds on the mother schizont (Fig. 15). The original schizont decreases in size as the merozoite buds grow and develop. Division with formation of fully developed daughter cells is completed as the posterior end of the merozoites constrict to the point where the advancing plasma membranes fuse, sealing off the newly formed cells from the bud residue (Figs. 15 and 17).

A small proportion of the intraerythrocytic parasites, rather than continuing to reproduce asexually, differentiate into sexual forms (Fig. 14). After formation both the microgametocyte and the macrogametocyte remain intraerythrocytic until they are released from the host cell within the gut of the mosquito. Morphologically the gametocytes are distinct from asexual forms in that they are usually larger parasites with single large nuclei. The gametocyte, like the merozoite, has a thick intermediate layer beneath a thin plasma membrane (Aikawa, 1971).

B. Stages in the Mosquito

The developmental cycle of the avian malaria parasite within the mosquito begins with the ingestion of blood from the infected bird. As the blood meal passes into the gut, the change in environmental factors induces gametocytes to emerge from the erythrocytes. The microgametocytes begin the prefertilization process by undergoing microgametogenesis or exflagellation.

Although there have been no ultrastructural studies made of microgametogenesis in any of the avian plasmodia, the process has been described for close haemosporidian relatives (*Leucocytozoon* sp.) (Aikawa *et al.*, 1970). Initially, the extracellular microgametocyte elon-

---

Fig. 18. A large *Plasmodium gallinaceum* oocyst developing in the midgut of an adult, female *Aedes aegypti* mosquito. Peripheral vacuolization (V) just beneath the capsule (C) is one of the earlier changes in oocyst structure which results in sporoblast formation. The number of lipid droplets (L) also increase at this stage, along with nuclei (N). 4500 ×. (Courtesy of Dr. John Terzakis; Terzakis *et al.*, 1966.)

Fig. 19. Mature *Plasmodium gallinaceum* sporozoites within a mature oocyst. Note the complex nature of the sporozoite's pellicle with its plasma membrane (PM), the thick underlying membrane, and the microtubules (MT). In the upper left, two sporozoites are shown in longitudinal sections through their anterior (A) and posterior (P) ends. The uppermost one shows the anterior end with microtubules radiating from the polar rings, micronemes, and also portions of the paired organelles (PO). (Courtesy of Dr. John Terzakis and taken from Terzakis, 1968.)

gates and chromatin masses condense at the periphery of the nucleus. The nuclear membrane breaks down and the clumps of chromatin are dispersed throughout the cytoplasm. These chromatin masses become membrane-bound and are associated with the elemental flagellar apparatus. The flagellar apparatus or axonemes, composed of microtubules in the conventional 9 + 2 arrangement, extend into the surface of the microgametocyte forming primary flagellar buds. These develop into larger secondary buds as chromatin masses migrate into them. The secondary buds, in turn, grow out and form threadlike structures, the microgametes.

As the microgametes become free of the residual gametocyte, they swim actively to the macrogametes. If a microgamete successfully attaches to and penetrates a macrogamete, fertilization occurs. Several hours after fertilization, the nuclei of the joined gametes fuse. The result of nuclear fusion is the production of the zygote which, in turn, elongates and undergoes a number of cytoplasmic changes forming the ookinete.

The ookinete is a cone-shaped organism about 15 μm in length. It has a large centrally positioned nucleus with a distinct nucleolus, tubular cristate mitochondria, and a variety of inclusions including membrane-bound crystalloid aggregates. The cytoplasm of the ookinete is bound by two membranes which cover an inner array of microtubules. Apical organelles, analogous to the paired organelles of the merozoite, are present (Garnham *et al.*, 1962).

The mature ookinete migrates to the gut wall and passes through the peritrophic membrane of the mosquito midgut. There it pushes the brush border of the mucosal cells aside and subsequently passes into a mucosal cell.

Once intracellular, the ookinete migrates through the cell to the hemocoel. At the hemocoel side of the midgut, the ookinete begins the formation of the oocyst (Fig. 18). The ookinete rounds up, increases in size, and lays down a thick surface capsule. The cytoplasm of the oocyst fills with free ribosomes. Mitochondria are present along with lipid droplets and a variety of other types of inclusions. At the beginning of the cytoplasmic segregation, the oocyst contains many nuclei. At the periphery of the oocyst the cytoplasm appears vacuolated and the numbers of lipid inclusions increase markedly. Peripheral vacuoles coalesce with segments of oocyst cytoplasm. Islands of cytoplasm form the primary sporoblasts. The mature sporoblasts produce the infective sporozoites through a budding process (Fig. 19) (Terzakis, 1971). The processes of sporoblast budding when viewed ultrastructurally resemble the segmentation of the mature intraerythrocytic schizont. When the oocyst becomes filled with sporozoites, the oocyst ruptures and releases the sporozoites into the hemocoel.

Sporozoites are spindle-shaped organisms measuring about 9 μm in length and a little over a micrometer in diameter (Fig. 19). Like the other motile forms (i.e., merozoites and ookinetes), the sporozoite has a complex surface architecture. The cytostome is a constant feature of the sporozoite and occurs in its midregion. The anterior portion of the sporozoite contains paired organelles whose ductules extend to the apical cup. These anterior organelles are thought to play a role in host cell penetration. The major organelles include a prominent, centrally located oblong nucleus and many small mitochondria (Garnham *et al.*, 1960).

VIII. Metabolism and Biochemistry

As in all cells, the avian plasmodia couple degradative processes to processes which generate high-energy compounds and essential metabolites for use in the synthesis of cellular materials.

A. Energy-Yielding Processes

The energy-yielding catabolic processes of the avian blood parasites are basically fermentative in nature, producing acids from simple sugar substrates. The parasite is unable to accumulate polysaccharides for later consumption and, as a result, must be constantly supplied with an exogenous source of sugar. The intracellular parasites appear to stimulate the host erythrocytes to take up greater amounts of glucose than they would normally require to support their own metabolism. This exogenously supplied glucose is utilized mainly as potential energy rather than for direct use in the biosynthesis of other compounds (Moulder, 1962). The Embden-Meyerhof pathway and the Krebs cycle provide, to a limited extent as they do in higher cell types, key intermediates for lipid and amino acid synthesis as well as providing a major source of adenosine triphosphate (ATP). The major portion of the supplied glucose is catabolized via the Embden-Meyerhof pathway producing lactic acid as the end product (Moulder, 1962; Sherman *et al.*, 1970). Through a substrate type of phosphorylation, high-energy ATP is produced from ADP, but in low yields. The remaining portion of the sugar, under aerobic conditions, is oxidized from pyruvate to CO_2 and water via the Krebs cycle. Even though the Krebs cycle seems to be completely functional in the avian blood parasites and is much more efficient in its ATP production (i.e., ATP produced per mole glucose consumed) than anerobic glycolysis, these microorganisms use the less efficient metabolic route to obtain chemical energy necessary for growth and development (Honigberg, 1967).

It has been demonstrated experimentally that extracellular parasites isolated from erythrocytes by disruptive methods lose key intermediates of the Krebs cycle through diffusion. From this it has been suggested that, *in vivo*, this leakage of highly polar compounds might be one reason for the inefficiency of the Krebs cycle in these parasites. It also is probable that the avian plasmodia with their highly permeable cytoplasmic membranes may utilize host cell pools of essential metabolites, including ATP (Moulder, 1962).

Other energy-yielding pathways, in addition to the Embden-Meyerhof and Krebs cycles, exist but probably play a minor role in overall energy production. Most are shunts which utilize portions of the Krebs cycle, such as the NADP-mediated oxidation of glutamate to α-ketoglutarate (Sherman *et al.*, 1971b).

Most knowledge about the energy-yielding mechanisms of the avian plasmodia was obtained through work done on the intraerythrocytic parasites. Little is known about the metabolism of the tissue forms or the various stages within the mosquito. It is possible that significant differences may occur in the relative use of these energy-yielding systems by the different developmental stages of the parasites.

The avian plasmodia apparently lack a functional pentose phosphate pathway (i.e., phosphogluconate oxidative pathway). This pathway usually provides cells with a source of pentoses for nucleotide biosynthesis and reduced coenzymes for fatty acid and steroid synthesis. The intraerythrocytic parasite compensates for this deficiency, however, by utilizing the products of the host cell shunt (Herman *et al.*, 1966). The pyridine nucleotide coenzymes required by the parasite are generally obtained directly from host cell pools. Parasitization results in significant increases in levels of many of these cofactors (e.g., NAD, NADP, and NADPH) and much of this increase is within the parasite. It is possible that these pyridine nucleotide cofactors function to regulate metabolism (Sherman, 1966).

B. Protein Metabolism

Most of the energy yielded by glycolysis is utilized in biosynthesis. Most avian blood parasites have a relatively short mean generation time, in the range of several hours, depending on the species and host involved. This must reflect an active production of cellular substance. Much of this newly synthesized matter is protein.

The avian plasmodia obtain free amino acids, the basic precursor material of protein, from at least three sources. The first, and probably the primary source of free amino acids, is from hydrolyzed host cell hemoglobin. The two major sources of free amino acids are the plasma and intraerythrocytic pools. The amount of a given amino acid within the plasma and its rate of intraerythrocytic accumulation are two factors thought to regulate plasmodial growth (Sherman *et al.*, 1971a).

Free amino acids are accumulated by erythrocytes by a mediated transport mechanism. Only relatively few types of amino acids enter by simple diffusion. Parasitic invasion and subsequent growth alters, in many cases, the rate of free amino acid accumulation by diffusion in the erythrocyte. In contrast, free amino acid accumulation by the intraerythrocytic parasite is mainly through diffusion. It has been suggested that this might be selectively advantageous to the parasite in that it can accumulate essential host-supplied nutrients without the expenditure of vast amounts of energy (Sherman and Tanigoshi, 1974).

C. Nucleic Acid Metabolism

The average *Plasmodium lophurae* blood parasite contains about one-tenth of a picogram of DNA (Bahr, 1966). This represents about 4% of the total DNA within the infected erythrocyte. The RNA content of this "average" parasite has been estimated to be about one to one and one-half times greater than the DNA content. During intraerythrocytic growth from young trophozoite to late segmenter there is about an eight-fold increase in DNA. Cellular RNA also increases dramatically during intraerythrocytic development (Clarke, 1952).

Some researchers have reported, contrary to what one might expect, little periodicity associated with DNA synthesis during a single growth cycle of intracellular avian parasites (Walsh and Sherman, 1968). However, this absence of stepwise synthesis of DNA by synchronous cultures of *P. lophurae* may be due to variation among individual parasites and the possibility that all parasite nuclei do not divide an equal number of times (Walsh and Sherman, 1968). The primate malarial parasites, in contrast, clearly synthesize DNA in periodic waves. In these species the period of DNA synthesis, the S phase of the cell cycle, shortly precedes nuclear division, the G and G_2 phases being of very short duration (Conklin *et al.*, 1973).

Chemically, the isolated parasite DNA has a guanine plus cytosine content (G + C) of 19 mole %. This is in contrast to the G + C content of DNA from the host erythrocyte which is 35 mole % (Walsh and Sherman, 1968). The avian plasmodia, in synthesizing nucleic acid, utilize intracellular pools of preformed bases and nucleosides. It has been demonstrated experimentally that exogenously supplied nucleosides are accumulated and incorporated to a greater extent than nucleotides

(Tracy and Sherman, 1972). Apparently, the parasites are able to phosphorylate nucleosides by specific kinase enzymes and, in turn, incorporate the nucleotides into nucleic acids. The avian plasmodia so far studied have a very limited capacity to synthesize purine nucleotides *de novo* and must rely on an exogenous source of preformed precursors, presumably derived from host cell pools. The parasites are incapable of synthesizing purine bases from simple organic precursors and have only a limited capacity to synthesize purine nucleotides from exogenously supplied purine bases. The avian blood parasites, for example, are capable of utilizing guanine to synthesize guanosine, but are not capable of carrying out adenine to adenosine transformations.

In contrast to their limited capabilities in purine biosynthesis, the avian parasites have a complete pathway for pyrimidine synthesis. A number of the essential enzymes in the pathway have been isolated and characterized from intraerythrocytic plasmodia (Walsh and Sherman, 1968). One of these is thymidylate synthetase. This enzyme initiates transformation of deoxyuridine 5′-monophosphate (dUMP) to deoxythymidylic acid (dTMP). The reaction is dependent on tetrahydrofolic acid which mediates the methyl group transfer. It is this type of participation in pyrimidine biosynthesis (i.e., thymidylate synthesis) which makes folate compounds key metabolites for plasmodial growth and development.

Trager (1958) demonstrated that *P. lophurae* grown in extracellular culture could utilize a reduced form of folic acid (folinic acid) but not folic acid itself or its biosynthetic precursors. In the intact erythrocyte, however, plasmodial growth was enhanced by folic acid and precursors such as *p*-aminobenzoic acid (PABA). Such findings led to speculation that the avian plasmodia were deficient in their biochemistry and relied on the host cell to convert folic acid to dihydrofolate. It is now known, however, that these organisms are capable of synthesizing *de novo* dihydrofolate from simple precursors, i.e., pteridine, *p*-aminobenzoic acid, and glutamate. Enzymes which mediate this complex synthesis have been isolated from the avian species *P. gallinaceum* and *P. lophurae* (Ferone, 1973).

The enzyme dihydropteroate synthetase, which catalyzes one of the steps in the conversion of pteridine and PABA to dihydropteroate, is sensitive to sulfonamides. The final product of the synthesis, dihydrofolate, is reduced by dihydrofolate reductase to the active coenzyme, tetrahydrofolate. At this point it enters the thymidylate synthesis cycle.

Dihydrofolate reductase, like dihydropteroate synthetase, has been the subject of a number of current investigations because of its inhibition by pyrimethamine. This enzyme increases in concentration during intracellular development of the small trophozoite to the large multinucleated trophozoite. Correlated with this increase is a shift in relative intracellular proportions of oxidized and reduced folates; that is, there is a higher proportion of reduced folates in infected cells containing large multinucleated trophozoites (Platzer, 1974).

D. Lipids

Little is known about the biosynthesis of lipids by avian malarial parasites. It has been shown by a number of investigators that the phospholipid content of both parasitized erythrocytes and tissue cells containing exoerythrocytic parasites increases greatly during infection. This finding has been interpreted to mean that late developmental stages have greater amounts of phospholipids than the earlier ones. The increase in parasite phospholipid is most likely associated with the synthesis of new membrane rather than with an increase in stored phospholipid. The plasma membranes of erythrocytic plasmodia seem to have a higher phospholipid to cholesterol ratio (P/C) than do erythrocyte membranes. The oocyst is the only developmental stage which stores phospholipids, although sporozoites and gametocytes seem to have minute lipid inclusions (DasGupta, 1960a). Chemical analysis of isolated erythrocytic parasites have revealed that phospholipids, sterols, and free fatty acids comprise the major lipid fractions. Only small amounts of triglycerides and sterol esters occur. A variety of fatty acids of various chain lengths and degrees of saturation are also found. Oleic acid comprises the major portion of the free fatty acids (Wallace *et al.*, 1965).

E. Carbohydrates

Little is known about the nature of the structural and storage carbohydrates of the avian plasmodia. For the most part, the various developmental stages occurring within the mosquito and the vertebrate are devoid of storage polysaccharides. The exception to this is the oocyst, whose thick capsule appears, by histochemical criteria, to contain polysaccharide (DasGupta, 1960b).

IX. Genetics

The genetics of the avian malarial parasite is not very well understood. It is presumed that the basic mechanism of genetic interchange is through recombination following gamete fusion. This sexual process occurs in the insect vector and results in the production of a diploid zygote, the ookinete. Meiosis occur early in the formation of the oocyst and thus all other stages are haploid. Genetic recombination has been experimentally demonstrated in the avian plasmodia by the transfer of drug resistance from one strain to another (Greenberg, 1956). Other types of directed genetic exchange such as transformation, transduction, and conjunction, such as occur in other microbial systems, have not been reported. Mutational change which results in phenotypical change surely occurs, for strain characteristics such as virulence and antigenicity change.

Not only are the genetic processes of the avian plasmodia unknown, but so is the nature of the basic genetic material, the chromosomes. Their nuclear material is distinctive in that it stains weakly with Feulgen (Chen, 1944). Although there have been a number of reported observations made by light microscopy on chromosome morphology, these observations have not been verified by electron microscopy. Bano (1959) reported seeing four diploid chromosomes in small oocysts in early phases of meiosis. Two of these "chromosomes" were dot-shaped, while two were large J-shaped structures. Similar dot-shaped "chromosomes" were observed in erythrocytic forms by Wolcott (1957); however, serious question has been raised as to whether or not these structures are actually the chromosomes of the avian malaria parasites (Huff *et al.*, 1960). Dividing nuclei of all developmental forms do, however, seem to go through the same mitotic phases and display the same mitotic apparatus found in other eukaryotes.

X. Cultivation

Cultivation of the various developmental stages of the avian plasmodia in defined systems has proved to be most difficult. The difficulties are presumably due to the inability to mimic *in vitro* the complex nutritional and environmental characteristics of the intracellular milieu. In spite of this, real progress has been made in culturing parasites in some of the different stages, and these successes, in turn, have yielded a great deal of basic information on their biochemistry, physiology, and chemotherapeutic sensitivity. However, apart from this, simple preservation of the parasites may be achieved over long periods of time by low-temperature freezing, as in liquid nitrogen.

A. Stages in the Vertebrate

1. Intraerythrocytic Stages

There has been less success in the cultivation of the intraerythrocytic stages of the avian parasites than in cultivation of any of the others. William Trager of Rockefeller University (e.g., Trager, 1943a, 1947,

1957, 1966) has done more work on developing culture methods for these stages than anyone else. Initially, he showed that intraerythocytic *Plasmodium lophurae* could be maintained viable at 40°C for several weeks in a complex balanced salts–erythrocyte extract media. His nutrient medium, which had a high potassium ion concentration, also contained glucose, blood serum, chick embryo extract, glutathione, and calcium pantothenate. Each of these ingredients was shown to have a definite beneficial effect on parasite survival. Such environmental factors as low parasite concentrations, frequent renewal of culture media, gentle agitation, and minimal aeration also prolonged survival. While the parasites remained viable for several weeks, actual growth and parasite multiplication were limited to the first day or so. Trager later obtained better growth of the cultured avian parasites in a medium which contained vitamins, purines, pyrimidines, and protein hydrolyzate in a modified salt solution which was high in sodium, low in potassium ions, lacked glutathione, but contained a small amount of glycerol. Even with his best medium, parasite growth generally slowed down and parasite density diminished after the second or third day of cultivation. By subculturing every second day, Trager was able to maintain continuous growth for 8 days. By the end of the period these parasites had produced some seventeen times, with a generation time of 11 hours.

Other intraerythrocytic avian parasites, besides *P. lophurae* and *P. gallinaceum,* have been successfully cultured for short periods utilizing similar culture techniques. These include *P. hexamerium, P. elongatum, P. circumflexum,* and *P. vaughani* (Glenn and Manwell, 1956).

2. Extracellular Erythrocytic Stages

It is clear that plasmodial nutritional requirements are extensive, indicating that the intracellular parasite is dependent on the host for many metabolites which many less dependent microorganisms can synthesize *de novo.* Over a period of years, Trager has demonstrated that the cultural requirements of extracellular erythrocytic parasites and intraerythrocytic ones are distinct in certain aspects. He has shown that the addition of pantothenic acid to culture media markedly increased the survival times of *P. lophurae,* although such additions to cultures of isolated extracellular parasites had little effect. However, the addition of the pantothenic acid containing factor, coenzyme A, had a beneficial effect on parasite viability, suggesting, therefore, that the host cell synthesizes coenzyme A from pantothenic acid which, in turn, is utilized by the parasite. A similar effect has been noted in the differential growth-stimulating effects of the vitamin folic acid and the related coenzyme tetrahydrofolic on intra- and extraerythrocytic forms. Therefore the role of the host cell is not merely to passively accumulate nutrients for the parasite but one of transformation and synthesis of new metabolites for it.

The axenic cultivation of intracellular avian plasmodia through one entire life cycle has not yet been achieved. Trager has, however, been able to get isolated extracellular parasites of *P. lophurae* to complete one growth cycle and to begin another. These extracellular forms of normal intraerythrocytic parasites appear morphologically well preserved and are infective after 5 days in culture.

3. Exoerythrocytic Forms

Attempts at propagation of the exoerythrocytic avian plasmodia *in vitro* have been much more successful than attempts at cultivation of intraerythrocytic parasites. A number of different culture methods, including roller tube, hanging drop procedures, fertile egg culture methods, and standard primary tissue culture methods have been used successfully over the years. Although the roller tube and the hanging drop methods are distinct, both have been used with great success in culturing the exoerythrocytic forms. Cultures of *P. gallinaceum* have been maintained without interruption and with little loss in virulence for several years by a method which basically involves the alternate subculturing of parasites in roller tubes and hanging drop cultures (Oliveira and Meyer, 1955). Both methods employ a thin plasma clot containing chick embryo extract as the solid phase of the culture medium and a chick serum–balanced salt solution–chick embryo extract solution as the fluid phase. The infected tissues are implanted in the solid phase and the cultures are sealed and incubated at 37°C. When the implanted tissues exhibit high concentrations of parasites, the infected material is transferred to hanging drop slide cultures. Continuity of the culture is maintained within the hanging drop chambers by placing noninfected embryonic tissues, either spleen, liver, or brain, in direct contact with the subcultured infected tissues.

Embryonated chicken or turkey eggs have also been used for propagation of exoerythrocytic parasites of *P. gallinaceum* and *P. fallax*. Continuous cultures can be maintained routinely by serially passing tissues from infected embryos onto the chorioallantoic membranes (CAM) of uninfected 1-week-old embryos (e.g., Huff *et al.*, 1960). The technique involved is similar to the standard methods utilized in cultivating other microbial agents.

Another cultivation method, that of primary tissue culture, has been developed mainly through the efforts of Drs. Frank Hawking and Clay Huff and colleagues, using *P. fallax* and *P. gallinaceum* (e.g., Hawking, 1945; Huff *et al.*, 1960; Davis *et al.*, 1966). The culture methods have been worked out to the extent that large numbers of exoerythrocytic parasites can be grown very rapidly on a routine basis. The technique utilizes tissues of infected embryos as both a source of cells to be cultured and as a source of infectious material. Brain or livers from the embryos are extracted, minced, and trypsinized. The resulting free cells are plated out in small volumes of tissue culture media, supplemented with 10% fetal calf serum, 1% folinic acid, 1% glutamine, and 1% minimal essential amino acid mixture and incubated at 37°C in a 5% CO_2 atmosphere. Subculturing allows the organisms to be maintained in culture for an indefinite period, although transfers must be made at relatively short intervals. Infection levels in these cultures can go as high as 70% with each cell containing multiple parasites.

Survival and multiplication of *Plasmodium elongatum* in tissue cultures derived from the duck, a host species not known to harbor this species of avian *Plasmodium* naturally, has also been achieved for limited lengths of time. Cultures were derived from the marrow, liver, and spleen, but only the first worked well (Manwell and Weiss, 1960).

B. Stages in the Mosquito

All stages of the sporogonic cycle from the gametocyte to infective sporozoite have been obtained by culturing infected mosquito tissues directly with actively growing mosquito cells. Both primary and established insect cell lines have been used successfully in stimulating development of the sporogonic stages. In one study, Ball and Chao (1971) reported that *Plasmodium relictum* oocysts developed much more readily when cultured in the presence of cells of Grace's *Aedes aegypti* cell line then when cultured alone. In the presence of *Aedes aegypti* cells, 3-day-old oocysts (8 μm in diameter) increased in size some two to three times during the first 7 days of culture, while 5-day-old oocysts matured fully to infective sporozoites with 6 to 8 days *in vitro*. The overall rate of *in vitro* development approximated that observed under normal *in vivo* conditions.

XI. Chemotherapy

Because for many years avian malaria was the laboratory model of choice for mass screening of potential antimalarials, the chemotherapy of avian malaria has been researched thoroughly. In two studies during and immediately preceding World War II alone, over 17,000 compounds were tested for effectiveness in controlling avian malarial infection (Wiselogle, 1946; Coatney *et al.*, 1953). It is in this way that the study of avian malaria has served man most profitably.

Indeed, without atebrin, the outcome of World War II in the Pacific theatre might well have been different. This drug replaced quinine, the supply of which had been cut off by the Japanese, and was largely the product of chemotherapeutic research using the avian malarias (Manwell, 1949).

The many effective chemotherapeutic agents may be grouped according to the developmental stage of the parasite which is affected during drug treatment. (1) There are schizontocidal drugs which act on the blood stages. Such drugs may be either natural extracts or synthetic in nature. (2) There are gametocytocidal drugs which directly kill or inhibit intraerythrocytic gametocytes. (3) Some of the potent antimalarial prophylactic drugs act against primary tissue schizonts. (4) A major group of drugs are the antirelapse drugs which inhibit the reproductive cycle of secondary tissue schizonts. (5) Last, there are sporontocidal drugs which block the development of sexual forms within the insect vector. Chemically, the most potent of the antimalarials generally fall into five distinct groups. These include: (1) the 4-aminoquinolines; (2) the 8-aminoquinolines; (3) the aminoacridines; (4) the antagonists of folic acid utilization; and (5) the inhibitors of dihydrofolate reductase, an enzyme functional in folic acid biosynthesis. Many other categories of compounds exhibit antiavian malarial activity and these range from simple metallic to complex organic compounds. However, the drugs of the five types listed have also enjoyed widespread use in the treatment of human malarias (Steck, 1971; Peters, 1970).

The natural, 350-year-old remedy, quinine, is a good example of a schizonticide. This alkaloid antimalarial originally was derived by extraction from the bark of the cinchona tree and later synthesized. Quinine, with its therapeutic effect on avian infections, served as a reference drug against which all other drugs were tested and compared. Quinine is active against blood schizonts but has little activity against tissue forms. Quinine exerts its chemotherapeutic effect by blocking parasite DNA synthesis.

The single nitrogen-containing double ring base of quinine is the structure upon which many potent antimalarials have been chemically constructed. This basic ring structure is referred to commonly as the quinoline base. Endochin, one such quinoline, exhibits marked schizontocidal activity as well as having a prophylactic effect, through an antisporozoite activity. This compound is interesting because it is highly active against avian plasmodia and relatively inactive against mammalian parasites.

A number of the 4-aminoquinolines are active schizontocides. One of the better known compounds within this group is chloroquine. Its use in the treatment of malaria does not result in radical cure because of its lack of activity against tissue forms. Like quinine, its mode of action involves the inhibition of DNA replication through the intercalation of the quinoline ring between base pairs of the parasites DNA. DNA polymerase is inhibited by the insertion of the quinoline ring which stabilizes DNA and prevents the final stages of assembly and polycondensation.

Quinacrine, or as it is often called, mepacrine or atebrin, is another synthetic heterocylic ring type antimalarial (i.e., aminoacridine) with potent blood schizontocidal activity and little tissue schizontocidal activity. This antimalarial is some seventeen times more effective than quinine in controlling ongoing *Plasmodium gallinaceum* infections. Like chloroquine and some of the other 4-aminoquinolines, it is relatively toxic at high doses. Its antiplasmodial action depends on the intercalation of acridine rings between the base pairs of DNA, thus inhibiting replication.

The 8-aminoquinolines are another major class of antimalarials. They were derived through carefully planned chemical modification of the basic quinoline structure. It is the only group of antimalarials tested in the avian system which demonstrated clear-cut effects on both tissue schizonts and blood gametocytes. Because of their anti-tissue-schizont activity, drugs of this group afford protection against relapse infection especially when used in combination with blood schizontocides, such as chloroquine. Many of the 8-aminoquinolines exhibit sporontocidal as well as gametocytocidal activities. For example, pamaquine inhibits exflagellation of microgametes within the mosquito vector while primaquine acts against sporozoites.

The general antimalarial activity of the 8-aminoquinolines is ascribed to the inhibition of mitochondria function (Aikawa and Beaudoin, 1970). Structural and functional analogies have been drawn between these antimalarials and the naphthoquinones, mitomycin, and streptomycin which inhibit electron transport, specifically, at the ubiquinone step. Due to the nature of the quinoline ring structure, it is believed that the 8-aminoquinolines, like the 4-aminoquinolines and the acridines, exert some antimalarial affect by binding to DNA and interfering with replication (Steck, 1971).

Another major group of chemicals which have potent antimalarial activity are those which interfere with folic acid metabolism. The avian plasmodia are nutritionally dependent on folic acid coenzymes for growth and reproduction. Biochemically, they depend on essential reactions mediated by the folates which include a one-carbon transfer reaction, production of high-energy reducing compounds, and pyrimidine biosynthesis. The antifolate antimalarials can be subdivided into two smaller groups, depending on whether they inhibit folic acid synthesis or inhibit its utilization. Those antimalarials which inhibit the synthesis of folic acid from the precursor pteridine and *p*-aminobenzoic acid (PABA) are usually chemical analogs of PABA which compete with PABA as substrate for the folic acid synthesizing enzyme. Sulfanilamide and related compounds serve as good examples of this class of antifolates. They are long-lasting schizontocides with limited potency. For example, sulfapyridine is effective against the avian parasite *Plasmodium circumflexum*.

Antifolate antimalarials of the second class exert their effect by inhibiting utilization of folic acid rather than, as in the case of the sulfa drugs, inhibiting folic acid synthesis. A good example of this group is pyrimethamine. Pyrimethamine is a potent, but slowly acting, blood and tissue schizontocide. Because of its marked suppressive activity against tissue forms, it is considered a rather good prophylactic drug. It also has marked sporontocidal activity. Its major weakness is that resistance to the drug by both erythrocytic and exoerythrocytic forms can be induced with relative ease. The drug, like others within the group, exerts its effect by inhibiting the activity of dihydrofolate reductase (Platzer, 1974).

XII. Resistance and Immunity

A. Constitutive Defense Mechanisms

Resistance mechanisms of birds to plasmodial infection generally fall into two overlapping categories, namely, constitutive and inducible types. The mechanism of constitutive or natural resistance usually involves an interplay of both physiological and biochemical factors which affect the host's susceptibility to parasitic invasion. For example, the nutrition of the host affects its natural resistance to infection. Undernutrition, as well as malnutrition (e.g., vitamin deficiencies and very high carbohydrate to protein ratio diets), will often increase the severity of the disease in birds as well as lessen their resistance to superinfection (Trager, 1943b; Brooke, 1945). Although these nutritional effects may be detrimental to the parasites as well, the adverse effect on the host is often such that the balance in the host–parasite relationship swings in favor of the parasite.

The nature of the physical barriers of the host through which the parasite must pass will often determine the outcome of a parasitization. Skin is one such barrier. Avian sporozoites may be uninfective following deposition in skin of resistant hosts. The limiting membrane of the host cell also acts as a physical barrier in controlling host susceptibility. Erythrocytes of different species have varying degrees of susceptibility to infection by certain plasmodial species. It is thought that this trait of cell susceptibility is, in part, hereditary (McGhee, 1957).

Age also is a predisposing factor which affects host resistance. At the cellular level, young, immature erythrocytes are often more susceptible to infection by certain plasmodial species than aged ones. This presumably reflects an alteration of the composition of cells as they age, which either enhances or retards parasite growth. Intracellular ATP concentrations which dwindle with age might be one such critical alteration. Change in ionic composition might be another, while loss of surface receptors may also affect the ability of the parasite to initiate infection of a cell.

The age of the host is often times critical in determining its susceptibility to infection or the severity of the infection (e.g., Manwell *et al.*, 1957). Young birds usually develop higher and more prolonged parasitemias with a greater mortality rate than do older birds. This type of "age resistance" is usually gained several weeks after hatching and is thought to be a consequence of a maturing immunological system.

Constitutive host resistance is, in part, an immunological function dependent on both humoral factors and the phagocytic activity of cells. The plasmodiacidal-plasmodiastatic activities of normal plasma of resistant species indicate that protective soluble factors exist. These might include natural antibodies which act as opsonins, or in conjunction with complement-produced plasmodial lysis, or would perhaps include substances such as properdin or lysins which are nonspecific proteins with antimicrobial activities. Certainly, in blood-induced infections, natural antibodies in the form of isoantibodies have been shown to have a protective effect in enhancing sequestration of parasitized erythrocytes.

While, no doubt, humoral factors exert a positive effect in natural resistance, still they are probably of minimal importance when compared to cellular elements. The reticuloendothelial system constitutes a major line of natural defense for the bird. Macrophages which line the vascular passages of organs like the spleen, liver, lymph nodes, and bone marrow, actively screen blood and phagocytose foreign material, such as extracellular plasmodia and *Plasmodium*-infected cells.

It should be noted, however, that in the avian malarias destruction of the parasite does not always occur following phagocytosis. Phagocytic cells commonly serve as host cells for exoerythrocytic parasites. In general, little species immunity to avian malaria exists, although there is some. Psittacines, for example, are in general insusceptible. This aspect of the subject is reviewed by Manwell (1963).

B. Inducible Defense Mechanisms

In addition to the constitutive resistance mechanisms, the host is also endowed with the capacity to mount a specific inducible response against invading pathogens.

Acquired immunity develops as a result of exposure to the pathogen itself through active infection or to its products via immunization. The immunity which develops during avian malaria is basically a cell-mediated event with antibody playing a supportive role. Due to the organism's intracellular habit, the classical protective effect exerted by specific antibody is limited in malarial infections. It is probable that extracellular merozoites are affected by specific antibody while in transit from one host cell to the next. In *in vitro* systems, free merozoites tend to aggregate in the presence of specific antibody and to collect an immune precipitate on their posterior ends (Graham *et al.*, 1973b). Similar effects have been noted with sporozoites. Immune precipitates and cell aggregation, however, have seemingly little effect on infectivity. In general, humoral immunity is incomplete, form-specific, and limited to given strains of parasites (e.g., Manwell and Goldstein, 1939, 1940). Immune plasma exhibits, through passive transfer or by *in vitro* incubation, little killing effect on erythrocytic parasites (this does not seem to be the case with exoerythrocytic forms) and generally only reduces the severity of infection without inhibiting it (Hegner and Eskridge, 1938; Graham *et al.*, 1973b).

Protective antibody is produced by plasma cells which reside in lymphoid tissues scattered throughout the body. These cells are formed by blastogenic transformation of antigen-stimulated lymphocytes within bursa-dependent lymphoid follicles of various reticuloendothelial system (RES) tissues. Experimentally by depleting plasma cell and B-cell populations within the adult, via neonatal bursectomy, the bird is rendered hypogammaglobulinemic and is incapable of mounting a specific humoral response to infection. The protective effect of the humoral response is realized when these birds are challenged and resultant infections are more severe than those in intact control birds. The infections have shorter prepatent periods, higher parasitemias, and greater mortality rates in the birds with deficient humoral responses (Stutz *et al.*, 1972).

The mechanisms of cell-mediated immunity in avian malaria have not been described as extensively as they have in other experimental infections. A functional thymus-derived lymphocyte-mediated immune system exists in birds as it does in mammals. Neonatal thymectomy of chicks results in loss of cell-mediated immune responses associated with such reactions as homograph rejection (Aspinall *et al.*, 1963). However, one report suggests that the effect of neonatal thymectomy on the acquisition of immunity to *Plasmodium lophurae* infections in adult chickens is minimal (Longenecker *et al.*, 1966). These researchers attempt to explain this curious finding by suggesting that birds are more fully developed at hatching than mammals are at birth and, as a result, peripheralization of immunocompetent lymphoid cells from the thymus to other lymphoid organs might occur before hatching.

Many years ago, Taliaferro and Cannon (1931) stated in reference to avian malaria that the "mechanism of immunity to superinfection is primarily cellular and consists of an increased rate of phagocytosis by the cells of the reticuloendothelial system, especially those of the spleen and liver. It is the product of two factors; (a) an increase in the number of phagocytic cells and (b) a greatly increased rate of phagocytosis by the individual phagocyte." Taliaferro was probably the first of the early immunologists to recognize that the mesenchymal reserves of the reticuloendothelial system (RES), with its reparative and antiparasitic functions, are an important component of the inducible defense mechanisms of the host. The system functions first through the mitotic proliferation of the RES cells and, second, through the transformation of reticular and monocytoid cells into actively engulfing macrophages (Taliaferro and Taliaferro, 1955). Early in the primary infection with more virulent species (e.g., *P. gallinaceum* and *P. lophurae*) there is lymphoid cell loss through destructive processes and cell transformation. These losses are compensated for by a hyperplastic response of the lymphoid tissues. The speed at which this recuperative response occurs is fundamental to the control of the infection. If the response is too slow, the bird will succumb. In contrast, superinfected immune birds quickly counterbalance the initial lymphoid depletion with an intensified hyperplastic response and thus control the infection. The hyperplastic condition regresses only when the infection has subsided. Immunity which depends on continued low-grade infection is termed "premunition" or "infection immunity" and seems to be a more effective form of protection against future infection than is a "sterile" type of immunity.

The spleen and, to a lesser extent, the liver seem to be most important organs in regulating the immune response to superinfection. In the ontogeny of the protective response there seems to be an interplay between the thymus and the bursa in seeding immunocompetent cells to the spleen (Longenecker *et al.*, 1966). As a result, the fully developed organ possesses phagocytic and antibody-forming capacities, as well as cytopoietic ability (Taliaferro and Taliaferro, 1955). The spleen seems to be much more important in host resistance after establishing itself during infection as a functional organ of defense. The effects of splenectomy are usually greater, in terms of increased parasitemias and greater mortality rates, if carried out during the acute phases of disease than when done prior to infection (Terzian, 1946; Longenecker *et al.*, 1966). This time-related effect of splenectomy on the course of infection may explain in part the varying results obtained by some investigators (e.g., Causey, 1939; Manwell *et al.*, 1957). Species of hosts and species and strains of parasites utilized, might also account for differences in the effects noted. Splenectomy during the recovery phase usually lessens the host resistance to reinfection. Relapse into acute infection is a common result of splenectomy during recovery (e.g., Al-Dabagh, 1960; El Nahal, 1966).

C. Active Immunization

Acquired immunity can be obtained by active immunization as well as by natural infection. A variety of immunogens have been tested for effectiveness in inducing a protective type of immunity. For example, sporozoites and erythrocytic parasites have been treated by ultraviolet light, drying, freeze-thawing, and formalinization to obtain antigens which have been successfully used to immunize birds against homologous challenge. The protection afforded by immunization is usually bolstered by the use of adjuvants (Richards, 1966).

Little is known, however, about the composition and the interrelationship of the specific antigens derived from the various developmental forms. The most thoroughly described avian plasmodial antigens are those derived from blood forms. This is because blood forms are relatively easy to collect and obtain in large amounts.

Like the antigens of so many of the blood parasites, plasmodial antigens can be either cell-associated or plasma-soluble. The soluble plasma antigens are called ectoplasmodial antigens and are detected in increasing quantity and number throughout the course of the acute infection (Todorovic *et al.*, 1967, 1968a,b). They are thought to originate from the parasite, but their relationship to the organism is not yet understood. In *P. gallinaceum* infection, three distinct proteinaceous ectoplasmodial antigens have been detected. The fluorescent antibody technique has indicated that these soluble antigens are associated with parasitized erythrocytes as well as extracellular merozoites. These antigens are said to stimulate a protective humoral response in nonimmune hosts. They seem to be genus-specific (parasite) and will confer protection, upon immunization, on many different hosts challenged with various species of plasmodia.

A second group of ectoantigens have been described. These are the nonspecific serum antigens found in the blood of animals with many types of acute haemosporidian infections. When used as immunizing agents, these soluble factors induce the host to produce opsonic antibodies which enhance erythrophagocytosis. These serum antigens are most effective when used to immunize birds against homologous infections, although, even heterologous serum antigens from widely different infections are reported to be slightly protective (Corwin and Cox, 1969).

Cell-associated plasmodial antigens seem to exist for the erythrocytic stages of a number of avian species. Some twelve to sixteen different proteinaceous antigens have been detected by electrophoretic techniques (Finerty and Dimopoullos, 1968). Some of these are thought to be antigens common to organisms of various genera, whereas others are probably species-, strain-, or stage-specific. By serological testing, using such methods as immuno-gel diffusion, immunofluorescence, and complement fixation, common antigens have been detected between the avian, human, primate, and rodent plasmodia. Because of the striking similarities in antigenic makeup between the avian and human organisms, *P. gallinaceum* blood smears have been utilized as antigen in the fluorescent antibody (FA) test in the successful diagnosis of human infections with all four species (Kielmann *et al.*, 1971). Species- and strain-specific antigens have also been demonstrated by gel diffusion for a variety of types of avian parasites. Common antigens might be expected to be shared by different developmental forms of a single strain of parasite; this appears to be the case. By the FA test, conjugated antierythrocytic parasite sera could detect exoerythrocytic forms in the brain and livers of infected animals, while antisporozoite conjugate reacted positively with erythrocytic forms (Voller and Taffs, 1963).

XIII. Pathogenicity

The pathogenicity of avian malaria varies considerably with the species (and even the strain) of parasite, and also with the species of host. Reasons for such differences are, in general, not known. Most of what is known is based on experimental infections (the majority blood-induced, and most in canaries and chickens) and it is unreliable to extrapolate from them to what may occur in nature. Occasionally, birds obviously ill are picked up and brought to the laboratory. Examination shows them to be suffering from an acute malaria infection, more often than not due to *Plasmodium cathemerium, P. relictum,* or *P. circumflexum.* It is thus very likely that malaria is a fairly frequent cause of death. Plasmodia of the *Novyella* group are seldom lethal.

In parts of the Far East, malaria may be a quite pathogenic infection in domestic fowls, although they are apparently not the natural hosts. There *Plasmodium gallinaceum* and *P. juxtanucleare* are the usual causes. The latter also occurs in Central and South America, and has been found in Mexico. Malaria has even been found in chickens in Wisconsin, but the species of parasite was uncertain. Malaria was one of the important causes of the decimation of indigenous bird species in Hawaii (Warner, 1968, 1973). Because birds of many species are commonly imported, together with their parasites, of course, it is also a very potential danger wherever this trade exists (Manwell and Rossi, 1975).

It is likely that malaria infection often occurs while birds are still nestlings and that it affects them more severely than adults though nestling often occurs early in the spring when the mosquito population is still low.

XIV. The Disease

A. Clinical Aspects

Human malaria and avian malaria are very similar in many ways, although the symptomatology of the two differs. The fever and paroxysms of human malaria are not seen in birds, except to some degree in pigeons with pinottii malaria (Fig. 20). With other host species, if the bird is acutely ill it will ruffle up its feathers and tuck its head under its wing, but there is no rise in body temperature. Often there are no visible symptoms.

If the course of parasitemia is graphed, there will be a prepatent period of a few days, a patent period in which parasitemia will reach a maximum and then decline (giving a bell-shaped curve), and then a subpatent period, at first often interrupted by small low peaks of parasitemia representing short-term relapses. The length of the prepatent period will depend on the duration of preerythrocytic schizogony, if the infection is mosquito-induced, and on the number of parasites in the inoculum, if blood-induced. It will also vary with the individual bird and the species of parasite.

Once infected, whether by the injection of malarious blood or the bite of a mosquito, birds remain infected for a long time, and probably for life (Manwell, 1934). After a clinical recovery there are usually so few parasites in the blood that a long search may be required to find any. The bird will also appear quite well, though occasional long-term relapses undoubtedly occur.

In laboratory-acquired infections, at least, relapses are seldom as severe as the original infection.

B. Pathology

Due to the hemotrophic nature of avian plasmodia, the major pathology in malaria is associated with the blood and the various organs of the circulatory system and is attributable to dysfunction in circulation. This is as true of the rodent, primate, and human malarias as it is of the avian disease.

Many of the harmful effects of the disease, such as the severe anemia, can be attributed directly to enhanced red cell destruction. The level of cell destruction is usually proportional to the parasitemia, although often

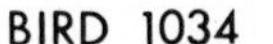

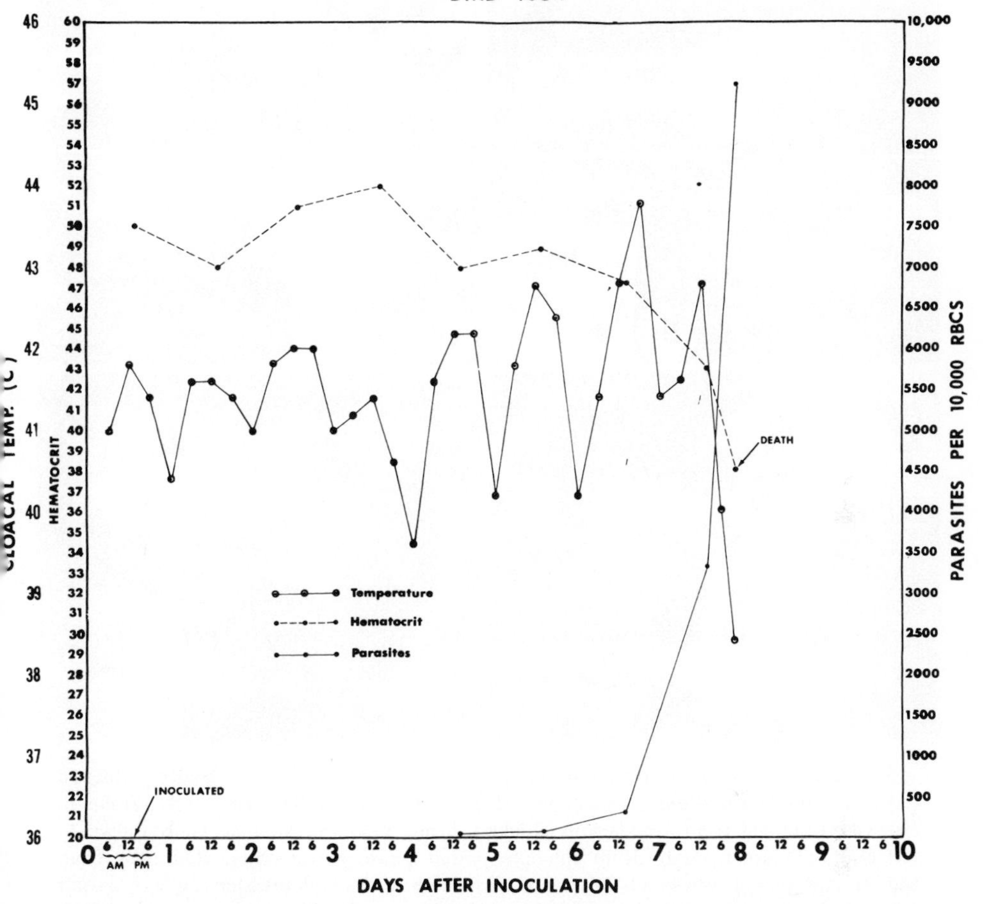

Fig. 20. Temperature chart of a pigeon infected with *Plasmodium pinottii.* As in human malaria there is fever when merozoite liberation is at its height. (*Plasmodium pinottii* has a reproductive cycle of about 24 hours.) Temperature remains nearly normal until almost the end of the prepatent period.

times in severe infections shortly following crisis it is disproportionately high (Wright and Kreier, 1969). Yet anemia is by no means always a chief cause of death. Anemia, more severe than ever occurs in the natural course of pinottii malaria of pigeons (which is lethal in almost 100% of cases), can be artificially and quickly recovered from. A characteristic symptom of this form of malaria is extreme water loss which may precipitate the fatal outcome (Manwell and Stone, 1966, 1967).

The mechanism of erythrocyte destruction is either by intravascular hemolysis or by phagocytosis. Parasitized erythrocytes are most suscepti-

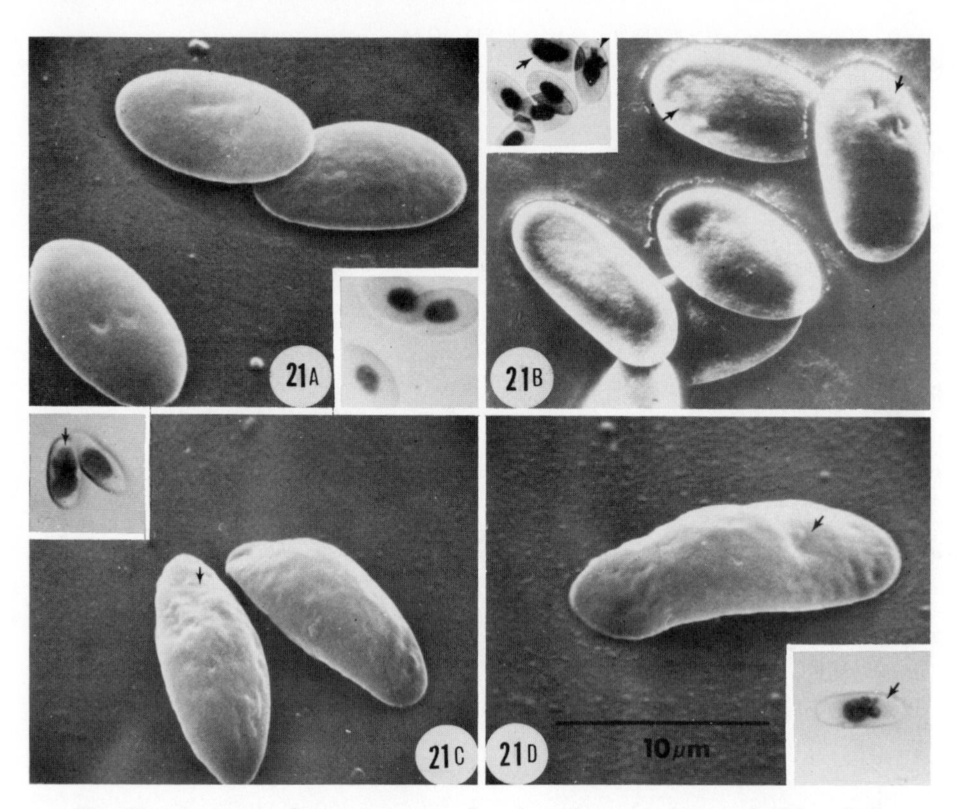

Fig. 21. Scanning electron micrographs of erythrocytes from a healthy chicken (A) and from *Plasmodium gallinaceum*-infected chickens (B–D). Red cells shown in (B) were obtained from a malarious chicken during peak parasitemia. Cells shown in (C) and (D) are from birds in the late patent phase of infection. Red cells from the middle patent phase of infection (B) have surface depressions where larger intracellular parasites are located (arrows). Abnormal cell shapes occur in both infected and noninfected erythrocytes in the late patent period (C and D). Arrows indicate location of intracellular plasmodia. 4500 ×. (From Seed and Kreier, 1972.)

ble to hemolysis, being made fragile by the growing intracellular parasites, especially during the late schizogonic stages (Fig. 21). Nonparasitized erythrocytes also become increasingly fragile during the acute phases of infection due to pathological changes in the suspending blood plasma (Seed and Kreier, 1972). In normal healthy birds, intravascular hemolysis as a mechanism to remove defective cells from the circulation is a rare occurrence; however, in malaria it seems to play an important role. The major portion of erythrocytes lost by means other than parasite emergence are, however, sequestered intact. This enhanced sequestration is thought to be the result of the combination of three factors: (1) an infection-induced physiochemical alteration of both parasitized and non-parasitized red cells which prematurely ages the cells (Seed and Kreier, 1972); (2) hypersplenism, that is, hyperphagotrophic activity of the spleen in which the selectivity (targeting specificity) of individual hyper-active macrophages in phagocytizing defective red cells is broadened (George *et al.*, 1966); and (3) antibody, in this case, the infection-induced opsonins with antiparasitic and autoantigenic specificity which coat antigen-modified cells therefore enhancing erythrophagocytosis (Zuckerman, 1964).

Plasma chemistry is also severely affected during certain types of avian malarial infections and this probably contributes to the pathogenesis of the disease. For example, anoxia, a common condition in malaria, results not only from massive red cell destruction but also from a drop of plasma pH which effectively reduces the oxygen-binding capacity of the hemoglobin (Rigdon and Rostorfer, 1946). Substantial increases in plasma proteins, particularly in the gamma and macroglobulin fractions, as well as increases in fibrin concentration contribute to the increased viscosity of plasma of critically ill birds. Such a thickening undoubtedly enhances sludging of blood cells in capillaries and contributes to a lessening of effective circulation within vascular beds of various organs and probably increases phagocytoses of red cells.

In addition to the many physical and chemical modifications of plasma which might have pathological consequences, the plasma may contain soluble substances (e.g., antigen–antibody complexes) which initiate pathological events in various target tissues (Soni and Cox, 1974).

During active infections of chickens with *Plasmodium gallinaceum* or *P. lophurae,* the bone marrow often becomes hyperplastic generating and releasing blood cells into the peripheral circulation to compensate for the massive numbers lost during infection. There is a massive replacement of fat by mitotically active cellular tissue. In extremely prolonged infections or in dying birds, the bone marrow may be stressed to the extent that it becomes aplastic with calcareous infiltration (Taliaferro and Taliaferro, 1955). In at least one type of avian infection *(Plasmodium elongatum),* stem cells of bone marrow may serve as host cells and, as a result, there is inhibition of the normal reparative functions of the bone marrow.

The enlargement of the spleen is a characteristic pathological finding in malaria. The enlargement may be as great as twentyfold by weight and is associated with the organ's hyperactive state. The spleen is largest during the crisis and gradually decreases in the recovery phase. This is not just an edematous type of enlargement but rather a cellular one. The hypercellularity is associated with reparative and immunological defense mechanisms (Taliaferro and Taliaferro, 1955) of the bird. Besides the increase in size, the spleen becomes very dark in color and somewhat friable. Splenic infarction often occurs. However, because of the rapid regeneration during the recovery phase, little permanent damage is done. Infarcts are usually seen as large thrombi in the central vein and to a lesser extent in the smaller arterioles and venules. At times when such infarcts are not repaired, necrotic areas develop (Hewitt, 1940).

The cellular reactions in the liver during infection are similar to those which occur in the spleen. Endothelial cells lining the capillaries become swollen with cell debris and malarial pigment (Hewitt, 1940). Initially, the pigment is scattered in masses throughout the liver while later in the infection it localizes in the periportal connective tissues or about the central vein or sinusoids leading to it. Focal necrotic areas occur but are infrequent. In terminal infections the hepatic cells become vacuolated and frequently show fatty degenerative changes. Liver cords become enlarged and engorged with lymphoid cells. In contrast to the lymphoid hyperplasia which occurs in the spleen during infection, there is only minimal hyperplasia in the liver and the lymphocytes which fill the hepatic sinuses are derived from the spleen (Taliaferro and Taliaferro, 1955).

Acute glomerulonephritis often occurs in *P. gallinaceum* malaria. Clinically, this is indicated by the dramatic increases in plasma blood urea nitrogen (BUN) levels which occur during the latter stages of infection. Nephritis usually develops during the patent phase of infection and increases in severity concomitant with the detection of circulating soluble antigen–antibody complexes (Soni and Cox, 1974).

A common cause of death in some types of avian malaria is a type of cerebral stroke. It is caused not by the constricting of capillaries as in most types of strokes but, rather, due to the extensive swelling of endothelial cells of the vessel walls which occlude the lumen and prevent normal blood flow (Hewitt, 1940; Graham *et al.*, 1973a). This type of blockage of brain capillaries, caused by concentration of exoerythrocytic stages in their endothelium, occurs almost regularly in chukars infected with *Plasmodium octamerium*, with lethal results (Manwell, 1968). Gliosis and granulomatalike lesions may also occur as a consequence of the anoxic condition which develops as a result of the restricted circulation (Al-Dabagh, 1961). Al-Dabagh (1966) has reviewed the subject, although he is chiefly concerned with *P. gallinaceum*.

REFERENCES

Aikawa, M. (1966). The fine structure of the erythrocytic stages of three avian malarial parasites, *Plasmodium fallax*, *P. lophurae* and *P. cathemerium*. *Am. J. Trop. Med. Hyg.* **15**, 449–471.

Aikawa, M. (1967). Ultrastructure of the pellicular complex of *Plasmodium fallax*. *J. Cell Biol.* **35**, 103–113.

Aikawa, M. (1971). *Plasmodium:* The fine structure of malarial parasites. *Exp. Parasitol.* **30**, 284–320.

Aikawa, M., and Beaudoin, R. L. (1970). *Plasmodium fallax:* High-resolution autoradiography of exoerythrocytic stages treated with primaquine *in vitro*. *Exp. Parasitol.* **27**, 454–463.

Aikawa, M., Hepler, P. K., Huff, C. G., and Sprinz, H. (1966). The feeding mechanism of avian malarial parasites. *J. Cell Biol.* **28**, 355–373.

Aikawa, M., Huff, C. G., and Strome, C. P. A. (1970). Morphological study of microgametogenesis of *Leucocytozoon simondi*. *J. Ultrastruct. Res.* **32**, 43–68.

Al-Dabagh, M. A. (1960). The effect of splenectomy on *Plasmodium juxtanucleare* infections in chicks. *Trans. R. Soc. Trop. Med. Hyg.* **54**, 400–405.

Al-Dabagh, M. A. (1961). Symptomatic partial paralysis in chicks infected with *Plasmodium juxtanucleare*. *J. Comp. Pathol.* **71**, 217–221.

Al-Dabagh, M. A. (1966). "Mechanisms of Death and Tissue Injury in Malaria." Shafik Press, Baghdad.

Aspinall, R. L., Meyer, B. K., Graetzer, M. A., and Wolfe, H. R. (1963). Effect of thymectomy and bursectomy on the survival of skin homografts in chickens. *J. Immunol.* **90**, 872–877.

Bahr, G. F. (1966). Quantitative cytochemical study of erythrocytic stages of *Plasmodium lophurae* and *Plasmodium berghei*. *Mil. Med.* **131**, 1064–1070.

Ball, G. H., and Chao, J. (1963). Contributions of *in vitro* culture towards understanding the relationships between avian malaria and the invertebrate host. *Ann. N.Y. Acad. Sci.* **113**, 322–331.

Ball, G. H., and Chao, J. (1971). The cultivation of *Plasmodium relictum* in mosquito cell lines. *J. Parasitol.* **57**, 391–395.

Bano, L. (1959). A cytological study of the early oocysts of seven species of *Plasmodium* and the occurrence of post-zygotic meisosis. *Parasitology* **49**, 559–585.

Brooke, M. M. (1945). Effect of dietary changes upon avian malaria. *Am. J. Hyg.* **41**, 81–108.

Causey, O. R. (1939). The effect of splenectomy on the course of malarial infection in canaries. *Am. J. Hyg.* **30**, 93–99.

Chen, T. T. (1944). The nuclei in avian malaria parasites. I. The structure of nuclei in *Plasmodium elongatum* with some considerations on technique. *Am. J. Hyg.* **40**, 26–34.

Clarke, D. H. (1952). The use of phosphorus 32 in studies on *Plasmodium gallinaceum*. I. The developmeint of a method for the quantitative determination of parasite growth and development *in vitro*. *J. Exp. Med.* **96**, 439–450.

Coatney, G. R., Cooper, W. C., Eddy, N. B., and Greenberg, J. (1953). Survey of antimalarial agents. *U.S., Public Health Serv., Public Health Monogr.* **193.**

Conklin, K. A., Chou, S. C., Siddiqui, W. A., and Schnell, J. V. (1973). DNA and RNA synthesis by intraerythrocytic stages of *Plasmodium knowlesi*. *J. Protozool.* **20**, 683–688.

Corwin, R. M., and Cox, H. W. (1969). The immunogenic activities of the nonspecific serum antigens of acute hemosporidian infections. *Mil. Med.* **134**, 1258–1265.

DasGupta, B. (1960a). Lipids in different stages of the life cycle of malaria parasites and some other protozoa. *Parasitology* **50**, 501–508.

DasGupta, B. (1960b). Polysaccharides in the different stages of the life cycles of malaria parasites and some other sporozoa. *Parasitology* **50**, 509–514.

Davis, A. G., Huff, C. G., and Plamer, T. T. (1966). Procedures for maximum production of exoerythrocytic stages of *Plasmodium fallax* in tissue culture. *Exp. Parasitol.* **19**, 1–8.

El-Nahal, H. S. (1966). Fluorescent antibody studies on infections with *Plasmodium gallinaceum.* Effect of splenectomy during latency on parasitemia and antibody titer. *J. Parasitol.* **52**, 570–572.

Ferone, R. (1973). The enzymatic synthesis of dihydropteroate and dihydrofolate by *Plasmodium berghei. J. Protozool.* **20**, 459–464.

Finerty, J. F., and Dimopoullos, G. T. (1968). Disk electrophoretic separation of soluble preparations of *Plasmodium lophurae* and *Anaplasma marginale. J. Parasitol.* **54**, 585–587.

Garnham, P. C. C., Bird, R. G., and Baker, J. R. (1960). Electronmicroscope studies of motile stages of malaria parasites. I. The fine structure of the sporozoites of Haemamoeba (=*Plasmodium*) *gallinacea. Trans. R. Soc. Trop. Med. Hyg.* **54**, 274–278.

Garnham, P. C. C., Bird, R. G., and Baker, J. R. (1962). Electronmicroscope studies of the motile stages of malaria parasites. III. The ookinetes of *Haemamoeba* and *Plasmodium. Trans. R. Soc. Trop. Med. Hyg.* **56**, 116–120.

George, J. N., Stokes, E. F., Wicker, D. J., and Conrad, M. E. (1966). Studies of the mechanism of hemolysis in experimental malaria. *Mil. Med.* **131**, 1217–1224.

Glenn, S., and Manwell, R. D. (1956). Further studies on the cultivation of the avian malaria parasites. II. The effects of heterologous sera and added metabolites on growth and reproduction *in vitro. Exp. Parasitol.* **5**, 22–23.

Graham, H. A., Stauber, L. A., Palczuk, N. C., and Barnes, W. D. (1973a). Immunity to exoerythrocytic forms of malaria. I. Course of infection of *Plasmodium fallax* in turkeys. *Exp. Parasitol.* **34**, 364–371.

Graham, H. A., Stauber, L. A., Palczuk, N. C., and Barnes, W. D. (1973b). Immunity to exoerythrocytic forms of malaria. II. Passive transfer of immunity to exoerythrocytic forms. *Exp. Parasitol.* **34**, 372–381.

Greenberg, J. (1956). Mixed lethal strains of *Plasmodium gallinaceum:* Drug-sensitive, transferable (SP) x drug-resistant, nontransferable (BI). *Exp. Parasitol.* **5**, 359–370.

Hawking, F. (1945). Growth of protozoa in tissue culture. I. *Plasmodium gallinaceum,* exoerythrocytic forms. *Trans. R. Soc. Trop. Med. Hyg.* **39**, 245–263.

Hegner, R., and Eskridge, L. (1938). Passive immunity in avian malaria. *Am. J. Hyg.* **28**, 367–376.

Herman, Y. F., Ward, R. A., and Herman, R. H. (1966). Stimulation of the utilization of 1-14C-glucose in chicken red blood cells infected with *Plasmodium gallinaceum. Am. J. Trop. Med. Hyg.* **15**, 276–280.

Hewitt, R. (1940). "Bird Malaria," Am. J. Hyg. Mongr. Ser. No. 15. Johns Hopkins Press, Baltimore, Maryland.

Honigberg, B. M. (1967). Chemistry of parasitemia among some protozoa. *Chem. Zool.* **1**, 695–814.

Huff, C. G. (1931). The inheritance of natural immunity to *Plasmodium cathemerium* in two species of *Culex. J. Prev. Med.* **5**, 249–259.

Huff, C. G. (1954). Changes in host-cell preferences in malaria parasites and their relation to splenic reticular cells. *J. Infect. Dis.* **94**, 173–177.

Huff, C. G. (1963). Experimental research on avian malaria. *Adv. Parasitol.* **1**, 1–65.

Huff, C. G. (1965). Susceptibilty of mosquitoes to avian malaria. *Exp. Parasitol.* **16**, 107–132.

Huff, C. G. (1968). Recent experimental research on avian malaria. *Adv. Parasitol.* **6,** 295–311.

Huff, C. G., Pipkin, A. C., Weathersby, A. B., and Jensen, D. V. (1960). The morphology and behavior of living exoerythrocytic stages of *Plasmodium gallinaceum* and *P. fallax* and their host cells. *J. Biophys. Biochem. Cytol.* **7,** 93–102.

James, S. P., and Tate, P. (1937). New knowledge of the life-cycle of malaria parasites. *Nature (London)* **139,** 545.

Kielmann, A., Weiss, N., and Sarasin, G. (1971). *Plasmodium gallinaceum* as antigen in the diagnosis of human malaria. *Bull. W.H.O.* **43,** 612–616.

Ladda, R., Aikawa, M., and Sprinz, H. (1969). Penetration of erythrocytes by merozoites of mammalian and avian malarial parasites. *J. Parasitol.* **65,** 633–644.

Longenecker, B. M., Breitenbach, R. P., and Farmer, J. N. (1966). The role of the bursa Fabricius, spleen and thymus in the control of a *Plasmodium lophurae* infection in the chicken. *J. Immunol.* **97,** 594–599.

Manwell, R. D. (1934). The duration of malarial infection birds. *Am. J. Hyg.* **19,** 532–538.

Manwell, R. D. (1949). Birds, malaria, and war. *Am. Sci.* **37,** 60–68.

Manwell, R. D. (1963). Factors making for host-parasite specificity; with special emphasis on blood protozoa. *Ann. N.Y. Acad. Sci.* **113,** 332–342.

Manwell, R. D. (1968). *Plasmodium octamerium,* n. sp., an avian malaria parasite from the pintail whydah, *Vidua macroura. J. Protozool.* **15,** 680–685.

Manwell, R. D., and Goldstein, F. (1939). Strain immunity in avian malaria. *Am. J. Hyg.* **30,** 115–122.

Manwell, R. D., and Goldstein, F. (1940). Passive immunity in avian malaria. *J. Exp. Med.* **71,** 409–422.

Manwell, R. D., and Rossi, G. (1975). Blood protozoa of imported birds. *J. Protozool.* **22,** 124–127.

Manwell, R. D., and Stone, W. (1966). The role of anemia in pinottii malaria of the pigeon. *J. Parasitol.* **52,** 1145–1149.

Manwell, R. D., and Stone, W. (1967). Fever and blood sugar in pinottii malaria of pigeons. *J. Protozool.* **16,** 99–102.

Manwell, R. D., and Weiss, M. (1960). *In vitro* growth of *Plasmodium elongatum* in duck tissues. *J. Protozool.* **7,** 342–346.

Manwell, R. D., Weiss, M. L., and Spandorf, A. A. (1957). Effects of splenectomy on *elongatum* and *circumflexum* malaria in ducks. *Exp. Parasitol.* **6,** 358–366.

McGhee, R. B. (1957). Comparative susceptibility of various erythrocytes to four species of avian plasmodia. *J. Infect. Dis.* **100,** 92–96.

Moulder, J. (1962). "The Biochemistry of Intracellular Parasitism," pp. 13–42. Univ. Chicago Press, Chicago, Illinois.

Oliveira, M. X., and Meyer, H. (1955). *Plasmodium gallinaceum* in tissue culture. Observations after one year of cultivation. *Parasitology* **45,** 1–4.

Peters, W., ed. (1970). "Chemotherapy and Drug Resistance in Malaria." Academic Press, New York.

Platzer, E. G. (1974). Dihydrofolate reductase in *Plasmodium lophurae* and duckling erythrocytes. *J. Protozool.* **21,** 400–405.

Richards, W. H. (1966). Active immunization of chicks against *Plasmodium gallinaceum* by inactivated homologous parasites. *Nature (London)* **212,** 1492–1494.

Rigdon, R., and Rostorer, H. H. (1946). Blood oxygen in ducks with malaria. *J. Nat. Malar. Soc.* **5,** 253–262.

Seed, T. M., and Kreier, J. P. (1972). *Plasmodium gallinaceum:* Erythrocyte membrane alterations and associated plasma changes induced by experimental infections. *Proc. Helminthol. Soc. Wash.* **39,** 387–411.

Sherman, I. W. (1966). Levels of oxidized and reduced pyridine nucleotides in avian malaria (*Plasmodium lophurae*). *Am. J. Trop. Med. Hyg.* **15,** 814–817.

Sherman, I. W., and Tanigoshi, L. (1974). Incorporation of 14-C-amino acids by malarial plasmodia (*Plasmodium lophurae*). VI. Changes in the kinetic constants of amino acid transport during infection. *Exp. Parasitol.* **35,** 369–373.

Sherman, I. W., Ting, I. P., and Tanigoshi, L. (1970). *Plasmodium lophurae:* Glucose-1-14C and glucose-6-14C catabolism by free plasmodia and duckling host erythrocytes. *Comp. Biochem. Physiol.* **34,** 625–639.

Sherman, I. W., Tanigoshi, L., and Mudd, J. B. (1971a). Incorporation of 14-C-amino-acids by malaria (*Plasmodium lophurae*). II. Migration and incorporation of amino-acids. *Int. J. Biochem.* **2,** 27–40.

Sherman, I. W., Peterson, I., Tanigoshi, L., and Ting, I. P. (1971b). The glutamate dehydrogenase of *Plasmodium lophurae* (Avian Malaria). *Exp. Parasitol.* **29,** 433–439.

Soni, J. L., and Cox, H. W. (1974). Pathogenesis of acute avian malaria. I. Immunologic reactions associated with anemia, splenomegaly and nephritis of acute *Plasmodium gallinaceum* infections of chickens. *Am. J. Trop. Med. Hyg.* **23,** 577–585.

Steck, E. (1971). Malaria: General. *In* "The Chemotherapy of Protozoan Diseases," Chapter 22–23, pp. 1–376. Div. Med. Chem., Walter Reed Army Inst. Res., Washington, D.C.

Stutz, D. R., Ferris, D. H., and Voss, E. W. (1972). Enhanced susceptibility of burseatomized chickens to *Plasmodium gallinaceum:* Comparison of three bursectomy methods. *Proc. Helminthol. Soc. Wash.* **39,** 460–464.

Taliaferro, W. H., and Cannon, W. (1931). Acquired immunity in avian malaria. *J. Prev. Med.* **5,** 37–64.

Taliaferro, W. H., and Taliaferro, L. G. (1955). Reactions of the connective tissue in chickens to *Plasmodium gallinaceum* and *P. lophurae.* I. Histopathology during initial infections and superinfections. *J. Infect. Dis.* **97,** 99–136.

Terzakis, J. A. (1968). Uranyl acetate, a stain and a fixative. *J. Ultrastruct. Res.* **22,** 168–184.

Terzakis, J. A. (1971). *Plasmodium gallinaceum:* Drug induced ultrastructural changes in oocysts. *Exp. Parasitol.* **30,** 260–266.

Terzakis, J. A., Spring, H., and Ward, R. A. (1966). Sporoblast and sporozoite formation in *Plasmodium gallinaceum* infection of *Aedes aegypti. Mil. Med.* **131,** 984–992.

Terzian, L. A. (1946). The effect of splenectomy on avian malarial infections. *J. Infect. Dis.* **79,** 215–220.

Todorovic, R., Ferris, D. H., and Ristic, M. (1967). Immunogenic properties of soluble serum antigens of *Plasmodium gallinaceum. Am. J. Trop. Med. Hyg.* **17,** 695–701.

Todorovic, R., Ferris, D. H., and Ristic, M. (1968a). Antigens of *Plasmodium gallinaceum.* I. Biophysical and biochemical characterization of explasmodial antigens. *Am. J. Trop. Med. Hyg.* **17,** 685–694.

Todorovic, R., Ferris, D. H. ,and Ristic, M. (1968b). Antigens of *Plasmodium gallinaceum.* II. Immunoserologic characterization of explasmodial antigens and their antibodies. *Am. J. Trop. Med. Hyg.* **17,** 695–699.

Tracy, S. M., and Sherman, I. W. (1972). Purine uptake and utilization by the avian malaria parasite *Plasmodium lophurae*. *J. Protozool.* **19**, 541–549.

Trager, W. (1943b). Further studies on the survival and development *in vitro* of a malarial parasite. *J. Exp. Med.* **77**, 411–420.

Trager, W. (1943b). The influence of biotin upon susceptibility to malaria. *J. Exp. Med.* **77**, 557–582.

Trager, W. (1947). The development of the malaria parasite *Plasmodium lophurae* in red blood cell suspensions *in vitro*. *J. Parasitol.* **33**, 345–350.

Trager, W. (1957). The nutrition of an intracellular parasite (avian malaria). *Acta Trop.* **14**, 289–301.

Trager, W. (1958). Folinic acid and non-dialyzable materials in the nutrition of malaria parasites. *J. Exp. Med.* **108**, 753–772.

Trager, W. (1966). Coenzyme A and the antimalarial action *in vitro* of antipanthothenate against *Plasmodium lophurae, P. coatneys'* and *P. falciparum*. *Trans. N.Y. Acad. Sci.* **28**, 1094–1103.

Voller, A., and Taffs, L. L. (1963). Fluorescent antibody staining of exoerythrocytic stages of *Plasmodium gallinaceum*. *Trans. R. Soc. Trop. Med. Hyg.* **57**, 32–33.

Wallace, W. R., Finerty, J. F., and Dimopoullos, G. T. (1965). Studies on the lipids of *Plasmodium lophurae* and *Plasmodium berghei*. *Am. J. Trop. Med. Hyg.* **14**, 715–718.

Walsh, C. J., and Sherman, I. W. (1968). Isolation, characterization and synthesis of DNA from a malarial parasite. *J. Protozool.* **15**, 503–508.

Warner, R. E. (1968). The role of introduced diseases in the extinction of endemic Hawaiian avifauna. *Condor* **70**, 101–120.

Warner, R. E. (1973). To kill a Honeycreeper. *Nat. Hist., N.Y.* **82**, 30–39.

Wiselogle, F. Y., ed. (1946). "A survey of Antimalarial Drugs, 1941–1945." Edwards, Ann Arbor, Michigan.

Wolcott, G. B. (1957). Chromosome studies in the genus *Plasmodium*. *J. Protozool.* **4**, 48–51.

Wright, R. H., and Kreier, J. P. (1969). *Plasmodium gallinaceum:* Chicken erythrocyte survival as determined by sodium radiochromate^{51} and di-isopropylfluorophosphate^{32} labeling. *Exp. Parasitol.* **25**, 339–352.

Zuckerman, A. (1965). Autoimmunization and other types of indirect damage to host cells as factors in certain protozoan diseases. *Exp. Parasitol.* **15**, 138–183.

8

Plasmodia of Rodents

Richard Carter and Carter L. Diggs

I. Introduction

Unlike the majority of protozoan groups discussed in this treatise the rodent malaria parasites are not of direct practical concern to man or his domestic animals. Their interest to us is twofold. Because of their ready adaptation to certain laboratory rodents, primarily mice and white rats, and to transmission through species of laboratory-reared mosquitoes, the murine plasmodia are easily the most practical models for the experimental study of mammalian malaria. Contrary to some early expectations, these parasites have proved to be truly analogous to the malarias of man and other primates in most essential aspects of structure, physiology, and life cycle. There are those, however, including without doubt their discoverer Ignace H. Vincke, for whom the interest of the rodent malarias transcends their convenience as laboratory tools. The complex of plasmodia of African murines, in particular, has captured the attention of many whose primary interests lay in the biology of the parasites themselves in their natural hosts and their home environment. In this survey of the rodent malarias, we have set out to give an account of both the general biology of the parasites and their role as models in experimental malariology.

Previous reviews on various aspects of the rodent malarias have appeared from time to time (Draper, 1953; Thurston, 1953; Vincke *et al.*, 1953; Fabiani, 1959; Sergent, 1959; Ciucă *et al.*, 1964; Garnham, 1966; Zuckerman, 1969, 1970). Symposia devoted to this subject were published in the *Ind. J. Malariol.* (Singh, 1954) and in the *Ann. Soc. Belge Med. Trop.* (Jadin, 1965a). A series of reviews dealing with all aspects of rodent malaria is in preparation by Academic Press in a single volume edited by Peters and Killick-Kendrick.

Among the parasites of rodents numerous species of blood protozoa have been described including representatives of four families of Sporozoa—the haemogregarine family Hepatozoidae, the haemosporidian families Haemoproteidae and Plasmodiidae, and piroplasms of the family Babesiidae. All are intracellular parasites of blood cells; all are transmitted by arthropod vectors. The majority of recorded species of blood sporozoa of rodents are classified as babesias or haemogregarines. In a comprehensive check list, Killick-Kendrick (1974a) cites 56 species of *Hepatozoa* in rodents and 53 species of *Babesia.* By contrast, only 6 species and subspecies of *Hepatocystis* and 2 of *Rayella* (Haemoproteidae) and 18 species and subspecies of *Plasmodium* (Plasmodiidae) are recorded. Nevertheless such a number of named plasmodia (20 species and subspecies by present reckoning) exceeds that in any other mammalian order with the exception of the primates. Landau *et al.* (1969) described an unusual parasite of the spiny mouse (*Acomys*). A new genus, *Anthemosoma,* was created to accommodate the organism whose characteristics bear a superficial resemblance both to the Plasmodiidae and to the Babesiidae. For want of a better solution, *Anthemosoma* has been placed in the family Dactylosomidae, a group of parasites otherwise known only in fish and amphibia. The vector(s) and full life cycle of *Anthemosoma* are unknown.

Of the three mammalian subgenera of *Plasmodium* the malaria parasites of rodents are placed in the subgenus *Vinckeia* (Garnham, 1966). This heterogeneous group comprises all mammalian plasmodia other than those of the higher primates. In addition to those of rodents, the *Vinckeia* include malaria parasites of lemurs and other lower primates, bats, and ungulates. There is, however, little to unify members of this group. The remarkably short duration of the preerythrocytic development in the vertebrate host of 2 to 3 days has been described as characteristic of the subgenus. However, since preerythrocytic development has been studied only in the murine malarias this cannot be taken as a general feature of the *Vinckeia.* The site of preerythrocytic development of the murine plasmodia is in the parenchymal cells of the liver. This fact, together with certain aspects of the biochemistry and fine structure of these parasites, is of special significance. It places the murine malaria parasites firmly beside those of man and other primates and not, as was at one time thought likely, with the avian malarias whose preerythrocytic development takes place in the cells of the reticuloendothelial system.

Rodents are the most numerous, diversified, and widely distributed of the mammalian orders. Nevertheless the geographical distribution and number of species known to harbor malaria parasites is highly restricted. Plasmodia are known from murine rodents (Murinae), scaly-tailed flying squirrels (Anomalurinae), and porcupines (Atherurinae) of tropical Africa, and from flying squirrels (Petauristinae) of Southeast Asia. Malaria parasites have never been found in rodents of the New World or from neotropical or palearctic regions of the Old World.

The natural vectors of rodent malarias are known only in the case of certain species parasitizing African murines and that of the African bush-tailed porcupine. In common with other mammalian plasmodia, the vectors are species of anopheline mosquitoes.

II. Malaria Parasites of Nonmurine Rodents

A. Malaria Parasites of Asian Flying Squirrels

Two malaria parasites of Asian flying squirrels have been named: *P. booliati* from the giant flying squirrel (*Petaurista petaurista*) of West Malaysia (Sandosham *et al.,* 1965) and *P. watteni* from *Petaurista petaurista grandis* of Taiwan (Lien and Cross, 1968). In addition, unclassified *Plasmodium* species have been recorded in the spotted giant flying squirrel *Petaurista elegans* of West Malaysia (Yap *et al.,* 1970) and in the black-winged flying squirrel *(Petinomys vordermanni)* and the dark-tailed flying squirrel *(Hylopetes spadiceus)* both of West Malaysia (L. F. Yap, unpublished, quoted by Killick-Kendrick, 1974a).

Only the blood stages of these parasites have been seen. The vectors are unknown.

B. Malaria Parasites of Nonmurine African Rodents

Four species of *Plasmodium* have been described in nonmurine rodents of Africa, three in the African scaly-tailed flying squirrels *Anomalurus derbianus* and *Anomalurus peli* and one in the brush-tailed porcupine *Atherurus africanus.*

Malaria parasites have been described in *A. derbianus* from the foothills of the Usambara mountains of Tanzania and from the Ivory Coast.

The parasite from Tanzania has been named *P. anomaluri* (Pringle, 1960). The parasite from the Ivory Coast (Killick-Kendrick and Bellier, 1970) has not been classified.

Plasmodium landauae and *P. pulmophilum* are two distinct species of malaria parasites from *A. peli* on the Ivory Coast (Killick-Kendrick, 1973b). Parasitemias in these animals are generally low while the incidence of infection in the scaly-tails is apparently high. Blood from scaly-tails infected with either *P. landauae* or *P. pulmophilum* failed to infect either intact or splenectomized mice. The vectors of these parasites have not been identified.

Unlike other rodents known to harbor malaria parasites, *A. africanus* (the African brush-tailed porcupine) is a wholly ground-living species. In Cameroon and the lower Congo basin west of Lake Kivu these animals have been found infected with the malaria parasite *P. atheruri* (van den Berghe *et al.*, 1958). The parasite is transmitted by *Anopheles smithii vanthieli* which inhabits the burrows used by the porcupines and feeds readily upon these animals. Sporozoite rates of 24% have been found in these mosquitoes. Patent blood infections occur in 50 to 60% of *A. africanus* in enzootic localities.

In sporozoite-induced infections in two young *Atherurus* parasitemia developed in 15 to 20 days. The preerythrocytic stage must, therefore, be no longer than this period and could be considerably less. The blood infection is described by Garnham (1966). The young rings are smoothly rounded with a small vacuole. As the trophozoite develops, the cytoplasm becomes amoeboid and vacuolated; pigment is not conspicuous. The mature schizont occupies about one-half the erythrocyte and produces four merozoites arranged around a clump of black pigment. Gametocytes are about 7 μm in diameter and contain dark pigment granules. The microgametocyte has brownish cytoplasm and a large irregular nucleus. In the macrogametocyte, the cytoplasm stains blue and the nucleus is compact. The asexual cycle lasts for 24 hours. Sporozoites of *P. atheruri* measure 10 μm in length.

Unique among malaria parasites of nonmurine rodents, *P. atheruri* has been adapted to laboratory mice, although splenectomized rats and hamsters are apparently refractory to infection. Attention is drawn to the similarity between the description of *P. atheruri* and that of *P. chabaudi* given below.

III. Malaria Parasites of African Murine Rodents

The malaria parasites of African murines are the rodent malarias familiar as models in laboratory research; the most widely studied is *Plasmodium berghei*. The discovery of *P. berghei* (Vincke and Lips, 1948) was an event of unusual consequence in the field of experimental malaria. By 1952 the literature on this parasite exceeded 120 references; Jadin (1965b) was able to cite almost 500 papers on the subject. Today, a rough survey of the literature suggests that in the order of 1500 articles have been written on topics relating to rodent malaria.

Vincke's isolation of *P. berghei* represented the culmination of 5 years of study of anopheline mosquitoes in the vicinity of Elisabethville (now Lubumbashi) in Katanga. In the course of his investigations, Vincke had been impressed by the annual appearance of high sporozoite rates in the glands of a silvatic species *Anopheles dureni millcampsi*. After a prolonged effort to identify the vertebrate host, he found a malaria parasite in the blood of a thicket rat, *Grammomys surdaster*, one of the rodent inhabitants of the gallery forests where *A. d. millcampsi* abound. Identical infections were produced in laboratory rats by inoculation of either the infected blood of *G. surdaster* or sporozoites from the glands of *A. d. millcampsi*. The *Plasmodium* of the mosquito and the thicket rat were thus proved to be one and the same. The parasite was named *Plasmodium berghei* (Vincke and Lips, 1948).

Mice as well as laboratory rats were found to be highly susceptible to infection by inoculation of blood forms of *P. berghei;* only the latter, however, succumbed readily to infection with sporozoites.

In 1952, Vincke isolated a second species of *Plasmodium* from *A. d. millcampsi*. This parasite readily infected mice, although white rats were refactory. Morphologically distinct from *P. berghei,* Rodhain (1952) named the parasite *Plasmodium vinckei.* In 1955, Bruce-Chwatt and Gibson made the first observation of a rodent *Plasmodium* outside Katanga. They described a *P. vinckei*-like parasite in a blood smear taken from a rodent identified as *Praomys jacksoni* captured in Western Nigeria. Landau and Chabaud's (1965) discovery of *P. berghei* and *P. vinckei*-like parasites in the blood of specimens of the shiny thicket rat, *Thamnomys rutilans,* in the Central African Republic (C.A.R.) marked the beginning of a period of rapid increase in the number of known murine plasmodia. *Plasmodium berghei*-like and *P. vinckei*-like parasites were discovered shortly after near Brazzaville (Adam *et al.*, 1966), while in 1967 Killick-Kendrick reconfirmed the existence of Bruce-Chwatt's *P. vinckei*-like parasite in Nigeria and isolated a *P. berghei*-like parasite from murines of the same region (Killick-Kendrick *et al.*, 1968).

Murine plasmodia were thus shown to extend across the Lower Guinea Forest from the highland regions of Katanga, through the lower Congo basin, and into the lowland regions of Western Nigeria (Fig. 1). It is very doubtful, however, whether the distribution of the parasite across this vast area is truly continuous. The attempts to isolate murine plasmodia in intermediate regions have not generally been successful, although at

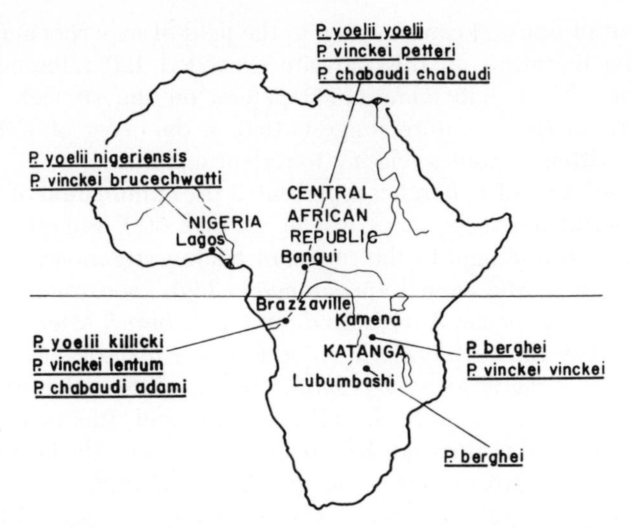

Fig. 1. The distribution of the murine malaria parasites in Africa.

the 9th International Congress on Tropical Medicine and Malaria, Athens, October 14–21, 1973, Bafort reported the presence of murine malaria parasites in the Cameroons. As will be apparent the parasites from the four major foci have clearly evolved in isolation from each other suggesting the existence of significant faunal barriers between the regions.

A. Taxonomy and Morphology

For about half a decade following the discoveries of Landau and others, the taxonomy of the murine malaria parasites lay in a state of some confusion. To the general satisfaction of most authorities, however, the problem has now been resolved with the recognition of four distinct species—*P. berghei, P. yoelii, P. vinckei,* and *P. chabaudi,* each, with exception of *P. berghei,* represented by two or more subspecies. The criteria upon which the current system of classification is based are (1) morphology of blood stage parasites, (2) characteristics of sporogony including temperature of transmission, size of mature oocysts, and sporozoite length, (3) growth rate and size of preerythrocytic stages, and (4) the electrophoretic mobility of enzymes of the blood stage parasites (Carter, 1973; R. Carter, unpublished, 1974; Killick-Kendrick, 1975).

The murine malarias are divided into two groups: the berghei group, comprising *P. berghei* and *P. yoelii,* and the vinckei group, comprising *P. vinckei* and *P. chabaudi.* The similarities within and distinctions between the two groups are manifest in a variety of aspects including the morphology and behavior of blood stage parasites, patterns of cross protec-

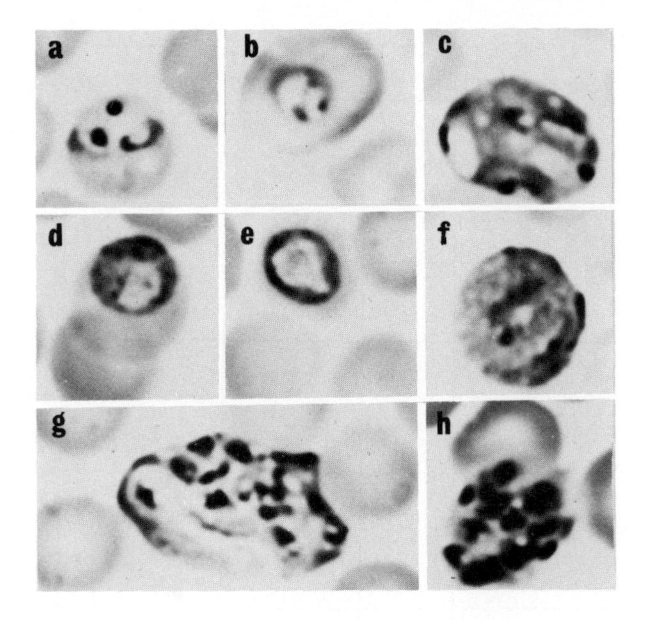

Fig. 2. Blood stages of a berghei group parasite (*P. yoelii*) on blood smears fixed with methanol and stained with Giemsa's stain × 10^3. (a) Young ring stage parasites in an immature red blood cell (RBC); (b) immature RBC parasitized by a single young trophozoite; (c) immature RBC multiply parasitized by young trophozoites; (d) and (e) mature RBC's each parasitized by a single trophozoite; (f) macrogametocyte; (g) marked hypertrophy in a multiply parasitized immature RBC; (h) mature schizont.

tion, serology, and enzyme type. Characteristics of members of each group are shown in Figs. 2, 3, and 4 and listed in Tables I and II.

1. The Berghei Group (Fig. 2)

In his taxonomic revision of "*P. berghei*," Killick-Kendrick (1974a) has given a full description of the berghei group. Members of this group, previously classified as the single species *P. berghei* and subspecies, are effectively indistinguishable upon the basis of their blood stage morphology. All show a marked intrinsic preference for reticulocytes and other young red blood cells, although in virulent infections this preference may not be readily apparent; as reticulocytes are depleted in virulent infections, mature erythrocytes become extensively invaded. Polyparasitism is common especially in reticulocytes and young red blood cells and is accompanied by marked hypertrophy of the infected cell. The very youngest trophozoites or rings are round, the central vacuole is not usually prominent, and the youngest rings have a single nucleus. Hairlike ring

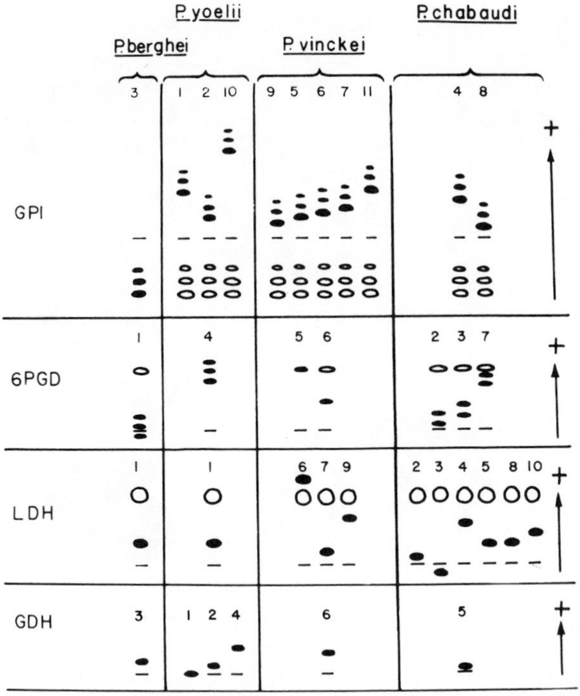

Fig. 3. Forms of enzymes of the murine malaria parasites as separated by electrophoresis on starch gel according to the technique of R. Carter (unpublished, 1974). Distinction between all enzyme variants is not possible on a single system of electrophoresis. Open circles, host enzyme; closed circles, parasite enzyme. For each enzyme the variants of the enzyme found in the parasites are designated by separate numbers.

forms of the type more common in vinckei group parasites are not seen. The older trophozoites, which may develop a large central vacuole, and young schizonts remain rounded, not amoeboid. In infected reticulocytes the malaria pigment appears as fine black granules while in infected normocytes the pigment has a brown or golden aspect. The schizonts produce generally between 10 and 18 merozoites, although the mean number and range are dependent upon the host species. The schizonts show a tendency to congregate in the internal organs.

Gametocytes appear in the circulation within 3 to 4 days after blood inoculation. These are large parasites filling the entire host cell which itself shows some hypertrophy. Fine pigment granules are distributed evenly through the cytoplasm. With Giemsa's stain the macrogametocytes stain blue and have a dark eccentrically placed nucleus surrounded by a pink aureola; the cytoplasm of the microgametocytes stains pink as does

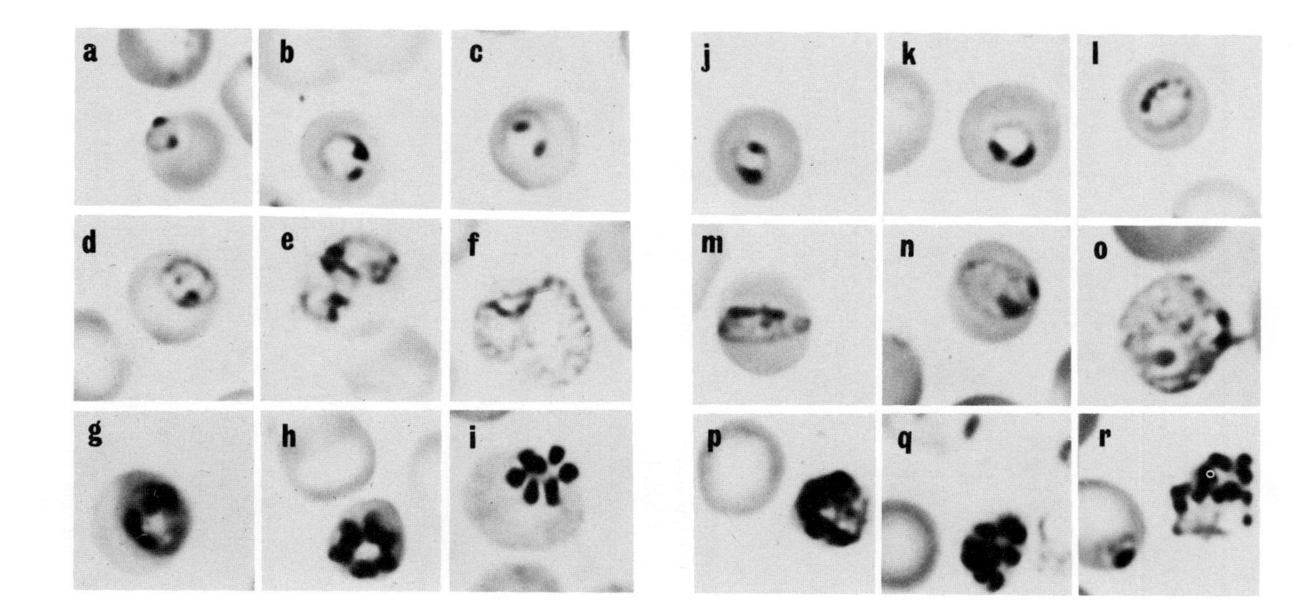

Fig. 4. Blood stages of vinckei group parasites on blood smears fixed with methanol and stained with Giemsa's stain $\times 10^3$. (a)–(i) *P. chabaudi;* (j)–(r) *P. vinckei;* (a)–(c) and (j)–(l) ring stage parasites in mature RBC's; (d), (e), (m), and (n) trophozoites in mature RBC's; (f) and (o) macrogametocytes; (g), (h), (p), and (q) immature schizonts; (i) and (r) mature segmenting schizonts.

Table I

Distribution of Enzyme Forms among Species and Subspecies of Murine Malaria Parasite

Plasmodium sp.	Enzyme forms			
	GPI	6PGD	LDH	GDH
P. berghei	3	1	1	3
P. y. yoelii	1,2,10	4	1	4
P. y. killicki	1	4	1	1
P. y. nigeriensis	2	4	1	2
P. v. vinckei	7	6	6	6
P. v. petteri	5,9	5	7	6
P. v. lentum	6,11	5	7,9	6
P. v. brucechwatti	6	6	9	6
P. c. chabaudi	4	2,3,7	2,3,4,5	5
P. c. adami	8	2	8,10	5

the nucleus, although the latter is of a deeper hue. According to Corradetti and Verolini (1951), only mature erythrocytes are parasitized by the sexual forms.

The blood infections of berghei group parasites are normally highly asynchronous in laboratory animals. In artifically synchronized infections the asexual division cycle has been reported to take 22–25 hours (Killick-Kendrick and Warren, 1968), although Walter (1968) found the cycle in *P. berghei* to be 18 hours, and Carter (unpublished, 1975) has demonstrated a cycle of 16 to 18 hours in *P. y. yoelii.* A variety of laboratory rodents including white rats, cotton rats, mice, hamsters, and laboratory-reared thicket rats *(G. surdaster)* are susceptible to blood-induced infections with berghei group parasites.

The member species of the berghei group, *P. berghei* and *P. yoelii,* are distinguished primarily by characteristics of sporogony and by the enzyme forms of the blood stage parasites. All berghei group parasites share the same form of lactate dehydrogenase, LDH-1; all are characterized by 6PGD forms with three subbands (Fig. 3; Table I).

A. PLASMODIUM (VINCKEIA) BERGHEI Vincke and Lips 1948. *Plasmodium berghei* is a malaria parasite of murines of Katanga. Its principle vertebrate host is the thicket rat *Grammomys surdaster;* its only known natural vector is *Anopheles dureni millcampsi.* The geographical distribution of *P. berghei* is apparently limited by the range of this mosquito which is restricted to the highland areas of Katanga at altitudes of not less than 900 m (Vincke, 1954).

Although attempts to establish laboratory colonies of *A. d. millcampsi* have met with consistent failure, *Plasmodium berghei* may be transmitted in the laboratory by several other species of anopheline of which *A.*

stephensi is the most suitable. Sporogony is extremely sensitive to temperature. For the optimum production of viable sporozoites, fed mosquitoes must be maintained at 19° to 21°C (Vanderberg and Yoeli, 1966). At maturity oocysts are generally not more than 45 μm in diameter (Yoeli and Most, 1960; Landau and Killick-Kendrick, 1966). Sporozoites first reach the salivary glands on the fourteenth day after a blood meal. Most measurements of sporozoite length fall between 10 and 12 μm (Garnham, 1966; Bafort, 1971a; Killick-Kendrick, 1973a).

In the liver of white rats, the preerythrocytic schizonts reach maturity in not less than 50 hours after inoculation of sporozoites (Yoeli, 1965; Wery, 1968). The mean diameter of liver schizonts of *P. berghei* is apparently very variable. Except in carefully controlled conditions, comparisons with other species of murine *Plasmodia* are unlikely to be reliable. Landau and Killick-Kendrick (1966) reported the mean diameter of mature schizonts of *P. berghei* to be 27 μm; this compared with a mean value for mature preerythrocytic schizonts of *P. y. yoelii* of about 35 μm measured under comparable conditions. Using the same strain of *P. berghei,* however, Yoeli and Most (1965) found schizonts of a wide range of sizes, the mean diameter being about 40 μm. The discrepancy between these and other results in the literature reflects the difficulty of standardizing such measurements. Nevertheless, under controlled conditions significant differences between the smaller preerythrocytic schizonts of *P. berghei* and the larger schizonts of *P. yoelii* can be demonstrated. Blood infections due to *P. berghei* in mice and young rats typically run a fulminating and fatal course.

The enzyme forms characterizing *P. berghei* are GPI-3, 6PGD-1, LDH-1, and GDH-3 (Table I).

B. PLASMODIUM (VINCKEIA) YOELII Landau and Killick-Kendrick 1966. *Plasmodium yoelii* is known as a parasite of murine rodents in three widely separated localities situated in lowland areas of the Lower Guinea forest (Fig. 1). The three geographical populations are recognized as separate subspecies: *P. y. yoelii* from the C.A.R., *P. y. killicki* from the vicinity of Brazzaville, and *P. y. nigeriensis* from Western Nigeria. In all three regions the only proved host is the shiny thicket rat *Thamnomys rutilans.* The vectors are not known in any instance, although *Anopheles cinctus* has been implicated in this role in Nigeria and Brazzaville (Killick-Kendrick, 1974b).

In the morphology of their blood stages the subspecies of *P. yoelii* are indistinguishable from each other as they are indistinguishable from *P. berghei.* The primary distinctions between *P. yoelii* and *P. berghei* lie in characteristics of the sporogonic and preerythrocytic stages and in the enzyme forms of the blood stages.

Table II
Biological Characteristics of the Murine Malaria Parasites [a]

	berghei group			
	P. berghei	*P. yoelii*		
		P. y. yoelii	*P. y. killicki*	*P. y. nigeriensis*
Morphology of blood stages	berghei-like	berghei-like	berghei-like	berghei-like
No. of merozoites per schizont	12–18	12–18	12–18	12–18
Blood infection synchronous	No	No	No	No
Upper limit to temperature of mosquito transmission (°C)	23	> 26	> 26	> 26
Optimum temperature range for mosquito transmission (°C)	19–21	24–26	24–26	24–26
Mean oocyst size at optimum temperature (μm)	45	75	60	60
Mean sporozoite length at optimum temperature (μm)	11–12	15–16	14–18	17
Sporozoites first reach glands at optimum temperature (days)	13–14	10	10	9–11
Mean diameter of mature preerythrocytic schizonts in mouse or white rat (μm)	27	35	35	42–50
Duration of preerythrocytic development (hours)	43–52	43–47	46	47–48
Susceptibility to blood infection				
Hamsters, white rats	Yes	Yes	Yes	Yes
Mice	Yes	Yes	Yes	Yes
Susceptibility to sporozoite-induced infection				
Hamsters, white rats	Yes	Yes	Yes	Yes
Mice	Variable	Yes	Yes	Yes

[a] Because of the problems of achieving standard conditions of measurement, values for certain of the parameters represented here have shown wide variations in the hands of different investigators. Values for the measurements, as presented here, must be accepted with a degree of caution. Rather than try to indicate the range of values found in the literature, we have attempted to give what we hope is a fair value, or natural range of values, for each parameter in each subspecies. In most cases, we feel that the measurements given lie close to the true values and are reliable reflections of differences between the species and subspecies of murine malaria parasite. Although we have little doubt that differences in the preerythrocytic development of the parasite do exist, we are less confident that the measurements quoted here are always reliable reflections of such differences.

[b] ND, no data.

vinckei group					
P. vinckeii				*P. chabaudi*	
P. v. vinckei	*P. v. petteri*	*P. v. lentum*	*P. v. brucechwatti*	*P. c. chabaudi*	*P. c. adami*
vinckei-like	vinckei-like	vinckei-like	vinckei-like	chabaudi-like	chabaudi-like
6–12	10–12	10–12	10–12	6–8	6–8
Yes	Yes	Yes	Yes	Yes	Yes
> 26	> 26	> 26	> 26	> 26	> 26
24–26	24–26	24–26	24–26	24–26	24–26
43	50	47	54	50	50
12–26	16	16–19	15	12	12
11	11	11	12–13	11	11
45	ND [b]	38	43	35	ND
52–61	ND	65–72	61–65	50–55	ND
No	ND	ND	No	No	ND
Yes	Yes	Yes	Yes	Yes	Yes
Yes	ND	ND	ND	ND	ND
Yes	Yes	Yes	Yes	Yes	Yes

Like *P. berghei,* the subspecies of *P. yoelii* may be transmitted in the laboratory using *A. stephensi* as vector. *Plasmodium yoelii* subspecies, however, undergo optimum transmission between 24° and 26°C, temperatures lethal to the sporogonic stages of *P. berghei.* At this temperature both oocysts and sporozoites of subspecies of *P. yoelii* are larger than those of *P. berghei.* Killick-Kendrick (1974b) states that the mean diameter of mature oocysts of *P. yoelii* is 60 μm or greater; the mean length of sporozoites of *P. yoelii* is not less than 14 μm. At 24°C, sporozoites first enter the glands of the mosquito on the tenth day after a blood meal.

Tissue schizonts of *P. yoelii* reach maturity more rapidly than those of *P. berghei.* The minimum maturation time in white rats is between 45 and 50 hours while the mean diameter of the mature schizonts is always greater than 30 μm (Killick-Kendrick, 1974b).

All members of *P. yoelii* so tested demonstrate a marked level of resistance to chloroquine and certain related drugs in the presence of *para* aminobenzoic acid (PABA) in the diet of the host (Warhurst and Killick-Kendrick, 1967; Carter, 1972). Strains of *P. berghei,* on the other hand, are normally highly sensitive to chloroquine with or without the presence of PABA. Blood infections in mice and rats are, in most cases, mild and rarely fatal.

The enzyme forms characterizing the subspecies of *P. yoelii* are GPI-1 or -2, 6PGD-4, LDH-1, and GDH-1, -2 or -4 (Killick-Kendrick, 1974b) (Table I).

Within the berghei group the subspecies of *P. yoelii* are united primarily by the higher optimum temperature of transmission (24°C), by the generally less virulent course of blood infection as compared to *P. berghei,* and by the shared forms of GPI and 6PGD.

i. *Plasmodium yoelii yoelii* Landau and Killick-Kendrick 1966. *Plasmodium yoelii yoelii* is a common malaria parasite of *T. rutilans* in the C.A.R. At 24°C, mature oocysts have a mean diameter of 75 μm (Landau and Killick-Kendrick, 1966); the mean length of sporozoites in crushed salivary glands is about 15 μm (Killick-Kendrick, 1973a). Under the standard conditions quoted by Killick-Kendrick (1974b), the mean diameter of mature tissue schizonts is given as 37 μm. *Plasmodium y. yoelii* is uniquely characterized by GDH-4 (Table I).

ii. *Plasmodium yoelii killicki* Landau, Michel and Adam 1968. *Plasmodium yoelii killicki* is found in *T. rutilans* from the vicinity of Brazzaville. At 22° to 24°C, mature oocysts have a mean diameter of about 60 μm; the mean length of sporozoites in crushed salivary glands is between 14 and 15 μm (Killick-Kendrick, 1973a). Under standard conditions (Killick-Kendrick, 1974b) the mean diameter of mature tissue schizonts is given as 35 μm. *Plasmodium y. killicki* is uniquely characterized by GDH-1 (Table I).

iii. *Plasmodium yoelii nigeriensis* Killick-Kendrick 1973. *Plasmodium y. nigeriensis* has been isolated only once from a specimen of *T. rutilans* captured in the Illobi Forest in Western Nigeria (Killick-Kendrick *et al.,* 1968). At 24°C, the mean diameter of mature oocysts is 60 μm and the mean length of salivary gland sporozoites about 17 μm (Killick-Kendrick, 1973a). The mean diameter of mature tissue schizonts measured under the standard conditions of Killick-Kendrick (1974b) is given as between 42 and 50 μm. *Plasmodium y. nigeriensis* is uniquely characterized by GDH-2 (Table I).

2. The Vinckei Group (Fig. 4)

The separate identity of the two species representing the vinckei group—*P. vinckei* and *P. chabaudi*—has only recently been clearly recognized (Carter and Walliker, 1975). The failure to appreciate the distinction between these two species in the C.A.R. and in Brazzaville obviously contributed to the difficulties of establishing a satisfactory system of classification in the murine malarias. Moreover, it must now be recognized that many of the earlier observations made upon vinckei group parasites from these two regions are subject to uncertain interpretation. The comparative antigenic studies of vinckei group parasite from the C.A.R. and Brazzaville with *P. vinckei* from other regions by Cox and Voller (1966), Bray and El-Nahal (1966), and El-Nahal (1967a) can no longer be taken as reliable analyses of the relationships between these parasites. In the light of present knowledge of enzyme forms of *P. vinckei* and *P. chabaudi*, the distinction between these species can be clearly recognized in the enzyme studies on the vinckei group made by Carter (1973). In the original accounts of both *P. chabaudi* (Landau, 1965) and *P. vinckei* (Landau *et al.*, 1970) it is clear that the parasites described possessed morphological characteristics of each of the two species now recognized as *P. vinckei* and *P. chabaudi*, indicating that these workers were almost certainly dealing with mixed infections.

The parasites of the vinckei group are most readily distinguished from those of the berghei group by their lack of preference for young erythrocytes. In the absence of such a preference, mature red blood cells are extensively invaded in infections with vinckei group parasites. Due to the absence of a restriction to a minor population of red blood cells (e.g., reticulocytes), multiple invasion of blood cells is comparatively rare except at high parasitemias. Hypertrophy of infected cells so characteristic of the berghei group is virtually never seen. In blood smears stained with Giemsa, the general impression of a vinckei group parasite is of a more delicate transparent organism than the heavily staining berghei group parasite. Ring stages may be compact with a small central vacuole or possess a hairlike circle of cytoplasm around a central vacuole. Unlike the member species of the berghei group, *P. vinckei* and *P. chabaudi* are clearly distinguished from each other by the morphology of the later trophozoites and schizonts of the blood infection (see below). The gametocytes completely fill the host cell and are of equal size. The cytoplasm of the macrogametocyte is pale blue and is filled with up to 20 grains of golden colored pigment. Microgametocytes possess deep pink cytoplasm and nucleus, the large nuclear area being distinguished by the absence of pigment.

Whereas blood infections of berghei group parasites are normally highly asynchronous, the asexual parasites in infections of vinckei group species are comparatively well synchronized with a 24-hour cycle. Schizogony is most common in the evening or early morning hours, although the time of the peak differs somewhat in the two species. Schizonts are never very numerous in the peripheral circulation, however; it is probable that significant sequestration of these forms occurs in the internal organs. Unless so adapted in the laboratory, vinckei group parasites do not infect white rats, cotton rats, or hamsters. Mice and laboratory-reared thicket rats are highly susceptible to infection.

None of the members of the vinckei group are restricted to mosquito transmission at low temperatures, i.e., less than 21°C. For some at least, e.g., *P. vinckei vinckei,* transmission may be achieved over a wide temperature range, 16°–28°C (Bafort, 1971a). Optimum production of infective sporozoites occurs at temperatures of 24° to 26°C.

A. PLASMODIUM (VINCKEIA) VINCKEI Rodhain 1952. *Plasmodium vinckei* is the most widely distributed of the species of murine *Plasmodium* and has been found in all four major geographic foci (Fig. 1). The parasites in each region are accorded subspecies status. In Katanga, the natural host of *P. vinckei* is unknown but is presumed to be *Grammomys surdaster;* its vector is *A. d. millcampsi,* the mosquito that transmits *P. berghei.* In other regions the rodent host is *T. rutilans;* in Nigeria *Praomys tullbergi* may also be infected. The vectors of *P. vinckei* outside Katanga are unknown although it is possible that *A. cinctus* may fulfill this role in some regions (Killick-Kendrick, 1971).

Plasmodium vinckei is readily distinguished from *P. chabaudi* on the basis of the morphology of the trophozoites and schizonts of the blood infection. As trophozoites develop beyond the ring stage, they become rounded and abundant golden brown pigment granules appear in the cytoplasm. The cytoplasm itself remains palely staining. The nucleus enlarges only slightly becoming oval in shape and may lie in a clear region of cytoplasm. Trophozoites, which frequently contain one or two small vacuoles, remain round or oval and occupy up to ¾ of the host cell. As schizogony commences, the trophozoites condense slightly and appear more deeply stained. The pigment coalesces into a single mass which usually remains centrally placed in the schizont. Normally, between ten and twelve merozoites are produced in infections in laboratory mice although the range commonly extends from eight to sixteen. The number of merozoites produced by an erythrocytic schizont are, however, said to vary according to host species. During the growth of the trophozoite, the host cell becomes very distorted and appears to lose hemoglobin. As schizogony nears completion, the infected erythrocyte is reduced to a pale and severely misshapen remnant of the original blood cell. In *P. vinckei* infections, schizogony reaches a peak in the early evening hours.

Within the vinckei group, extensive enzyme variation occurs both between *P. vinckei* and *P. chabaudi* and within each species. No enzyme forms are shared between subspecies of *P. vinckei* and those of *P. chabaudi*. All *P. vinckei* are characterized by GDH-6, while all *P. chabaudi* share GDH-5. *Plasmodium vinckei* subspecies are characterized by a single major band of 6PGD, *P. chabaudi* by two (Fig. 3).

Sporozoites of *P. vinckei* are characteristically slender with pointed ends in contrast to blunt-ended stumpy sporozoites of *P. chabaudi*.

i. *Plasmodium vinckei vinckei* Rodhain 1952. *Plasmodium vinckei vinckei* is a rare parasite from the Katanga plateau. It has been isolated only twice, by Vincke in 1952 (Rodhain, 1952) and by Bafort in 1967 (Bafort, 1969a) in each instance from the glands of *A. d. millcampsi*. Its natural host is presumed to be *G. surdaster*.

In many ways the least typical of the subspecies of *P. vinckei, P. v. vinckei* most closely resembles its Nigerian counterpart *P. v. brucechwatti* with which it shares the enzyme forms 6PGD-6 in addition to GDH-6. Unlike the benign infections associated with *P. v. petteri* and *P. v. lentum* in mice and laboratory thicket rats, laboratory-induced infections of *P. v. vinckei* and *P. v. brucechwatti* are usually fulminating and fatal in these hosts. According to Rodhain (1952) and Bafort (1971a), however, *P. v. vinckei* gives rise to a comparatively mild infection in mice upon first isolation. Bafort (1971a) records that the mean number of merozoites released by the blood schizonts of *P. v. vinckei* is about eight; in our experience ten to twelve are more commonly found.

At 24° to 25°C, Bafort (1971a) found that the mean size of mature oocysts of *P. v. brucechwatti* in *A. stephensi* varied between 42 and 65μm. Killick-Kendrick (1975) gives a mean value of 40 μm. According to Killick-Kendrick (1975), the mean length of sporozoites of *P. v. vinckei* is about 12 μm, although Bafort (1971a) found values ranging from 11 to 19 μm with a mean of 15 to 16 μm. In mosquitoes maintained at 24° to 25°C, sporozoites reach the glands on the tenth day after a blood meal.

The minimum maturation time of the preerythrocytic schizonts of *P. v. vinckei* is 53 hours according to Bafort (1969b), although he (1971a) mentions 61 hours as the minimum time, the mean diameter of mature schizonts being approximately 45 μm.

In addition to GDH-6 and 6PGD-6, *P. v. vinckei* is uniquely characterized by GPI-7 and LDH-6 (Table I).

ii. *Plasmodium vinckei petteri* Carter and Walliker 1975. Among populations of the shiny thicket rat, *T. rutilans,* in the vicinity of Bangui in the C.A.R., almost 100% of adult animals are infected with parasites of the vinckei group. The majority of these infections are due to *P. c. chabaudi* while about one-third have infections of *P. vinckei petteri.* Mixed infections of these two parasites and *P. y. yoelii* are common in various combinations, the latter being found in about one-half the adult rats.

Plasmodium v. petteri is very similar to *P. v. lentum* from the Brazzaville region. In the laboratory, both give rise to benign infections in mice and thicket rats. In addition to GDH-6, common to all members of the species, these two subspecies share the enzyme forms LDH-7 and 6PGD-5. The only reliable means of distinguishing these parasites is on the basis of the forms of GPI, *P. v. petteri* being characterized by GPI-6 or -11 and *P. v. lentum* by GPI-5 or -9 (Table I).

At 24° to 26°C, the mean diameter of mature oocysts of *P. v. petteri* in *A. stephensi* is 50 μm. The mean length of the sporozoites is reported as 16–17 μm (Carter and Walliker, 1975). Sporozoites appear in the glands on the eleventh day after a blood meal. The preerythrocytic stages have not been described.

iii. *Plasmodium vinckei lentum* Landau, Michel, Adam and Boulard 1970. *Plasmodium vinckei lentum* is a common parasite of *T. rutilans* from the vicinity of Brazzaville. Adam *et al.* (1966) found 10 out of 24 specimens of *Thamnomys* infected with parasites of the vinckei group. The majority of such infections are due to *P. v. lentum,* although *P. c. adami* is also found. Although the natural vector of these parasites is unknown, *A. cinctus* has been implicated on indirect evidence (Killick-Kendrick, 1971).

At the time of the original description of *P. v. lentum* Landau *et al.* (1970), the distinction between *P. vinckei* and *P. chabaudi* was not appreciated. From the extreme length of the sporozoites, mean length 21 μm, there is little doubt that their description does relate to the parasite now recognized as *P. v. lentum.* Their value for sporozoite length (in *A. stephensi* at 23° to 25°C) is significantly greater than that given by either Killick-Kendrick (1975), 19 μm, or Carter and Walliker (1976), 18–19 μm. It is clear, therefore, that the measurements of Landau *et al.* (1970) do not reflect the presence of *P. c. adami,* the mean length of whose sporozoite is about 13 μm (Carter and Walliker, 1976). The sporozoites of *P. v. lentum* first reach the glands of experimentally infected *A. stephensi* maintained at 23° to 25°C on the tenth or eleventh day after a blood meal. Mature oocysts had a mean diameter of 47 μm (Landau *et al.,* 1970), or 45 μm (Carter and Walliker, 1976).

According to Landau *et al.* (1970), the mean diameter of the mature preerythrocytic schizont is about 38 μm and the minimum time to maturation between 65 and 72 hours. This is a considerably longer prepatent period than that recorded for most other murine malarias including *P. v. vinckei* whose preerythrocytic schizonts reach maturity between 45 and 55 hours after inoculation of sporozoites. It is, however, comparable to that of *P. v. brucechwatti,* the only other lowland subspecies of *P. vinckei* for which such figures are available. Bafort (1971b) recorded a prepatent period of 61 to 65 hours after the inoculation of sporozoites of *P. v. brucechwatti* in mice.

In our experience, blood infections of *P. v. lentum* in mice and thicket rats have remained comparatively mild during the course of a number of passages in the laboratory. The morphology of the blood stages is typical of the species.

Plasmodium v. lentum is characterized by the enzyme forms GPI-6 or -11, 6PGD-5 and LDH-7 being shared with *P. v. petteri* and GPI-6 and LDH-9 with *P. v. brucechwatti* (Table I).

iv. *Plasmodium vinckei brucechwatti* Killick-Kendrick 1975. In 1967, Killick-Kendrick found 6 out of 18 specimens of *T. rutilans* captured in Western Nigeria infected with a malaria parasite of the vinckei group (Killick-Kendrick *et al.,* 1968). This parasite is now recognized as a subspecies of *P. vinckei* by the name *P. vinckei brucechwatti.* Its natural vector is unknown, although *A. cinctus* may be involved in this role.

As in the case of *P. v. vinckei,* isolates of *P. v. brucechwatti* initially gave rise to benign blood infection in mice. However, within the space of a few blood passages in laboratory hosts, the infections developed fulminating and frequently fatal courses in mice and thicket rats. The morphology of the blood stage parasites is typical of the species.

Sporogonic development has been achieved in a variety of laboratory anophelines including *A. quadrimaculatus, A. stephensi,* and *A. labranchiae atroparvus.* Each of these species develop light sporozoite infections in the salivary glands at 24° to 26°C. There is only one record of successful transmission of *P. v. brucechwatti* from the mosquito to a rodent host (Bafort, 1971b).

The mean diameter of mature oocysts in *A. stephensi* at 24° to 26°C is about 48 μm and the mean length of sporozoites about 15 μm (Killick-Kendrick, 1975). Sporogonic development is rather slow, the first major sporozoite invasion of the glands not occurring until the thirteenth day after a blood meal.

In white mice the minimum time to maturity of the preerythrocytic stages was found to be between 61 and 65 hours. The mean diameter of preerythrocytic schizonts is 43 μm (Bafort, 1971b).

Plasmodium v. brucechwatti is characterized by GPI-6, 6PGD-6, LDH-

9, and GDH-6 (Table I). None of these enzyme forms is itself unique to the subspecies. The combination of forms is uniquely characteristic of *P. v. brucechwatti.*

B. PLASMODIUM (VINCKEIA) CHABAUDI Landau 1965. *Plasmodium chabaudi* has been found only in *T. rutilans* of the C.A.R. and Brazzaville. The parasites in each region are accorded subspecies status as *P. chabaudi chabaudi* and *P. chabaudi adami,* respectively. The natural vectors are not known in either region.

Plasmodium chabaudi is distinguished from *P. vinckei* by the morphology of the trophozoites and schizonts of the blood infection and by the morphology of the sporozoites; no enzyme forms are known to be shared between *P. chabaudi* and *P. vinckei.*

One of the most striking aspects of the morphology of the blood stages of *P. chabaudi* is the virtual absence of discernible malaria pigment in the growing trophozoite. Only as the blood schizont reaches maturity does the pigment become apparent as a single large granule of black pigment at the center or edge of the dividing parasite.

As the young trophozoite develops beyond the ring stage, the central vacuole is lost and the parasite becomes feebly ameboid having a characteristically ragged outline and several small vacuoles within the cytoplasm. The trophozoites never become as large as those *P. vinckei,* seldom exceeding one-half the diameter of the host cell. As the parasites enter the early stages of schizogony, an intense reddening of the cytoplasm of the host cell may be seen. Although a highly unpredictable phenomenon, the reddening of the host cell is uniquely characteristic of *P. chabaudi* subspecies.

As schizogony progresses the parasite becomes condensed and darkly staining. In contrast to *P. vinckei, P. chabaudi* usually has little obvious disruptive effect on its host cell. On fixed blood smears the merozoites often appear to separate from the schizont leaving the host cell intact. The schizonts are the smallest of any species of murine *Plasmodium* normally producing between four and eight merozoites. Schizogony reaches a peak in the peripheral circulation in the late evening and early hours of the morning. The similarity between this description and that of *P. atheruri* has been noted above.

Both subspecies of *P. chabaudi* may be transmitted in the laboratory in *A. stephensi* maintained at 24° to 26°C. Mature oocysts reach a diameter of about 50 μm (Carter and Walliker, 1975, 1976). Sporozoites which first reach the salivary glands on the eleventh day following a blood meal are characteristically short with blunt ends. The mean lengths of sporozoites of *P. chabaudi* have been variously reported as 13 μm (Landau and Killick-Kendrick, 1966) and 11 to 13 μm (Carter and Walliker, 1975, 1976).

The preerythrocytic stages of *P. c. chabaudi* have been described by Landau and Killick-Kendrick (1966). The mean diameter of the mature schizont was recorded by these workers at about 37 μm. The prepatent period was between 50 and 55 hours.

The two subspecies of *P. chabaudi* are virtually identical by most criteria; they are distinguished only by certain enzyme forms of the blood stage parasites. All *P. chabaudi* are characterized by GDH-5 and isolates of either subspecies may be characterized by LDH-2.

i. *Plasmodium chabaudi chabaudi* Landau 1965. Almost 100% of adult specimens of *T. rutilans* from the field station at Bukoko in the C.A.R. harbor infections of *P. chabaudi chabaudi* (Killick- Kendrick, 1971). The original description of this parasite (Landau, 1965) includes characteristics of *P. v. petteri* as well as those of *P. c. chabaudi.* However, the measurements of sporogony given by Landau and Killick-Kendrick (1966) and Wery (1968) clearly indicate that the organism referred to in these studies was that now recognized as *P. c. chabaudi.*

In the laboratory, blood infections in mice and thicket rats are generally fulminating and frequently fatal.

Plasmodium c. chabaudi is distinguished from *P. c. adami* by its enzyme forms of which GPI-4 is found in all isolates of the *P. c. chabaudi* and GPI-8 in isolates of *P. c. adami.* The enzymes LDH and 6PGD are remarkably polymorphic among isolates of *P. c. chabaudi;* 6PGD-2, -3, and -7 and LDH-2, -3, -4, and -5 are all found (Table I). There is no evidence from the distribution of these forms that *P. c. chabaudi* is represented by genetically isolated subpopulations in the C.A.R. On the contrary, the random reassortment of the enzyme forms among the isolates of *P. c. chabaudi* strongly indicates a single breeding population of organisms.

ii. *Plasmodium chabaudi adami* Carter and Walliker 1977. *Plasmodium chabaudi adami* is a parasite of *T. rutilans* from the region of Brazzaville. As in the case of the vinckei group parasites from the C.A.R., the distinction between *P. chabaudi* and *P. vinckei* from Brazzaville has only recently been recognized (Carter and Walliker, 1977). Early allusions to vinckei group parasites of this region clearly refer to characteristics of both species (Garnham *et al.,* 1967b; Wery, 1968).

Of the two existing isolates of *P. c. adami,* one runs a fulminating course of infection in mice and thicket rats in the laboratory. The other runs a benign course of infection not experienced with *P. chabaudi* unless superinfected with organisms such as *Eperythrozoon coccoides* (Ott and Stauber, 1967).

Plasmodium c. adami is otherwise indistinguishable from *P. c. chabaudi* in the morphology of sporozonic and blood stage parasites. The preerythrocytic stages of *P. c. adami* have not been studied.

The subspecies is uniquely characterized by GPI-8 and LDH-8 or -10 (Table I).

3. Comments on the Taxonomy of the Murine Plasmodia

The literature on the murine plasmodia has, in the course of time, accumulated a variety of reports which in one way or another defy acceptance in the context of the taxonomic scheme advocated here.

The earliest such report involved a malaria parasite isolated from a rat caught near Lierre in Belgium. The parasite was named "*P. inopinatum*" (Resseler, 1956). According to Garnham (1966) the infection was probably *P. berghei* acquired during the period of captivity of the rat in the Institute of Tropical Medicine in Antwerp.

Raffaele (1965) claimed to have transmitted a strain of *P. berghei* through *Aedes aegypti* and infected chickens. During the course of passage in chickens the morphology of the parasites closely resembled that of *P. gallinaceum,* a malaria parasite of chickens. Believing that he had induced some form of transformation in the original *P. berghei* Raffaele named the novel parasite "*P. metastaticum.*" The true nature of this parasite has never been fully investigated. Although no strain of *P. gallinaceum* was maintained in the Instituto di Malariologia, where these observations were made, their most likely explanation is as a result of cross-infection at some stage.

Neither "*P. inopinatum* or "*P. metastaticum*" can be regarded as valid species.

With the discovery of murine malarias in regions of Africa outside Katanga, the taxonomy of these parasites entered a state of uncertainty. The problems arose initially from the natural variation among isolates from a single geographical locus and the difficulty of compiling comparable measurements in different laboratories. Discussion centered primarily on the question of the validity of distinctions being made between berghei group parasites from different parts of Africa (Bafort, 1970c). Eventually even the distinction between berghei group and vinckei group parasites was held in question (Bafort, 1971a). The case for abandoning the distinction between the parasites finally rested upon the interpretation of two observations.

Since the time of its isolation in 1948, one of the original strains of Katangan *P. berghei,* K173, has been passaged by blood inoculation in laboratory rodents. In 1969 this strain had long since failed to produce gametocytes, the only stage infective to mosquitoes. Nevertheless, in that year, Peters *et al.* (1969) reported that various sublines of K173 derived from stock material preserved under liquid nitrogen had reacquired the ability to produce gametocytes and could be cyclically transmitted through mosquitoes. This in itself was remarkable. Equally perplexing was the fact that the line of *P. berghei* K173 (referred to as NS, NS/L3, BS, etc.) could be transmitted at temperatures of 24° to 26°C. Previous evidence had indicated that such temperatures were outside the range at which *P. berghei* could be transmitted but coincided with the optimum temperatures for transmission of *P. yoelii.* Moreover, measurements of sporogonic stages in these lines (oocyst diameter 35–70 μm and sporozoite length 18–22 μm) exceeded those previously recorded for other isolates of *P. berghei* but fell within the range characteristic of *P. yoelii* subspecies. On these and other grounds, Bafort (1970c) argued that the validity of the distinctions between *P. yoelii* and *P. berghei* must be seriously doubted.

In the same year, Bafort reported that lines of a strain of *P. v. vinckei* which he had isolated in 1967 had spontaneously transformed into a type almost indistinguishable from one of the *P. yoelii* subspecies in the morphology and behavior of all stages of its life cycle (Bafort, 1970a).

The observations on the "NS" lines of *P. berghei* K173 and the "berghei-like" lines of *P. v. vinckei* 67 led Bafort (1971a) to propose that the murine malarias could reasonably be considered a single "polymorphic species in which variants or mutations, spontaneous or induced, occur."

The origins of these lines have never been satisfactorily explained. There is no doubt, however, that the "berghei-like" lines are not *P. v. vinckei* or are the "NS" lines *P. berghei.* Significantly or coincidentally, both sets of anomalous lines have identical enzyme types GPI-2, 6PGD-4, LDH-1, and GDH-2. Only one other murine malaria, the single isolate of *P. y. nigeriensis,* is characterized by this enzyme combination. The spontaneous transformation of all four enzymes from the enzyme type of the supposed parental lines to that of *P. y. nigeriensis* is virtually inconceivable. Whatever the origin of the "NS" and "berghei-like" lines, maybe we can only conclude that they did not arise by transformation of either *P. berghei* K173 or *P. v. vinckei* 67. Thus evidence for transformations between the morphological types representing the recognized species of murine malaria does not exist.

The instances quoted above are only a few of the many reported morphological and biological transformations in murine malarias in the laboratory. Parasites may be adapted to hosts not normally susceptible to infection, e.g., the adaptation of *P. v. vinckei* to rats (Rodhain, 1954; Adler and Foner, 1961); spontaneous increase in virulence may occur either gradually in the course of laboratory passage or suddenly and spontaneously in a single animal (Yoeli *et al.,* 1975). Conversely, virulence may be attenuated by passage under unusual conditions such as tissue culture (Weiss and deGiusti, 1964) or in hosts with a dietary deficiency such as the absence of PABA (Bafort, 1971a). Marked changes in morphology and behavior may be associated with the acquisition of resistance to certain drugs (Peters, 1965a).

Two general points must be made regarding the taxonomic significance of such observations. First, it must be recognized that cross-contamination of strains in the laboratory can and does occur even under apparently improbable circumstances. A supposed transformation may be due in reality to such contamination with parasites of a different type. It is also possible that subpatent infections of one parasite species passaged in the presence of another may, under certain conditions, attain patency in preference over the previously predominant species. Reports claiming the transformation of parasites from one recognized phenotype to another, as opposed to the differential selection of distinct lines of parasite already present, cannot, therefore, be readily accepted in the absence of sound experimental verification. Second, even genuine changes in phenotype arising in the laboratory should not of themselves be taken as grounds for modifying the supposed taxonomic status of either the parental or altered daughter lines.

Taxonomic statements about malaria parasites, as about other organisms, generally attempt to reflect evolutionary relationships. Biological evolution proceeds by the accumulation of genetic differences between populations of organisms derived from an original stock. A direct manifestation of such differences at the level of the gene product is provided by the demonstration of enzyme types in the murine malarias. Enzyme differences between the murine malaria parasites are the most direct available measures of their natural genetic relationships and the most reliable guide to their closer taxonomic relationships.

4. Criteria for the Identification of the Murine Plasmodia

Within each of the species and subspecies of murine plasmodia, as defined by enzyme type, there is no doubt that natural variation of biological characteristics does occur. Moreover, parasites may, as already indicated, undergo major morphological and physiological changes in the laboratory. Consequently, the problem of identifying a line of murine malaria can be far from easy.

It is generally possible to recognize the species or species group of a murine malaria parasite on fixed blood smears stained with Giemsa's stain. It may be necessary to examine several such smears taken at different times in the infection to ensure that all the stages of the blood infection have been seen. Having tentatively or definitely identified the species of parasite in this way, the only reliable way to determine the subspecies and even strain is to conduct an enzyme analysis. A scheme recommended in making such a diagnosis is given in Table III.

Table III

Scheme for Identification of Murine Malaria Parasites by Enzyme Type

Test for GPI	Test for GDH	
GPI-1 = *P. y. yoelii*	GDH-4 = *P. y. yoelii*	(confirmation = 6PGD-4)
or		
P. y. killicki	GDH-1 = *P. y. killicki*	
	Test for GDH	
GPI-2 = *P. y. yoelii*	GDH-4 = *P. y. yoelii*	(confirmation = 6PGD-4)
or		
P. y. nigeriensis	GDH-2 = *P. y. nigeriensis*	(confirmation = 6PGD-4)
GPI-10 = *P. y. yoelii*		(confirmation = 6PGD-4)
GPI-3 = *P. berghei*		(confirmation = 6PGD-1)
GPI-5		
GPI-9 = *P. v. petteri*		(confirmation = 6PGD-5)
GPI-11 = *P. v. lentum*		(confirmation = LDH-9 or -7 and 6PGD-5)
	Test for 6PGD	
GPI-6 = *P. v. lentum*	6PGD-5 = *P. v. lentum*	(confirmation = LDH-9 or -7)
or		
P. v. brucechwatti	6PGD-6 = *P. v. brucechwatti*	(confirmation = LDH-9)
GPI-7 = *P. v. vinckei*		(confirmation = LDH-6)
GPI-4 = *P. c. chabaudi*		(confirmation = GDH-5)
GPI-8 = *P. c. adami*		(confirmation = GDH-5)

B. Natural History

1. Epizootiology

The natural history of murine malaria parasites has been most extensively studied in Katanga, the region of their original isolation. These parasites have been collected in about six localities distributed over a wide area of Katanga from Kamena in the north to Lubumbashi (Elisabethville) in the south and from Sandoa in the west almost as far as Lake Tanganyika in the east (Fig. 1). According to Bafort (1971a), over 400 isolates of *P. berghei* have been collected since its discovery in 1948, almost all from the glands of *A d. millicampsi;* the great majority of these strains have since been lost. The remaining isolations have been made on comparatively rare occasions by subinouclation of blood from naturally infected rodents. *Plasmodium v. vinckei* has, by contrast, been isolated only twice, on each occasion from the glands of specimens of *A. d. millicampsi* captured near Kamena in the northern part of Katanga. As Bafort (1971a) points out, *P. v. vinckei* may not be quite as rare a parasite in Katanga as these figures indicate. Most of the early isolations from *A. d. millicampsi* were made by inoculation of the infected glands into white rats, a host which is insusceptible to infection with *P. vinckei.*

No vectors other than *A. d. millicampsi* are known to be involved in the transmission of murine malarias in Katanga. Within Katanga, this anopheline is restricted to areas at altitudes of not less than 900 m; the mosquito is not found anywhere outside the province. Three rodent species may be infected with *P. berghei:* the thicket rat, *Grammomys surdaster, Praomys jacksoni,* and the pygmy mouse, *Leggada bella.* Natural infection of *P. v. vinckei* have not been encountered in rodents.

The habitats of the rodent hosts and anopheline vector of these parasites are found in the so-called gallery forests lining the streams and rivers which traverse the savannas of the Katanga plateau (Vincke, 1954; Yoeli, 1965; Bafort, 1969a). Among the trees and bushes of these narrow swaths of forest the mean daily temperature is around 20°C, the daily maximum rarely rising above 22°C (Bafort, 1971a; Yoeli, 1965). In this shaded environment, the thicket rats build their nests of leaves at heights of 2 to 3 m above the ground. Other than an occasional *Anopheles implexus, A. d. millicampsi* is the only anopheline to frequent the forest galleries at this level. These mosquitoes lay their eggs in the fast flowing streams running between the tree roots, and the adults spend their lives entirely within the forest galleries feeding exclusively upon the local rodents.

Outside of Katanga the known foci of transmission of murine malaria are all in low lying areas. Ambient temperatures are markedly higher than those in Katanga. *Anopheles d. millicampsi* and *G. surdaster* are not found; the universal and virtually the only known host of the lowland parasites is the shiny thicket rat *(Thamnomys rutilans)*, although in Nigeria a single specimen of *Praomys tulbergi* was found infected with *P. v. brucechwatti* (Bruce-Chwatt and Gibson, 1955). Several thousand specimens of other murine species have failed to reveal plasmodial infections in the lowland areas.

The habitats of the shiny thicket rats in the lowland regions are not unlike those of *G. surdaster* in Katanga. In the C.A.R. the animals were caught at the edge of a forest clearing (Landau and Chabaud, 1965), in Nigeria in open secondary forest (Killick-Kendrick *et al.*, 1968), and near Brazzaville in areas described as much degraded gallery forest (Adam *et al.*, 1966). As in Katanga, the thicket rats of these regions build nests of leaves in bushes and trees at heights of 2 to 3 m above the ground.

In the C.A.R., the trapping areas lie at an altitude of 600 m, the highest of the lowland foci (Landau and Chabaud, 1965); the mean annual temperature is 24°C. At Brazzaville, where the trapping areas lie at an altitude of 340 m, the mean anuual temperature is 25°C (Adam *et al.*, 1966). In the Ilobi forest, the locality from which the Nigerian isolations were made, the altitude varies from 70 to 130 m. The rainy season here lasts from April to October, a lull occurring around the month of August. As judged by the incidence of malaria in the local rodents, the transmission appears to reach a peak in November, 1 month after the end of the rainy season. The ambient temperatures in the Ilobi forest fluctuate between a mean daily maximum of 31° and a minimum of 22°C (Killick-Kendrick *et al.*, 1968).

The only two anopheline species found in the trapping area in the C.A.R. were *A. obscurus* and *A. coustani* (Landau and Chabaud, 1965); only the larval forms were seen. In Brazzaville, *A. cinctus* has been found infected with very long sporozoites similar to those of *P. v. lentum* and *P. y. killicki* (J. P. Adam, unpublished). In the vicinity of the Ilobi forest in Nigeria, *A. cinctus, A. coustani,* and *A. obscurus* are all found, although only the larval stages were seen (Killick-Kendrick *et al.*, 1968). None of these species, however, has been directly incriminated as the vector of a murine malaria parasite.

The incidence of infection in *T. rutilans* in the lowland areas is much greater than in *G. surdaster* in Katanga. In the C.A.R., 100% of adult thicket rats (42 out of 42) were infected with malaria parasites (Landau and Chabaud, 1965), in Brazzaville about 40% (10 out of 24) (Adam *et al.*, 1966), and in Western Nigeria a similar proportion (6 out of 18) (Killick-Kendrick *et al.*, 1968). Whereas, in Katanga, berghei group parasites greatly outnumber those of the vinckei group, in the lowland regions the latter are more common. In the C.A.R. the vinckei group parasites are about twice as prevalent as *P. yoelii* and are represented primarily by *P. chabaudi*. In Brazzaville and Nigeria, *P. yoelii* is apparently rather rare in comparison to the vinckei group parasites. In these areas, *P. chabaudi* is less common than *P. vinckei* and may be absent altogether from Nigeria. In view of the geographical distribution of the parasites of the two major groups, it is tempting to speculate that the berghei group malarias are better adapted than are the vinckei group parasites to transmission in the cool environment of the Katanga biotopes while the reverse is true in the torrid conditions of the lowland regions.

Reflecting their zoogeographical isolation, the murine plasmodia of Katanga are unlike those of the lowland areas. *Plasmodium berghei* itself is restricted to Katanga, the berghei group being represented in the lowland regions by the subspecies of *P. yoelii*. *Plasmodium chabaudi* is unknown in Katanga, the vinckei group being represented there by *P. v. vinckei*. This parasite, however, is itself somewhat atypical of the species as represented by the lowland forms of *P. vinckei*.

In spite of their similarities, each of the lowland populations of murine malaria parasites is nevertheless unique and in apparent isolation from the others. Such isolation implies the existence of past or present faunal barriers separating the regions. Western Nigeria is cut off from the relevant areas of Brazzaville and the C.A.R. by a succession of geological barriers represented by the Niger and Cross rivers, the Cameroon Highlands, and the Sanager river, each of which is known to represent a significant faunal barrier (Killick-Kendrick, 1973a). Other than sheer distance (about 600 miles), there are no obvious barriers between Brazzaville and the C.A.R. The absence of such barriers between them is reflected in the similarity between the populations of murine malaria parasites in these two regions. The murine plasmodia of Brazzaville and the C.A.R. are undoubtedly the most closely related of those of any two foci of transmission.

2. Natural Infections

The levels of parasitemia in naturally infected thicket rats are almost always very low regardless of the species of *Plasmodium* involved. In Katanga, thick blood smears consistently failed to reveal the presence of parasites in specimens of *G. surdaster* shown, by subinoculation, to be infected. Patent parasitemias on blood smears were more frequently seen in infected animals from the lowland regions. In Nigeria, parasites were visible on thick smears in four out of six infected animals (Killick-

Kendrick *et al.*, 1968). In Brazzaville, all of ten infected animals had parasitemias detectable on blood smear (Adam *et al.*, 1966). In the C.A.R., all of 42 adult thicket rats were found to be infected (Landau and Chabaud, 1965). Of these animals, 13 were negative on thick blood smears, 24 showed low levels of parasitemia, and 5 had elevated parasitemias. Thicket rats less than 2 months old were never found to be infected. Since the mothers suckle their young over a long period of time, it is possible that the young animals acquire effective protection against infection via antibodies transferred in their mother's milk. Unfortunately, no information is available on the immunity of the naturally infected rodents.

After an animal has been isolated from the wild, natural infections in thicket rats from the C.A.R. persist until death up to 2½ years later. During this time, parasites are rarely seen in the blood, although their presence may be demonstrated by subinoculation. Such infections pass through occasional intermittent periods of latency.

In addition to their chronic blood infections, thicket rats from the C.A.R. have been found to harbor an unusual form of liver schizont (Landau, 1973). Such schizonts persist at least 8 months after the isolation of an animal from the wild. Cross-reactivity of schizonts with antisera to *P. chabaudi* and *P. yoelii* has demonstrated their plasmodial nature (El-Nahal, 1971b). The schizonts slowly increase in size reaching a maximum diameter of about 32 μm at which stage they contain about 250 nuclei; this compares with 10,000 to 20,000 nuclei in liver schizonts of murine malarias following sporozoite inoculation in the laboratory.

Forms similar to such naturally occurring chronic schizonts were produced in the laboratory in thicket rats by Landau (1973) directly following the inoculation of sporozoites of *P. y. yoelii.* The finding that the experimental schizonts developed directly from sporozoites led her to propose that the chronic schizonts found in nature are of similar origin and do not arise from the secondary invasion of the liver by merozoites. The existence of the chronic schizonts could account for the persistence of naturally acquired infections.

C. Life Cycle and Fine Structure (Figs. 5 and 6)

As we have already indicated, the malaria parasites of murine rodents are, in all essential respects, typical mammalian malaria parasites. This generalization is clearly demonstrated in the life cycle and fine structure of these parasites, while the distinction between the rodent and primate *Plasmodia* and those of birds and reptiles is apparent at several stages.

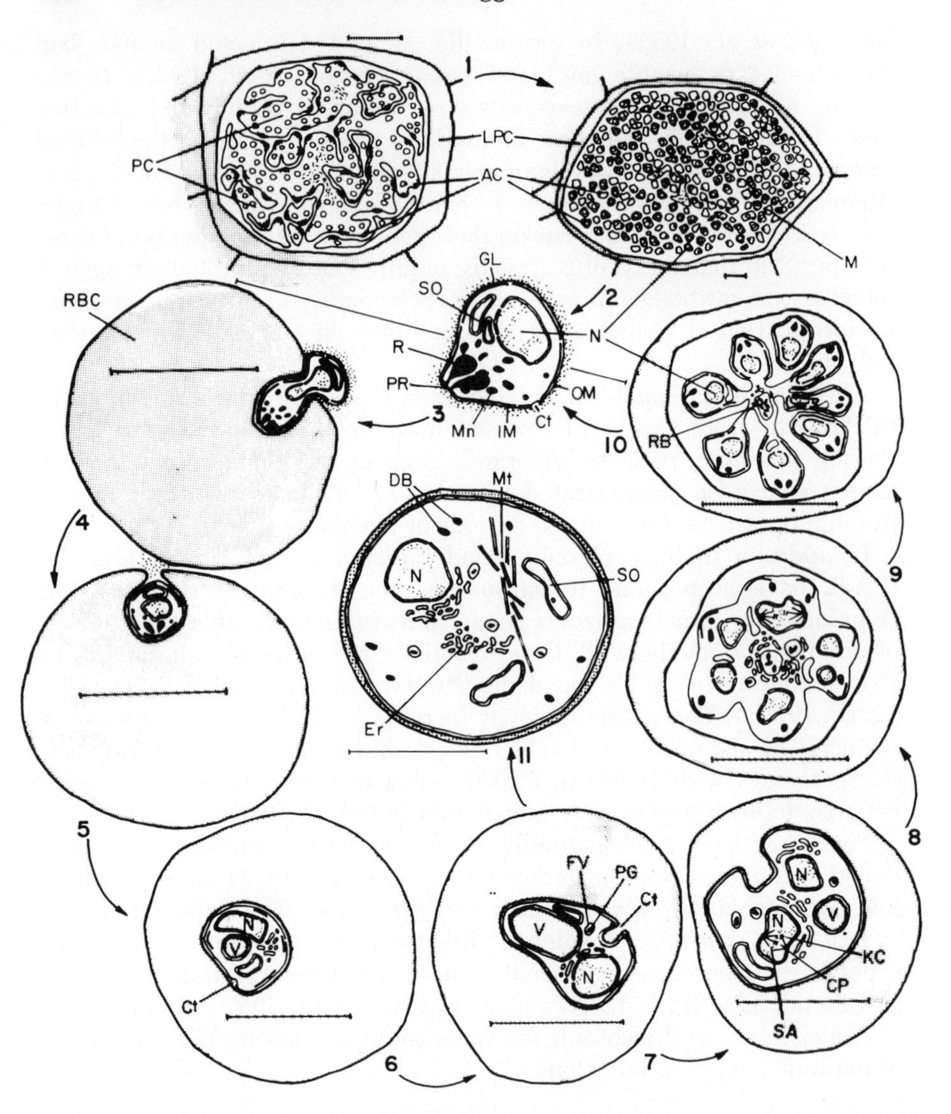

Fig. 5. Semidiagrammatic representation of the development of a murine malaria parasite in the rodent host, as revealed by the electron microscope. (1) Early preerythrocytic schizont in the liver parenchyma cell; (2) mature preerythrocytic schizont; (3) merozoite in the blood stream; (4) merozoite in the process of entering a red blood cell; (5) merozoite in the final stages of entering a red blood cell; (6) merozoite, or young ring, stage trophozoite, in the final stages of dedifferentiation; (7) ring stage trophozoite with functional cytostome; (8) early schizont; (9) late schizont; (10) mature segmenter with merozoites in the process of separation from the residual body; (11) mature macrogametocyte. Bar represents 5 μm. AC, apical complex; CP, centriolar plaque; Ct, cytostome; DB, dense body; Er, endoplasmic

1. Preerythrocytic Stages (Desser *et al.*, 1972; Bafort and Howells, 1970; Garnham *et al.*, 1969b; Terzakis *et al.*, 1974)

The first developmental stage in the rodent host after the inoculation of sporozoites into the bloodstream is the preerythrocytic stage in the parenchyma cells of the liver. Depending upon the species of murine *Plasmodium* involved, the preerythrocytic stage lasts from 2 to 3 days.

The manner of entry of a sporozoite into a parenchymal cell is entirely unknown as are the earliest events of the development of the preerythrocytic parasite. It is known, however, that the sporozoites are completely cleared from the circulation within 30 minutes to 1 hour of their inoculation. Within the host parenchyma cell, the young preerythrocytic schizont is surrounded by two membranes of which the inner is the limiting membrane or plasmalemma of the parasite. As it develops, the schizont is seen to be within a vacuole in the host cell cytoplasm. Within the cytoplasm of the parasite are found numerous double-membraned organelles, seen in all stages in the rodent host, and thought to be nonfunctional mitochondria. These structures have been observed in rare instances to contain occasional tubular cristae. Aggregations of smooth concentric membranes are also seen in the parasite cytoplasm. Found in all stages in the rodent host, it has been suggested that these concentric membranes, or membrane whorls, may play the role of mitochondria; it has also been proposed that they may represent a reserve of membrane for the growing parasite. There is, however, a distinct possibility that the membrane whorls may be no more than artifacts arising in the handling of the material. Numerous ribosomes are present throughout the cytoplasm.

A cytostome, probably the main organ of ingestion in the blood stage parasites (see below), has not been seen in preerythrocytic schizonts. The method by which the schizont acquires its nutrition is uncertain. It has been suggested that the contents of the host vacuole are engulfed by pinocytosis. The schizonts do not actively ingest host cell cytoplasm.

As the schizont grows within the cytoplasm of a parenchymal cell, the nucleus divides many times. The nuclei are, at first, distributed throughout an undivided cytoplasm. As the schizont nears maturity the cytoplasm becomes divided into lobed or island structures called "pseudocytomeres." At this stage, the whole schizont lies within a vacuole occupy-

---

reticulum; FV, food vacuole of parasite; GL, granular layer; IM, inner membrane of parasite; KC, kinetochore; LPC, liver parenchyma cell; M, merozoite; Mn, microneme; Mt, microtubule; N, nucleus; OM, outer membrane or plasmalemma of parasite; PC, pseudocytomere; PG, pigment granule; PR, polar ring; R, rhoptry; RB, residual body; RBC, red blood cell; SA, spindle apparatus; SO, smooth membraned organelle; V, "vacuole" continuous with host cell membrane.

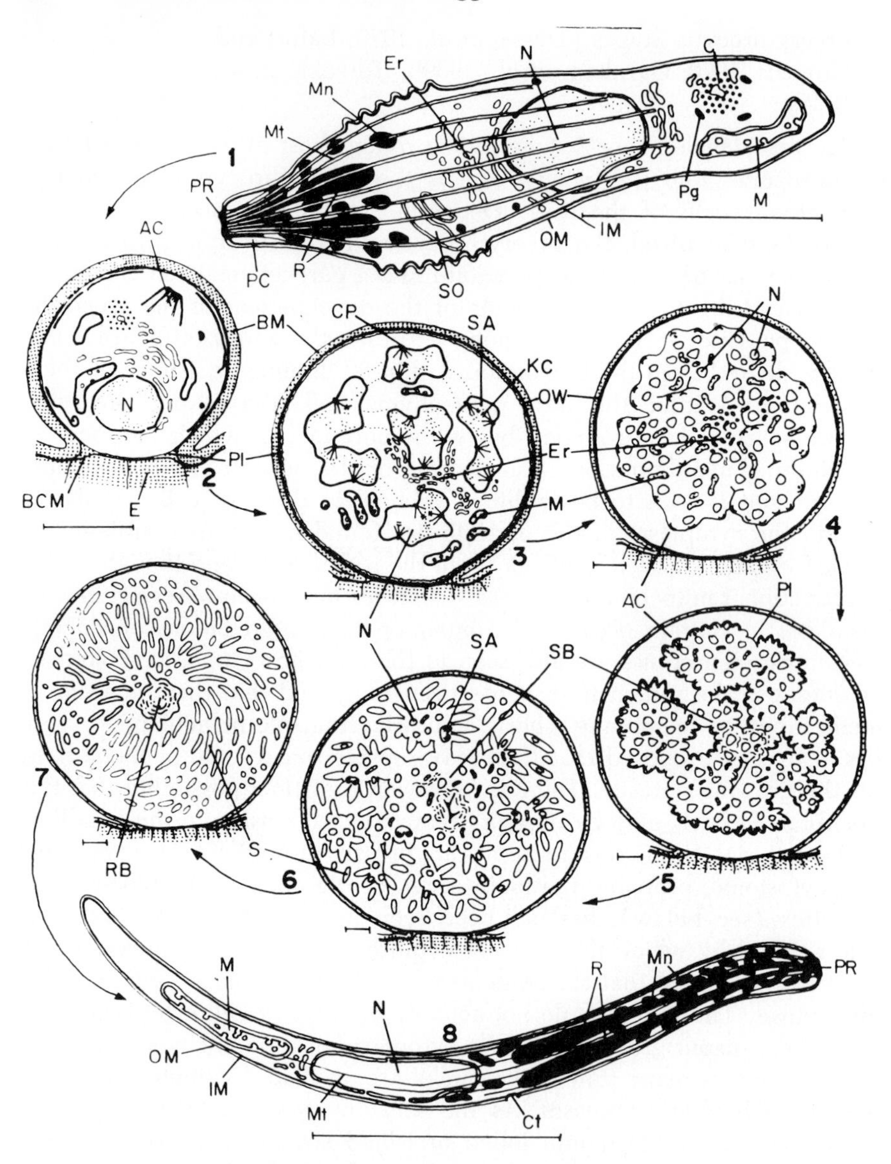

Fig. 6. Semidiagrammatic representation of the development of a murine malaria parasite in the mosquito from the ookinete to the sporozoite as revealed by the electron microscope. (1) Fully differentiated ookinete at 12 to 18 hours after fertilization; (2) early oocyst with structure of the ookinete in the process of dedifferentiation (24–48 days); (3) oocyst with polyploid but undivided nucleus; the capsule or oocyst wall is now recognizable (2–5 days); (4) early sporoblastoid formation; the plasmalemma has separated from the oocyst wall, numerous individual nuclei are present forming a syncitium in the undivided cytoplasm of the oocyst; thickenings beneath

ing almost the entire volume of the now much enlarged host cell. The individual nuclei lie along the periphery of the pseudocytomeres and budding of merozoites begins. Just beneath the outer membrane of the maturing schizont, a thick inner membrane is laid down in the region of each developing merozoite bud. The tip of each bud thickens to form the apical rings and rhoptries (characteristic organelles of the extracellular stages of the parasite), and the nuclei enter the merozoites. Upon completion of schizogony, the merozoites are fully separated from the residual parent body.

Each merozoite has a double-membraned pellicle consisting of a thin outer membrane and a thicker inner membrane. According to Desser *et al.* (1972), the pellicle of the preerythrocytic merozoite is surrounded by tiny microfibrillae; neither subpellicular microtubules or the pellicular labyrinth, found in merozoites of avian and reptilian plasmodia, have been described in the preerythrocytic merozoites of rodent malaria parasites. A nonfunctioning cytostome of internal diameter 75 mμ is present. At the anterior end of the pear-shaped merozoite are two apical rings behind which the rhoptries extend into the cytoplasm. The double-membraned nucleus lies adjacent to the base of the merozoite. Ribosomes and aggregations of smooth concentric membranes are found in the cytoplasm, but cristate mitochondria have not been seen. The merozoite is approximately 1.5 mμ in length.

On release of the merozoites (between 8,000 and 20,000 may be produced by a single schizont depending on the species), the original site of the schizont is quickly infiltrated by leukocytes. Once shed into the bloodstream, the preerythrocytic merozoite speedily enters a red blood cell. The thick inner membrane of the merozoite and the structures of the apical complex, polar rings and rhoptries, are rapidly resorbed in a process of dedifferentiation. The young trophozoite, as it exists at this

---

the plasmalemma represent early stages of formation of the apical complexes of the developing sporozoites (5–8 days); (5) sporoblastoid formation at a later stage (7–9 days); (6) late sporoblastoid formation with nearly mature sporozoites beginning to bud off from the sporoblastoid body; at this stage each nucleus appears to undergo a final division, the daughter nuclei each entering a separate sporozoite (8–10 days); (7) mature oocyst with sporozoite fully differentiated and separated from the residual body (9–12 days); (8) sporozoite. Bar represents 5 μm. AC, apical complex; BCM, basal cell membrane; C, crystalloid; Ct, cytostome; CP, centriolar plaque; E, epithelial cell of mosquito midgut; Er, endoplasmic reticulum; IM, inner membrane of parasite; KC, kinetochore; M, mitochondrion; Mn, microneme; Mt, microtubule; N, nucleus; OM, outer membrane, or plasmalemma, of parasite; OW, oocyst wall; PC, pellicular cavity; PG, pigment granule; Pl, plasmalemma of parasite; PR, polar ring; R, rhoptry; RB, residual body; S, sporozoite; SA, spindle apparatus; SB, sporoblastoid body; SO, smooth membraned organelle.

stage, is surrounded by two membranes of which the outer is derived from the membrane of the host cell during the entry of the merozoite while the inner membrane of the trophozoite represents the original plasmalemma of the merozoite.

2. Blood Stages

A. ASEXUAL DEVELOPMENT (Bafort, 1970b; Kilby and Silverman, 1969; Ladda, 1969; Rudzinska and Trager, 1959; Scalzi and Bahr, 1968; Theakston *et al.*, 1968; Jadin *et al.*, 1968; Aikawa, 1971; Fulton and Flewett, 1956; Howells *et al.*, 1968a,b; Rudzinska, 1969; Arnold *et al.*, 1969; Howells, 1970; Cox and Vickerman, 1966; Ladda *et al.*, 1969; Sead *et al.*, 1973; Vickerman and Cox, 1967). The growth and development of the erythrocytic trophozoite and schizont is presumed to be much the same whether it originates from an erythrocytic or a preerythrocytic merozoite. Following its entry into the host red blood cell, the surface area of the young trophozoite expands greatly as the parasite develops from the compact form of the merozoite into an extended crescentic or ringlike body. At the same time, the densely packed ribosomes of the merozoite become more thinly dispersed in the cytoplasm of the early trophozoite.

Unlike the sporogonic and preerythrocytic parasites, the blood stage trophozoites actively ingest and digest the cytoplasm of the host cell. The mechanism by which the trophozoites ingest the host cytoplasm has been greatly discussed. The following account represents our interpretation of the balance of evidence in the literature.

Within the cytoplasm of the erythrocytic trophozoites are found unit-membraned vesicles containing host cell cytoplasm. These vesicles or phagosomes, are the sites of digestion of the host hemoglobin. On completion of the partial digestion process, one or two crystalline granules of the so called "malaria pigment" or hemozoin remain in each phagosome. The phagosomes themselves are formed by the activity of the cytostome which is present and fully functional in the trophozoites. This structure is a pore in the inner membrane of the trophozoite surrounded by two electron-dense rings. The internal diameter of the orifice varies from 50 to 80 mμ. The cytoplasm of the host cell is ingested through the cytostome to a depth of up to 400 mμ. Within the cytostome the host cell membrane is lost and the invagination is nipped off to form a unit-membraned vesicle within the cytoplasm of the trophozoite. Because the process is repeated, numerous such vesicles or phagosomes accumulate, each of which will produce one or two crystals of hemozoin. It is not clear whether or not phagosomes are formed solely by this mechanism or by others as, for example, by direct phagocytosis of host cytoplasm at other regions of the parasite membrane but this seems unlikely.

The large double-membrane bounded areas enclosing regions of host cell cytoplasm were thought at one time to represent the "food vacuoles." These structures seen with the electron microscope correspond to the central vacuole seen by light microscopy in the ring stage parasite and to the vacuoles often seen in trophozoites of a later stage. Such structures have never been observed to contain malaria pigment nor do they contain the enzyme acid phosphatase, shown to be present in the phagosomes. Acid phosphatase is present in lysosomes in other organisms and is associated with digestive activity. The large double-membraned vacuoles are not, therefore, believed to be true food vacuoles. Indeed, most, or all, are probably continuous with the cytoplasm of the host cell and are thus invaginations of the surface of the trophozoite. They are, therefore, not even true vacuoles in that they are not fully enclosed bodies lying free within the cytoplasm of the trophozoite. Such invaginations presumably serve to increase the surface area available for the exchange of metabolites between the parasite and the host cell.

In addition to the phagosomes and the large double-membraned "vacuoles," the cytoplasm of the trophozoite contains a variety of other structures. The young trophozoite has an extensive vesicular endoplasmic reticulum but comparatively few ribosomes; the ribosomes, however, increase in number considerably as the parasite grows. Cristate mitochondria are absent from the trophozoite at all stages of growth, although the double-membraned organelles resembling acristate mitochondria are present. All available evidence indicates that these organelles are devoid of any normal mitochondrial activity.

Throughout the development of the trophozoite the membrane whorls, similar to those seen in the merozoites and preerythrocytic stages, are present in the cytoplasm. As already suggested, these may be no more than artifacts of the process of preparation for electron microscopy.

The nucleus of the young trophozoite is surrounded by a double membrane and contains numerous nuclear pores; a nucleolus has not been reported. The dense chromatin-packed nucleus of the merozoite swells in the trophozoite and the chromatin material becomes dispersed as maturation begins. A few thinly dispersed ribosomes are present within the nucleus.

The beginning of schizogony is marked by the division of the nucleus and the differentiation of cytoplasmic organelles. Nuclear division is characterized by the appearance of spindle structures composed of microtubules radiating into the nucleoplasm from points on the surface of the nuclear membrane. The points of attachment of the spindle to the nuclear membrane are known as the "centriolar plaques" and are seen as electron-dense regions occluding one or two pores in the nuclear membrane and extending from the cytoplasm through and into the nucleoplasm. The spindle microtubules radiate from a centriolar plaque through the nucleoplasm. They are seen in appropriate sections to make contact with microtubules of another spindle attached to a centriolar plaque on the nuclear membrane. Invaginations in the surface of the nuclear membrane occur in the region of each centriolar plaque. At the midpoint of the complete spindle apparatus, electron-dense structures on and perpendicular to the microtubules may be seen. Each of these structures, referred to as a kinetochore, is composed of three electron-dense bands separated by two light bands. The kinetochores are considered to represent the points at which the chromosomes are attached to the spindle microtubules during the segregation of the genetic material. At no stage, however, have the chromosomes themselves been seen as organized structures. Nuclear division is apparently achieved by a fission process which is repeated until the full complement of schizont nuclei is attained. The nuclear membrane remains intact throughout the division process.

Accompanying the division of the nucleus, the cytoplasm of the schizont undergoes differentiation. Merozoite buds develop just beneath the plasmalemma, or inner membrane, of the trophozoite with the formation of a thick inner membrane in the region of each growing bud. The plasmalemma separates from the outer membrane so that the schizont lies in a single-membraned vacuole within the cytoplasm of the host cell or its remnants. Polar rings form at the surface of each bud while the rhoptries enter below them and the thick inner membrane extends around the developing merozoite. Organelles, including endoplasmic reticulum, ribosomes, double-membraned organelles, dense bodies or micronemes, and finally the nucleus, migrate into the merozoite which separates from the residual body of the schizont. The residual body, surrounded by a unit membrane, now contains the pigment granules aggregated into a single vesicle, together with some remnants of endoplasmic reticulum and sometimes concentric membranes whorls.

The free erythrocytic merozoite is essentially similar to that of preerythrocytic origin. Microfibrils, such as those observed surrounding the preerythrocytic merozoites, have not been seen in erythrocytic merozoites. Subpellicular microtubules have rarely been observed in the merozoites of rodent plasmodia. During their brief journey in the bloodstream, the merozoites acquire a granular coat surrounding the plasmalemma. The origin of the surface coat is unknown but probably derives, at least in part, from materials in the plasma.

As it reenters a red blood cell, the merozoite approaches with its apex making direct contact with the host cell membrane. The membrane of the host cell invaginates, the invagination widening and deepening as the merozoite enters. Finally the host membranes close behind it and the merozoite lies within a tight fitting vacuole in the cytoplasm of the host cell. At no stage during the penetration is the host cell membrane broken by the merozoite. In the process of entry, the outer granular coat of the merozoite is sluffed off and remains outside the parasitized cell. Having entered the host cell, the organs of the apical complex, polar rings, rhoptries, and micronemes and the thick inner membrane of the merozoite are resorbed in the process of dedifferentiation. The cycle of trophozoite growth and schizogony is thus reinitiated.

B. SEXUAL DEVELOPMENT (Ladda, 1969; Aikawa *et al.*, 1969; Garnham *et al.*, 1967a; Sinden *et al.*, 1976). Certain merozoites, of either erythrocytic or preerythrocytic origin (Killick-Kendrick and Warren, 1968), are not destined to become schizonts but gametocytes. The early stages of gametocyte development in murine malarias as in most other plasmodia are not well understood. Based primarily on the study of avian systems Aikawa *et al.* (1969) have postulated that the early events of dedifferentiation determine whether a merozoite becomes a gametocyte or a schizont. These workers proposed that only the structures of the apical complex and the rhoptries are dedifferentiated, while the thick inner membrane of the merozoite remains and grows with the developing gametocyte. In avian and reptilian systems, the gametocytes at maturity are seen to be surrounded by three membranes: the outermost being of host cell origin, the middle being the outer membrane proper or plasmalemma of the parasite, and the inner membrane being the thick pellicular membrane supposedly derived from the merozoites.

In the murine malarias, and generally in the mammalian parasites, only two membranes may be readily seen to surround the mature gametocyte. Sinden *et al.* (1976) report the presence of a discontinuous third inner membrane in gametocytes of murine malaria parasites. It would seem reasonable to suppose that the relationship of the two outer membranes is the same as that seen in the trophozoites. The cytoplasm of the gametocytes is filled with ribosomes and electron-dense or osmophilic bodies similar to the micronemes. The osmophilic bodies are bounded by a unit membrane and are occasionally seen to connect with the plasmalemma. Both the microneme-like elements and the ribosomes are considerably fewer in the microgametocyte than in the macrogametocyte. The comparative absence of ribosomes in the microgametocyte is thought to account for the pink color of the cytoplasm as seen by light microscopy with Giemsa's stain. The cytostome is fully functional and is probably the sole source of phagosomes. Invaginations in the membrane of the gametocytes are rare. Consequently, the large double-membraned "vacuoles" containing host cytoplasm are not usually seen. The digestive activity of the phagosomes leads in the mature gametocyte to the formation of 20 or more pigment granules scattered evenly throughout the cytoplasm of the gametocyte. In some gametocytes, somewhat irregular arrangements of microtubules may be seen traversing the cytoplasm. Double-membraned organelles are commonly found in the macrogametocytes; these rarely contain a few tubular cristae. The gametocytes are characterized by a single large double-membraned nucleus. The nuclear envelope contains pores, some of which appear to be plugged by structures consisting of eight tubules arranged symmetrically around a core. A dense granular cytoplasmic mass lying in a diverticulum of the nuclear envelope is referred to by Sinden *et al.* (1976) as a microtubule organizing center (see below). Other than growth in size, no evidence of nuclear reorganization or nuclear division is seen at any stage during the development of the gametocytes. The mature gametocytes occupy the entire volume of the host cell; only a small rim of host cytoplasm remains surrounding the parasite.

3. Development in the Mosquito

A. GAMETE FORMATION AND FERTILIZATION (Sinden and Croall, 1974; Sinden *et al.*, 1976; Garnham *et al.*, 1967a). On the ingestion of the mature gametocytes in a blood meal by the mosquito vector a complicated and rapidly evolving sequence of events is initiated in the final development and release of the gametes. At the level of the light microscope these events are seen as follows. Within a few minutes the membranes of the host cell of each gametocyte begin to degenerate as the gametocytes become extracellular in the process of "emergence." On completion of emergence the macrogametocyte is in effect a *macrogamete* ready for fertilization. The microgametocyte, however, undergoes further development in the process of "exflagellation." Within ten minutes of ingestion the *microgametes* may be seen to thrust their way out of the microgametocyte with rapid and extremely vigorous movements. As each microgamete breaks free of the parent body it appears, by light microscopy, as a single membrane bounded flagellum. Usually eight microgametes are produced from a single microgametocyte. The residual body of the exflagellating microgametocyte consists largely of pigment granules with some nuclear material all enclosed by the remains of the gametocyte membrane.

At the ultrastructural level the events of emergence and exflagellation are as follows. Within one or two minutes the gametocytes round off and the osmophilic bodies migrate toward the surface. In the macrogametocyte, or macrogamete as it is after shedding the surrounding host cell, no further ultrastructural changes are apparent prior to fertilization. In the microgametocyte eight kinetosomes develop in association with the microtubule organizing center near the nuclear envelope. Shortly thereafter a spindle forms within the nucleus, the kinetosomes being distributed in two sets of four at the poles of the nuclear spindle. As at other stages of nuclear division in the life cycle of these parasites the nuclear membrane remains intact, the poles of the spindle being connected to centriolar plaques in the nuclear membrane. During the three divisions of the genetic material in the microgametocyte the kinetosomes and the axonemes segregate in close association with the centriolar plaques. Kinetochores are present in the nuclear spindle but, as at other stages in the life cycle, chromatin is not seen associated with them. Of the three segregations of the genetic material the third is closely linked to the release of the microgametes.

The kinetosomes, lying on the cytoplasmic side of the centriolar plaques, are each surrounded by a network of tubules forming the "perikinetosomal basket." Each perikinetosomal basket contains a spherical structure within which lies a central dense granule. Growing out from the kinetosomes the axonemes elongate in the cytoplasm around the contour of the nucleus. The axonemes are constructed on the familiar 9 + 2 configuration of microtubules. During exflagellation the kinetosome and juxtakinetosomal sphere and granule are passed through the perikinetosomal basket followed by the axoneme to form the emerging microgamete. The plasmalemma of the microgametocyte becomes closely applied to the emerging axoneme and associated organelles to become the cell membrane of the microgamete. The nuclear bud, its centriolar plaque now separated from the kinetosome, passes through the perikinetosomal basket to become incorporated as the microgamete nucleus.

On their release the microgametes swim vigorously and rapidly through the blood meal. Fertilization of a high proportion of macrogametes probably ensues within minutes. There is no detailed ultrastructural record of this event in any *Plasmodium* species. In the light microscope, however, one end of the microgamete is seen to attach to the membrane of the macrogamete; after a few moments of energetic motion the microgamete appears to swim right inside the macrogamete. In reality the membranes of male and female gametes probably undergo fusion resulting in the release of the contents of the microgamete, including the axoneme and nucleus, into the cytoplasm of the macrogamete. The axoneme remains motile for several moments after entering the macrogamete causing a vigorous commotion within the cytoplasm of the fertilized cell, or zygote. Following fertilization the gamete nuclei fuse and during the next 12–18 hours the zygote undergoes reorganization into the elongated and weakly motile cell known as the ookinete.

B. OOKINETE DEVELOPMENT (Canning and Sinden, 1973; Davies, 1974; Garnham *et al.*, 1969a; Bafort, 1971a). The ookinete develops all those structures characteristic of the extracellular and motile stages of the parasites (gametes excepted). It is surrounded by two membranes, a thin outer plasmalemma and a thick inner membrane. At the anterior end of the ookinete are found the organelles of the apical complex, the polar rings and the rhoptries. Below the polar rings an electron-dense collar has been described, while extending from the polar rings beneath the inner membrane of the pellicle is a layer of microtubules. Cristate mitochondria are occasionally seen within the cytoplasm of the ookinete; the acristate double-membraned organelles are, however, more common. Abundant smooth endoplasmic reticulum is present; ribosomes, however, are less numerous than they become in the young oocyst. A crystalloid resembling an aggregate of virus particles is occasionally found associated with the endoplasmic reticulum. Concentric membrane whorls may be present. Pigment granules from the parental macrogamete remain scattered in the cytoplasm of the ookinete. A single double-membraned nucleus is present within which nuclear spindles are occasionally seen. The first, possibly meiotic, postzygotic segregation of the genetic material probably occurs in the ookinete.

The development from fertilized zygote to a recognizable motile vermiform ookinete takes 12 to 18 hours. The fully developed ookinete transverses the wall of the mosquito midgut to take up a position on the external wall of the gut about 24 hours after the blood meal. The ookinete usually comes to rest between the basal lamina of the midgut wall and the epithelial basement cell membrane. Some evidence, however, suggests that early oocyst development may also take place deeper within the epithelial cells of the midgut wall. Having reached its final position, dedifferentiation of the ookinete proceeds rapidly with resorbtion of the subpellicular microtubules and apical complex and disintegration of the inner cell membrane.

C. DEVELOPMENT OF THE OOCYST AND SPOROZOITE FORMATION (Davies, 1974; Canning and Sinden, 1973; Howells and Davies, 1971; Bafort, 1971a; Vanderberg *et al.*, 1967). Within the young oocyst, division of the genetic material proceeds with the appearance of many spindle apparatus within the nucleus. The process of spindle formation is essentially the same as that seen in the erythrocytic stages. Centriolar plaques form in the pores of the nuclear membrane. From each centriolar plaque microtubules radiate inward through the nucleoplasm toward a similar apparatus derived from an opposing section of nuclear membrane. Kinetochores are found attached to the spindle tubules; the chromosomes themselves never appear as organized structures.

Between 3 and 7 days after a blood meal, the nucleus of the oocyst develops numerous spindles and its membrane becomes highly irregular and invaginated. Although several separate nuclear elements are usually seen in cross sections in the electron microscope at this stage, narrow connections between such elements are also frequently observed. Several spindles may be seen in any area of the nucleus. It is, therefore, quite likely that the nuclear membrane remains undivided up to this point as the nucleus becomes highly polyploid.

As the young oocyst develops, cristate mitochondria increase in numbers in the cytoplasm. At this stage they can be shown to contain enzymes such as succinate dehydrogenase and isocitrate dehydrogenase, characteristic of functioning mitochondria. The endoplasmic reticulum of the young oocyst is more fully developed than it is in the ookinete; numerous ribosomes are present. Granules of malaria pigment remain scattered through the cytoplasm of the oocyst throughout its development. Membrane whorls may also be found.

During the first 4 days of its development, no indication of a wall is seen surrounding the oocyst. As it matures, however, an amorphous layer develops outside the plasmalemma of the oocyst and beneath the basal lamina of the midgut. This layer entirely surrounds the oocyst and becomes the capsule or oocyst wall. Prior to the beginning of sporozoite development, or sporoblastoid formation, the plasmalemma forms numerous invaginations in the capsule giving it a placentalike appearance. A cytostome has not been observed in the oocyst; it is possible that nutrients are supplied directly through, or by, the capsule which is readily permeable to certain substances, e.g., amino acids.

The first stages of sporoblastoid formation begin around the seventh day after a blood meal. At this time, the nucleus has probably divided into numerous discrete nuclei to form a syncytium in the oocyst cytoplasm. Spindles are no longer present and connections between the individual ovoid nuclei cannot be found. The periphery of the oocyst becomes vacuolated in appearance as the plasmalemma separates from the outer capsule. The sporoblastoid body, as it is now termed, lies in a vacuole formed by the capsule of the oocyst. As sporozoite buds appear in the outer membrane of the sporoblastoid body, a thick inner membrane is laid down within each bud. Polar rings, rhoptries, micronemes, and subpellicular microtubules are formed and endoplasmic reticulum, ribosomes, and a cristate mitochondrion migrate into each developing sporozoite. The nuclei undergo one final division at this stage. Each nucleus comes to lie beneath a pair of adjacent sporozoite buds, a typical spindle is formed, and the nucleus divides, one-half entering each sporozoite. The fully formed sporozoites separate from the sporoblastoid body and now lie free within the oocyst capsule. During sporogony several sporoblastoid bodies may ultimately be formed in the continuing course of sporozoite formation. The fully mature oocyst eventually contains several thousand sporozoites lying within the cavity of the oocyst capsule.

D. THE SPOROZOITES (Vanderberg *et al.*, 1967; Sinden and Garnham, 1973; Bafort, 1971a; Sinden, 1974; Howells, 1970a). The mechanism by which the sporozoites leave the capsule is not entirely clear. It has generally been supposed that a complete rupture of the oocyst wall occurs releasing sporozoites into the hemocoel. Sinden (1974), however, has presented evidence that in some instances the sporozoites may excyst via holes formed by small ruptures in the oocyst wall. The sporozoites migrate through the hemocoelomic fluid to the salivary glands which they penetrate coming to rest in the lumen of the glands. They may remain here in viable condition apparently indefinitely until inoculated into the bloodstream of the rodent host during the next blood meal of the mosquito vector.

Although sporozoites from the salivary glands are apparently more infective than those still within the oocyst, no differences in their ultrastructure have been recorded. The sporozoites are characterized by all those structures associated with motile and invasive stages of the parasite. The pellicle consists of two membranes, a thick inner membrane and a thin outer plasmalemma. Below the inner membrane are rows of microtubules running longitudinally from the anterior end. Depending upon the species, fourteen to sixteen microtubules traverse two-thirds of the periphery of the sporozoite. A single microtubule runs longitudinally down the remaining one-third of the periphery. The microtubules are no longer seen toward the posterior end of the sporozoite.

The apical complex of the sporozoite consists of the polar rings and possibly a collar; several long rhoptries and numerous micronemes extend halfway down the length of the parasite. The cytostome is not commonly seen in the sporozoites of murine plasmodia although it is more common in those of some other species of mammalian malaria parasite. When seen the cytostome is in a nonfunctional state. The cytoplasmic components include cristate mitochondria apparently in functional condition, as demonstrated by the presence of succinate dehydrogenase activity. Ribosomes are also present; membrane whorls have not been seen. Most sporozoites contain a single nucleus although, rarely, two may be present.

4. Comments on the Life Cycle and Fine Structure

The foregoing account of fine structural events in the life cycle of the murine plasmodia represents our attempt to synthesize the information available in the literature. In addition to our own fallibility, our reconstruction of these events suffers from the inevitable uncertainties involved in interpreting some of the relevant data. We have tried to indicate those areas in which interpretation may be controversial. We wish now to point to some aspects of special interest in comparison with other groups of *Plasmodium* species. In particular, we hope to underline the basic similarity between the murine and primate malaria parasites and to indicate the points of distinction between these and the plasmodia of birds and reptiles.

The fine structure of the parasites in the vertebrate host reveals several points of difference between avian and mammalian plasmodia. We have been impressed by the rarity with which subpellicular microtubules have been observed in merozoites of murine malaria parasites, nor do they appear to be more commonly encountered in primate malarias. By contrast, the subpellicular microtubules are apparently regularly encountered in merozoites of avian plasmodia. Likewise, in avian systems, a "spherical body" is frequently reported in close association with the mitochrondrion of the merozoite; the organelle is lost in the trophozoite. We have found no reference to such an organelle in the literature on the murine malarias. Indeed, in these parasites, as in most primate malarias, functional cristate mitochondria are not found in the blood stages. Gametocytes of the avian plasmodia are characterized by a very clear triple membrane surrounding the parasite cell. The composition of the cell membranes of the gametocytes of murine and primate malaria parasites is less clear; generally a double membrane appears to be present, although a triple third inner membrane has been reported in several species. The cytostome of the blood stages of murine and primate malarias is about one-half the diameter of that of the avian parasites. Notable exceptions to these generalizations are the quartan malarias of man and primates, *P. malariae* and *P. brasilianum* (Sterling *et al.*, 1972). In these parasites, subpellicular microtubules are clearly present in the blood stage merozoites. Cristate mitochondria are present at all stages in the life cycle and are associated in the merozoite with the spherical body. Gametocytes are clearly surrounded by three membranes.

The most striking difference between the malaria parasites of birds and reptiles, on one hand, and those of primates and murine rodents, on the other, is in the site of the preerythrocytic development. Among the avian parasites this occurs in the fixed tissue cells of the reticuloendothelial system and involves repeated cycles of schizogony and even reinvasion by blood stage parasites. In primates and murines, however, preerythrocytic development takes place in the parenchyma cells of the liver. Reinvasion of the liver from the blood does not occur, nor is there any evidence that reinvasion by preerythrocytic merozoites is possible. Instead, evidence from both murine and primate systems suggests that the liver infection may be maintained by long-lasting slow-growing schizonts derived directly from sporozoites.

Finally, we would like to point out that the structure known as the conoid has not been found in any stage of any malaria parasite or piroplasm. The term conoid was introduced by Gustafson *et al.* (1954) to describe the truncated spiral conelike structure found just below the polar rings in *Toxoplasma*. It is found in all toxoplasmatid species and coccidian parasites and is apparently present in the ookinetes of *Haemoproteus*. The term should not be used in describing any part of the apical complex of malaria parasites.

D. Genetics (Walliker *et al.*, 1973, 1975; Yoeli *et al.*, 1969; Diggens *et al.*, 1970; Schoenfeld *et al.*, 1974; Morgan, 1974; Carter, 1973; R. Carter, unpublished, 1974; Carter and Walliker, 1975; Canning and Sinden, 1973; Oxbrow, 1973; Aikawa *et al.*, 1972)

1. Mechanisms of Inheritance

The murine plasmodia, like all other groups of malaria parasites and haemosporidia, are sexual organisms. Male and female gametocytes arise in the blood of the rodent host which, on ingestion by a mosquito, give rise to free gametes. The mechanism by which the sex of a gametocyte is determined is unknown. The following lines of evidence, however, indicate that it is not under genetic control: (i) cloned lines of blood parasites can produce both male and female gametocytes, and (ii) all blood stage parasites are derived from a haploid stage probably arising during development in the mosquito. The formation of male and female gametocytes by segregation of sex-determining genes or chromosomes is, therefore, unlikely. It is more probable that the differentiation of the gametocytes is controlled by physiological mechanisms.

Following the release of male and female gametes in the stomach of the mosquito, fertilization rapidly ensues; replication and/or segregation of the genetic material, as manifest by the formation of nuclear spindles, begins very shortly thereafter. It is not at all clear at which point the diploid complement of genetic material present in the zygote is reduced to the haploid complement found in the blood stage parasites. This event, the reduction division or meiosis, is all important in determining the genetic fate of the progeny of the zygote. The studies of Walliker *et al.* (1973, 1975) have demonstrated that recombination and segregation of genetic markers (drug resistance and enzyme variants) takes place at some stage during development in the mosquito (or, as a remote possibility, during preerythrocyte development in the liver). Following cross-fertilization between parental lines of parasite in the mosquito, parasites showing recombinant characteristics are readily demonstrated in the subsequent blood infection. Thus both recombination and segregation of genetic markers occur in a meiotic, or reduction, division prior to the formation of the blood stage parasites. Cloning studies confirm that the blood parasites are always endowed with a *haploid* complement of genetic markers. It is pointed out that phenomena deriving from dominance or recessivity of genetic traits cannot arise in the blood stage parasites in view of their haploid state.

There is little doubt that conventional genetic mechanisms play an important and probably major role in the inheritance of murine, and presumably other, malaria parasites. The chromosomes themselves have not, however, been recognized as organized structures. Their existence is, nevertheless, indicated by the cytological evidence of the kinetochores. These structures appear to segregate in association with the spindle microtubules seen in dividing nuclei at various stages in the life cycle. The kinetochores themselves do not contain DNA; the presence of this material has, however, been demonstrated in avian malaria (Aikawa *et al.*, 1972) close to the kinetochores during spindle formation. It is reasonable to suppose that the kinetochores serve as centromeres to which individual chromosomes are attached during segregation of the genetic material. Based on this assumption, Canning and Sinden (1973) have estimated that between five and ten chromosomes may be present on each half of a spindle in the young oocyst. It is not clear, however, whether this represents the diploid or the haploid number of chromosomes. The genetic evidence of Walliker *et al.* (1973, 1975) indicates that at least three unlinked units of segregation (chromosomes) are represented in the parasites. Linkage between markers has not thus far been demonstrated.

As already indicated, the precise point at which the reduction division occurs is not clear. It is generally supposed to occur in the ookinete during the first divisions of the genetic material following fertilization, although concrete evidence for this is lacking. Further insights into the mechanisms of chromosomal inheritance in the murine malaria parasites will be largely dependent upon the use of greater numbers of genetic markers and on further efforts to clarify our understanding of these events at the cytological level.

Evidence for extrachromosomal and other forms of non-Mendelian inheritance in murine *Plasmodia* is lacking at present. Yoeli *et al.* (1969) conducted experiments in which they claimed to have demonstrated an episomal type of exchange of genetic material between blood stage parasites. Their results, however, have proved to be unreproducible and can, in any case, be explained on the basis of spontaneous mutation.

2. Mutation and Transformation

On the basis of both biochemical and genetic evidence, it is now clear that mutation in a single Mendelian factor may account for the emergence of resistance in murine malaria to certain drugs of the antimetabolite class (see below). In the presence of drug pressure, lines of parasite resistant to pyrimethamine may arise in the course of a single blood infection in all species of murine *Plasmodium*. Such resistant traits exhibit single gene inheritance; resistant lines have been shown to be characterized by structurally altered forms of dihydrofolate reductase which, unlike the enzyme in drug-sensitive lines of parasite, are exceptionally resistant to inhibition by pyrimethamine (Diggens *et al.*, 1970). Estimates of the rate of spontaneous mutation to pyrimethamine resistance in blood parasites are remarkably low, being in the order of 1 in 10^{10}–10^{11} (Morgan, 1974; Schoenfeld *et al.*, 1974).

Resistance to certain drugs such as those of the 4-aminoquinoline class probably has a quite different genetic basis to that involving pyrimethamine. High levels of resistance to drugs such as chloroquine have so far been impossible to induce in vinckei group parasites (Powers *et al.*, 1969; Rosario, 1976). Comparatively low levels of chloroquine resistance have been achieved in both *P. vinckei* and *P. chabaudi*. Chloroquine resistance in *P. chabaudi* (toleration of about twice the minimum effective drug dose) is inherited in Mendelian fashion as a single factor (Rosario, 1976).

Lines resistant to maximum tolerated doses of chloroquine in mice have been derived from the normally highly sensitive *P. berghei* on several occasions (Peters, 1970). Unfortunately, nothing is known of the inheritance of such resistance. Generally prolonged serial blood passage under gradually increasing drug doses is required to elicit high levels of resistance. Emergence of resistance has rarely been achieved in a single step. Initially, lines of *P. berghei* resistant to chloroquine are almost always unstable in the absence of drug pressure, unlike pyrimethamine-resistant lines which are normally stable over an indefinite period in the absence of the drug. Stability of resistance to drugs such as chloroquine has been claimed after periods of prolonged passage under drug pressure. The genetic basis of this type of drug resistance is unlikely to be as simple as it is in the case of pyrimethamine resistance. Such resistance could arise from the gradual accumulation of single mutations each contributing in small degree to the overall level of drug resistance in the parasite. Alternatively, resistance to drugs such as chloroquine may be, in part, due to inducible systems activated by the presence of the drug.

Changes in parasite virulence, like drug resistance, could be accounted for on the basis of single mutations, multiple mutations, or induced transformation. D. Walliker, M. Yoeli, and A. Sanderson (personal communication) have recently demonstrated that a trait conferring fulminating virulence in *P. y. yoelii* is inherited as a single Mendelian factor. The trait arose in a single step during passage of a mild strain of *P. y. yoelii* (Yoeli *et al.*, 1975). Some other types of modification of virulence, other than those mediated by extraneous factors such as diet or mixed infection, could be due to selection of multiple mutations or physiologically mediated transformations. One or other of these mechanisms probably explains the increases in virulence frequently experienced during prolonged serial blood passage of lines of murine plasmodia in the laboratory.

3. Natural Variations

In addition to traits such as drug resistance and changes in virulence arising in the laboratory, the murine malaria parasites may exhibit considerable variation within a single wild population. An example of such natural variation is the extensive polymorphism of their enzymes. There can be little doubt that variation among these organisms is equally prevalent in other characteristics less readily recognized or analyzed than the enzyme forms.

Experiments in the laboratory indicate that genetic recombination among murine plasmodia of a single subspecies takes place readily. The analysis of enzyme variation in parasite populations of single locality indicates that genetic recombination among members of a single species takes place equally readily in the wild. By contrast, members of different *Plasmodium* species parasitizing the same local rodent population remain in total genetic isolation from each other (Carter and Walliker, 1975).

The concept has thus arisen that each regional subspecies of murine *Plasmodium* is a genetically heterogeneous group of organisms and that within each population genetic variation is readily and continuously reassorted by cross-fertilization among the parasites.

E. Biochemistry and Physiology

The distinction between biochemistry and physiology in unicellular organisms is, more than usual, a semantic question. In the sense that its biochemistry deals with the basic chemical events of the cell (enzymology, intermediary metabolism, and molecular biology), the physiology of an organism such as the malaria parasite may be defined as that which concerns its development and functions at the level of the organelle and the whole cell and its relationship with its environment. Defined in this way it can be said that most general areas of biochemistry have been examined to one degree or another in murine, as in other, plasmodia. Other than on a descriptive basis, however, aspects of cellular development and function have been largely ignored in these organisms.

1. Respiration

The murine plasmodia, like other malaria parasites, depend upon the metabolism of glucose for their main supply of energy. In blood stage parasites most of the glucose is metabolized no further than lactate; only a very small amount is converted to carbon dioxide (Bowman *et al.*, 1961). Substrate level phosphorylation during glycolysis is probably the main source of ATP formed within the parasite (Carter *et al.*, 1972). Nagaragan (1968a), however, indicated that at least a part of the ATP requirment of the parasite may be supplied by the host cell. This concept is also supported by the results of Brewer and Coan (1969).

There is now a considerable amount of evidence indicating that the asexual parasites of the blood stage do not possess conventional mitochondrial metabolic activity and do not phosphorylate ATP by oxidative processes. As already discussed, cristate mitochondria are not found in parasites of any of the blood stages with the possible exception of the mature macrogametocyte (Howells, 1970a). Citric acid cycle activity cannot be demonstrated nor can the presence of such enzymes as succinate dehydrogenase (SDH) and NAD- and NADP-dependent isocitrate dehydrogenase (IDH) be shown (Howells, 1970a; Howells and Maxwell, 1973a). Various workers, however, have shown the presence in the asexual parasites of a cyanide-sensitive cytochrome oxidase, found in functional mitochondria as the terminal component of the electron transport chain (Theakston *et al.*, 1969; Schiebel and Miller, 1969) and of both NADH and NADPH oxidizing activities (diaphorases) (Theakston *et al.*, 1970). Both cyanide and antimycin A (an inhibitor of electron transport) inhibit oxygen consumption by the parasites by almost 100%. This oxygen-dependent electron transport is not involved in energy production. Homewood *et al.* (1972b) obtained evidence for a cyanide-insensitive oxygen-independent electron transport system involved in energy production. Energy production by this system was uncoupled by rotenone and inhibited by antimycin A. In accordance with the inability of other workers to demonstrate SDH activity or a functioning citric acid cycle in the parasites, malonate (an inhibitor of SDH) had no effect on either the energy production or the oxygen consumption of the parasites. In the absence of dehydrogenases of the citric acid cycle, the sources of reducing equivalents for the oxygen-independent electron transport system are obscure although the diaphorases may well be involved in catalyzing their entry into the electron transport chain.

In both parasite and mature red blood cell respiration is, therefore, largely anaerobic. In parasitized red blood cells, glucose is metabolized at a much higher rate than it is in uninfected cells. Reticulocytes unlike mature erythrocytes, possess functional mitochondria supporting citric acid cycle activity and oxidative phosphorylation. Both glycolytic and oxidative activity of the host cell appear to be greatly stimulated in parasitized reticulocytes (Bryant *et al.*, 1964; Howells and Maxwell, 1973b).

Respiratory metabolism of the parasites at stages outside the blood has been studied only indirectly by cytochemical techniques. Cristate mitochondria are found in the ookinete, the oocyst, and the sporozoites. SDH activity has been demonstrated in mitochondria of the parasites at all these stages (Howells, 1970a) probably indicating the presence of citric acid cycle activity and oxidative phosphorylation throughout the sporogonic development. As in the blood parasites, cytochrome oxidase activity is found in all sporogonic stage parasites, both in the mitochondria and elsewhere in the cytoplasm.

Sparsely distributed cristae are apparently present in the mitochrondria of the preerythrocytic parasites (Terzakis *et al.*, 1974). As usual, cytochrome oxidase is present, but Howells and Bafort (1970) could not demonstrate SDH activity.

2. Intermediary Metabolism

A scheme relating the known aspects of intermediary metabolism in murine plasmodia is proposed in Fig. 7.

A. CARBOHYDRATES, ORGANIC ACIDS, LIPIDS, AND AMINO ACIDS. The blood stages of the murine malaria parasites undoubtedly metabolize larger amounts of glucose than any other substance. As already indicated the greater part of this is converted during glycolysis to lactate in which form it is presumably returned to the plasma to be disposed of by the host. Because of an inability to store carbohydrate as glycogen or any other polysaccharide (Sen Gupta *et al.*, 1955; Ciucă *et al.*, 1963), the blood stage parasites are dependent upon a continuous supply of glucose from the plasma.

In addition to being catabolized to lactate, glucose is apparently metabolized by the blood parasites by other pathways including the pentose phosphate pathway, or pathway of direct oxidation of glucose. Although the amount of glucose oxidized by this pathway is certainly small (Bowman *et al.*, 1961), Langer *et al.* (1967) claimed to have demonstrated most of the necessary enzymes including glucose-6-phosphate dehydrogenase (G6PD) and 6-phosphogluconate dehydrogenase (6PGD) in parasites of the blood stages. There is no doubt that the parasites possess their own form of 6PGD (Theakston and Fletcher, 1973; Carter, 1973).

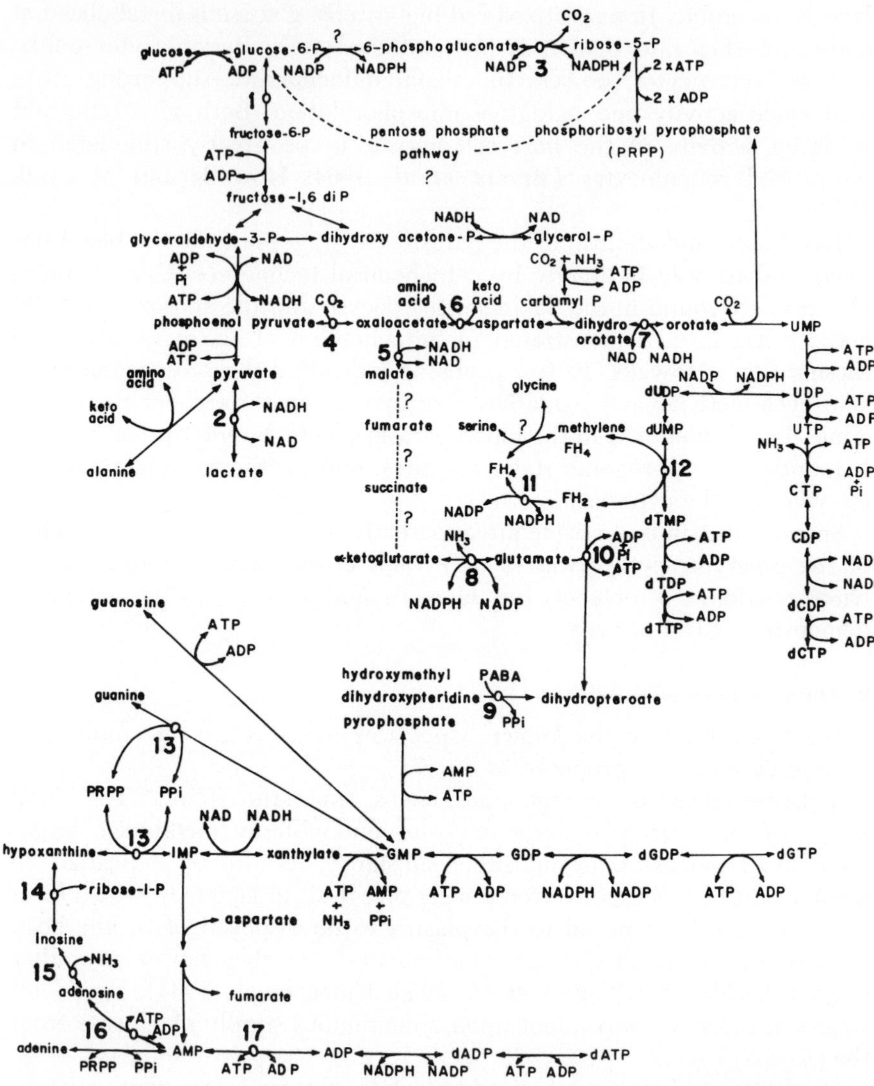

Fig. 7. Pathways of intermediary metabolism in the blood stage parasites of murine malaria. Numbers indicate parasite enzyme whose presence has been directly demonstrated. (1) Glucosephosphate isomerase; (2) lactate dehydrogenase; (3) 6-phosphogluconate dehydrogenase; (4) pyruvate carboxylase; (5) NAD-dependent malate dehydrogenase; (6) aspartate aminotransferase; (7) dihydroorotate dehydrogenase; (8) NADP-dependent glutamate dehydrogenase; (9) dihydropteroate synthetase; (10) dihydrofolate synthetase; (11) dihydrofolate reductase; (12) d-thymidylate synthetase; (13) hypoxanthine and guanine phosphoribosyltransferase; (14) purine nucleoside (inosine) phosphorylase; (15) adenosine deaminase; (16) adenosine kinase; (17) adenylate kinase.

The presence of G6PD of parasite origin has never been convincingly confirmed.

The operation of the pentose phosphate pathway is important primarily as a source of reduced NADP, required in many general synthetic reactions, in maintaining the level of cellular reduction, and as a source of ribose phosphates required in nucleic acid synthesis. The extent to which the parasites are able to meet these requirements by means of this pathway is unknown.

Murine malaria parasites have been shown to be capable of fixing small quantities of carbon dioxide into intermediary metabolites (Siu, 1966; Nagaragan, 1968b). Carbon dioxide condenses with phosphoenolpyruvate to form the four-carbon dicarboxylic acids, malate, succinate, and fumarate and the related amino acids, aspartate and glutamate. The process is absolutely dependent upon the presence of glucose, probably mainly as a source of pyruvate. Exogenous pyruvate, which does not enter the parasite (Nagaragan, 1968b), cannot be substituted for glucose to support carbon dioxide fixation. Inhibition of lactate dehydrogenase (LDH) with oxamate, however, prevents the removal of pyruvate and greatly increases the amount of carbon dioxide fixed.

Products of glucose metabolism are incorporated into the lipids of the blood stage parasites mainly as phospholipid (Cenedella, 1968; Cenedella *et al.*, 1969). Almost all is incorporated as the three-carbon glycerol moiety of the triglycerides, less than 5% being converted to fatty acid. The fatty acids of the parasite appear to be supplied largely from the pool in the host plasma.

The complex pathways of amino acid metabolism have not been widely investigated in the murine malarias. Langer *et al.* (1969) studied the metabolism of methionine in *P. berghei*. The parasites were able to synthesize the methyl group of methionine *de novo* but were unable to store the methyl group or exchange it with choline or phosphatidylcholine (lecithin), normally reservoirs of methyl groups. The inability of the parasites to store methyl groups probably means that they are dependent upon a continuous supply of methionine.

The presence of an FMN-dependent L-amino-acid oxidase was demonstrated in *P. berghei* by Langer and Phisphumvidhi (1971). It was estimated that the activity of this enzyme might account for up to one-third of the oxygen consumption of the parasites.

The essential amino acid requirements of the murine malarias are unknown. We have already seen that glutamate and aspartate may be synthesized from glucose, as can alanine (Bryant *et al.*, 1964). Most of the remaining amino acids are probably supplied by the digestion of hemoglobin or by the pool of free amino acids in the plasma.

Cytochemical and electrophoretic techniques have been used by several investigators to demonstrate directly the presence of parasite enzymes. The most active enzymes of the blood stage parasites are glucose phosphate isomerase (GPI) (Carter, 1970) and LDH (Phisphumvidhi and Langer, 1969). Both of these glycolytic enzymes are also highly active in the host cell. The enzymes of host and parasite are, however, readily distinguished by electrophoresis. 6PGD is the only enzyme of the pentose phosphate pathway whose presence has been directly demonstrated in the parasite (Theakston and Fletcher, 1973; Carter, 1973). The activity of the parasite enzyme is, in this case, much weaker than that of the enzyme of the host red blood cell. The presence of G6PD of parasite origin is in some doubt although this enzyme is active in the host cell.

Both parasite and host cell possess NAD-dependent malate dehydrogenase (MDH) activity (Carter, 1970; Tsukamoto, 1974), although the activity of the parasite enzyme is much weaker that that of the host cell form. The presence of NADP-dependent MDH (malic enzyme or decarboxylating MDH) cannot be demonstrated in the parasites (R. Carter, unpublished observation). The carbon dioxide fixing enzyme phosphoenolpyruvate carboxylase (PEPC) has been demonstrated in *P. berghei* (Siu, 1967; Forrester and Siu, 1971). The enzyme has both carboxylase and carboxykinase activities. Langer *et al.* (1970) demonstrated the existence of a parasite NADP-dependent glutamate dehydrogenase; this enzyme is absent from the host cell. Tsukamoto (1974) has identified a parasite-specific aspartate aminotransferase.

B. PYRIMIDINES AND PURINES. According to most reports the murine malarias are entirely dependent upon an exogenous supply of purines but are unable to utilize exogenous pyrimidines. These are synthesized by the parasites *de novo.* Whereas the parasites recycle purines extensively by salvage pathways (see below) they appear to lack the necessary pathways for recycling pyrimidines, although these molecules apparently enter the parasitized blood cells with ease (Neame *et al.,* 1974).

i. Pyrimidines. The central intermediate in *de novo* synthesis of pyrimidines is uridylic acid (UMP) formed in several steps from carbamyl phosphate, aspartate and phosphoribosyl pyrophosphate (PRPP) (Fig. 7). One of the enzymes of this pathway, dihydroorotate dehydrogenase, has been demonstrated in the blood stage parasites by Krooth *et al.* (1969). Formation of other ribose and deoxyribose nucleotides proceeds by a series of conversions starting from UMP.

The steps leading to the synthesis of deoxythymidylic acid (dTMP) have been studied in some detail in murine malaria parasites. The conversion of dUMP to dTMP is mediated by the enzyme deoxythymidylate synthetase catalyzing the transfer of the methyl moiety from N^5, N^{10}-methylene tetrahydrofolate to dUMP. This enzyme has been demonstrated in *P. chabaudi* by Walter and co-workers (1970) and in *P. berghei* by Reid and Freidkin (1973). The all important synthesis of dTMP is thus achieved in a single step from dUMP. The reformation of the methylated tetrahydrofolate cofactor is less simple. As a result of the deoxythymidylate synthetase reaction, methylene tetrahydrofolate is oxidized to dihydrofolate. Dihydrofolate is then reduced to tetrahydrofolate by the enzyme dihydrofolate reductase whose presence in *P. berghei* was first demonstrated by Ferone *et al.* (1969) and in *P. chabaudi* by Walter and Königk (1971). The reaction catalyzing the methylation of tetrahydrofolate to methylene tetrahydrofolate has not been studied but probably involves the donation of a methylene group by serine.

In accordance with the inability of the parasites to utilize preformed pyrimidines, Walter and co-workers (1970) were unable to demonstrate the presence of deoxythymidine kinase, one of the enzymes of the pyrimidine salvage pathway, in *P. chabaudi.*

Like the pyrimidines themselves, tetrahydrofolic acid must be synthesized by the parasites. Exogenous supplies of either folate or folinic acid are not utilized. The synthesis of dihydrofolate by condensation of glutamate with dihydropteroate is catalyzed by dihydrofolate synthetase whose presence in the parasites was demonstrated by Ferone (1973). Dihydropteroate synthetase has been demonstrated in *P. chabaudi* by Walter and Königk (1973). This enzyme catalyzes the condensation of *p*-aminobenzoate (PABA) with hydroxymethyl dihydropteridine pyrophosphate (HMDPP) to form dihydropteroate. HMDPP is itself derived from guanylic acid (GMP). PABA, on the other hand, is an essential vitamin for murine plasmodia.

Unlike the parasites, the host itself is unable to synthesize dihydrofolic acid *de novo;* an exogenous source of folate or folinic acid (one of the derivatives of folic acid) is necessary to meet its requirement for these cofactors. Folate or folinate, on the other hand, are not able to supplement the requirements of the parasite. The differences in folate metabolism in the host and the parasites expose the latter to attack by some of the most effective antimalarial agents—the "antimetabolites" (see Section III, F,1,b). Drugs of the sulfonamide class act by competitive inhibition of the condensation of PABA with HMDPP (Walter and Königk, 1973). The parasites are also unfortunate in being endowed with a dihydrofolate reductase highly sensitive to inhibition by antimalarials such as pyrimethamine and cycloguanil. At therapeutic concentrations these drugs do not significantly affect the activity of the host enzyme.

The *de novo* synthesis of pyrimidines and the absence of a pyrimidine salvage pathway is probably a feature of all stages of the life cycle of the murine malaria parasites. Thus Davies and Howells (1973) found that exogenous pyrimidines failed to become incorporated into the nucleic acids of oocysts of *P. yoelii*. Jacobs *et al.* (1974) were unable to demonstrate incorporation of pyrimidines into sporozoites of *P. berghei* but found that purines were readily incorporated when these substances were included in the diet of mosquitos during sporogonic development.

ii. Purines. In contrast to the inability of the blood stage parasites to utilize exogenous pyrimidines, exogenously supplied purines both enter the parasite cell and are incorporated into its nucleic acids (Van Dyke *et al.*, 1970). In common with most other protozoa, the murine plasmodia appear to lack the *de novo* pathway of purine synthesis but sustain an effective salvage pathway for the utilization of preformed purine bases and nucleosides. In this respect, the parasites differ from most mammalian cells which synthesize purines *de novo*. A notable exception, however, is the mature erythrocyte which possesses only the salvage pathway (Lowry *et al.*, 1962).

A wide variety of exogenously supplied purines, purine nucleosides, and purine nucleotides are incorporated into the nucleic acids of murine malaria parasites. These include the purines hypoxanthine (Büngener and Nielsen, 1968) and adenine (Lucknow *et al.*, 1973), the nucleosides adenosine and guanosine (Neame *et al.*, 1974), and the nucleotides AMP, ADP, ATP, and GTP (Lantz *et al.*, 1971). Since none of their phosphate is incorporated into nucleic acid the nucleotides presumably undergo some rearrangements in the course of which they are catabolized at least to the nucleoside level. AMP is the nucleotide most readily incorporated by the parasites.

Red blood cells contain high levels of ATP. Adenine is not found in the plasma and the parasites and red blood cells are thought to obtain this purine during their passage through the liver. In the course of malaria infection both infected and uninfected red blood cells of the rodent host become steadily depleted of ATP. The parasite appears to deprive its host of its normal requirement of adenine by competing for this substance.

The scheme in Fig. 7 for the purine salvage pathways inside the parasite is based on the proposals of Luckow *et al.* (1973) and Jaffé and Gutteridge (1974). Adenine and adenosine may either be converted directly to AMP and then to ADP, ATP, and dATP, or via inosine and hypoxanthine to IMP. IMP is a common intermediate from which both ATP and GTP and their deoxyribose derivatives can be derived. Guanine and guanosine may be converted directly to GMP and then to GDP, GTP and dGTP.

Of the enzymes required to fulfill this scheme, adenosine kinase (Büngener and Nielsen, 1968), adenosine deaminase (Büngener, 1967), adenylate kinase (Carter, 1970), purine nucleoside (inosine) phosphorylase, and hypoxanthine and guanine phosphoribosyltransferase (Büngener and Nielsen, 1968) have been demonstrated in the murine malaria parasites. As already indicated the purine salvage pathway appears to operate in the sporogonic as well as in the blood stages of the parasites' life cycle.

3. Nucleic Acids

According to Bahr and Mikel (1972), the amount of DNA present in the murine malaria parasites far exceeds their probable genetic requirements. On cesium chloride density gradient centrifugation the DNA of murine plasmodia generally bands as a single component with base composition about 20% G + C (Gutteridge *et al.*, 1971). This is an exceptionally low G + C content and contrasts with about 40% G + C in the primate malarias *P. knowlesi* and *P. falciparum*. In some lines of berghei group parasites, Warhurst *et al.* (1971a) found a rapidly renaturing (highly repetitive) band of satellite DNA.

The DNA of rodent malaria parasites is associated with about four times its own mass in proteins of both basic histone and nonhistone types and is arranged in chromatin fibers accommodating about 32 strands of double helical DNA per fiber. The association of DNA with basic proteins such as histone is typical of the organization of the genetic material in eukaryotic cells. DNA synthesis continues throughout the growth of the blood stage trophozoite, but is largely completed by the early schizont stage.

In blood stage trophozoites the DNA apparently concentrates near the nuclear membrane in merozoites and young trophozoites; at the metaphase stage in nuclear division the DNA appears to occupy an ill-defined area in the equatorial region of the mitotic spindle. Condensed chromosomes have not been seen at any stage of the life cycle.

The behavior of RNA in murine malaria parasites is hard to reconstruct from currently available information. The parasites possess fairly typical eukaryotic ribosomes consisting of ribonucleoprotein in two subunits, one 60, the other 40 S. In higher organisms the RNA components of these subunits are derived from a precursor synthesized in the nucleolus, a structure not so far recognized in murine plasmodia. It has been suggested that the host may itself synthesize one of the RNA subunits of the parasite ribosomes (Tokuyasu *et al.*, 1969). The origin of ribosomal and other classes of RNA in the parasites is, however, far from clear. Activation of amino acids with tRNA for protein synthesis has been described (Ilan *et al.*, 1969). RNA is actively synthesized in parasitized cells throughout the division cycle. According to Whitfield (1953) the blood stage parasites contain twice as much RNA as DNA.

4. Nutrition and Digestion

Like all other plasmodia, blood stage murine malaria parasites are dependent upon a continuous supply of glucose derived from the plasma. They also appear to be dependent upon their host for the greater part of their fatty requirement. There is, however, surprisingly little known of the amino acid requirements of the parasites. There can be little doubt that these are in large part met by digestion of hemoglobin. In other species of malaria (*P. knowlesi*) it has been found that amino acids poorly represented or absent in the hemoglobin of the host must be supplied to the parasite from the plasma or culture medium. In addition to hemoglobin, the host red blood cell, itself rich in ATP, is able to supply the parasite with its purine requirements.

PABA is an essential vitamin of the blood stage parasites. The suppression of malarial infections in animals on a milk diet (Maegraith *et al.*, 1952) is due the absence of PABA from milk (Hawking, 1953).

The oxygen consumption of the blood stage parasites is low (Coombs and Gutteridge, 1975) and is hardly, if at all, involved in the respiratory processes of the parasite cell. Indeed, high oxygen tensions are not conducive to optimum growth of the parasites. Carbon dioxide is utilized in the synthesis of metabolic intermediates via the CO_2 fixation pathways.

Gases and other small molecules utilized by the blood stage parasites presumably enter the parasite cell by diffusion or other mechanisms of direct transport across the cell membrane. The host cell proteins, mainly hemoglobin, must, however, be digested before they can be utilized by the parasites. Digestion of hemoglobin occurs inside food vacuoles which accumulate in the cytoplasm of the trophozoite. As previously discussed, the food vacuoles arise by the activity of the cytostome through which invaginations of the host cytoplasm develop. The vesicles so formed are pinched off to form the food vacuoles lying free within the cytoplasm of the parasite. In avian malarias it has been shown that the food vacuoles contain acid phosphatases, enzymes normally associated with hydrolytic activity in lysosomes; these enzymes are not found elsewhere in the parasite or in the host cell and are clearly of parasite origin. The residual product of hemoglobin digestion is hemozoin, the substance forming the malaria pigment. Hemozoin contains the iron and porphyrin ring of hemoglobin but little is otherwise known of its structure.

Cook *et al.* (1961) isolated two proteases from *P. berghei.* These enzymes hydrolyzed denatured, but not native, mouse hemoglobin; reticulocyte protein was digested more rapidly than normocyte hemoglobin, an observation which might, in part, explain the preference of these parasites for young red blood cells. Chan and Lee (1974) isolated three distinct proteases from *P. berghei.* Two of these enzymes had much greater reactivity with mouse than with human hemoglobin. These workers also suggested that the specificity of the enzymes involved in digestion of hemoglobin might affect the host specificity of the parasites.

F. Chemotherapy

A complete discussion of the extensive literature on chemotherapy and drug resistance in murine malaria is not within the scope of this review. For a fuller account of this subject up to 1970 the reader is referred to the monograph by Peters (1970). Since their discovery, the murine *Plasmodia* have been exploited extensively in antimalarial screening programs and in basic studies on the mechanisms of drug action and drug resistance.

1. Antimalarial Action

The plasmodicidal activity of antimalarials of the four major classes—the 8-aminoquinolines (e.g., primaquine), the 4-aminoquinolines (e.g., chloroquine), the quinines, the dihydrofolate reductase inhibitors (e.g., pyrimethamine) and antagonists of PABA metabolism, and the sulfones and sulfonamides—has been extensively studied. The antimalarial activity of antibiotics such as the tetracyclines and lincomycin derivatives has also been examined. Members of each of the four major classes of antimalarial drugs share general structural similarities and mechanisms of antimalarial action and area of attack in the parasite life cycle (see Fig. 8).

A. 8-AMINOQUINOLINES. The 8-aminoquinolines, including pamaquine and primaquine, are effective against all stages of the parasite life cycle: blood forms, both sexual and asexual, sporogonic, and preerythrocytic stages. The preerythrocytic stages are, however, the most susceptible to the action of these drugs. The mechanism of action of the 8-aminoquinolines is not understood.

Drugs of this class are comparatively toxic, their effect on the oxidation-reduction system of the erythrocytes rendering these cells susceptible to lysis. The toxicity of 8-aminoquinolines for erythrocytes and parasites is similar. For this reason, the drugs are only useful for their prophylactic activity against the liver stages and in effecting radical cure.

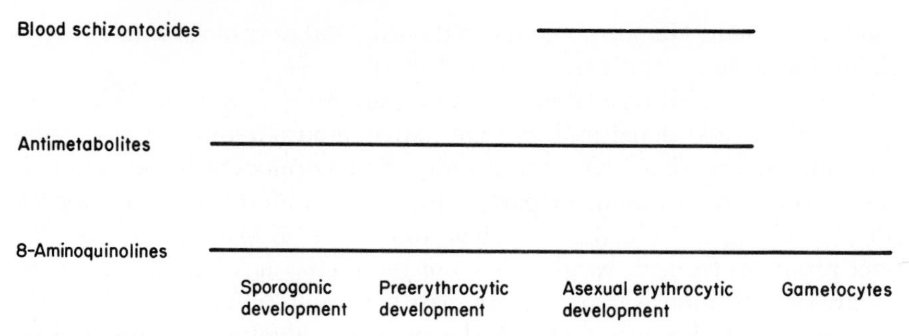

Fig. 8. The areas of antimalarial activity of three classes of drug in the life cycle of the murine malaria parasites.

B. ANTIMETABOLITES. Drugs of the two classes represented by the dihydrofolate reductase inhibitors, e.g., pyrimethamine and proguanil and the PABA antagonists, sulfones such as dapsone, and sulfonamides such as sulfadiazine, are collectively referred to as the antimetabolites. These drugs are effective against all actively dividing parasites and thus act against all stages except the gametocytes. The general mechanisms of action of these drugs in preventing thymidylic acid synthesis has already been discussed. The efficiency of the antimalarial activity of the sulfonamides, which competitively inhibit incorporation of PABA into dihydrofolate, is reduced in inverse proportion to the amount of PABA in the host diet. Reflecting the close sequential relationship of the metabolic steps attacked, the dihydrofolate reductase inhibitors act synergistically in combination with the sulfa drugs.

The dihydrofolate reductase inhibitors, pyrimethamine and proguanil, are two of the most effective antimalarials known. However, they act less rapidly in reducing parasitemia in established infections than do the 4-aminoquinolines (see below) and resistance is easily developed. They are, nevertheless, highly effective prophylactic drugs. Resistance to the sulfa-based drugs also arises easily and these drugs are handicapped by the antagonistic effect exerted by PABA against their antimalarial activity.

C. BLOOD SCHIZONTOCIDES (4-AMINOQUINOLINES AND QUININE). Drugs of the 4-aminoquinoline and quinine class are known as blood schizontocides; they include the oldest (quinine) and most effective (chloroquine) therapeutic antimalarials in existence. Almost entirely without effect against the sporogonic and preerythrocytic stages, these drugs act with great speed against the asexual blood parasites, which, combined with the beneficial antiinflammatory and antipyretic side effects of quinine and chloroquine, results in rapid and effective therapy.

Although studied with some intensity, real clues to the mechanism of antimalarial action of chloroquine and other 4-aminoquinolines have remained tantalizingly elusive. These drugs are rapidly concentrated by the parasitized blood cells. Macomber *et al.* (1966) found that mouse erythrocytes infected with *P. berghei* concentrate chloroquine 100-fold above plasma levels, while Fitch (1969) demonstrated concentrations of up to 600 times those in the plasma. Within minutes of administration of the drugs, the parasites undergo striking morphological changes. The fine pigment granules of the trophozoites begin to aggregate forming a single large lump of pigment 1–2 hours after administration. Pigment aggregation is due to the fusion of the small cytoplasmic food vacuoles into a single large "autophagic" vacuole. This process, similar to the aggregation of the food vacuoles during schizogony, has been studied by Warhust *et al.* (1971b, 1972) and Warhurst and Baggaley (1972). Homewood *et al.* (1972a) have reported that autophagic vacuole formation is an energy-dependent process involving electron transport; it is also dependent upon active RNA and protein synthesis.

In 1969, Fitch demonstrated chloroquine binding sites of three different affinities in *P. berghei.* The most avid had an intrinsic association constant of about 10^8 *M*. Normal plasma concentrations of chloroquine in mammals of about 10^{-6} *M* (Berliner *et al.*, 1968) will readily saturate this site. In chloroquine-resistant *P. berghei,* the high-affinity binding sites are absent or only present in very small numbers (Fitch, 1969). Such parasites concentrate less than one-half as much chloroquine as do chloroquine-sensitive parasites. In addition to the high-affinity sites, two other types of chloroquine binding site were demonstrated with association constants of about 10^5 *M* and 10^3 *M*. These sites are apparently unaltered in resistant parasites and are probably not involved in the antimalarial activity of chloroquine. Negligible amounts of chloroquine accumulate in uninfected red blood cells at an external chloroquine concentration.

At one time it had been thought that the chloroquine binding affinity of DNA might be involved in the antiplasmodial activity of the drug. However, as the chloroquine binding association constant of DNA corresponds to the binding sites of lowest affinity in the parasites, it seems unlikely that this property is involved in its initial concentration within the parasites. The high-affinity binding site has been found to be associated with a membrane component in *P. berghei* while the low affinity sites are freely soluble (Kramer and Matusik, 1971).

Results of Warhurst *et al.* (1972) and Fitch (1972) show that quinine and chloroquine compete for the same high-affinity binding site in *P. berghei.* Quinine, however, does not induce pigment clumping as does chloroquine. Pursuing this observation, Warhurst (1973) has suggested that at least some of the high-affinity binding sites of Fitch (1969) are associated with the digestive vacuoles possibly with receptor molecules within the membrane of the vacuole. Homewood *et al.* (1972a) have postulated that chloroquine-induced autophagic vacuole formation is the result of neutralization of the acidic content of the vacuole by the presence of the basic chloroquine molecule. These workers reported that a similar type of pigment clumping could be induced simply by raising the pH of the external medium. Hence, it is suggested that drugs such as quinine which apparently bind to the same site as chloroquine fail to cause clumping as they are much less basic than chloroquine.

The chloroquine binding site postulated to exist in the food vacuole membrane may not be the site of antimalarial action of the drug but only represents the means by which it is concentrated to high levels. Pigment clumping, therefore, may not be, in itself, a pathological manifestation. On subsequent release of the highly concentrated drug into the cytoplasm, however, general toxic effects including disruption of glycolysis and nucleic acid and protein synthesis (Gutteridge and Trigg, 1972) may be manifest. The schizontocides, to which malaria strains show cross-resistance, may not necessarily act on the same final target(s). The binding site by which they are initially concentrated, however, is probably the same.

D. ANTIBIOTICS. Until recently the antibacterial antibiotics were rarely considered for use as antimalarials. In some recent studies, however, various such compounds, including erythromycin, tetracyclines, and lincomycin derivatives, have been found to exert marked antimalarial activity against blood stage parasites (Robinson and Warhurst, 1972; Colwell *et al.*, 1972).

2. Parasite Susceptibility

Blood stages of *P. berghei, P. vinckei,* and *P. chabaudi* are naturally sensitive to antimalarials of each of the four main classes (see Table IV). One of the distinguishing characteristics of *P. yoelii* and its subspecies, however, is their relative insensitivity to the blood schizontocidal action of 8-amino- and 4-aminoquinolines and of quinine (Warhurst and Killick-Kendrick, 1967; Peters *et al.*, 1970). The sensitivity of the preerythrocytic stages of *P. yoelii* to the 8-aminoquinolines is, nevertheless, similar to that of *P. berghei* (Most *et al.*, 1967; Gregory and Peters, 1970; King *et al.*, 1972). In comparison to *P. berghei, P. yoelii* subspecies are hypersensitive to drugs of the antimetabolite class (Warhurst and Killick-Kendrick, 1967; Peters, 1968, 1970; Diggens and Gregory, 1969; Peters *et al.*, 1970; King *et al.*, 1972).

The resistance of *P. yoelii* subspecies to chloroquine is conditional upon the presence of PABA in the diet of the host (Carter, 1973). In the absence of PABA or when it is present in amounts just sufficient to maintain

Table IV

Normal Sensitivity to Antimalarial Drugs of Blood Stages of *Plasmodium* sp.[a]

Drug	Route of adminsstration of drug[b]	*P. berghei*	*P. yoelii*	*P. chabaudi*
8-Aminoquinolines				
Primaquine phosphate	sc	5.8 ± 1.1 [c]		6.4 ± 1.45 [i]
	sc	2.7 ± 0.3 [d]	6.6 ± 1.6 [d]	
Pamaquine	or	3 to 15 [e]	20 to 40 [g]	
Blood schizontocides				
4-Aminoquinolines				
Chloroquine phosphate	sc	3.9 ± 0.3 [c]		3.7 ± 0.8 [i]
	sc	4.0 ± 0.9 [d]	8.5 ± 3.3 [d]	
	ip		> 60 [g]	
Chloroquine sulfate	sc	2–4 [f]	1.9 to 16 [f]	
	ip		> 60 [h]	
Mepacrine methanesulfonate	sc	6.6 ± 1.25 [c]		6.6 ± 1.4 [i]
	sc	5.5 ± 1.6 [d]	8.2 ± 3.0 [d]	
Mepacrine hydrochloride	ip		5 to 10 [g]	
Cinchona derivatives				
Quinine bisulfate	sc	200 ± 35 [c]		160 ± 25 [i]
	sc	95 ± 28 [d]	140 ± 111 [d]	
Quinine hydrochloride	or		150 to 300 [g]	
Antimetabolites				
DHFR inhibitors				
Pyrimethamine	ip	0.2 ± 0.05 [c]		0.064 ± 0.23 [i]
	ip	0.1 ± 0.03 [d]	0.06 ± 0.01 [d]	
Cycloguanil hydrochloride	sc	9.0 ± 7.0 [c]		8.0 ± 5.8 [i]
	sc	11.0 ± 6.3 [d]	4.6 ± 2.2 [d]	
PABA antagonists				
Sulfone				
Dapsone	sc	1.0 ± 0.6 [c]		0.042 ± 0.022 [i]
	sc	0.41 ± 0.17 [d]	0.09 ± 0.02 [d]	
Sulfonamide				
Sulfadiazine	sc	2.7(0.7 to 7.0) [c]		0.054 ± 0.03 [i]
	sc	0.25 ± 0.04 [d]	0.08 ± 0.06 [d]	

[a] The data are presented as the number of milligrams of drug per kilogram of body weight per day required to reduce the parasitemia to 10% of that of untreated animals (ED_{99}) on the fifth day after four daily administrations of the drug. All tests were performed in mice (strain of mouse and dietary and other conditions not necessarily specified). The accuracy of the figures varies according to the source of the data; in some cases, the values of the measurements quoted are "guesstimates" based on available information.

It must be assumed that the conditions of drug assay represented in this table varied considerably. Among the most important variables governing drug response is their normal rate of growth, the parasites are highly susceptible to chloroquine. At sufficiently high levels of PABA *P. yoelii* is unaffected by the maximum dose of the drug tolerated by the host.

Blood stage parasites of *P. berghei* and *P. chabaudi* are of comparable sensitivity to the 8-amino- and 4-aminoquinolines, to quinine, and to the dihydrofolate reductase inhibitors. *Plasmodium chabaudi,* however, is hypersensitive to sulfones and sulfonamides (Peters, 1967) and more sensitive than *P. berghei* to dietary levels of PABA.

Plasmodium vinckei has not been used extensively in chemotherapy studies. The blood stage parasites of this species are normally sensitive to chloroquine, pyrimethamine, and primaquine (Powers *et al.,* 1969).

3. Drug Resistance

Resistance to drugs of each of the main classes of antimalarials has been demonstrated in experimental infections of murine *plasmodia.* The manner in which drug resistance arises, however, differs profoundly according to the class of drug and the species of *Plasmodium* involved.

A. 8-AMINOQUINOLINES. There are few reports of resistance to 8-aminoquinolines by murine malaria parasites. Peters (1966) described a line of *P. berghei* whose blood stages were resistant to primaquine. The blood parasites were also resistant to the dihydrofolate reductase inhibitors pyrimethamine and cycloguanil, to quinine, and to the sulfone dapsone;

---

the composition of the diet of the host. As indicated in the text, variations in dietary levels of PABA may drastically change the response of all murine malaria species to the PABA antagonists (sulfones and sulfonamides) and also to the response of representatives of *P. yoelii* to 4-aminoquinolines. Much of the variation in drug response represented in the table undoubtedly reflects these relationships.

The data quoted from Peters *et al.* (1970) are derived from studies made on *P. berghei* strain K173 and on lines of the "NS" series (see Section III,A,3). Although these workers recognize the "NS" lines as representatives of *P. berghei* they are, according to the criteria followed here, members of *P. yoelii* and are, therefore, included as representatives of this species.

Fully curative drug doses can generally be expected to be not more than twice the doses indicated here. Conditions which modify the effectiveness of a drug should, however, always be taken into account.

[b] sc, Subcutaneous; or, oral; ip, intraperitoneal.

[c] Peters (1965c).

[d] Peters *et al.* (1970).

[e] Hill (1950); Thurston (1950).

[f] Peters (1968).

[g] Warhurst and Killick-Kendrick (1967).

[h] Carter (1973).

[i] Peters (1967).

the parasites were sensitive to 4-aminoquinolines and to sulfonamides. Unlike the chloroquine-resistant lines of *P. berghei* (described below), the primaquine-resistant line did not lose its ability to form malaria pigment granules. Resistance was, however, associated with a benign course of infection as compared to that of the virulent parental line of parasites. In the absence of drug pressure, serial blood passage resulted in loss of resistance. Since the line of parasites used in this study was incapable of being transmitted through mosquitoes, it is not known whether the drug resistance of the blood stages would have been shown by the preerythrocytic parasites as well.

Blood stage parasites of *P. yoelii* exhibit a certain degree of natural resistance to pamaquine and primaquine (Warhurst and Killick-Kendrick, 1967; Peters *et al.*, 1970). The liver stages of these parasites are normally sensitive to the 8-aminoquinolines (Gregory and Peters, 1970).

B. ANTIMETABOLITES. Resistance of blood stage parasites to drugs of the dihydrofolate reductase inhibitor class and to sulfones and sulfonamides is readily developed in all four species of murine malaria. Resistance to these drugs may be selected after a single course of drug treatment in an infected animal. Parasites selected in a single step may show high levels of drug resistance; such resistance is usually stable through mosquito transmission and through multiple serial blood passages in the absence of drug pressure.

In studies on inheritance in *P. yoelii* and *P. chabaudi,* Walliker and his colleagues (1973, 1975) have shown that pyrimethamine resistance is inherited as a single Mendelian factor. The biochemical studies of Diggens *et al.* (1970) have shown that in pyrimethamine-resistant lines of *P. berghei* the dihydrofolate reductase has a much lower affinity for pyrimethamine than has the enzyme of the sensitive parental strain. From these two lines of evidence there is little doubt that the resistance of the parasites to this drug is due to mutation in a gene coding for dihydrofolate reductase. Such mutant enzymes also show reduced affinity for their normal substrate and are, therefore, more sensitive than the wild type to decreased levels of dihydrofolate in the parasite cell. This, no doubt, largely accounts for the increased susceptibility of many pyrimethamine-resistant parasites to reduced levels of PABA (Jacobs, 1964), and their hypersensitivity to sulfonamides.

It is likely that resistance to other drugs of the dihydrofolate reductase inhibitor class, e.g., proguanil and cycloguanil, commonly arises by similar mutations in the gene coding for dihydrofolate reductase. Cross-resistance between drugs of this class is usually, but not always, found (Peter, 1970). The preerythrocytic stages of parasites whose blood stages are resistant to dihydrofolate reductase inhibitors are themselves resistant to the same drugs and share the same pattern of cross-resistance and sensitivity (Diggens and Gregory, 1969). Morphological or behavioral changes in the parasites, such as loss of virulence, are not normally associated with resistance to the dihydrofolate reductase inhibitors. In one instance, however, pyrimethamine resistance in *P. yoelii* was accompanied by a dramatic increase in virulence (Morgan, 1974). Genetic studies have subsequently shown that the drug resistance and virulence characters in this line are independently inherited and do not therefore have any direct relationship to each other (D. Walliker, personal communication).

Lines selected for resistance to the sulfa-based drugs are generally cross-resistant to the dihydrofolate reductase inhibitors, in contrast to lines selected for resistance to the dihydrofolate reductase inhibitors which usually demonstrate hypersensitivity to the sulfa drugs. Sulfadiazine-resistant parasites lose their requirement for PABA and may even be developed by selecting for independence of the PABA requirement. Resistance of this drug is inherited in simple Mendelian fashion (R. A. F. Macleod, personal communication) and probably arises as a mutation in a single gene.

The mechanisms by which parasites resistant to the sulfa drugs circumvent their requirement for PABA and also, apparently, the necessity for the reduction of dihydrofolate by the parasite, is not known. The evidence implies, however, that such parasites may have abandoned the usual pathways for furnishing themselves with tetrahydrofolate in favor of an alternative, as yet unknown, source of supply.

Parasites resistant to sulfa drugs tend to be less virulent than the parental lines.

C. BLOOD SCHIZONTOCIDES. *Plasmodium berghei* has the dubious distinction of being the first mammalian malaria parasite to exhibit resistance to chloroquine (Ramakrishnan *et al.*, 1957).

There are striking differences between the species of murine plasmodia both in the natural response of the parasites to the blood schizontocides and in their potential to acquire resistance. In general, resistance to quinine and the 4-aminoquinolines, such as chloroquine, is less easily elicited than is resistance to other antimalarials. *Plasmodium yoelii,* however, has a naturally high level of resistance to these drugs conditional upon the level of PABA to which the parasites are exposed. The induction of resistance of the naturally sensitive *P. berghei* to 4-aminoquinolines generally requires prolonged exposure of the parasites to gradually increasing levels of drug. Nevertheless, under such drug pressure highly resistant lines of *P. berghei* may be selected which survive the maximum doses of drug tolerated by the host.

Induction or selection of chloroquine-resistant parasites in the vinckei group is apparently even more difficult. Only comparatively low levels of resistance have been achieved in these parasites. Powers *et al.* (1969) were unable to select a chloroquine-resistant line from a parental line of *P. v. vinckei* sensitive to the drug at a normal level. These workers were, however, able to select a chloroquine-resistant line of this species using a line of *P. v. vinckei,* already resistant to pyrimethamine, as the starting material. Efforts to achieve chloroquine resistance in *P. chabaudi* have met with only limited success. A line of this species able to tolerate about three times the level of chloroquine effective against the parental lines, itself resistant to pyrimethamine, was obtained by selection under gradually increasing drug pressure (Rosario, 1976).

Chloroquine-resistant lines of *P. berghei* and *P. vinckei* are initially unstable in the absence of drug pressure and may revert rapidly to sensitivity in the course of a few blood passages in the absence of drug. After prolonged maintenance under drug pressure, however, several lines are now reputed to be stably resistant in the absence of the drug. In some instances, resistance once lost may be rapidly recovered by renewed drug pressure (Peters, 1965c); in other instances, resistance once lost has been apparently irrecoverable (Hawking, 1966).

In contrast to the sudden emergence of high levels of resistance to drugs of the antimetabolite class, resistance to chloroquine and other 4-aminoquinolines has rarely been achieved in a single step. It seems unlikely that high resistance to these drugs normally arises as the result of a mutation in a single gene. Studies on the inheritance of high levels of resistance to blood schizontocides are lacking, although Rosario (1976) has demonstrated that a low level of resistance in *P. chabaudi* is inherited as a single Mendelian factor. Some aspects of the patterns of acquisition and loss of chloroquine resistance suggest that it may be inducible under the direct influence of the drug itself. Alternatively, resistance may arise by successive mutations each contributing in small degree to the level of resistance finally achieved; or resistance could be achieved by selecting for cytoplasmically inherited components which increase in concentration in the cytoplasm under drug pressure thereby confering increasing levels of resistance on the parasites.

The morphology and behavior of chloroquine-resistant lines of *P. berghei* and *P. vinckei* are highly characteristic. Infections with the resistant lines of *P. berghei* run much milder courses and develop lower parasitemias than do infection with the parental lines (Peters, 1965a). The resistant parasites are confined almost exclusively to reticulocytes and young red blood cells, while parasites of the sensitive parental lines may be commonly found in mature cells. Multiple invasion of reticulocytes is very common. Schizonts are virtually absent in the peripheral blood being sequestered in the veins of the internal organs (Miller and Frémont, 1969). Under the light microscope, pigment granules are not seen in the resistant parasites. The cytoplasm contains numerous round clear "vacuoles."

The moderately chloroquine-resistant line of *P. v. vinckei* (Powers *et al.*, 1969) was only slightly less virulent than its parental line. As in the parental line, the resistant parasites continued to develop predominantly in mature red blood cells; schizonts were commonly found in the peripheral circulation. Like the chloroquine-resistant lines of *P. berghei* the resistant *P. v. vinckei* lacked conspicuous pigment granules and was characterized by numerous round cytoplasmic "vacuoles."

The slightly chloroquine-resistant line of *P. chabaudi* is morphologically indistinguishable from its parental line.

The morphology of the blood stages of the naturally resistant subspecies of *P. yoelii* is virtually indistinguishable from those of the naturally chloroquine-sensitive *P. berghei.* Under chloroquine pressure, however, the blood parasites of *P. yoelii* are found almost exclusively in reticulocytes, the pigment granules disappear, and vacuolization of the cytoplasm occurs.

Lines of *P. berghei* selected for resistance to quinine or to mepacrine behave in most essential respects like the chloroquine-resistant parasites of this species (Jacobs, 1964; Peters, 1966). Benign infections with "vacuolated," pigment-less parasites occur with such lines. There is complete cross-resistance between quinine- and 4-aminoquinoline-resistant lines, while all are normally sensitive or hypersensitive to pyrimethamine, cycloguanil, dapsone, and sulfadiazine. Most such lines are also resistant to primaquine (8-aminoquinoline), although the mepacrine-resistant line of Peters (1966) and a chloroquine-resistant line of Hawking and Gammage (1962) responded with normal sensitivity to this drug.

Even less is known of the mechanisms of resistance of the parasites to the 4-aminoquinolines than is understood of mechanisms of action of these drugs. Fitch (1969), however, has found that the high-affinity chloroquine binding site of the sensitive parasites, thought to be involved in initial concentration of the drug within the parasites, is absent in resistant lines of *P. berghei.*

G. Murine Malarias in the Laboratory

1. Laboratory Transmission

For many years after its first isolation, attempts to transmit *P. berghei* through laboratory-bred mosquitoes met with almost invariant failure. The natural vector, *A. d. millcampsi,* was not generally available for such studies because of consistent failure to maintain this species in the laboratory. In spite of occasional and transient successes, such as those of Yoeli and Wall (1951), Box *et al.* (1953), and Perez-Reyes (1953), it was not until 1965 that Yoeli and his colleagues at the New York University School of Medicine realized that the secret of successful transmission of *P. berghei* lay in maintaining the vector at temperatures of 21°C or less throughout the period of sporogonic development (Vanderberg and Yoeli, 1966). The temperatures used in previous attempts at transmission were generally in the range of 24° to 26°C, temperatures lethal to the late sporogonic development of *P. berghei.* It is ironic that *P. berghei* is the only murine *Plasmodium* suffering this temperature restriction. Had the many different murine malaria parasites which became available in and after the year 1965 been discovered first, it is probable that sporogony would have been achieved in the laboratory and the demonstration of the preerythrocytic stages of these parasites would have taken place many years earlier.

A. TEMPERATURE DEPENDENCE OF SPOROGONY. Sporogonic development of *P. berghei* is generally restricted to temperatures between 16° and 23°C, optimum transmission occurring between 18° and 21°C (Vanderberg and Yoeli, 1966; Bafort, 1970c). At temperatures below 14°C or above 24°C, transmission is completely inhibited. In contrast to *P. berghei,* all other murine malaria species have been shown to be capable of complete sporogonic development at temperatures at least as high as 28°C (e.g., Wery, 1968; Bafort, 1971a). Optimum development in *P. yoelii, P. vinckei,* and *P. chabaudi* is generally considered to occur between 24° and 26°C.

B. VECTOR CHARACTERISTICS. Numerous species of mosquito have been tested as potential laboratory vectors of murine malaria parasites. None but a small number of anopheline species have proved to be capable of supporting cyclical transmission. Not all anophelines capable of sustaining oocyst formation are able to supply infective sporozoites. Among laboratory-reared species *A. stephensi, A. quadrimaculatus, A. atroparvus, A. sundaicus, A. aztecus, A. freeborni, A. gambiae, A.,* and *A. albimanus* have all at one time or another been recorded to be capable of acquiring oocyst infections with one or more murine malaria species. Among these, *A. albimanus, A. freeborni,* and *A. atroparvus* achieved only very light and rare oocyst infections and have never sustained cyclical transmission (Yoeli, 1965; Wery, 1968; Killick-Kendrick, 1975). Cyclical transmission has occasionally been achieved with one or another of the murine malarias with *A. gambiae, A. aztecus,* and *A. sundaicus* (Bafort, 1971a; Perez-Reyes, 1953; Wery, 1968).

Two species, *A. quadrimaculatus* and *A. stephensi,* rank as the most successful laboratory vectors of murine malaria. *Anopheles quadri-*

maculatus supports heavy gut infections with *P. berghei* and *P. vinckei* subspecies (up to several hundred oocysts per gut and infection rates of 50% or greater) (Yoeli, 1965; Bafort, 1971a; Killick-Kendrick, 1975). However, conversion to infective sporozoites is not efficient in the case of *P. berghei* (not more than one to two hundred sporozoites per mosquito with a conversion rate from gut to gland infection of 75%) and is very poor, indeed, for *P. vinckei.*

Anopheles stephensi is without doubt the most useful laboratory vector and is capable of transmitting members of all species of murine malaria parasite. Although oocyst counts with *P. berghei* infections of *A. stephensi* may be slightly lower than those in *A. quadrimaculatus* (100–200 oocysts per gut with a 40% infection rate compared to 100–500 oocysts per gut with a 40–80% infection rate), the conversion to infective sporozoites is many times more efficient. Thus Yoeli (1965) records that 95% of *A. stephensi* carrying gut infections of *P. berghei* acquired sporozoite infected glands with mean numbers of 7000 to 8000 sporozoites per mosquito. Wery (1968) recorded slightly lower oocyst counts in *P. yoelii* infections of *A. stephensi,* 10–150 oocysts per gut. *Plasmodium chabaudi,* which has not been transmitted by any laboratory raised mosquito other than *A. stephensi,* gave gut infections in 30 to 70% of mosquitoes fed, with a range of 20 to 50 oocysts per gut (Wery, 1968). Conversion to infective sporozoites is comparatively more efficient for *P. chabaudi* than it is for *P. yoelii.* Wery (1968) estimated that 25–35% of *P. chabaudi* sporozoites were infective compared to 2–25% of sporozoites of *P. yoelii* in the *A. stephensi* system. After a certain amount of adaptation, Bafort (1971a) was able to achieve gut infections with *P. v. vinckei* in almost 100% of *A. stephensi;* gland infections of several thousand sporozoites were attained in some mosquitoes.

C. PARASITE CHARACTERISTICS. It is generally true that the berghei group parasites are more readily transmitted through *A. stephensi* than are members of the vinckei group. In the berghei group, there is a tendency for oocysts to degenerate before reaching maturity and for sporozoites reaching the salivary gland to be of low infectivity. In compensation, both oocyst and gland sporozoite counts of berghei group species are generally high.

The transmission potential of the subspecies of *P. vinckei* is highly variable. High oocyst counts in *A. stephensi* have been achieved for both *P. v. vinckei* and *P. v. brucechwatti.* Sporozoites of both subspecies may reach the salivary glands in high numbers. However, while sporozoites of *P. v. vinckei* readily infect a suitable host, those of *P. v. brucechwatti* are almost always totally sterile. Both *P. v. lentum* and *P. v. petteri* are satisfactorily transmitted by *A. stephensi.* Little information is available on the characteristics of the mosquito infection with these two subspecies.

The comparatively poor infectivity of *P. chabaudi* subspecies transmitted by *A. stephensi* is due to the rather low-grade gut infections (associated with generally very low levels of exflagellation in the blood of the host). As we have already seen, however, the sporozoites that reach the glands are of high infectivity.

D. CHARACTERISTICS OF THE BLOOD INFECTION. As important for successful transmission as the vector–parasite combination and the temperature of sporogony is the nature of the blood infection in the host upon which the mosquitoes feed. It is a characteristic of murine malaria parasites, not shared with some avian and primate species, that multiple blood passages reduce their infectivity to mosquitoes. As few as ten serial blood passages may render a line of parasites incapable of infecting mosquitoes while continued passage without transmission results in total loss of gametocytes. This situation certainly contributed to the difficulties met with by early investigators in their attempts to transmit *P. berghei*. Sporozoite-induced infections, however, are not themselves ideal material for reinfecting mosquitoes. It is generally considered that infections arising in the first blood passage after a sporozoite-induced infection are the most suitable for transmission through mosquitoes (Wery, 1968).

Infectivity to mosquitoes is also strongly dependent upon the time point in the course of an infection at which the mosquitoes are allowed to feed. The berghei and vinckei group parasites exhibit two, more or less distinct, patterns for the optimum time in an infection for successful mosquito transmission. For *P. berghei* and *P. yoelii* the best time to feed mosquitoes is on the third and fourth days of a blood-induced infection. Beyond the fifth day, infectivity falls virtually to zero for the remainder of the infection in spite of the fact that gametocytes continue to be present and may even increase in number.

Plasmodium chabaudi is most infective to mosquitoes just after the crisis of infection, around the seventh to tenth day, although transmission may be achieved rather more irregularly before crisis during the rising phase of the parasitemia (Wery, 1968). *Plasmodium vinckei* is also infective to mosquitoes during the postcrisis period (Bafort, 1969c), but transmits to mosquitoes rather more readily than *P. chabaudi* during the early stage of a blood infection.

2. Host Susceptibility

A. SUSCEPTIBILITY TO INFECTION BY BLOOD STAGE PARASITES. The blood stages of murine malaria parasites, mainly *P. berghei*, have, at various times, been inoculated into a wide variety of vertebrates. With the exception of the successful infection with *P. berghei* of splenectomized primates, *Macaca mulatta, Macaca speciosa,* and marmosets, reported by Wellde *et al.* (1966), these parasites have been found to cause blood infections only in certain species of rodents and bats and some avian embryos.

Man (Bafort, 1971a), ungulates, such as lambs, pigs, and calves (Durbin, 1951), carnivores, such as puppies, kittens (Durbin, 1951), and mongoose (Satya Prakash *et al.*, 1952), marsupials, such as the bandicoot (Satya Prakash *et al.*, 1952) and the opossum (Wellde *et al.*, 1966), and chicks (Durbin, 1951) have been found to be totally insusceptible to infection with murine malaria. Adult rabbits and guinea pigs are likewise almost totally refractory to infection with any species of murine malaria parasite (Garnham, 1966; Bafort, 1971a). McGhee (1954) succeeded in infecting the erythrocytes of duck and goose, but not chick, embryos with *P. berghei.* Cox (1964) found that a large number of different rodent species, mainly *Myomorphs,* e.g., murine rodents (rats and mice) and a few cricetine rodents (hamsters and cotton rats), but also some *Sciuromorphs* (squirrels), are susceptible in various degrees to blood-induced infections of *P. berghei.* A few species of megachiropteran bats, including fruit bats, have been successfully infected.

Among myomorphan rodents, the host range of berghei group parasites is apparently somewhat wider than for the vinckei group species. Whereas the five rodent species used most commonly in malaria research, the murine species *Mus musculus* (the laboratory mouse), *Rattus norvegicus* (the laboratory rat) and *Grammomys surdaster* (the African thicket rat), and the cricetine species *Mesocricetus auratus* (the golden hamster) and *Sigmodon hispidus* (the cotton rat), are all susceptible to blood-induced infections with berghei group parasites, only *M. musculus* and *G. surdaster* readily succumb to blood-induced infections with parasites of the vinckei group. By adaptation, however, these parasites may become infective to rats and hamsters (see Section III,G,2,c below). Unlike berghei group parasites those of the vinckei group do not give rise to blood infections in *Microtis agrestis* (the field vole), *Clethrionomys glareolus* (the field mouse), *Acomys cahirinus* (the spiny mouse), or fruit bats (Bafort, 1971a).

B. SUSCEPTIBILITY TO SPOROZOITE-INDUCED INFECTION OF THE LIVER. The pattern of susceptibility in different rodents is not necessarily the same for a blood-induced infection as it is for a sporozoite-induced infection of the liver. A wider range of rodent species is susceptible to sporozoite-induced liver infection by a wider range of parasite species than is the case for blood-induced infections. Those species which sustain blood infections are usually also susceptible to sporozoite-induced infection. Thus *M. musculus* and *G. surdaster* acquire blood infections following inoculation of either sporozoites or blood of parasites of the vinckei group. Likewise, normal blood infections with parasites of the berghei group follow inoculation of either sporozoites or blood stages into *R. norvegicus, M. auratus, G. surdaster,* or *S. hispidus.* Many strains of laboratory mouse, however, are relatively refractory to infection with sporozoites of berghei group parasites, although all are highly susceptible to blood infection.

Such refractoriness to sporozoite-induced infection in an animal highly susceptible to blood-induced infection is unusual. By contrast, a number of instances have been recorded in which animals refractory to blood-induced infection have acquired liver infection following the inoculation of sporozoites. Thus, Bafort (1971a) found that *Mastomys natalensis, M. agrestis, C. glareolus, Peromyscus maniculatus, A. cahirinus, M. auratus,* and *R. norvegicus* all developed apparently normal liver infections following inoculation of sporozoites of *P. v. vinckei.* None of these animals was capable of sustaining a blood infection of this parasite. Likewise, guinea pigs and rabbits, which are essentially refractory to blood-induced infections of both *P. berghei* and *P. yoelii,* were found to devêlop liver schizonts (but not blood infections) following inoculation with sporozoites of *P. yoelii* (Wery, 1968).

As in the case of susceptibility to blood stage infection, the vinckei group parasites appear to be more restricted than those of the berghei group in the range of animals susceptible to sporozoite-induced infection. Thus guinea pigs, rabbits, and cotton rats are not only insusceptible to blood-induced infection with *P. vinckei* but also fail to develop liver schizonts following inoculation of sporozoites of this parasites (Bafort, 1971a).

C. FACTORS AFFECTING HOST SUSCEPTIBILITY. Not all members of a particular species of rodent may be equally susceptible to infection with a given malaria parasite. In a series of papers by Greenberg and his colleagues (Greenberg *et al.*, 1953; Greenberg, 1956; Greenberg and Kendrick, 1958; Nadel *et al.*, 1955) different inbred strains of mice were reported to differ markedly and consistently in the virulence of infection sustained with *P. berghei.* In cross-breeding studies these workers demonstrated major hereditary differences between the mice in their susceptibility to infection with this parasite. Most *et al.* (1966) and Vincke and Bafort (1968) showed that inbred strains of mouse differed widely in their susceptibility to infection with sporozoites of *P. berghei;* while some strains were almost totally refractory, others were highly susceptible. The relative susceptibility of the different mouse strains was not affected by the species of anopheline used to supply the sporozoites.

The patterns of host susceptibility indicated above are not immutable. Splenectomized rats are susceptible to infection with either *P. vinckei* or *P. chabaudi.* After several serial passages in splenectomized animals, these parasites become so adapted that they are able to infect and develop high parasitemias in intact laboratory rats. Bafort (1971a) was able to infect golden hamsters with *P. v. vinckei* after passage through a chinese hamster, *Gricetulus griseus.* Adler and Foner (1961) adapted *P. v. vinckei* to splenectomized hamsters. Animals refractory to sporozoite infection, however, have never been rendered susceptible by splenectomy.

The age of the host is often of great importance in determining the susceptibility of an animal to infection. Adult rats, more than 7 or 8 weeks of age, normally survive *P. berghei* infection (Singer *et al.*, 1955) and are completely refractory to infection with *P. vinckei.* Young rats, on the other hand, succumb to fulminating infections and may show low-grade parasitemias with *P. vinckei.* Whereas adult rabbits are quite refractory to infection with *P. berghei,* newborn rabbits develop parasitemias which may rise as high as 5% (Garnham, 1966).

3. Characteristics of the Blood Infection

A. COURSE OF INFECTION. The course of blood infections of murine plasmodia is determined by both the parasite species and strain and its host. In the most susceptible hosts, e.g., most strains of mice and young rats, *P. berghei* generally runs a fulminating and fatal course. In the earlier stages of the infection, the parasites invade mainly the reticulocytes and polychromatophilic erythrocytes as a result of which there is a rapid depletion of immature red blood cells. In the most susceptible animals, young rats and certain strains of mice, parasitemias in blood-induced infections may reach 80–90% within 5 to 6 days. By this time, following the depletion of the immature red cells, mainly mature cells are infected, a dramatic fall in hematocrit is encountered, and death rapidly supervenes. In the most resistant strains of mice and in adult rats, an entirely different course of infection is presented. Parasitemias rise much more slowly, erythropoiesis is able to replenish reticulocytes as they are eliminated, and the parasites remain in immature red cells throughout the course of the infection. High parasitemias of 30 to 50% may not be reached until the second or third week of infection. Shortly after this, a wave of deaths may occur in the more resistant mouse strains although adult rats almost invariably recover fully.

Most strains and subspecies of *P. yoelii* run a comparatively benign course in laboratory rodents similar to that seen with *P. berghei* in adult rats and resistant mouse strains. Like *P. berghei* these parasites show a strong preference for immature red cells. Because of the slow rise in parasitemia, the young blood cells are not depleted but are able to rise steadily in response to the erythropoietic stimulus provided by the general destruction of red blood cells. Parasitemia closely follows reticulocytemia and rises steadily from 1 to 5% on the third or fourth day of blood-induced infection to 20–60% after 2 to 3 weeks. At this time some deaths may occur; generally, however, the parasites clear rapidly in the third or fourth week and full recovery is achieved.

Certain strains of *P. yoelii* have become highly virulent under various circumstances (e.g., Yoeli *et al.*, 1975). These parasites multiply at almost exponential rates through most of the period of infection usually killing mice within 5 to 8 days. Although retaining their intrinsic preference for immature blood cells, these parasites rapidly deplete such cells from the circulation and enter mature erythrocytes. As in the case of other strains and species of murine plasmodia, the degree of virulence of the "virulent" strains of *P. yoelii* is governed to a considerable extent by the genetic constitution of the host.

Plasmodium vinckei infections vary in virulence according to the strain and subspecies of the parasite. Infections of *P. v. vinckei* and *P. v. brucechwatti* are highly virulent in mice and thicket rats; parasitemias rise rapidly to high levels infecting up to 100% of red cells by the end of the first week of infection. Mortality rates of between 70 and 100% within 6 to 10 days are common. Lines of *P. v. lentum,* by contrast, run a rather mild course, parasitemias rarely rising above 20–30%; deaths are rare. *Plasmodium v. petteri* is somewhat more virulent but deaths are also rare. Following the peak of infection, parasitemias fall again rapidly and are generally no longer detectable on thin smears beyond the third week of infection.

The virulence of *P. chabaudi* infections is variable and is dependent upon the strain of parasite and presumably also upon the genetic constitution of the host. Parasitemias generally rise to high levels in mice and thicket rats reaching a peak of 60 to 100% of blood cells infected between 6 and 8 days after inoculation of blood stage parasites. In the more virulent infections many animals may die at or shortly after this time. Following the peak of infection, the parasitemias in the survivors decline rapidly. The parasites are not, however, immediately cleared from the blood and may persist at low levels for several weeks. In milder infections the peak parasitemia may not exceed 20–30% of blood cells infected. Recovery from such an infection is almost invariable.

During the period after the parasites cease to be detectable on blood smears, the host is immune to reinfection with a homologous parasite. For at least 1 or 2 months the parasites persist in the circulation at low levels so that the host is said to be in a state of premunition (infection immunity) (Zuckerman, 1970). Premunition immunity is apparently dependent upon the presence of an intact spleen. Removal of the spleen at this stage of an infection leads to a frequently severe relapse and death ensues in some cases. Although some animals appear to remain in a state of premunition for the rest of their lives (up to 2 years) (Sergent and Poncet, 1955; Weiss, 1965), others clear their parasites within 1 or 2 months and a period of apparently "sterile" immunity follows during which the animals are able to resist reinfection or contain a challenge infection at low levels even following splenectomy (Zuckerman, 1953; Corradetti and Verolini, 1957; Kretschmar, 1963b). It is not clear how long such a state of sterile, as opposed to premunition immunity may last.

B. FACTORS AFFECTING VIRULENCE. The virulence of blood infections is frequently greatly enhanced by serial passage following the isolation of a strain of parasite from the wild. In some cases transmission through mosquitoes attenuates the level of virulence in such parasites. The parasitemia may sometimes be profoundly affected by concomitant infection with another organism. The presence of *Eperythrozoon coccoides* leads to delayed and much reduced parasitemias in infections with *P. berghei* (Peters, 1965b) and *P. chabaudi* (Ott and Stauber, 1967). Hsu and Geiman (1952), on the other hand, found that the presence of *Haemobartonella muris* caused heightened virulence of *P. berghei* infections. The effects of concomitant infection of murine malaria with various agents including *Trypanosoma lewisi* (Jackson, 1959), *Leishmania donovani* (Ebert, 1970), *Trichuris muris* (Phillips *et al.*, 1974), viruses (Cox *et al.*, 1974; Yoeli *et al.*, 1955), and schistosomes (Yoeli, 1956) have been recorded. The diet of the host is of great importance in determining the virulence of an infection. Thus, even intrisically highly virulent parasites run a mild course of infection when inadequate levels of PABA are present in the diet of the host. Other extrinsic factors such as environmentally induced stress may affect the resistance of a host to plasmodial infection (Kretschmar, 1965). As already indicated the genetic constitution of the rodent host may be of great significance in determining the virulence of an infection.

H. Pathology

The pathological changes which occur in rodents infected with malaria parasites have been rather extensively studied. In this section, those features of the infection which are commonly observable in the course of experimental work as well as some more subtle changes which have been the subject of specific investigations will be discussed. Not surprisingly, most of the pathological changes are quantitatively related to the intensity of parasitemia (Kreier and Laste, 1967; Kreier *et al.*, 1972a,b) which, in turn, is a function of the precise host–parasite system involved. Some of the more general aspects to be discussed will primarily concern *P. berghei* in mice, the most commonly studied fulminant murine malaria. Intense parasitemia of the degree observed in this host–parasite system results in readily appreciable morphological changes in erythrocytes (Kreier *et al.*, 1972a). In contrast to the blood stage infection, the pre-erythrocytic stages result in no apparent disease and minimal pathological changes; very little will, therefore, be said about them in this context.

1. Gross Pathology

The mouse with terminal malaria due to *P. berghei* exhibits a characteristic appearance. Perhaps the most striking feature is the extreme pallor due to severe anemia. The skin and mucous membranes are of a brown tinge because of the wide spread dissemination of malaria pigment. The animals have the universal ruffled fur of severe disease, and are cachectic in appearance, perhaps due in large part to dehydration resulting from low fluid intake in the terminal period. The appearance of the abdominal viscera is striking; the spleen, especially, is greatly enlarged and of a deep dark brown to maroon color due to the concentration of pigment within it. The liver is also of a dark brown color due to pigment deposits. This discoloration is, indeed, present throughout all the abdominal viscera and serosal surfaces as well as the lungs and brain.

2. Hematological Changes

The most obvious microscopic change in the blood is the presence of parasites within the erythrocytes. There are also striking changes in the cellular elements of the blood as infection proceeds. *Plasmodium yoelii* has been particularly well studied in this regard (Topley *et al.*, 1970). Coincident with the rise in parasitemia there is a progressive fall in both hematocrit and red cell count. There is a slight decrease in the mean corpuscular hemoglobin concentration which is coincident with peak parasitemia. The morphology of the erythrocytes is also altered, both microcytic spherocytes and macrocytes being readily apparent. The latter cells are quite often basophilic, reflecting the intense erythropoietic stimulus of the anemia. In animals that survive the infection, reticulocytosis follows peak parasitemia. The interrelationships between the dynamics of parasitemia and reticulocytosis are complex. The berghei group of parasites are strikingly reticulocytotropic whereas parasites of the vinckei group prefer mature erythrocytes. Many malariologists consider it likely that the availability of reticulocytes is a limiting factor in the development of parasitemia in some host–parasite systems such as, for example, *P. berghei* in the rat. In this situation, essentially all basophilic erythrocytes are parasitized during the early stages of infection and, in fact, the rate of increase and extent of parasitemia can be enhanced by artificial induction of reticulocytosis prior to inoculation with *P. berghei* (Singer, 1954a). The degree of tropism for reticulocytes is variable among the various members of the berghei group and, in general, is less pronounced in the more virulent host–parasite systems. In the most fulminant infections, reticulocytosis does not have time to occur prior to the death of the host. In some systems, however, artificially induced polycythemia can completely block an otherwise lethal infection (Ladda and Lalli, 1966).

Leukocytosis is also a feature of murine malaria and is predominantly due to an increase in monocytes associated with an increase in atypical mononuclear cells (Topley *et al.*, 1970). Platelets are also influenced by malaria. There is a modest thrombocytopenia in the *P. yoelii* model, but more striking is a pronounced enlargement of platelets in a number of malarias including *P. berghei.* A very surprising feature is the presence of parasites *(P. berghei)* within platelets (Fajardo, 1973). It is not clear whether the parasites gain entry into platelets by active penetration or by phagocytosis. Parasites have been observed within platelets both with and without evidence of degenerative changes.

Although the destruction of erythrocytes accompanying the development and release of malaria parasites would appear to be an obvious cause of anemia in malaria, a number of investigators have collected evidence that the degree of anemia may be out of proportion to that which can be explained by direct destruction by the parasites (Greenberg and Coatney, 1954; Zuckerman, 1966). A large body of literature has accumulated on other possible mechanisms for the destruction of erythrocytes during malarial infection. In some of these studies, inadequate evidence for the existence of alternative mechanisms of erythrocyte destruction is presented. However, it is clear that erythrophagocytosis of uninfected cells is a very common phenomenon in murine malaria (Zuckerman, 1966), although its impact on the overall status of the erythron is not clear. Other modes of destruction of erythrocytes, such as direct intravascular extracellular hemolysis, have not been ruled out. In any case, the underlying mechanisms by which erythrophagocytosis and/or extracellular hemolysis may be promoted have not been elucidated. Immune reactions against parasitized cells will not be considered in this discussion since such reactions would result, for most part, in the same degree of erythrocyte destruction as that due to direct parasitization; this topic is discussed below in relationship to immunity.

There are three major categories of pathways by which erythrophagocytosis and/or hemolysis of uninfected erythrocytes would be promoted. The first of these is a specific immune response against antigen residing on the surface of the erythrocytes. Three categories of antigens can be envisaged. (a) Antigen derived from parasites could arrive at the surface of the erythrocyte through the plasma and become attached to it thus providing a passively sensitized target for antibodies or cells directed against the parasite antigen. (b) If there were cross-reactions between erythrocytes and parasites and an immune response to these cross-reacting antigens occurred, sensitization of erythrocytes for subsequent destruction could be explained. (c) A true autoimmune hemolytic reaction could also result in red cell destruction; in this case, either a change in the antigen or in the immune system itself would be required. The data relevant to these points are still not definitive. Coomb's reactions have been reported with erythrocytes from animals infected with *P. berghei* (Zuckerman, 1960) but might well have been false positive due to agglutination of reticulocytes, as was demonstrated in a later study (Zuckerman and Spira, 1961). However, there is evidence for agglutination of erythrocytes obtained from uninfected animals and treated with trypsin prior to testing with serum from animals with *P. berghei* malaria (Cox *et al.*, 1966; Kreier *et al.*, 1966). Trypsin treatment may uncover reactive sites not otherwise exposed, such an alteration in the erythrocyte surface being accomplished during the course of malaria through enzymes derived from the parasites. Normal red cells are apparently not affected by serum factors; neither acute nor immune serum from *P. berghei*-infected mice influences survival of normal autologous erythrocytes in syngeneic recipients (George *et al.*, 1966; Kreier and Leste, 1968). Studies of the erythrocyte membrane during *P. berghei* infections suggest that there may be changes in protein constituents which could conceivably alter their antigenicity thus allowing the development of autoimmunity (Weidekamm *et al.*, 1973).

A second possible mechanism for the preparation of erythrocytes for subsequent destruction during malaria is similar to the first but involves damage by a toxic substance(s) rather than a specific immune reaction. It has been show that a substance is elaborated during the course of primate malaria which increases the osmotic fragility of erythrocytes (Fife *et al.*, 1972). Although no such factor has been demonstrated in murine malaria, the increased osmotic fragility of uninfected erythrocytes has been well documented (Fogel *et al.*, 1966). If such a toxic factor is produced, the phenomenon offers an additional potential explanation for excessive red cell destruction during malaria.

The third major category of effects which can be postulated to increase destruction of erythrocytes during malaria is hyperfunction of phagocytic elements. Macrophage activation apparently does take place during malaria as discussed in Section III,I,3 and could well result in erythrophagocytosis as observed in the spleen of animals with *P. berghei* malaria. This increased phagocytic capacity, which, in the context of splenomegaly such as that seen in malaria, constitutes the hypersplenic state, could conceivably be active against completely normal erythrocytes but would be expected to be even more effective against erythrocytes which have been damaged in one or other of the ways described above (George *et al.*, 1966).

Another alteration of erythrocytes which might result in greater fragility, and thus destruction, is the pitting or removal of parasites from infected red cells in the spleen during the passage of the cells into the venous sinuses from the splenic cords (Schnitzer *et al.*, 1972). This curious phenomenon results in small spherocytes which are more susceptible to hemolysis than red cells of normal size.

A somewhat trivial explanation of anemia to related splenomegaly is the fact that in the presence of an expanded blood volume, due to the dilation of splenic venous sinuses with secondary hemodilution, the erythrocyte concentration declines, a phenomenon which would be interpreted as a loss of erythrocytes as estimated from peripheral blood samples.

The fact that deep vascular schizogony can occur in murine malaria (Miller and Frémont, 1969) might lead to an erroneously low estimate of the extent of parasitization of an animal, so that blood destruction by parasites in deep vasculature might lead to significant anemia but prevent an accurate interpretation of its cause.

3. Changes in Blood Chemistry

A number of changes occur in blood chemistry during the course of experimental murine malaria. Among the most profound is the hypoglycemia which develops almost certainly as a result of rapid consumption of glucose by the parasites (Sadun *et al.*, 1965). Other changes include a moderate decrease in albumin and alkaline phosphatase. Increases in transaminase are probably due to liver damage, since blood levels of ornithine carbamoyltransferase, a highly specific indicator of liver damage, are also elevated (Einheber *et al.*, 1967). An increase in Bromsulphalein retention (Sadun *et al.*, 1965) also favors this interpretation.

4. Pathology of the Liver

Although the liver is the site of the preerythrocytic stages of murine malaria parasites, no discernible disease is caused by this stage of the infection; there is little histopathology to be observed except for the presence of the parasites themselves within the hepatic parenchymal cells. However, an inflammatory response to preerythrocytic stages, appearing as small grossly observed foci, has been described.

During the blood stages of the infection with *P. berghei* in mice a number of hepatic changes occur. Of interest is the hypertrophy of the

Kupffer cells with striking collection of malaria pigment both in mice (Singer, 1954a) and hamsters (MacCallum, 1969a). There is an early inflammatory response in the interlobular areas which converts to areas of erythropoiesis somewhat later in the infection. There is depletion of liver glycogen early in the infection (Mercado and von Brand, 1954), followed by extensive fatty infiltration in depleted areas (von Brand and Mercado, 1958). Administration of adrenal corticosteroids can partially reverse the glycogen depletion indicating that impaired synthesis may be at least in part responsible for the phenomenon (Mercado and von Brand, 1954). The extensive consumption of glucose mentioned above may also play a role.

Mice with *P. berghei* infections contain a factor which inhibits liver mitochondrial function *in vitro* (Riley and Maegraith, 1962). Although this factor is presumably not specific for the liver enzymes, it might explain some of the malfunction of this organ during the disease.

5. Renal Pathology

A variety of types of renal abnormality have been described in rodent malaria. Infection with *P. berghei* results in retention of blood urea nitrogen as well as a defect in the excretion of phenolsulfonphthalein. Hemoglobinuria occurs in mice and hamsters (Sesta *et al.*, 1968). Histologically there is an accumulation of pigment-laden macrophages and parasitized cells in the veins of the organ, as well as fatty change of the parenchymal cells. In addition to these changes, which are explicable in terms of the presence of intravascular hemoglobulin due to hemolysis and anoxia, respectively, there is evidence for immune complex nephritis as well. Mouse immunoglobulin, β-1-C globulin, and malaria antigen have been detected in complexes localized in the glomerular capillary walls (Boonpucknavig *et al.*, 1972). Globulin with specificity for *P. berghei* was eluted from these kidneys. Proteinuria was demonstrated in these animals as well, although the contribution of hemoglobin to the measurements is not clear.

6. Pulmonary Changes

Plasmodium berghei infection results in a striking accumulation of pigment-laden phagocytic cells within the pulmonary alveolar capillaries. This is true both in mice (Singer, 1954a) and in hamsters (MacCallum, 1968), but, in the latter case, there is apparently more extensive obstruction of the pulmonary vasculature and more profound sequelae of the phenomenon. In the hamster, endothelial overgrowth by masses of macrophages with partial venous occlusion occurs. Subsequently, there is lymphatic dilation and stasis with coagulation of lymph and resulting obstruction of the lymphatic channels followed by pulmonary edema. After chemotherapy, macrophage migration to the mediastinal lymph nodes brings about complete resolution of the pulmonary lesions (MacCallum, 1969b).

7. Central Nervous System Pathology

Typically, few specific changes are seen in the brain in murine malaria. There are three host–parasite systems, however, in which striking central nervous system changes can be seen. Young hamsters infected with *P. berghei* die approximately 1 week after infection with relatively low parasitemias. Cerebral hemorrhages similar to those found in human cerebral malaria are seen in these animals. Grossly, the brains show wide spread petechial hemorrhages and, both grossly and microscopically, congestion of the cerebral capillaries by aggregates of parasitized as well as nonparasitized cells are seen. Necrosis of neurons is associated with these areas of hemorrhage and occlusion. A striking feature of this syndrome is the fact that it can be inhibited by neonatal thymectomy or by the administration of anti-thymocyte serum. It has been postulated that the aggregates which occlude the cerebral capillaries are immune agglutinates and that interference with the immune response inhibits the production of the antibodies involved (Wright *et al.*, 1971).

A lethal line of the 17X strain of *P. yoelii* also causes striking intravascular sequestration of erythrocytes in the brain. In this situation, however, death occurs with high parasitemia so the influence of the cerebral lesions on mortality is not clear (Yoeli and Hargreaves, 1974).

A third central nervous system syndrome is the paralysis observed in rats infected with a line of *P. berghei* (Mercado, 1965). This syndrome is associated with extensive focal cerebral hemorrhages occurring approximately 1 week after inoculation and is fatal in most animals. The parasitemia is usually less than 10% in animals which succumb. Curiously, this syndrome is inhibited by prior splenectomy of the animals (Mercado, 1973). The possibility that this syndrome is due to a contaminating infection has not been excluded.

8. Response of the Spleen, Thymus, and Lymphocytes

Grossly, the most dramatic pathological event in murine malaria is splenomegaly, as mentioned above. In addition to the great enlargement of this organ, there may be frank infarcts of the spleen, particularly in hamsters. The microscopic appearance of the organ is equally striking. Dynamic changes occur in the spleen which are as marked as those observed in the peripheral blood (Singer, 1954a). In *P. berghei* infections there is an increase in both white and red pulp with the hyperplasia of the white pulp consisting primarily of an increase in mononuclear cells, probably both lymphocytes and macrophages. The increase in size of the red pulp is due primarily to congestion in the early stages of the infection. As the infection develops there is onset of active erythropoiesis within the splenic cords and simultaneous decrease in the size of the white pulp. Phagocytosis by splenic macrophages begins early in the infection, as indicated by extensive erythrophagocytosis and deposits of malaria pigment within these cells, and continues throughout the course of the infection.

Changes in the lymph nodes are much less pronounced in keeping with the fact that these organs, unlike the spleen, serve the extravascular spaces rather than the blood. However, there is a response which includes swelling of the medullary cords due to increases in a variety of cell types, among them granulocytes. Later, there is a generalized hyperplasia of lymphatic tissue with a relative depletion of small lymphocytes and the development of large lymphoid cells in the germinal centers. Macrophages with ingested malarial pigment can also be seen within the lymph nodes later in the course of the infection.

The thymus changes little in early infection, but late in the course of the disease there is involution due to loss of thymocytes; coincident with this is an accumulation of cellular debris.

In all of these organs there is evidence for a loss of small lymphoid cells. This has been studied further from the point of view of the subpopulations of cells involved in mice (Krettli and Nussenzweig, 1974) and rats (Gravely *et al.*, 1976). Both T cells and complement receptor lymphocytes are markedly depleted in both thymus and lymph nodes, but there is a simultaneous increase in lymphocytes which bear neither of these markers in mice and young cats. T cell depletion is not severe in mature rats, however. The possible relationship between these phenomena and the immunosuppression and age-related immunity observed in malaria is of interest.

9. Immunosuppression

A number of workers have reported the suppression of the immune response to a variety of antigens, including erythrocytes, bacteria, viruses, and soluble proteins, during the course of *P. berghei* and *P. yoelii* infections. Although cell-mediated responses generally tend to be spared (Greenwood *et al.*, 1971), graft rejection may also be impaired (Sengers *et al.*, 1971). The spontaneous development of autoimmune disease in NZB mice can be prevented by concurrent malaria infection (Greenwood and Greenwood, 1971). Similarly, murine lymphomas due to the Maloney lymphomagenic virus develop earlier and more frequently in mice infected with *P. yoelii* as compared with controls which receive only the virus (Wedderburn, 1974).

The mechanisms of this immunosuppression by malaria infection are unclear. The depletion of lymphocytes described above may be germane. The report of a remarkable proliferative response of T cells in mice which, although infected with *P. yoelii,* were not immunosuppressed, supports this view (Jayawardena *et al.*, 1975). Proliferation of T cells was not found in mice infected with *P. berghei* in which immunosuppression occurred. It has been suggested, however, that the defect may lie not at the level of the lymphocyte but rather with the processing of antigen by macrophages (Greenwood *et al.*, 1971) since it has been observed that those antigens to which the response is not suppressed are those that do not depend on macrophage processing. Evidence interpreted to support this hypothesis has been obtained (Loose *et al.*, 1972); it was observed that macrophages from animals with malaria could phagocytize sheep red blood cells, but the plaque-forming cell response to intravenous injections of these macrophages with their enclosed erythrocyte antigen into normal recipients was less than that when the macrophages were obtained from normal donors. In these experiments, interpretation of the data is limited since it is not clear how many lymphocytes were also transferred.

I. Immunity to Murine Malaria

1. Immune Response during Active Infection

The immune response to murine malaria parasites is distinctive for each particular host–parasite combination. As in other aspects of murine malariology, *P. berghei* is the most studied species. In its natural host, *G. surdaster,* there appears to be an excellent host–parasite equilibrium with relatively low parasitemias maintained over a long period of time favoring the chronic coexistence of host and parasite. In most experimental hosts, on the other hand, this balance has not been achieved and there is usually a relatively rapid resolution of the infection terminating either in the eradication of the parasite or the death of the host. Although, in mice, death usually supervenes, there are differences in the response of various strains of mice to *P. berghei* infection as studied in detail by Greenberg and his associates (Greenberg *et al.*, 1953, 1954; Greenberg and Coatney, 1954; Greenberg, 1956; Greenberg and Kendrick, 1958). They observed a spectrum of responses, as reflected by the extent of parasitemia and times of death, in a series of inbred mouse strains and their hybrids. In certain strains and hybrids, survival was considerably prolonged as compared to that in the most susceptible mice, but almost all eventually succumbed to the infection. Mice are, therefore, essentially incompetent with respect to their ability to mount an effective immune response to challenge with *P. berghei.* As will be detailed below, however, this can be overcome by artificial immunization

Rats have a quite different response to *P. berghei.* Whereas young rats uniformly succumb to the infection, older animals develop an increasingly greater immune competency until, at maturity, there is only a mild and transient parasitemia in response to challenge (Zuckerman and Yoeli, 1954; Singer *et al.*, 1955). The animals are also solidly immune to rechallenge. The reasons for this age dependence of susceptibility in rats has not been definitely clarified. Two main hypotheses have been elaborated to explain it. On the one hand, it has been suggested that old rats are not stimulated to produce reticulocytes to as great an extent as younger rats in malarial infection; since the parasite is primarily limited to reticulocytes, the paucity of these cells maintains the infection in older rats at a low level (Zuckerman and Yoeli, 1954). This hypothesis is supported by the fact that a significant enhancement in parasitemia can be induced in old rats by the artificial induction of reticulocytosis with phenylhydrazine or by bleeding (Singer, 1954a). However, Singer *et al.* (1955) have expressed the view that these effects influence the parasitemia only before the onset of active immunity. Other workers (Gravely *et al.*, 1976) argue that age immunity is based on a more fundamental immunological defect in young animals and support this contention with the fact that severe malaria can be induced in old rats by treatment with antilymphocyte serum (Smalley, 1975). The transfer of spleen cells from old to young rats has thus far failed to modify the course of subsequent *P. berghei* infection in the recipients. However, Folch and Waksman (1974) report that spleens of young rats contain large numbers of suppressor "T" cells. Age immunity has also been described in mice and hamsters (Hsu and Geiman, 1952) but is not so pronounced as in rats. The immune response to *P. berghei* in adult hamsters is intermediate between that seen in mice and rats; the animals usually die after a protracted infection.

Of other murine malarias, *P. yoelii* is generally the least virulent in most strains of mice. After a comparatively low-grade parasitemia, commonly rising no higher than about 20%, the animals usually make a complete recovery associated with solid immunity to challenge. This generalization does not always hold, however; the 17X strain of *P. yoelii* has been observed to undergo a spontaneous shift to a virulent state in which 100% of mice may die with overwhelming parasitemias within a week to 10 days (Yoeli and Hargreaves, 1974). Moreover, continous passage of the so-called nonlethal form of *P. yoelii* in Balb/c mice can result in a host–parasite situation involving 100% mortality of host animals. *Plasmodium chabaudi* and *P. vinckei* are similar to the berghei group of parasites with respect to the immune response during infection, although significant quantitative differences are present (Cox, 1966, 1970).

Very little data are available regarding the immune response to moderate numbers of viable sporozoites. This is in sharp contrast to the situation with respect to artificially acquired immunity to sporozoites, as will be discussed below.

Aspects of the immune response during active infection other than resistance to challenge have been documented in some detail. Peripheral blood leukocytes undergo a brisk proliferative response during malaria infection, as demonstrated by increased incorporation of tritiated thymidine (Golenser *et al.*, 1975). Even more striking is the proliferative response of splenic T lymphocytes (Jayawardena *et al.*, 1975). The almost universal splenomegaly which is observed in acute malaria is, in part, due to the proliferation of lymphoid cells and/or the recruitment of phagocytic cells from the periphery. Antibody production has been demonstrated by a variety of tests and is present in high titers even when there is a fatal termination to the infection (Finerty *et al.*, 1973). The lack of correlation between the immune response as measured by fluorescent antibody, complement fixation, precipitation or hemagglutination tests with resistance to challenge is an important fact. Protective humoral antibody can, however, be demonstrated by appropriate *in vivo* tests (Diggs *et al.*, 1972; Hamburger and Kreier, 1975, 1976). Presumably the predominant antigens reactive in serological tests are somatic antigens unassociated with the mechanisms of immunity to challenge.

2. Artificially Acquired Immunity

Artificially acquired immunity to murine malarias is a subject which has been given a great deal of attention during the last decade. Rodent models have provided much of the information responsible for current understanding of immunity to malaria, in general. Artificial immunization against *P. berghei* has been demonstrated with both sporozoites and blood stage immunogens inactivated with ionizing radiation. Similar procedures have subsequently been used against human malaria parasites.

A. ARTIFICIAL IMMUNIZATION WITH SPOROZOITES. Sporozoites have been used with particular effectiveness as immunogenic agents. Sporozoites of *P. berghei* irradiated with 17,000 rads from an X-ray source and then injected into A/J mice gives rise to a striking immunity to challenge with nonirradiated sporozoites (Nussenzweig *et al.*, 1969a). A similar protection can be induced by the use of nonirradiated sporozoites followed by the administration of a chemotherapeutic agent which is effective against the blood forms of the parasites; the recipient animals thus undergo a full preerythrocytic infection but experience no blood invasion (Verhave, 1975).

On subsequent sporozoite challenge, animals immunized with sporozoites demonstrate either complete immunity or behave as unimmunized animals. Thus, although a high proportion of animals demonstrate no parasitemia on challenge, those that break through develop a fatal infection. There is no cross-protection against blood parasites; if animals fully immune to sporozoite infection are challenged with blood forms, the animals exhibit typical fatal disease. Animals immunized against sporozoites of *P. berghei* are, however, also immune to challenge with those of *P. vinckei* (Nussenzweig *et al.*, 1969b) thus indicating a higher degree of cross-reactivity between the sporozoites of these species than among the blood forms. This method of immunization has been experimentally applied to man with successful immunization against *P. falciparum* and *P. vivax*, thus demonstrating the utility of the rodent model as a predictor of immunological phenomena in man (Clyde *et al.*, 1973).

B. ARTIFICIAL IMMUNIZATION WITH PREERYTHROCYTIC FORMS. Very little work has been done with preerythrocytic forms of murine malaria parasites as immunizing agents. In one study, however, the preerythrocytic forms of *P. fallax,* a bird malaria parasite, were found to protect mice from subsequent challenge with sporozoites of *P. berghei* (Holbrook *et al.*, 1976). This striking finding, if repeatable, might signal an important new area for investigation.

C. ARTIFICIAL IMMUNIZATION WITH BLOOD FORMS. Murine rodents can also be artificially immunized against the blood forms of malarial parasites. As in the case of sporozoites, among the most successful immunogens are irradiated parasites. Both rats and mice have been immunized against *P. berghei* with irradiated parasites (Wellde and Sadun, 1967). Other maneuvers which resulted in immunization include the simultaneous administration of parasitized erythrocytes and a chemotherapeutic agent (Wellde *et al.*, 1972), infection of animals on a milk diet (a maneuver which deprives the parasite of PABA and thus suppresses its growth) (Gilbertson *et al.*, 1970), the use of parasite extracts (Zuckerman *et al.*, 1965), and the immunization of animals against virulent strains by prior infection with attenuated parasites (Weiss and deGiusti, 1966). Prior exposure to an avirulent form of *P. yoelii* strain 17X provided striking protection against a virulent line of this parasite (Hargreaves *et al.*, 1975).

In all these experiments the immunity obtained differs from that obtained with sporozoites in that it is relative, usually resulting in animals which sustain at least a low-grade parasitemia on challenge, but typically are able to rapidly clear the infection. In the case of rats, in which the infection is normally self-limiting, immunization causes a further moderation in the intensity of parasitemia due to challenge and in some cases entirely prevents detectable parasitemia. This relative immunity is similar to that encountered after spontaneous self-cure of murine malaria. Immunization with blood forms, like sporozoite immunization, has been extended beyond the rodent models. *Aotus* monkeys can be immunized against *P. falciparum* with irradiated parasitized erythrocytes (Sadun *et al.*, 1969).

3. Mechanisms of Immunity

The mechanisms operative in immunity against malaria are not completely understood. Figure 9 summarizes currently entertained concepts. In the case of sporozoite-induced immunity two *in vitro* tests for antibody may be relevant. The first is the circumsporozoite precipitin test which consists of the development of an immune precipitate around the sporozoites on incubation with immune serum (Spitalny and Nussenzweig, 1973). The second is the sporozoite neutralization test; this is performed by incubation of sporozoites in immune serum followed by injection of

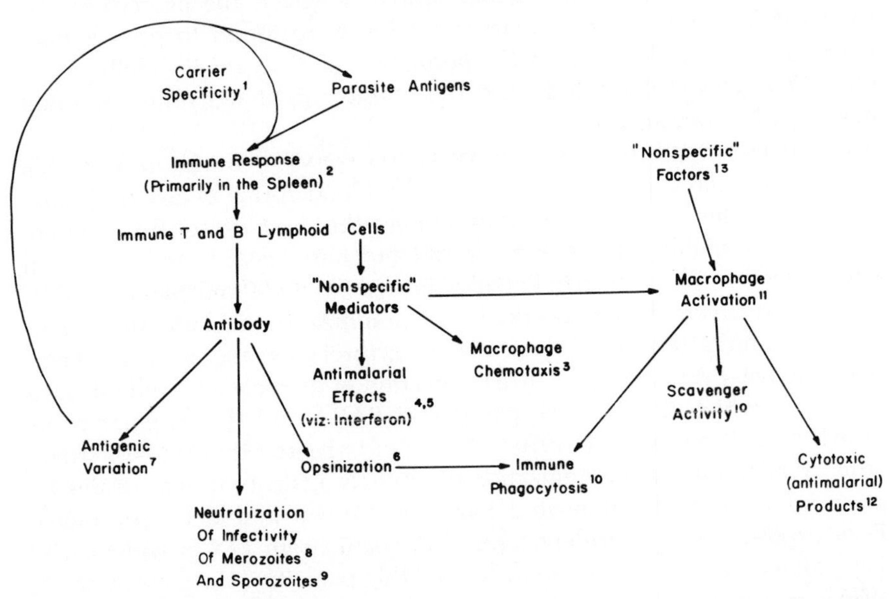

Fig. 9. Putative events in the protective immune response to rodent malaria. Key to the superscript numbers: (1) Brown, 1971; (2) Brown and Phillips, 1974; (3) Wyler and Gallin, 1975; (4) Schultz *et al.*, 1968; (5) Jahiel *et al.*, 1968; (6) Criswell *et al.*, 1971; (7) Briggs *et al.*, 1968; (8) Diggs and Osler, 1973; (9) Vanderberg *et al.*, 1969; (10) Chow and Kreier, 1972; (11) Cox *et al.*, 1964; (12) Clark *et al.*, 1976; (13) Martin *et al.*, 1967.

the sensitized sporozoites into mice for infectivity testing (Vanderberg *et al.*, 1969). It is not known if the two tests involved the same antigen(s) or the same immunoglobulin class(es). Although sporozoite neutralization, by definition, measures antibody with activity against sporozoite infectivity, it is not clear to what extent this is operative *in vivo;* serum which can neutralize sporozoites under *in vitro* conditions cannot completely neutralize the infectivity of sporozoites when passively transferred to mice prior to the administration of infective sporozoites. Thus, there is clearly some escape mechanism operative *in vivo* which allows the sporozoites to survive antiserum which would otherwise render them noninfectious. The *in vitro* phase of sporozoite neutralization is not dependent on heat-labile serum factors. Regarding the *in vivo* phase, the A/J mouse, which has been used for the majority of the studies with sporozoites, is deficient in the fifth component of complement, a component required for most complement-mediated cytotoxic reactions. Just in the month prior to submission of this manuscript has it been found possible to transfer immunity against sporozoites to nonimmune mice with immune lymphoid cells. Similarly, recent data indicate that nonantibody effectors are probably active *in vivo.*

Unlike the situation with sporozoites, *in vitro* serological tests thus far devised give no clue as to the mechanisms of action of immunity against blood forms. Murine malaria parasites have not been as susceptible to analysis as has *P. knowlesi,* a monkey malaria parasite, regarding the effector mechanisms of antiparasitic immunity. Schizonts have not been isolated from murine malaria parasites under conditions which allow a determination of the reactivity of antibody with the surface of these infected cells (Prior and Kreier, 1972; Kreier *et al.*, 1976), nor have viable merozoites been available until recently. Serum transfer experiments in rats and mice indicate that antibody is quite effective in the murine systems and a number of lines of evidence suggest that the merozoite is the target cell which is attacked by antibody (Chow and Kreier, 1972; Diggs and Olser, 1973; Hamburger and Kreier, 1975). Immunity is also transferred via the colostrum and transplacentally from immune pregnant female rats to their offspring (Bruce-Chwatt, 1963). The antiparasitic effect of immune serum is also complement independent in the case of blood forms as determined by the lack of effect of cobra venom factor on serum-mediated protection (Diggs *et al.*, 1972). It is also possible to immunize C5-deficient mice with irradiated blood forms of *P. berghei* and to demonstrate passive transfer of immunity to C5-deficient recipient mice with immune serum from C5-deficient donors.

A possible explanation for the fact that immunity against blood forms is usually of the relative type involves the phenomenon of antigenic variation. Although not as clear as for trypanosomes or *P. knowlesi,* there is good evidence for this phenomenon in murine malaria. If mice are immunized against *P. berghei* and then challenged, the parasites which develop in the immunized animals can be shown to have a different specificity from those in the inoculum as demonstrated by passive transfer experiments with immune serum directed against parasites homologous with the parasites used for inoculation. Whereas parasitemia due to these homologous organisms is delayed by the immune serum, that due to the "variant" organism is not (Briggs *et al.,* 1968).

Induction of immunity to the blood forms of *P. berghei* is thymus-dependent (Brown *et al.,* 1968) and its expression is transferred by cells more efficiently than by serum, at least in the case of the rat–*P. berghei* model (Stechschulte, 1969). This could be interpreted to suggest that cells and serum mediate protection through different mechanisms with the cell-dependent mechanisms being more efficient. An alternative explanation is that both maneuvers result in protection mediated by antibody and that the cells are more efficient because they provide a continuous supply of antibody rather that the limited amount made available in serum transfer experiments. This view is supported by cell transfer studies (Gravely and Kreier, 1976) which show that immunity is transferred with "B" cells and by the kinetics of parasitemia after challenge of animals passively protected by serum or by cells. In animals immunized by passive transfer of immune serum, the onset of parasitemia is delayed; a patent infection eventually appears, however, but then declines, presumably as a result of the onset of active immunity (Diggs and Osler, 1969). In the case of animals immunized by transfer of cells, however, the parasitemia, if present at all, usually rises at a rate similar to that in the controls but is suppressed more rapidly (Stechschulte, 1969). This is consistent with the production of protective antibodies by the transferred cells, as it would be expected that the activity of such antibodies, as expressed by a lowered parasitemia, would be delayed in time.

Antigenic variation is a second factor which probably limits the efficiency of serum-mediated passive immunity. Unless passively supplied antibody is of very broad specificity for many different antigenic variants, it is to be expected that the parasites will escape from the effects of the antibody. In the same way, immune lymphocytes of restricted specificity would have only limited potential for antiparasitic effects either through antibody production or through any other immune effector mechanisms dependent on antigen recognition. However, in the case of cell transfer experiments, the transplant, as well as the lymphoid tissues of the recipient, can actively respond to new antigenic variants as they arise. The subpopulation of lymphocytes responsible for transfer of protection is currently under intense study; while there are suggestions from different model systems that either T or B cells may be active, recent work (Gravely and Kreier, 1976) indicates a major role for B cells. Brown (1971) has proposed that an immune response to a given antigenic variant may prime the immune system for a secondary response to a second variant. This hypothesis presumes that a carrier portion of the molecule is shared by the two immunogens. If this is the case, such carrier specificity would be expected to be transferred with the lymphoid cells but not with serum, thus offering an additional explanation for the greater efficiency of cells.

T cells may have a variety of roles in addition to the helper function during antibody production. Interferon has been shown to be produced during the course of malaria (Huang *et al.*, 1968) and also to exert an antiparasitic effect (Schultz *et al.*, 1968; Jahiel *et al.*, 1968); although other sources are possible, the T lymphocyte is a good candidate for the source of this material. Monocyte chemotactic factor has been demonstrated in the spleen of mice with *P. berghei* malaria, thus suggesting a mechanism for the great influx of mononuclear cells into the spleen during the course of the disease (Wyler and Gallin, 1975). This material may also be derived from T lymphocytes.

In addition to the effector roles of antibody and lymphocytes the system of phagocytes is clearly involved in the response of the host to malaria. This is abundantly evident in the histological response to the infection and is quite likely to be, at least partly, responsible for the very dramatic splenomegaly which accompanies this disease. The central role of the spleen in malaria (Brown and Phillips, 1974) is emphasized by the fact that splenectomy predisposes animals to more severe disease on challenge with malaria parasites and can exacerbate chronic malaria, sometimes to the extent of converting an asymptomatic infection to a fatal one. The system of phagocytes has been studied in malaria by carbon clearance experiments designed to test that ability of the system to take up a putatively inert particle from the blood. In experiments with *P. vinckei*, it has been shown that there is a stimulation of the ability to remove carbon from the circulation during the course of malaria although during terminal malaria there is effective blockade of this ability (Cox *et al.*, 1964). This would suggest that malaria induces a nonspecific hyperphagocytic condition. That malaria is influenced by nonspecific stimulation of the phagocytic system is also suggested by a number of studies in which administration of endotoxin (Martin *et al.*, 1967) or killed organisms such as *Cornybacterium parvum* (Nussenzweig, 1967) result in a lower parasitemia on challenge with malaria parasites. In a similar way, there are data with respect to protection of animals from malaria by infection with *Babesia* (Cox, 1968) in the assumed absence of immunological cross-reactivity between the parasites.

The targets of these activated phagocytes would appear to be diverse. Clearly there is phagocytosis of pigment during malaria but it is not clear to what extent pigment found within phagocytic cells results from phagocytosis of free pigment after schizogony and to what extent it represents the residuum left after phagocytosis and digestion of parasites or parasitized erythrocytes. Most likely, both phenomena occur. In the first case, this phagocytosis is a scavenger activity which, although protecting the host from the possible detrimental effects of circulating pigment, cannot be thought of as a primary immune effector mechanism against viable parasites. In the second situation, however, there is direct protection against replicating organisms. Phagocytosis of parasites and parasitized cells is not well understood from the point of view of specificity. A synergistic effect of immune serum with macrophages against *P. berghei* has been shown in experiments in diffusion chambers implanted in the peritoneal cavities of mice (Criswell *et al.*, 1971). A more pronounced effect was obtained with macrophages from mice with chronic malaria. Whether the targets of these effects are free merozoites or infected erythrocytes was not studied in these experiments.

Although merozoites are the most likely targets of antibody (Chow and Kreier, 1972), it is not clear whether phagocytosis of such sensitized merozoites is a prominent event *in vivo;* evidence of this is not conspicuous in the literature. Phagocytosis of antibody-sensitized schizonts is another potentially important effector function of phagocytes, at least by analogy with the *P. knowlesi* system (Brown, 1971). Again, however, there are little data to suggest that this is an important mechanism. In fact, the data which are available would suggest that phagocytosis is largely nonspecific; even nonparasitized erythrocytes are phagocytized to an impressive extent during the course of malaria infection. The extent of phagocytosis of normal and parasitized cells appears to be in proportion to their numbers, although the data to support this are inevitably subject to considerable error. This apparently nonspecific hyperphagocytosis of erythrocytes might be expected in view of the demonstrated activation of the phagocytic system for test materials such as carbon particles. It may also be influenced by the fact that fatty acids with hemolytic activity are increased in the circulation during the course of malaria, apparently influencing the osmotic fragility of the erythrocytes and thereby possibly also increasing their susceptibility to phagocytosis (Fife *et al.*, 1972). This is evidenced by the fact that immunologically detectable erythrocyte membrane damage occurs during infection (Kreier *et al.*, 1966), and by the fact that many of the reticulocytes produced during infection are very short-lived (Kreier *et al.*, 1972).

Clark and his collaborators (1976) have recently demonstrated dramatic immunity to both *P. yoelii* and *P. vinckei* in animals immunized with BCG (bacillus Calmette Gûerin). In such animals, intraerythrocytic parasites can be observed to be undergoing degeneration in a manner similar to that observed in animals undergoing spontaneous cure. It has been suggested that the mediator of this effect is a cytotoxic product of macrophages and that this substance gains entry into the intact parasitized erythrocyte in order to exert its parasiticidal activity.

IV. Conclusion

During almost three decades since their discovery, the murine *Plasmodia* have probably contributed more to our understanding of basic malariological phenomena than any other group of malaria parasites. From its silvatic haunts in the highlands of Katanga, the descent of *P. berghei* into a laboratory in Antwerp was, perhaps, "a small step for a malaria parasite, but a giant step for malariology."

REFERENCES

Adam, J. P., Landau, I., and Chabaud, A. G. (1966). Découverte dans la région de Brazzaville de rongeurs, infectés par des *Plasmodium*. *C. R. Hebd. Seances Acad. Sci.* **263,** 140–141.

Adler, S., and Foner, A. (1961). Observations on *Plasmodium vinckei* before and after adaptation to splenectomized hamsters. *Bull. Res. Counc. Isr., Sect. E* **9,** 1–23.

Aikawa, M. (1971). Parasitological review: *Plasmodium:* The fine structure of malaria parasites. *Exp. Parasitol.* **30,** 284–320.

Aikawa, M., Huff, C. G., and Sprinz, H. (1969). Comparative fine structure study of the gametocytes of avian, reptilian and mammalian malarial parasites. *J. Ultrastruct. Res.* **26,** 316–331.

Aikawa, M., Sterling, C. R., and Rabbege, J. (1972). Cytochemistry of the nucleus of malaria parasites. *Proc. Helminthol. Soc. Wash.* **39,** 174–194.

Arnold, J. D., Balcerzak, S. P., and Martin, D. C. (1969). Studies on the red cell-parasite relationship. *Mil. Med.* **134,** 962–971.

Bafort, J.-M. (1969a). Etude du cycle biologique du *P. vinckei.* I. Isolement de la nouvelle souche du *P. vinckei. Ann. Soc. Belge Med. Trop.* **49,** 533–544.

Bafort, J.-M. (1969b). Etude du cycle biologique du *P. vinckei.* II. Schizogonie erythrocytaire et pre-erythrocytaire. *Ann. Soc. Belge Med. Trop.* **49,** 545–570.

Bafort, J.-M. (1969c). Etude du cycle biologique du *P. vinckei.* III. Sporogonie. *Ann. Soc. Belge Med. Trop.* **49,** 571–609.

Bafort, J.-M. (1970a). Transient morphological and biological characters of rodent malaria. *J. Protozool.* **17,** Suppl., 26.

Bafort, J.M. (1970b). On the fine structure of *Plasmodium v. vinckei J. Parasitol.* **56,** Suppl. 4, 15.

Bafort, J.-M. (1970c). The variability of *Plasmodium berghei* vincke and Lips 1948. *Ann. Soc. Belge Med. Trop.* **50,** 247–262.

Bafort, J.-M. (1971a). The biology of rodent malaria with particular reference to *Plasmodium vinckei vinckei* Rodhain 1952. *Ann. Soc. Belge Med. Trop.* **51,** 1–204.

Bafort, J.-M. (1971b). Le cycle biologique de *Plasmodium vinckei* de Nigeria. *Proc. Multicoll. Eur. Parasitol., 1st, 1970* pp. 235–237.

Bafort, J.-M., and Howells, R. E. (1970). Electron microscope studies on the pre-erythrocytic stages of rodent malaria. *Trans. R. Soc. Trop. Med. Hyg.* **64,** 467.

Bahr, G. F., and Mikel, U. (1972). The arrangement of DNA in the nucleus of rodent malaria parasites. *Proc. Helminthol. Soc. Wash.* **39,** 361–372.

Berliner, R. W., Earle, D. P., Taggart, J. V., Zubrod, C. G., Welch, W. J., Conan, N. J., Bauman, E., Scudder, S. T., and Shannon, J. A. (1968). Studies on the chemotherapy of the human malarias. VI. The physiological disposition, antimalarial activity and toxicity of several derivatives of 4-aminoquinolines. *J. Clin. Invest.* **27,** 98–107.

Boonpucknavig, S., Boonpucknavig, V., and Bhamarapravat, N. (1972). Immunopathological studies of *Plasmodium berghei* infected mice. *Arch. Pathol.* **94,** 322–330.

Bowman, I. B. R., Grant, P. T., Kermack, W. O., and Ogston, D. (1961). The metabolism of *Plasmodium berghei* the malaria parasite of rodents. 2. The effect of mepacrine on the metabolism of glucose by the parasite separated from its host cell. *Biochem. J.,* **78,** 472–478.

Box, E. D., Celaya, B. L., and Gingrich, W. (1953). Development of *Plasmodium berghei* in *Anopheles quadrimaculatus. Am. J. Trop. Med. Hyg.* **2,** 624–627.

Bray, R. S., and El-Nahal, H. M. S. (1966). Indirect haemagglutination test for malarial antibody. *Nature (London)* **212,** 83.

Brewer, G. J., and Coan, C. C. (1969). Interaction of red cell ATP levels and malaria, and the treatment of malaria with hyperoxia. *Mil. Med.* **134,** 1056–1067.

Briggs, N. T., Wellde, B. T., and Sadun, E. H. (1968). Variants of *Plasmodium berghei* resistant to passive transfer of immune serum. *Exp. Parasitol.* **22,** 338–345.

Brown, I. N., and Phillips, R. S. (1974). Immunity to *Plasmodium berghei* in rats, passive serum transfer and the role of the spleen. *Infect. Immun.* **10,** 1213–1218.

Brown, I. N., Allison, A. C., and Taylor, R. B. (1968). *Plasmodium berghei* infection in thymectomized rats. *Nature (London)* **219,** 292–293.

Brown, K. N. (1971). Protective immunity to malaria parasites, a model for the survival of cells in an immunologically hostile environment. *Nature (London)* **230,** 163–167.

Bruce-Chwatt, L. J. (1963). Congenital transmission of immunity in malaria. *In* "Immunity to Protozoa" (P. C. C. Garnham, A. E. Pierce, and I. Roitt, eds.), pp. 89–108. Blackwell, Oxford.

Bruce-Chwatt, L. J., and Gibson, F. D. (1955). A *Plasmodium* from a Nigerian rodent (laboratory demonstration) *Trans. R. Soc. Trop. Med. Hyg.* **49,** 9.

Bryant, C., Voller, A., and Smith, M. J. H. (1964). The incorporation of radioactivity from C^4 glucose into the soluble metabolic intermediates of malaria parasites. *Am. J. Trop. Med. Hyg.* **13,** 515–519.

Büngener, W. (1967). Adenosine deaminase and nucleoside phosphorylase in malaria parasites. *Z. Tropenmed. Parasitol.* **18,** 48–52.

Büngener, W., and Nielsen, G. (1968). Nucleic acid metabolism in experimental malaria. 2. The utilization of adenosine and hypoxanthine for the synthesis of nucleic acids in malaria parasites *Plasmodium berghei* and *P. vinckei*. *Z. Tropenmed. Parasitol.* **19**, 185–197.

Canning, E. U., and Sinden, R. E. (1973). The organization of the öokinete and observations on nuclear division in öocysts of *Plasmodium berghei*. *Parasitology* **67**, 29–40.

Carter, G., Van Dyke, K., and Mengoli, H. F. (1972). Energetics of the malaria parasite. *Proc. Helminthol. Soc. Wash.* **39**, 241–243.

Carter, R. (1970). Enzyme variation in *Plasmodium berghei*. *Trans. R. Soc. Trop. Med. Hyg.* **65**, 586–590.

Carter, R. (1972). Effect of PABA on chloroquine resistance in *Plasmodium berghei yoelii*. *Nature (London)* **238**, 98–99.

Carter, R. (1973). Enzyme variation in *Plasmodium berghei* and *Plasmodium vinckei*. *Parasitology* **66**, 297–307.

Carter, R., and Walliker, D. (1975). New observations on the malaria parasites of rodents of the Central African Republic—*Plasmodium vinckei petteri* subsp. nov. and *Plasmodium chabaudi* Landau 1965. *Ann. Trop. Med. Parasitol.* **69**, 187–196.

Carter, R., and Walliker, D. (1976). Malaria parasites of Rodents of the Congo (Brazzaville). *Plasmodium chabaudi adami* subsp. nov. and *Plasmodium vinckei lentum* Landau, Michel, Adam and Boulard 1970. *Ann. Parasitol. Hum. Comp.* (in press).

Cenedella, R. J. (1968). Lipid synthesis from glucose carbon by *Plasmodium berghei in vitro*. *Am. J. Trop. Med. Hyg.* **17**, 680–684.

Cenedella, R. J., Jarrell, J. J., and Saxe, L. H. (1969). Lipid synthesis *in vivo* from 1-^{14}C- oleic acid and 6-^{3}H- glucose by intra-erythrocytic *Plasmodium berghei*. *Mil. Med.* **134**, 1045–1055.

Chan, V. L., and Lee, P. Y. (1974). Host cell specific proteolytic enzymes in *Plasmodium berghei* infected erythrocytes. *Southeast Asian J. Trop. Med. Publ. Health* **5**, 447–449.

Chow, J., and Kreier, J. P. (1972). *Plasmodium berghei;* adherence and phagocytosis by rat macrophages *in vitro*. *Exp. Parasitol.* **31**, 13–18.

Ciucă, M., Cipiea, A. G., Bona, C., Pozsgi, N., Isfan, T., and Iuga, G. (1963). Etudes cytochimiques sanguines dans l'infection expérimentale avec *Plasmodium berghei* de la souris blanche. I. Structure cytochimique du parasite, des globules rouges et observations effectuées au microscope à contraste de phase. *Arch. Roum. Pathol. Exp. Microbiol.* **22**, 503–514.

Ciucă, M., Radovici, E., Cipiea, A., Isfan, T., and Ianco, L. (1964). Aspects cellulaires et sérobiochimiques de l'immunité du rat blanc (*Rattus norvegicus*) infecté au *Plasmodium berghei*. *Arch. Roum. Pathol. Exp. Microbiol.* **23**, 5–22; in *Trop. Dis. Bull.* **62**, 13–14 (1965).

Clark, I. A., Allison, A. C., and Cox, F. E. G. (1976). Protection of mice against *Babesia* and *Plasmodium* with B. C. G. *Nature (London)* **259**, 309–311.

Clyde, D. F., McCarthy, V. C., Miller, R. M., and Hornick, R. B. (1973). Specificity of protection of man immunized against sporozoite induced *falciparum* malaria. *Am. J. Med. Sci.* **266**, 398–403.

Colwell, E. J., Hickman, R. L., Intraprasert, R., and Tirabutana, C. (1972) Minocycline and tetracycline treatment of acute *falciparum* malaria in Thailand. *Am. J. Trop. Med. Hyg.* **21**, 144–149.

Cook, L., Grant, P. T., and Kermack, W. O. (1961). Proteolytic enzymes of the erythrocytic forms of rodent and simian species of malarial *Plasmodia*. *Exp. Parasitol.* **11**, 372–379.

Coombs, G. H., and Gutteridge, W. E. (1975). Growth *in vitro* and Metabolism of *Plasmodium vinckei chabaudi*. *J. Protozool.* **22**, 555–560.

Corradetti, A., and Verolini, F. (1951). Relazione tra *Plasmodium berghei* e cellule dellaserie rossa durante l'attacco primario nel ratto albino. *Riv. Parassitol.*, **12**, 69–84.

Corradetti, A., and Verolini, F. (1957). Dimostrazione che l'immunità acquista del ratto albino al *Plasmodium berghei* é una immunità assoluta. *Riv. Parassitol.* **18**, 65–68.

Cox, F. E. G. (1964). Studies on the host-parasite relationships of Haemosporidia. Ph.D. Thesis, University of London.

Cox, F. E. G. (1966). Acquired immunity to *Plasmodium vinckei* in mice. *Parasitology* **56**, 719–732.

Cox, F. E. G. (1968). Immunity to malaria after recovery from piroplasmosis in mice. *Nature (London)* **219**, 646.

Cox, F. E. G. (1970). Acquired immunity to *Plasmodium chabaudi* in Swiss T O mice. *Ann. Trop. Med. Parasitol.* **64**, 304–314.

Cox, F. E. G., and Vickerman, K. (1966). Pinocytosis in *Plasmodium vinckei*. *Ann. Trop. Med. Parasitol.* **60**, 293–296.

Cox, F. E. G., and Voller, A. (1966). Cross-immunity between the malaria parasites of rodents *Ann. Trop. Med. Parasitol.* **60**, 297.

Cox, F. E. G., Bilbey, D. L. J., and Nicol, T. (1964). Reticulo-endothelial activity in mice infected with *Plasmodium vinckei*. *J. Protozool.* **11**, 229–236.

Cox, F. E. G., Wedderbern, N., and Salaman, M. H. (1974). The effect of Manson-Barr virus on the severity of malaria in mice *J. Gen. Microbiol.* **85**, 338–364.

Cox, H. W., Schroeder, W. F., and Ristic, M. (1966). Hemagglutination and erythrophagocytosis associated with the anemia of *Plasmodium berghei* infections in rats. *J. Protozool.* **13**, 327–332.

Criswell, B. S., Butler, W. T., Rossen, R. D., and Knight, V. (1971). Murine malaria, the role of humoral factors and macrophages in destruction of parasitized erythrocyte. *J. Immunol.* **107**, 212–221.

Davies, E. E. (1974). Ultrastructural studies on the early öokinete stage of *Plasmodium berghei nigeriensis* and its transformation into an öocyst. *Ann. Trop. Med. Parasitol.* **68**, 283–290.

Davies, E. E., and Howells, R. E. (1973). Uptake of ^{3}H-adenosine and ^{3}H-thymidine by oocysts of *P. berghei berghei*. *Trans. R. Soc. Trop. Med. Hyg.* **67**, 20.

Desser, S. S., Weller, I., and Yoeli, M. (1972). An ultrastructural study of the pre-erythrocytic development of *Plasmodium berghei* in the tree rat *Thamnomys surdaster*. *Can. J. Zool.* **50**, 821–25.

Diggens, S. M., and Gregory, K. G. (1969). Comparative responses of various rodent malarias to chemotherapy. *Trans. R. Soc. Trop. Med. Hyg.* **63**, 7.

Diggens, S. M., Gutteridge, W. E., and Trigg, P. I. (1970). Altered dihydrofolate reductase associated with a pyrimethamine-resistant *Plasmodium berghei berghei* produced in a single step. *Nature (London)* **228**, 579–580.

Diggs, C. L., and Osler, A. G. (1969). Humoral immunity in rodent malaria. II Inhibition of parasitaemia by serum antibody. *J. Immunol.* **102**, 298–305.

Diggs, C. L., and Osler, A. G. (1973). Humoral immunity in rodent malaria. III. Studies on the site of antibody action. *J. Immunol.*, **114**, 1243–1247.

Diggs, C. L., Skin, H., Briggs, N. T., Laudenslayer, K., and Weber, R. M. (1972).

Antibody mediated immunity to *Plasmodium berghei* independent of the third component of complement. *Proc. Helminthol. Soc. Wash.* **39**, 456–459.

Draper, C. C. (1953). "*Plasmodium berghei:* A Bibliography of Published Papers up to April 1953," pp. 1–10. Ross Institute, London.

Durbin, C. G. (1951). Attempts to transfer *Plasmodium berghei* to domesticated animals. *Proc. Helminthol. Soc. Wash.* **18**, 108.

Ebert, F. (1970). Effects of *Leishmania donovani* infection on the course of *Plasmodium vinckei* infection in the mouse. *Z. Tropenmed. Parasitol.* **21**, 252–260

Einhaber, A., Wren, R. E., Rosen, H., and Martin, L. K. (1967). Ornithine carbamoyltransferase activity in plasma of mice with malaria as an index of liver damage. *Nature (London)* **215**, 1489–1491.

El-Nahal, H. M. S. (1967a). Serological cross reaction between rodent malaria parasites as determined by the indirect immunofluorescent technique. *Trans. R. Soc. Trop. Med. Hyg.* **61**, 7.

El-Nahal, H. M. S. (1967b). Studies on humoral immunity in malaria. Ph.D. Thesis, University of London.

Fabiani, G. (1959). Le paludisme expérimental (à *Plasmodium berghei* et à *Plasmodium vinckei*). Son intérêt en immunologie générale. *Alger. Med.* **63**, 793–799.

Fajardo, L. F. (1973). Malaria parasites in mammalian platelets. *Nature (London)* **243**, 298–299.

Ferone, R. (1973). The enzymic synthesis of dihydropteroate and dihydrofolate by *Plasmodium berghei. J. Protozool.* **20**, 459–464.

Ferone, R., Burchall, J. J., and Hitchings, G. H. (1969). *Plasmodium berghei* dihydrofolate reductase. Isolation, properties and inhibition by antifolates. *Mol. Pharmacol.* **5**, 49–59.

Fife, E. H., Jr., von Doenhoff, A. E., and D'Antonio, L. E. (1972). *In vitro* and *in vivo* studies on a lytic factor isolated from *Plasmodium knowlesi. Proc. Helminthol. Soc. Wash.* **39**, 373–382.

Finerty, J. F., Evans, C. B., and Hyde, C. L. (1973). *Plasmodium berghei* and *Eperythrozoon coccoides* antibody and immunoglobulin synthesis in germ free and conventional mice simultaneously infected. *Exp. Parasitol.* **34**, 76–84.

Fitch, C. D. (1969). Chloroquine resistance in malaria a deficiency of chloroquine binding. *Proc. Natl. Acad. Sci. U.S.A.* **64**, 1181–1187.

Fitch, C. D. (1972). Chloroquine resistance in malaria: Drug binding and cross resistance patterns. *Proc. Helminthol. Soc. Wash.* **39**, 265–270.

Fogel, B. J., Shields, C. E., and von Doenhoff, A. E. (1966). The osmotic fragility of erythrocytes in experimental malaria. *Am. J. Trop. Med. Hyg.* **15**, 269–275.

Folch, H., and Waksman, B. H. (1974). The splenic suppressor cell. I. Activity of thymus-dependent adherent cells: Changes with age and stress. *J. Immunol.* **113**, 127–139.

Forrester, L. J., and Siu, P. M. L. (1971). P-enol-pyruvate carboxylase from *Plasmodium berghei. Comp. Biochem. Physiol.* **38**, 73–85.

Fulton, J. D., and Flewett, T. H. (1956). The relation of *Plasmodium berghei* and *Plasmodium knowlesi* to their respective red-cell hosts. *Trans. R. Soc. Trop. Med. Hyg.* **50**, 150–156.

Garnham, P. C. C. (1966). "Malaria Parasites and other Haemosporidia. Blackwell. Oxford.

Garnham, P. C. C., Bird, R. G., and Baker, J. R. (1967a). Electron microscope studies on motile stages of malaria parasites. V. Exflagellation of *Plasmodium, Hepatocystis* and *Leucocytozoon. Trans. R. Soc. Trop. Med. Hyg.* **61**, 58–68.

Garnham, P. C. C., Landau, I., Killick-Kendrick, R., and Adam, J.-P. (1967b). Repartition et caracteres differentiels des *Plasmodium* de murides. *Bull. Soc. Pathol. Exot.* **60**, 118–127.

Garnham, P. C. C., Bird, R. G., Desser, S. S., and El-Nahal, H. M. S. (1969a). Electron microscopic studies on motile stages of malaria parasites. VI. The ookinete of *Plasmodium berghei yoelii* and its transformation into the early oocyst. *Trans. R. Soc. Trop. Med. Hyg.* **63**, 187–194.

Garnham, P. C. C., Bird, R. G., Baker, J. R., and Killick-Kendrick, R. (1969b). Electron microscope studies on motile stages of malaria parasites. VII. The fine structure of merozoites of exoerythrocytic schizonts of *Plasmodium berghei yoelii. Trans. R. Soc. Trop. Med. Hyg.* **63**, 328–332.

George, J. N., Stakes, E. F., Wicker, D. J., and Conrad, M. D. (1966). Studies on the mechanisms of hemolysis in experimental malaria. *Mil. Med.* **131**, Suppl., 1217–1224.

Gilbertson, M., Maegraith, B. G., and Fletcher, K. A. (1970). Resistance to superinfection with *Plasmodium berghei* in mice in which the original infection was suppressed by a milk diet. *Ann. Trop. Med. Parasitol.* **64**, 497–512.

Golenser, J., Spira, D. T., and Zuckerman, A. (1975). Dynamics of thymidine incorporation by spleen cells from rats infected with *Plasmodium berghei. Clin. Exp. Immunol.* **22**, 364–371.

Gravely, S. M., and Kreier, J. P. (1976). Adoptive transfer of immunity to *Plasmodium berghei* with immune T and B lymphocytes. *Infect. Immun.* **14**, 184–190.

Gravely, S. M., Hamburger, J., and Kreier, J. P. (1976). T and B cell population changes in young and adult rats infected with *Plasmodium berghei. Infect. Immun.* **14**, 178–183.

Greenberg, J. (1956). Differences in the course of *Plasmodium berghei* infections in some hybrid and backcross mice. *Am. J. Trop. Med. Hyg.* **5**, 19–28.

Greenberg, J., and Coatney, G. R. (1954). Some host-parasite relationships in *Plasmodium berghei* infections. *Ind. J. Malariol.* **8**, 313–325.

Greenberg, J., and Kendrick, L. P. (1958). Parasitaemia and survival in mice infected with *Plasmodium berghei.* Hybrids between Swiss (high parasitaemia) and Str (low parasitaemia) mice. *J. Parasitol.* **44**, 492–498.

Greenberg, J., Nadel, E. M., and Coatney, G. R. (1953). The influence of strain, sex and age of mice on infection with *Plasmodium bergehi. J. Infect. Dis.* **93**, 96–100.

Greenberg, J., Nadel, E. M., and Coatney, G. R. (1954). Differences in survival of several inbred strains of mice and their hybrids infected with *Plasmodium berghei. J. Infect. Dis.* **95**, 114–116.

Greenwood, B. M., and Greenwood, A. M. (1971). Malaria infection in adult NZB mice and adult (NZBxNZW) F1 hybrid mice. *Trans. R. Soc. Trop. Med. Hyg.* **65**, 581–585

Greenwood, B. M., Playfair, J. H., and Torrigani, G. (1971). Immunosuppression in murine malaria. I. General characteristics. *Clin. Exp. Immunol.* **8**, 467–478.

Gregory, K. G., and Peters, W. (1970). The chemotherapy of rodent malaria. IX. Causal prophylaxis. Part 1. A method for demonstrating drug action on exoerythrocytic stages. *Ann. Trop. Med. Parasitol.* **64**, 15–24.

Gustafson, P. V., Agar, V. D., and Cramer, D. J. (1954). An electron microscope study of *Toxoplasma. Am. J. Trop. Med. Hyg.* **3**, 1008–1021.

Gutteridge, W. E., and Trigg, P. I. (1972). Some studies on the DNA of *Plasmodium knowlesi. In* "Comparative Biochemistry of Parasites" (H. Van den Bossche, ed.), pp. 199–218. Academic Press, New York.

Gutteridge, W. E., Trigg, P. I., and Williamson, D. H. (1971). Properties of DNA from some malarial parasites. *Parasitology* **62**, 209–219.

Hamburger, J., and Kreier, J. P. (1975). Antibody mediated elimination of malaria parasites (*Plasmodium berghei*) *in vivo*. *Infect. Immunol.* **12**, 339–345.

Hamburger, J., and Kreier, J. P. (1976). The demonstration of protective humoral activity in serum from recovered rats by use of free blood stage parasites. *Exp. Parasitol.* (in press).

Hargreaves, B. J., Yoeli, M., Nussenzweig, R. S., Walliker, D., and Carter, R. (1975). Immunological studies in rodent malaria. I. Protective immunity induced in mice by mild strains of *Plasmodium berghei yoelii* against a virulent and fatal line of this plasmodium. *Ann. Trop. Med. Parasitol.* **69**, 289–299.

Hawking, F. (1953). Milk diet p-aminobenzoic and malaria (*Plasmodium berghei*). *Br. Med. J.* **1**, 1201–1202.

Hawking, F. (1966). Chloroquine resistance in *Plasmodium berghei*. *Am. J. Trop. Med. Hyg.* **15**, 287–293.

Hawking, F., and Gammage, K. (1962). Chloroquine resistance produced in *Plasmodium berghei*. *Trans. R. Soc. Trop. Med. Hyg.* **56**, 263.

Hill, J. (1950). The schizontocidal effect of some antimalarials against *Plasmodium berghei*. *Ann. Trop. Med. Parasitol.* **44**, 291–297.

Holbrook, T. W., Spitalny, G. L., and Palczuk, N. C. (1976). Stimulation of resistance in mice to sporozoite-induced *Plasmodium berghei* malaria by injections of avian exo-erythrocytic forms. *J. Parasitol.* (in press).

Homewood, C. A., Warhurst, D. C., Peters, W., and Baggaley, V. C. (1972a). Lysosomes, pH and the antimalarial action of chloroquine. *Nature* (*London*) **235**, 50–52.

Homewood, C. A., Warhurst, D. C., Peters, W., and Baggaley, V. C. (1972b). Electron transport in intra erythrocytic *Plasmodium berghei*. *Proc. Helminthol. Soc. Wash.* **39**, 382–386.

Howells, R. E. (1970a). Mitochondrial changes during the life cycle of *Plasmodium berghei*. *Ann. Trop. Med. Parasitol.* **64**, 181–187

Howells, R. E. (1970b). Electron microscope observations on the development and schizogony of the erythrocytic stages of *Plasmodium berghei*. *Ann. Trop. Med. Parasitol.* **64**, 305–307.

Howells, R. E., and Bafort, J. (1970). Histochemical observations on the pre-erythrocytic schizont of *Plasmodium berghei*. *Ann. Soc. Belge Med. Trop.* **50**, 587–594.

Howells, R. E., and Davies, E. E. (1971). Nuclear division in the oocyst of *Plasmodium berghei*. *Ann. Trop. Med. Parasitol.* **65**, 451–459.

Howells, R. E., and Maxwell, L. (1973a). Further studies on the mitochondrial changes during the life cycle of *Plasmodium berghei*: Electrophoretic studies on isocitrate dehydrogenase. *Ann. Trop. Med. Parasitol.* **67**, 279–283.

Howells, R. E., and Maxwell, L. (1973b). Citric acid cycle activity and chloroquine resistance in rodent malaria parasites: The role of the reticulocyte. *Ann. Trop. Med. Parasitol.* **67**, 285–300.

Howells, R. E., Peters, W., and Thomas, E. A. (1968a). Host parasite relationships. Part 3. The relationship between haemozoin formation and the age of the host cell. *Ann. Trop. Med. Parasitol.* **62**, 267–270.

Howells, R. E., Peters, W., and Thomas, E. A. (1968b). Host parasite relationships. Part 4. The relationship between haemozoin formation and host cell age in chloroquine and primaquine resistant strains of *Plasmodium berghei*. *Ann. Trop. Med. Parasitol.* **62**, 271–276.

Hsu, D. Y. M., and Geiman, Q. M. (1952). Synergistic effect of *Haemobartonella muris* on *Plasmodium berghei* in white rats. *Am. J. Trop. Med. Hyg.* **1**, 747–760.

Huang, K., Schultz, W. W., and Gordon, F. B. (1968). Interferon induced by *Plasmodium berghei. Science* **162**, 123–124.

Ilan, J., Ilan, J., and Tokuyasu, K. (1969). Amino acid activation for protein synthesis in *Plasmodium berghei. Mil. Med.* **134**, 1026–1031.

Jackson, G. J. (1959). Simultaneous infections with *Plasmodium berghei* and *Trypanosoma lewisi* in the rat (Research note). *J. Parasitol.* **45**, 94.

Jacobs, R. L. (1964). Role of p-aminobenzoic acid in *Plasmodium berghei* infection in the mouse. *Exp. Parasitol.* **15**, 213–255.

Jacobs, R. L., Miller, L. H., and Koontz, L. C. (1974). Labeling of sporozoites of *Plasmodium berghei* with tritiated purines. *J. Parasitol.* **60**, 340–343.

Jadin, J. (1965a). Colloque international sur le *Plasmodium berghei. Ann. Soc. Belge Med. Trop.* **45**, 251–496.

Jadin, J. (1965b). Bibliographie du *Plasmodium berghei* I. H. Vincke et M. Lips (1948–1964) *Ann. Soc. Belge Med. Trop.* **45**, 473–496.

Jadin, J. M., Creemers, J., and LeRay, D. (1968). Ultrastructure et biologie de *Plasmodium vinckei.* I. Absorbtion and digestion of haemoglobin in the trophozoite. *Ann. Soc. Belge Med. Trop.* **48**, 497–501.

Jaffé, J. J., and Gutteridge, W. E. (1974). Purine and pyrimidine metabolism in protozoa. *Actual. Protozool; Resume Discuss. Tables Rondes Congr. Int. Protozool. 4th, 1973* pp. 23–35.

Jahiel, R. I., Nussenzweig, R. S., Vanderberg, J., and Vilćek, J. (1968). Antimalarial effect of interferon inducers at different stages of development of *Plasmodium berghei* in the mouse. *Nature* (*London*) **220**, 710–711.

Jayawardena, A. N., Targett, G. A. T., Leuchars, E., Carter, R. L., Roenhoff, M. J., and Davies, A. J. S. (1975). T-cell activation in murine malarias. *Nature* (*London*) **258**, 149–151.

Kilby, V. A. A., and Silverman, P. H. (1969). Fine structural observations of the erythrocytic stages of *Plasmodium chabaudi* Landau 1965. *J. Protozool.* **16**, 354–370.

Killick-Kendrick, R. (1971). The collection of strains of murine malaria parasites in the field and their maintenance in the laboratory by cyclical passage. *Symp. Br. Soc. Parasitol.* **9**, 39–64.

Killick-Kendrick, R. (1973a). Parasitic protozoa of the blood of rodents. I. The life cycle and zoogeography of *Plasmodium berghei nigeriensis* subsp. nov. *Am. Trop. Med. Parasitol.* **67**, 261–277.

Killick-Kendrick, R. (1973b). Parasitic protozoa of the blood of rodents. III. Two new malaria parasites of anomaliurine flying squirrels of the Ivory coast. *Ann. Parasitol. Hum. Comp.* **48**, 639–651.

Killick-Kendrick, R. (1974a). Parasitic protozoa of the blood of rodents. II. Haemogregarines, malaria parasites and piroplasms: An annotated check list and host index. *Acta Trop.* **31**, 28–69.

Killick-Kendrick, R. (1974b). Parasitic protozoa of the blood of rodents. IV. A revision of *Plasmodium berghei. Parasitology* **69**, 225–237.

Killick-Kendrick, R. (1975). Parasitic protozoa of the blood of rodents. V. *Plasmodium vinckei brucechwatti* subsp. nov. A malaria parasite of the thicket rat, *Thamnomys rutilans,* in Nigeria. *Ann. Parasitol. Hum. Comp.* **50**, 251–264.

Killick-Kendrick, R., and Bellier, L. (1970). Blood parasites of scaly-tailed flying squirrels in the Ivory Coast (laboratory demonstration) *Trans. R. Soc. Trop. Med. Hyg.* **65**, 430–431.

Killick-Kendrick, R., and Warren, M. (1968). Primary exoerythrocytic schizonts of a mammalian *Plasmodium* as a source of gametocytes. *Nature (London)* **220,** 191–192.

Killick-Kendick, R., Shute, G. T., and Lambo, A. O. (1968). Malaria parasites of *Thamnomys rutilans* (Rodentia, Muridae) in Nigeria. *Bull. W.H.O.* **38,** 822–824.

King, M. E., Shefner, A. M., and Schneider, M. D. (1972). Utilization of a sporozoite induced rodent malaria system for assessment of drug activity. *Proc. Helminthol. Soc. Wash.* **39,** 288–291.

Kramer, P. A., and Matusik, J. E. (1971). Location of chloroquine binding sites in *Plasmodium berghei. Biochem. Pharmacol.* **20,** 1619–1626.

Kreier, J. P., and Leste, J. (1967). Relationship of parasitemia to erythrocyte destruction in *Plasmodium berghei* infected rats. *Exp. Parasitol.* **21,** 78–83.

Kreier, J. P., and Leste, J. (1968). Parasitemia and erythrocyte destruction in *Plasmodium berghei*-infected rats. II. Effect of infected host globulin. *Exp. Parasitol.* **23,** 198–204.

Kreier, J. P., Shapiro, H., Dilley, D. A., Szilvassy, I., and Ristic, M. (1966). Autoimmune reactions in rats with *Plasmodium berghei* infection. *Exp. Parasitol.* **19,** 155–162.

Kreier, J. P., Seed, T., Mohan, R., and Pfister, R. (1972a). *Plasmodium* spp.: The relationship between erythrocyte morphology and parasitization in chickens, rats and mice. *Exp. Parasitol.* **31,** 19–28.

Kreier, J. P., Mohan, R., Seed, T., and Pfister, R. (1972b). Studies of the morphology and survival characteristics of erythrocytes from mice and rats with *Plasmodium berghei* infection. *Z. Tropenmed. Parasitol.* **23,** 245–255.

Kreier, J. P., Hamburger, J., Seed, T. M., Saul, K., and Green, T. (1976). *Plasmodium berghei:* Characteristics of a selected population of small free blood stage parasites. *Z. Tropenmed. Parasitol.* **27,** 82–88.

Kretschmar, W. (1963a). Abhängigkeit des Verlaufs der Nagetier malaria (*Plasmodium berghei*) in der Maus von exogen Faktoren unter der Wahl des Mausestammes. I. Interferienende Bartonellosen. *Z. Versuchstierkd.* **3,** 151–166.

Kretschmar, W. (1963b). Weitere Untersuchungen über die Immunität bei der Nagetiermalaria. *Z. Tropenmed. Parasitol.* **14,** 41–48.

Kretschmar, W. (1965). The effect of stress and diet on resistance to *Plasmodium berghei* and malaria immunity in the mouse. *Ann. Soc. Belge Med. Trop.* **45,** 325–344.

Krettli, A. U., and Nussenzweig, R. (1974). Depletion of T and B lymphocytes during malarial infections. *Cell. Immunol.* **13,** 440–446.

Krooth, R. S., Wuu, K., and Ma, R. (1969). Dihydroorotic acid dehydrogenase introduction into erythrocyte by the malaria parasite. *Science* **164,** 1073–1075.

Ladda, R. L. (1969). New insights into the structure of rodent malaria parasites. *Mil. Med.* **134,** 825–864.

Ladda, R. L., and Lalli, F. (1966). The course of *Plasmodium berghei* infection in the polycythémic mouse. *J. Parasitol.* **52,** 383–385.

Ladda, R. L., Aikawa, M., and Sprinz, H. (1969). Penetration of erythrocytes by merozoites of mammalian and avian malarial parasites. *J. Parasitol.* **55,** 633–644.

Landau, I. (1965). Description de *Plasmodium chabaudi* n. sp. parasite de rongeurs africaine. *C. R. Hebd. Seances Acad. Sci.* **260,** 3758–3761.

Landau, I. (1973). Diversité des méchanismes assurant la pérennité de l'infection chez les sporozoires coccidiomorphes. *Mem. Mus. Natl. Hist. Nat. Ser. A* **77,** 1–62.

Landau, I., and Chabaud, A. G. (1965). Infection naturelle par deux *Plasmodium* du rongeurs *Thamnomys rutilans* en Republique Centrafricaine. *C. R. Hebd. Seances Acad. Sci.* **26**, 230–232.

Landau, I., and Killick-Kendrick, R. (1966). Rodent *Plasmodia* of the Republique Centrafricaine: The sporogony and tissue stages of *Plasmodium chabaudi* and *P. berghei yoelii*. *Trans. R. Soc. Trop. Med. Hyg.* **60**, 633–649.

Landau, I., Michel, J.-C., and Adam, J.-P. (1968). Cycle biologique au laboratoire de *Plasmodium berghei killick* n. subsp. sp. *Ann. Parasitol. Hum. Comp.* **43**, 545–550.

Landau, I., Boulard, Y., and Houin, R. (1969). *Anthemosoma garnhami* n.g.n. sp., premier *Dactylosomidae* connu chez un mammifère. *C. R. Hebd. Seances Acad. Sci.* **26**, 230–232.

Landau, I., Michel, J.-C., Adam, J.-P., and Boulard, Y. (1970). The life cycle of *Plasmodium vinckei lentum* subsp. nov. in the laboratory; comments on the nomenclature of the murine malaria parasites. *Ann. Trop. Med. Parasitol.* **64**, 315–323.

Langer, B. W., Jr., and Phisphumvidhi, P. (1971). The L-amino acid oxidase of *Plasmodium berghei*. *J. Parasitol.* **57**, 677–678.

Langer, B. W., Jr., Phisphumvidhi, P., and Friedlander, Y. (1967). Malaria parasite metabolism: The pentose cycle in *Plasmodium berghei*. *Exp. Parasitol.* **20**, 68–76.

Langer, B. W., Jr., Phisphumvidhi, P., Jiampermpoon, D., and Weidhorn, R. P. (1969). Malarial parasite metabolism: The metabolism of methionine by *Plasmodium berghei*. *Mil. Med.* **134**, 1039–1044.

Langer, B. W., Jr., Phisphumvidhi, P., and Jiampermpoon, D. (1970). Malarial parasite metabolism: The glutamic acid dehydrogenase of *Plasmodium berghei*. *Exp. Parasitol.* **28**, 298–303.

Lantz, C. H., Van Dyke, K., and Carter, G. (1971). *Plasmodium berghei: In vitro* incorporation of purine derivatives into nucleic acids. *Exp. Parasitol.* **29**, 402–416.

Lien, J. C., and Cross, J. H. (1968). *Plasmodium* (*Vinckeia*) *watteni* sp. n. from the Formosan flying squirrel, *Petaurista petaurista grandis*. *J. Parasitol.* **54**, 1171–1174.

Loose, L. D., Cook, J. A., and DiLuzio, N. R. (1972). Malarial immunosuppression: A macrophage mediated defect. *Proc. Helminthol. Soc. Wash.* **39**, 484–491.

Lowry, B. A., Williams, M. K., and London, I. M. (1962). Enzymatic deficiencies of purine nucleotide synthesis in the human erythrocyte. *J. Biol. Chem.* **237**, 1622–1625.

Lucknow, I., Schmidt, G., Walter, R. D., and Königk, E. (1973). Adenosine monophosphate salvage synthesis in *Plasmodium chabaudi*. *Z. Tropenmed. Parasitol.* **24**, 500–504.

MacCallum, D. K. (1968). Pulmonary changes resulting from experimental malaria infections in hamsters. *Arch. Pathol.* **122**, 681–688.

MacCallum, D. K. (1969a). Time sequence study on the hepatic system of macrophages in malaria infected hamsters. *J. Reticuloendothel. Soc.* **6**, 232–252.

MacCallum, D. K. (1969b). A study of macrophage-pulmonary vascular bed interactions in malaria infected hamsters. *J. Reticuloendothel. Soc.* **6**, 253–270.

Macomber, P. B., O'Brien, R. L., and Hahn, F. E. (1966). Chloroquine: Physiological basis of drug resistance in *Plasmodium berghei*. *Science* **152**, 1374–1375.

McGhee, R. B. (1954). The infection of duck and goose erythrocytes by the mammalian parasite *Plasmodium berghei*. *J. Protozool.* **1**, 145–148.

Maegraith, B. G., Deegan, T., and Jones, E. S. (1952). Suppression of malaria (*Plasmodium berghei*) by milk. *Br. Med. J.* **2**, 1382–1384.

Martin, L. K., Einhaber, A., Sadun, E. H., and Wren, R. E. (1967). Effect of bacterial endotoxin on the course of *Plasmodium berghei* infection. *Exp. Parasitol.* **20**, 186–199.

Mercado, T. (1965). Paralysis associated with *Plasmodium berghei* malaria in the rat. *J. Infec. Dis.* **115**, 465–572.

Mercado, T. (1973). *Plasmodium berghei:* Inhibition by splenectomy of a paralyzing syndrome in infected rats. *Exp. Parasitol.* **34**, 142–147.

Mercado, T., and von Brand, T. (1954). Glycogen studies on white rats infected with *Plasmodium berghei. Exp. Parasitol.* **3**, 259–266.

Mercado, T. and von Brand, T. (1957). The influence of some steroids on glycogenesis in the liver of rats infected with *Plasmodium berghei. Am. J. Hyg.* **66**, 20–28.

Miller, L. H., and Frémont, H. N. (1969). The sites of deep vascular schizogony in chloroquine-resistant *Plasmodium berghei* in mice. *Trans. R. Soc. Trop. Med. Hyg.* **63**, 195–197.

Morgan, S. (1974). The genetics of malaria parasites: Studies on pyrimethamine resistance. Ph.D. Thesis, University of Edinburgh.

Most, H., Nussenzweig, R. S., Vanderberg, J., Herman, R., and Yoeli, M. (1966). Susceptibility of genetically standardized (JAX) mouse strains to sporozoite and blood-induced *Plasmodium berghei* infections. *Mil. Med., Suppl.* **131**, 915–918.

Most, H., Herman, R., and Schoenfeld, D. (1967). Chemotherapy of sporozoite and blood induced *Plasmodium berghei* infections with selected antimalarial agents. *Am. J. Trop. Med. Hyg.* **16**, 572–575.

Nadel, E. M., Greenberg, J., Jay, G. E., and Coatney, G. R. (1955). Back-cross studies on the genetics of resistance to malaria in mice. *Genetics* **40**, 620–626.

Nagarajan, K. (1968a). Metabolism of *Plasmodium berghei.* II. ^{32}Pi incorporation into high-energy phosphates. *Exp. Parasitol.* **22**, 19–26.

Nagarajan, K. (1968b). Metabolism of *Plasmodium berghei.* III. Carbon dioxide fixation and role of pyruvate and dicarboxylic acids. *Exp. Parasitol.* **22**, 33–42.

Neame, K. D., Brownbill, P. A., and Homewood, C. A. (1974). The uptake and incorporation of nucleosides into normal erythrocytes and erythrocytes containing *Plasmodium berghei. Parasitology* **69**, 329–335.

Nussenzweig, R. S. (1967). Increased nonspecific resistance to malaria produced by administration of killed *Corynebacterium parvum. Exp. Parasitol.* **21**, 224–231.

Nussenzweig, R. S., Vanderberg, J., and Most, H. (1969a). Protective immunity produced by the injection of X-irradiated sporozoites of *Plasmodium berghei.* IV. Dose response, specificity and humoral immunity. *Mil. Med.* **134**, 1176–1182.

Nussenzweig, R. S., Vanderberg, J., Most, H., and Orton, C. (1969b). Specificity of protective immunity produced by X-irradiated *Plasmodium berghei* sporozoites. *Nature (London)* **222**, 488–489.

Ott, K. J., and Stauber, L. A. (1967). *Eperythrozoon coccoides:* Influence on course of infection of *Plasmodium chabaudi* in mouse. *Science* **155**, 1546–1548.

Oxbrow, A. I. (1973). Strain specific immunity to *Plasmodium berghei:* A new genetic marker. *Parasitology* **67**, 17–27.

Perez-Reyes, R. (1953). *Anopheles aztecus* (Hoffman, 1935) a new definitive host for the cyclical transmission of *Plasmodium berghei* Vincke and Lips, 1948. *J. Parasitol.* **39**, 603–604.

Peters, W. (1965a). Morphological and physiological variations in chloroquine-resistant *Plasmodium berghei,* Vincke and Lips, 1948. *Ann. Soc. Belge Med. Trop.* **45,** 365–378.

Peters, W. (1965b). Competitive relationship between *Eperythrozoon coccoides* and *Plasmodium berghei* in the mouse. *Exp. Parasitol.* **16,** 158–166.

Peters, W. (1965c). Drug resistance in *Plasmodium berghei,* Vincke and Lips, 1948. I. Chloroquine resistance. *Exp. Parasitol.* **17,** 80–89.

Peters, W. (1966). Drug response of mepacrine- and primaquine-resistant strains of *Plasmodium berghei,* Vincke and Lips, 1948. *Ann. Trop. Med. Parasitol.* **60,** 25–30.

Peters, W. (1967). Chemotherapy of *Plasmodium chabaudi* infection in albino mice. *Ann. Trop. Med. Parasitol.* **61,** 52–56.

Peters, W. (1968). The chemotherapy of rodent malaria. I. Host-parasite relationship. Part 1. The virulence of infection in relation to drug resistance and time elapsed since isolation of the wild strain. *Ann. Trop. Med. Parasitol.* **62,** 238–245.

Peters, W., ed. (1970). "Chemotherapy and Drug Resistance in Malaria." Academic Press, New York.

Peters, W., Robinson, B., Ramkaran, A., and Portus, J. (1969). A virulent chloroquine resistant strain of *Plasmodium berghei* transmitted through *Anopheles stephensi. Trans. R. Soc. Trop. Med. Hyg.* **63,** 8.

Peters, W., Bafort, J., and Ramkaran, A. E. (1970). The chemotherapy of rodent malaria. XI. Cyclically transmitted chloroquine-resistant variants of the Keyberg 173 strain of *Plasmodium berghei. Ann. Trop. Med. Parasitol.* **64,** 41–52.

Phillips, R. S., Selby, G. R., and Wakelin, D. (1974). The effect of *Plasmodium berghei* and *Trypanosoma brucei* infections on the immune expulsion of the nematode *Trichuris muris* from mice. *Int. J. Parasitol.* **4,** 409–415.

Phisphumvidhi, P., and Langer, B. W., Jr. (1969). Malarial parasite metabolism: The lactic acid dehydrogenase of *Plasmodium berghei. Exp. Parasitol.* **24,** 37–41.

Powers, K. G., Jacobs, R. L., Good, W. C., and Koontz, L. C. (1969). *Plasmodium vinckei:* Production of chloroquine-resistant strain. *Exp. Parasitol.* **26,** 193–202.

Pringle, G. (1960). Two new malaria parasites from East African vertebrates. *Trans. R. Soc. Trop. Med. Hyg.* **54,** 411–414.

Prior, R. B., and Kreier, J. P. (1972). Isolation of *Plasmodium berghei* by use of a continuous-flow ultrasonic system: A morphological and immunological evaluation. *Proc. Helminthol. Soc. Wash.* **39,** 563–574.

Raffaele, G. (1965). Adattamento allo suilluppo nel pollo di *Plasmodium berghei. Riv. Malariol.* **44,** 1–8.

Ramarkrishnan, S. P., Satya Prakash, and Choudhury, D. S. (1957). Selection of a strain of *Plasmodium berghei* highly resistant to chloroquine (Resochin). *Nature (London)* **179,** 975.

Reid, V. E., and Freidkin, M. (1973). Thymidylate synthetase in mouse erythrocytes infected with *Plasmodium berghei. Mol. Pharmacol.* **9,** 74–80.

Resseler, R. (1956). Un nouveau plasmodium de rat en Belgique: *Plasmodium inopinatum* n. sp. *Ann. Soc. Belge Med. Trop.* **36,** 259–263.

Riley, M. V., and Maegraith, B. G. (1962). Changes in the metabolism of liver mitochondria of mice infected with rapid acute *Plasmodium berghei* malaria. *Ann. Trop. Med. Parasitol.* **56,** 473–482.

Robinson, B. L., and Warhurst, D. C. (1972). Antimalarial activity of erythromycin. *Trans. R. Soc. Trop. Med. Hyg.* **66,** 525.

Rodhain, J. (1952). Un deuxième *Plasmodium* parasite de rongeurs sauvages au Katanga. *Ann. Soc. Belge Med. Trop.* **32,** 275–280.

Rodhain, J. (1954). Essai d'adaptation du *Plasmodium vinckei* au rat blanc. *Ann. Soc. Belge Med. Trop.* **34,** 217–228.

Rosario, V. E. (1976). Genetics of chloroquine resistance in malaria parasites. *Nature* **261,** 585–586.

Rudzinska, M. A. (1969). The fine structure of malaria parasites. *Int. Rev. Cytol.* **25,** 161–199.

Rudzinska, M. A., and Trager, W. (1959). Phagotrophy and two new structures in the malaria parasite, *Plasmodium berghei. J. Biophys. Biochem. Cytol.* **6,** 103–112.

Sadun, E. H., Williams, J. S., Meroney, F. C., and Hutt, G. (1965). Pathophysiology of *Plasmodium berghei* infections in mice. *Exp. Parasitol.* **17,** 277–286.

Sadun, E. H., Wellde, B. T., and Hickman, R. L. (1969). Resistance produced in owl monkeys (*Aotus trivirgatus*) by inoculation with irradiated *Plasmodium falciparum. Mil. Med.* **134,** 1165–1182.

Sandosham, A. A., Yap, L. F., and Omar, I. (1965). A malaria parasite, *Plasmodium* (*vinckeia*) *boolati* sp. nov. from a Malayan giant flying squirrel. *Med. J. Malaya* **20,** 3–7.

Satya, Prakash, Krishnaswami, A. K., and Ramakrishnan, S. P. (1952). Studies on *Plasmodium berghei.* V. On the host range in blood-induced infections. *Indian J. Malariol.* **6,** 175–182.

Scalzi, H. A., and Bahr, G. (1968). An electron microscopic examination of erythrocytic stages of two rodent malarial parasites, *Plasmodium chabaudi* and *Plasmodium vinckei. J. Ultrastruct. Res.* **24,** 116–133.

Schiebel, L. W., and Miller, J. (1969). Glycolytic and cytochrome oxidase activity in plasmodia. *Mil. Med.* **134,** 1074–1080.

Schnitzer, B., Sodeman, T., Mead, M. L., and Contacos, P. G. (1972). Pitting function of the spleen in malaria; ultrastructural observations. *Science* **177,** 175–177.

Schoenfeld, C., Most, H., and Entner, N. (1974). Chemotherapy of rodent malaria: Transfer of resistance vs mutation. *Exp. Parasitol.* **67,** 17–27.

Schultz, W. W., Huang, K., and Gordon, F. B. (1968). Role of interferon in experimental mouse malaria. *Nature* (*London*) **220,** 709–710.

Seed, T. M., Aikawa, M., Prior, R. B., Kreier, J. P., and Pfister, R. (1973). *Plasmodium* sp.: Topography of intra- and extracellular parasites. *Z. Tropenmed. Parasitol.* **24,** 525–535.

Sengers, R. C. A., Jerusalem, C. R., and Doesburg, W. H. (1971). Disturbed immunological responsiveness during *Plasmodium berghei* infection. *Exp. Parasitol.* **30,** 41–53.

Sen Gupta, P. C., Ray, H. N., Dutta, B. N., and Chaudhuri, R. N. (1955). A cytochemical study of *Plasmodium berghei* Vincke and Lips, 1948. *Ann. Trop. Med. Parasitol.* **49,** 273–277.

Sergent, E. (1959). Réflexion sur l'épidémiologie et l'immunologie du paludisme. *Arch. Inst. Pasteur Algér.* **37,** 1–52.

Sergent, E., and Poncet, A. (1955). Etude expérimentale du paludisme des rongeurs á *Plasmodium berghei.* II. Stade d'infection latente métacritique. *Arch. Inst. Pasteur Algér.* **33,** 195–222.

Sesta, J. J., Rosen, S., and Sprinz, H. (1968). Malarial nephropathy in the golden hamster. *Arch. Pathol.* **85,** 663–668.

Sinden, R. E. (1974). Excystment by sporozoites of malaria parasites. *Nature* (*London*) **252,** 314.

Sinden, R. E., Canning, E. U., and Spain, B. (1976). Gametogenesis and fertilization in *Plasmodium yoelii nigeriensis:* a transmission electron microscope study. *Proc. R. Soc. Lond. B.* **193**, 55–76.

Sinden, R. E., and Croall, N. A. (1974). Cytology and kinetics of microgametogenesis and fertilization in *Plasmodium yoelii nigeriensis. Parasitology* **70**, 53–65.

Sinden, R. E., and Garnham, P. C. C. (1973). A comparative study on the ultrastructure of Plasmodium sporozoites within the oocyst and salivary glands, with particular reference to the incidence of the micropore. *Trans. R. Soc. Trop. Med. Hyg.* **67**, 631–637.

Singer, I. (1954a). The effect of splenectomy or phenylhydrazine on infections with *Plasmodium berghei* in the white mouse. *J. Infect. Dis.* **94**, 159–163.

Singer, I. (1954b). The cellular reaction to infections with *Plasmodium berghei* in the white mouse. *J. Infect. Dis.* **94**, 241–261.

Singer, I., Hadfield, R., and Lakonen, M. (1955). The influence of age on the intensity of infection with *Plasmodium berghei* in the rat. *J. Infect. Dis.* **97**, 15–21.

Singh, J. (1954). Symposium on *Plasmodium berghei. Indian J. Malariol.* **8**, 237–394.

Siu, P. M. L. (1966). The mechanism of action of chloroquine on CO_2 fixation in *Plasmodium berghei. Fed. Proc., Fed. Am. Soc. Exp. Biol.* **25**, 756.

Siu, P. M. L. (1967). Carbon dioxide fixation in plasmodia and the effect of some antimalarial drugs on the enzyme. *Comp. Biochem. Physiol.* **23**, 785–795.

Smalley, M. E. (1975). The nature of age immunity to *Plasmodium berghei* in the rat. *Parasitol.* **71**, 337–347.

Spitalny, G. L., and Nussenzweig, R. S. (1973). *Plasmodium berghei:* Relationship between protective immunity and anti-sporozoite (CSP) antibody in mice. *Exp. Parasitol.* **33**, 168–178.

Stechschulte, D. J. (1969). Cell-mediated immunity in rats infected with *Plasmodium berghei. Mil. Med.* **134**, 1147–1152.

Sterling, C. R., Aikawa, M., and Nussenzweig, R. S. (1972). Morphological divergence in mammalian malaria parasites: The fine structure of *Plasmodium brasilianum. Proc. Helminthol. Soc. Wash.* **39**, 109–129.

Terzakis, J. A., Vanderberg, J. P., and Hutter, R. M. (1974). The mitochondria of pre-erythrocytic *Plasmodium berghei. J. Protozool.* **21**, 251–253.

Theakston, R. D. G., and Fletcher, K. A. (1973). An electron cytochemical study of 6-phosphogluconate dehydrogenase activity in infected erythrocytes during malaria. *Life Sci.* **13**, 4005–4010.

Theakston, R. D. G., Fletcher, K. A., and Maegraith, B. G. (1968). The fine structure of *Plasmodium vinckei,* a malaria parasite of rodents. *Ann. Trop. Med. Parasitol.* **62**, 122–134.

Theakston, R. D. G., Howells, R. E., Fletcher, K. A., Peters, W., Fullard, J., and Moore, G. A. (1969). The ultrastructural distribution of cytochrome oxidase activity in *Plasmodium berghei* and *P. gallinaceum. Life Sci.* **8**, Part 2, 521–529.

Theakston, R. D. G., Fletcher, K. A., and Maegraith, B. G. (1970). Ultrastructural localization of NADH- and NADPH-dehydrogenase in the erythrocytic stages of the rodent malaria parasite, *Plasmodium berghei. Life Sci.* **9**, Part 2, 421–429.

Thurston, J. P. (1950). The action of antimalarial drugs in mice infected with *Plasmodium berghei. Br. J. Pharmacol. Chemother.* **5**, 409–416.

Thurston, J. P. (1953). *Plasmodium berghei. Exp. Parasitol.* **2**, 311–332.

Tokuyasu, K., Ilan, J., and Ilan, J. (1969). Biogenesis of ribosomes in *Plasmodium berghei. Mil. Med.*, **134**, 1032–1038.

Topley, E., Bruce-Chwatt, L. J., and Dorrell, J. (1970). Haemotological study of a rodent malaria model. *J. Trop. Med. Hyg.* **75**, 1–8.

Tsukamoto, M. (1974). Differential detection of soluble enzymes specific to a rodent malaria parasite *Plasmodium berghei,* by electrophoresis on polyacrylamide gels. *Trop. Med.* **16,** 55–69

van den Berghe, L., Peel, E., Chardome, M., and Lambrecht, F. L. (1958). Le cycle asexué de *Plasmodium atheruri* n. sp. du porc-épic *Atherurus africanus centralis* au Congo belge. *Ann. Soc. Belge Med. Trop.* **38,** 971–976.

Vanderberg, J. P., and Yoeli, M. (1966). Effects of temperature on sporogonic development of *Plasmodium berghei. J. Parasitol.* **52,** 559–564.

Vanderberg, J. P., Rodhain, J., and Yoeli, M. (1967). Electron microscopical and histochemical studies of sporozoite formation in *Plasmodium berghei. J. Protozool.* **14,** 82–103.

Vanderberg, J. P., Nussenzweig, R. S., and Most, H. (1969). Protective immunity produced by the injection of X-irradiated sporozoite of *Plasmodium berghei. Mil. Med.* **134,** 1183–1190.

Van Dyke, K., Tremblay, G. C., Lantz, C. H., and Szustkiewicz, C. (1970). The source of purines and pyrimidines in *Plasmodium berghei. Ann. J. Trop. Med. Hyg.* **19,** 203–208.

Verhave, J. P. (1975). "Immunization with Sporozoites." Krips Repro, Meppel, The Netherlands.

Vickerman, K., and Cox, F. E. G. (1967). Merozoite formation in the erythrocytic stages of the malaria parasite *Plasmodium vinckei. Ann. Trop. Med. Parasitol.* **60,** 293–296.

Vincke, I. H. (1954). Natural history of *Plasmodium berghei. Indian J. Malariol.* **8,** 245–256.

Vincke, I. H., and Bafort, J. (1968). Methodes de standardization de l'inoculum de sporozoites de *Plasmodium berghei. Ann. Soc. Belge Med. Trop.* **48,** 181–194.

Vincke, I. H., and Lips, M. (1948). Un nouveau *Plasmodium* d'un rongeur sauvage du Congo, *Plasmodium berghei* n. sp. *Ann. Soc. Belge Med. Trop.* **28,** 97–104.

Vincke, I. H., Peeters, E. M. E., and Frankie, G. (1953). Essai d'étude d'ensemble sur le *Plasmodium berghei. Inst. R. Colon. Belge, Bull. Seances* **24,** 1364–1406.

von Brand, T., and Mercado, T. (1958). Quantitative and histochemical studies on liver lipids of rats infected with *Plasmodium berghei. Am. J. Hyg.* **67,** 311–320.

Walliker, D., Carter, R., and Morgan, S. (1973). Genetic recombination in *Plasmodium berghei. Parasitology* **66,** 309–320.

Walliker, D., Carter, R., and Sanderson, A. (1975). Genetic studies on *Plasmodium chabaudi:* Recombination between enzyme markers. *Parasitology* **70,** 19–24.

Walter, R. D. (1968). Untersuchungen über die entwicklung freier, erythrozytärer schizonten bei infektionen mit *Plasmodium berghei* und *Plasmodium chabaudi. Z. Tropenmed. Parasitol.* **19,** 415–426.

Walter, R. D., and Königk, E. (1971). Synthesis of desoxythymidylate synthetase and dihydrofolate reductase in *Plasmodium chabaudi* during synchronous schizogony. *Z. Tropenmed. Parasitol.* **22,** 250–255.

Walter, R. D., and Königk, E. (1973). *Plasmodium chabaudi:* Enzymic synthesis of dihydropteroate and inhibition of the process by sulphonamides. *Z. Tropenmed. Parasitol.* **22,** 256–259.

Walter, R. D., Mühlpfordt, H., and Königk, E. (1970). Comparative studies on desoxythymidylate synthesis in *Plasmodium chabaudi, Trypanosoma gambiense* and *T. lewisi. Z. Tropenmed. Parasitol.* **21,** 347–357.

Warhurst, D. C. (1973). Chemotheraputic agents and malaria research. *Symp. Br. Soc. Parasitol.* **11,** 1–28.

Warhurst, D. C., and Baggaley, V. C. (1972). Autophagic vacuole formation in *P. berghei. Trans. R. Soc. Trop. Med. Hyg.* **66,** 5.

Warhurst, D. C., and Killick-Kendrick, R. (1967). Spontaneous resistance to chloroquine in a strain of rodent malaria (*Plasmodium berghei yoelii*). *Nature* (*London*) **213**, 1048–1049.

Warhurst, D. C., Baggaley, V. C., and Robinson, B. L. (1971a). A technique for preparation of DNA from *P. berghei*. *Trans. R. Soc. Trop. Med. Hyg.* **65**, 9.

Warhurst, D. C., Robinson, B. L., Howells, R. E., and Peters, W. (1971b). The effect of cytotoxic agents on autophagic vacuole formation in chloroquine-treated malaria parasite (*Plasmodium berghei*). *Life Sci.* **10**, 761–771.

Warhurst, D. C., Homewood, C. A., and Baggaley, V. C. (1972). Observations *in vitro* on the mode of action of chloroquine and quinine in blood stages *Plasmodium berghei*. *J. Protozool.* **19**, Suppl., 53.

Wedderburn, N. (1974). Immunodepression produced by malarial infection in mice. *Parasites Immunized Host: Mech. Surv., Ciba Found. Symp., 1974* pp. 123–135.

Weidekamm, E., Walloch, D. F. H., Lin, P. S., and Hendricks, J. (1973). Erythrocyte membrane alterations due to infection with *Plasmodium berghei*. *Biochim. Biophys. Acta* **323**, 539–546.

Weiss, M. L. (1965). Development and duration of immunity to malaria (*Plasmodium berghei*) in mice. *Prog. Protozool. Int. Conf. Protozool., 2nd, 1965* Int. Congr. Ser. No. 91, p. 168.

Weiss, M. L., and deGiusti, D. L. (1964). Modification of a malaria parasite (*Plasmodium berghei*) following passage through a tissue culture. *Nature* (*London*) **201**, 731.

Weiss, M. L., and deGiusti, D. L. (1966). Active immunization against *Plasmodium berghei* malaria in mice. *Am. J. Trop. Med. Hyg.* **15**, 472–482.

Wellde, B. T., and Sadun, E. H. (1967). Resistance produced in rats and mice by exposure to irradiated *Plasmodium berghei*. *Exp. Parasitol.* **21**, 310–324.

Wellde, B. T., Briggs, N. T., and Sadun, E. H. (1966). Susceptibility to *Plasmodium berghei:* Parasitological, biochemical and haemotological studies in laboratory and wild animals. *Mil. Med.* **131**, 859–869.

Wellde, B. T., Diggs, C. L., Rodriguez, E., Jr., Briggs, N. T., Weber, R. M., and von Doenhoff, A. E., Jr. (1972). Requirements for induction of immunity to *Plasmodium berghei* by irradiated parasitized erythrocytes. *Proc. Helminthol. Soc. Wash.* **39**, 529–537.

Wery, M. (1968). Studies on the sporogony of rodent malaria parasites. *Ann. Soc. Belge Med. Trop.* **48**, 1.

Whitfield, P. R. (1953). Studies on the nucleic acids of the malaria parasite *Plasmodium berghei* (Vincke and Lips). *Aust. J. Biol. Sci.* **6**, 234–243.

Wright, D. H., Masembe, R. M., and Bazira, E. R. (1971). The effect of antithymocyte serum on golden hamsters and rats infected with *Plasmodium berghei*. *Br. J. Exp. Pathol.* **52**, 405–476.

Wyler, D. J., and Gallin, J. I. (1975). Mechanism of splenic hypertrophy in malaria. *Clin. Res.* **23**, 313A.

Yap, L. F., Muul, I., and Lim Boo Liat. (1970). A *Plasmodium* sp. from the spotted giant flying squirrel in W. Malaysia (Laboratory demonstration). *Southeast Asian J. Trop. Med. Publ. Health* **1**, 418.

Yoeli, M. (1956). Some aspects of concomitant infections of plasmodia and schistosomes. I. The effect of *Schistosoma mansoni* on the course of infection of *Plasmodium berghei* in the field vole. *Am. J. Trop. Med. Hyg.* **5**, 988–999.

Yoeli, M. (1965). Studies on *Plasmodium berghei* in nature and under experimental conditions. *Trans. R. Soc. Trop. Med. Hyg.* **59**, 255–276.

Yoeli, M., and Hargreaves, B. J. (1974). Brain capillary blockage produced by a virulent strain of rodent malaria. *Science* **184**, 572–573.

Yoeli, M., and Most, H. (1960). The biology of a newly isolated strain of *Plasmodium berghei* in a rodent host and in experimental mosquito vectors. *Trans. R. Soc. Trop. Med. Hyg.* **54,** 549–555.

Yoeli, M., and Most, H. (1965). Studies on sporozoite-induced infections of rodent malaria 1. The pre-erythrocytic tissue stage of *Plasmodium berghei. Am. J. Trop. Med. Hyg.* **14,** 700–714.

Yoeli, M., and Wall, W. J. (1951). Complete sporogonic development of *Plasmodium berghei* in experimentally infected *Anopheles* sp. *Nature* (*London*) **168,** 1078–1080.

Yoeli, M., Becker, Y., and Benkopf, H. (1955). The effect of West Nile virus on experimental malaria (*Plasmodium berghei*) in mice. *Harefuah* **49,** 116–119.

Yoeli, M., Upmanis, R. S., and Most, H. (1969). Drug resistance transfer among rodent plasmodia. I. Acquisition of resistance to pyrimethamine by a drug-sensitive strain of *Plasmodium berghei* in the course of its concomitant development with a pyrimethamine-resistant *P. vinckei* strain. *Parasitology* **59,** 429–447.

Yoeli, M., Hargreaves, B., Carter, R., and Walliker, D. (1975). Sudden increase in virulence in a strain of *Plasmodium berghei yoelii. Ann. Trop. Med. Parasitol.* **66,** 173–178.

Zuckerman, A. (1953). Residual immunity following radical cure of *Plasmodium berghei* in intact and splenectomized voles (*Microtus guentheri*). *J. Infect. Dis.* **92,** 205–223.

Zuckerman, A. (1960). Auto-antibody in rats with *Plasmodium berghei. Nature* (*London*) **185,** 189–190.

Zuckerman, A. (1966). Recent studies on factors involved in malarial anemia. *Mil. Med.* **131,** Suppl., 1201–1216.

Zuckerman, A. (1969). Current status of the immunology of malaria and of the antigenic analysis of plasmodia. *Bull. W.H.O.* **40,** 55–66.

Zuckerman, A. (1970). Malaria of lower mammals. *In* "Immunity to Parasitic Animals" (G. J. Jackson, R. Herman, and I. Singer, eds.), pp. 793–829. Appleton, New York.

Zuckerman, A., and Spira, D. (1961). Blood loss and replacement in malarial infections. V. Positive anti-globulin tests in rat anemias due to the rodent malarias (*Plasmodium berghei* and *Plasmodium vinckei*) to cardiac bleeding and to treatment with phenylhydrazine hydrochloride. *J. Infect. Dis.* **108,** 339–348.

Zuckerman, A., and Yoeli, M. (1954). Age and sex as factors influencing *Plasmodium berghei* infections in intact and splenectomized rats. *J. Infect. Dis.* **94,** 225–236.

Zuckerman, A., Hamburger, J., and Spira, D. (1965). Active immunization of rats against rodent malaria with a non-living plasmodial product. *Prog. Protozool. Int. Conf. Protozool. 2nd, 1965* Int. Congr. Ser. No. 91, pp. 50–51.

9

Plasmodia of Nonhuman Primates

William E. Collins and Masamichi Aikawa

I. Introduction

The plasmodia of the nonhuman primates are very similar to the malaria parasites that infect man. The relationships are so close that biologic and morphologic separation is, at times, quite difficult. However, the isolation of these parasites within their primate hosts has led to a considerable amount of diversity which has resulted in the present recognition of twenty distinct species. Two of these, *Plasmodium girardi* and *P. lemuris,* are only found in members of the family Lemuridae and are confined to the Malagasy Republic. In fact, so little is known of these parasites, other than the blood-infecting stages in naturally infected lemurs, that they are normally excluded from the discussion on primate malarias. They are the only malaria parasites found in the Prosimii, all others being in the Anthropoidea. Two species (*P. brasilianum* and *P. simium*) have been reported from the New World monkeys of the family Cebidae; eight (*P. coatneyi, P. cynomolgi, P. fieldi, P. fragile, P. gonderi, P. inui, P. knowlesi,* and *P. simiovale*) from the Old

World monkeys, family Cercopithecidae; four (*P. eylesi, P. hylobati, P. jefferyi,* and *P. youngi*) from gibbons, family Hylobatidae; and four (*P. pitheci, P. reichenowi, P. schwetzi,* and *P. sylvaticum*) from the great apes, family Pongidae.

The chronology of the discovery of these parasites is presented in Table I. It is surprising that, in spite of the great flurry of activity on the human malarias in the 1890's, it was not until 1907 that the first distinct nonhuman primate malaria was described. Other workers had reported malaria parasites in African monkeys which they called *P. kochi.* Later studies revealed, however, that they were probably reporting infections with *Hepatocystis.* Although Laveran (1905) saw the parasite, it was for Halberstaedter and von Prowazek (1907) to describe *P. pitheci* from the orangutan from Borneo. In the same article, they also briefly described *P. inui* from *Macaca cynomolgus* (= *M. fascicularis*) from Java, and *M. nemestrina* from Sumatra and Borneo. In the same year, Mayer (1907) discovered and described *P. cynomolgi* from a *M. fascicularis* monkey from Java; the following year, Gonder and von

Table I
Chronology of the Discovery of the Nonhuman Primate Malaria Parasites

Species	Original host	Described by
P. pitheci	*Pongo pygmaeus*	Halberstaedter and von Prowazek, 1907
P. inui	*Macaca fascicularis*	Halberstaedter and von Prowazek, 1907
P. cynomolgi	*Macaca fascicularis*	Mayer, 1907
P. brasilianum	*Cacajoa calvus*	Gonder and von Berenberg-Gossler, 1908
P. reichenowi	*Pan troglodytes* *Gorilla gorilla*	Sluiter *et al.*, 1922
P. knowlesi	*Macaca fascicularis*	Sinton and Mulligan, 1932
P. gonderi	*Cercocebus atys*	Sinton and Mulligan, 1933
P. schwetzi	*Pan troglodytes* *Gorilla gorilla*	Brumpt, 1939
P. hylobati	*Hylobates lensciscis*	Rodhain, 1941
P. simium	*Alouatta fusca*	Fonseca, 1951
P. girardi	*Lemur fulvus rufus*	Bück *et al.*, 1952
P. fieldi	*Macaca nemestrina*	Eyles *et al.*, 1962a
P. coatneyi	*Anopheles hackeri* [a]	Eyles *et al.*, 1962b
P. lemuris	*Lemur collaris*	Huff and Hoogstraal, 1963
P. youngi	*Hylobates lar*	Eyles *et al.*, 1964
P. eylesi	*Hylobates lar*	Warren *et al.*, 1965
P. fragile	*Macaca sinica*	Dissanaike *et al.*, 1965a
P. simiovale	*Macaca sinica*	Dissanaike *et al.*, 1965b
P. jefferyi	*Hylobates lar*	Warren *et al.*, 1966
P. sylvaticum	*Pongo pygmaeus*	Garnham *et al.*, 1972

[a] Only species of primate malaria isolated initially from its mosquito host.

Berenberg-Gossler (1908) described the *P. malariae*-like parasite, *P. brasilianum* from a cacajao [*Brachyurus calvus* (=*Cacajao calvus*)] from Brazil. There were many reports on the prevalence of malaria parasites in African and Asian monkeys during the following years, but many of the identifications were questionable or the descriptions were later shown to be invalid.

In the meantime, workers in Africa became very interested in the malaria parasites of the higher apes. Reichenow (1920) found parasites which he considered identical with the human parasites, *P. falciparum*, *P. vivax*, and *P. malariae*. However, further studies indicated that at least two of these were distinct. Sluiter *et al.* (1922) gave the name *P. reichenowi* to the *P. falciparum*-like parasite of the chimpanzees and gorillas in the Cameroons. It was not until 1939 that the vivax-like parasite was given the name *P. schwetzi* by Brumpt. Although the *P. malariae*-like parasite was named *P. rodhaini* by Brumpt in 1939, it is now generally considered that this is probably *P. malariae* in the nonhuman primate host.

In 1932–1933, Sinton and Mulligan reviewed the current knowledge concerning the monkey-infecting parasites and managed to bring some semblance of order to the subject. In the process, they renamed a parasite (*P. inui gonderi*) from the blood of a mangabey, *Cercocebus fuliginosus*, which Gonder and von Berenberg-Gossler (1908) had called *P. kochi*. The original authors had been working with a mixed infection of what was eventually to be called *P. gonderi* and *Hepatocystis kochi*. It had been shown as early as 1910 (Gonder and Rodenwaldt, 1910) that the African parasite had a tertian periodicity, whereas *P. inui* had a quartan periodicity. Therefore, *P. gonderi* was elevated to specific rank by Rodhain and van den Berghe (1936). Sinton and Mulligan also described *P. knowlesi*, the only primate malaria with a 24-hour schizogonic cycle.

There was again a period during which there was little activity on the isolation and description of new species. Rodhain (1941) described *P. hylobati* from the gibbon, *Hylobates moloch*, and da Fonseca (1951) described *P. simium* from the howler monkey, *Alouatta fusca*. *Plasmoduim girardi* was described from a lemur, *Lemur fulvus rufus*, by Bück *et al.*, (1952). In the 1960's, a great deal of interest was generated, primarily associated with the accidental infection of laboratory workers with *P. cynomolgi*. Concern was expressed that the involvement of man with the nonhuman primate malarias may complicate malaria eradication efforts where man and other primates cohabit the same environment. As a result, the National Institutes of Health (United States) established a field laboratory in collaboration with the Institute for Medical Research in Kuala Lumpur, Malaysia in 1960. These studies resulted in the discovery and description of five new species: two (*P. fieldi* and *P. coatneyi*) from monkeys and three (*P. youngi, P. eylesi,* and *P. jefferyi*) from gibbons. At approximately the same time, Dissanaike and his coworkers in Sri Lanka described two additional species (*P. fragile* and *P. simiovale*) from monkeys. Huff and Hoogstraal (1963) described *P. lemuris* from a black lemur, *Lemur collaris,* housed in the Tananarive Zoo.

The last species to be described, *P. sylvaticum,* was described by Garnham *et al.* (1972) from the orangutan. This relatively rare host from a remote area thus has the distinction of being the source for the first and also the most recent of this group of malaria parasites.

II. Life Cycles and Morphology

The life cycles and general morphology of all the nonhuman primate malaria parasites very closely resemble those of their related parasites in man. In fact, for each of the human malarias, there are one or more counterpart parasites found in monkeys. For example, *P. cynomolgi* of monkeys resembles, both morphologically and biologically, *P. vivax*; *P. brasilianum* resembles *P. malariae*; *P. fieldi* and *P. simiovale* resemble *P. ovale*; and, except for the morphology of the gametocyte, *P. coatneyi* and *P. fragile* have many similarities to *P. falciparum.* This close relationship between the parasites of man and the nonhuman primates has led to extensive investigations of these parasites as to their pathology, physiology, and chemotherapy so as to better understand the host–parasite relationships.

The life cycle of the primate malarias is diagrammatically presented in Fig. 1. After the mosquito injects sporozoites into the primate host, there is a period of from 5 to 15 days prior to the appearance of infected erythrocytes. This preerythrocytic period remained a mystery, however, until Shortt and Garnham (1948) and Shortt *et al.* (1948) demonstrated the exoerythrocytic (EE) schizogony of *P. cynomolgi* in the liver of *M. mulatta* monkeys. This was followed by the discovery of similar forms for the human-infecting parasites. The sporozoites leave the blood stream soon after injection and enter the parenchymal cells of the liver. As the parasite grows, it assumes different shapes (oval, round, or lobulate) and may contain one or more vacuoles and from several to many flocculi. Normally, the EE body develops within a single parenchymal cell of the liver, displacing the host cell nucleus. *Plasmodium brasilianum,* however, characteristically produces a marked enlargement of the host cell nucleus (Sodeman *et al.,* 1969), a feature found in none of the other nonhuman

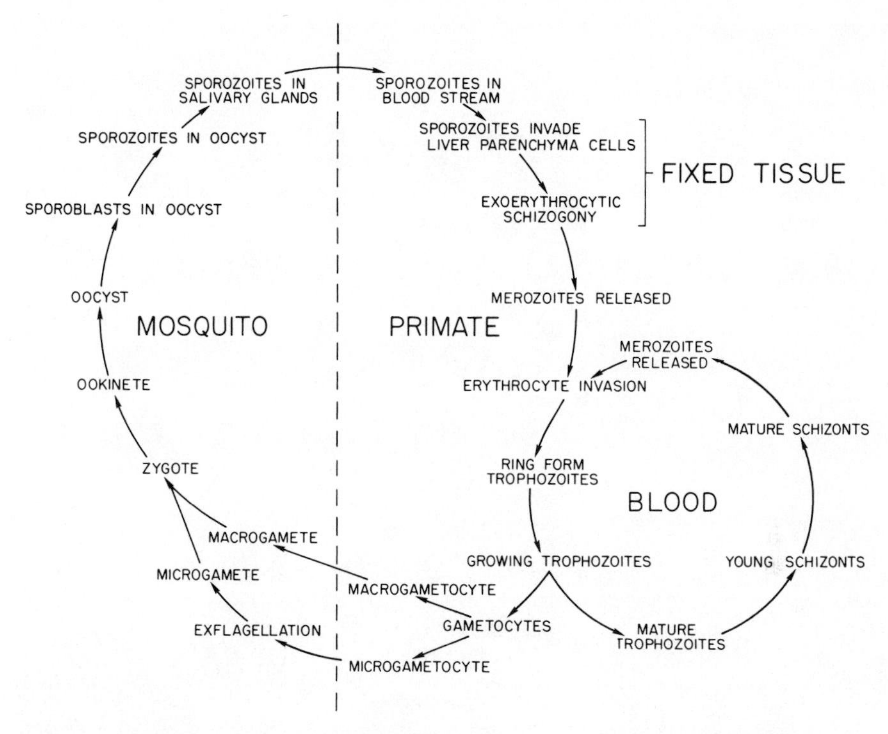

Fig. 1. Diagrammatic presentation of the life cycle of the primate malaria parasite (from Coatney *et al.*, 1971).

primate malarias but found in liver cells infected with *P. malariae* and *P. ovale*. In the older forms, distinct clefts appear in some of the schizonts. The appearance of the EE bodies of *P. knowlesi* is illustrated in Figs. 2–13. There are differences such as growth rate, presence or absence of vacuoles, inclusions, and clefts between the different species of *Plasmodium* which facilitate their identification. The development in the size of the EE schizonts of *P. knowlesi* is presented in Table II. The EE body of this parasite is almost indistinguishable from the host cell nucleus at 40 hours after sporozoite inoculation. However, by 48 hours, it can be readily identified. The mean diameter at this time ranges from 6 to 8 μm. The *P. knowlesi* EE body grows rapidly and development is completed in from 112 to 120 hours. At this time, very young trophozoites are found in host erythrocytes. In the 3-day period between 48 and 120 hours, the mean diameter of the EE body increases almost sixfold. At this time, thousands of merozoites are released to begin the erythrocytic cycle.

The morphology of the parasites in the red blood cells is similar to

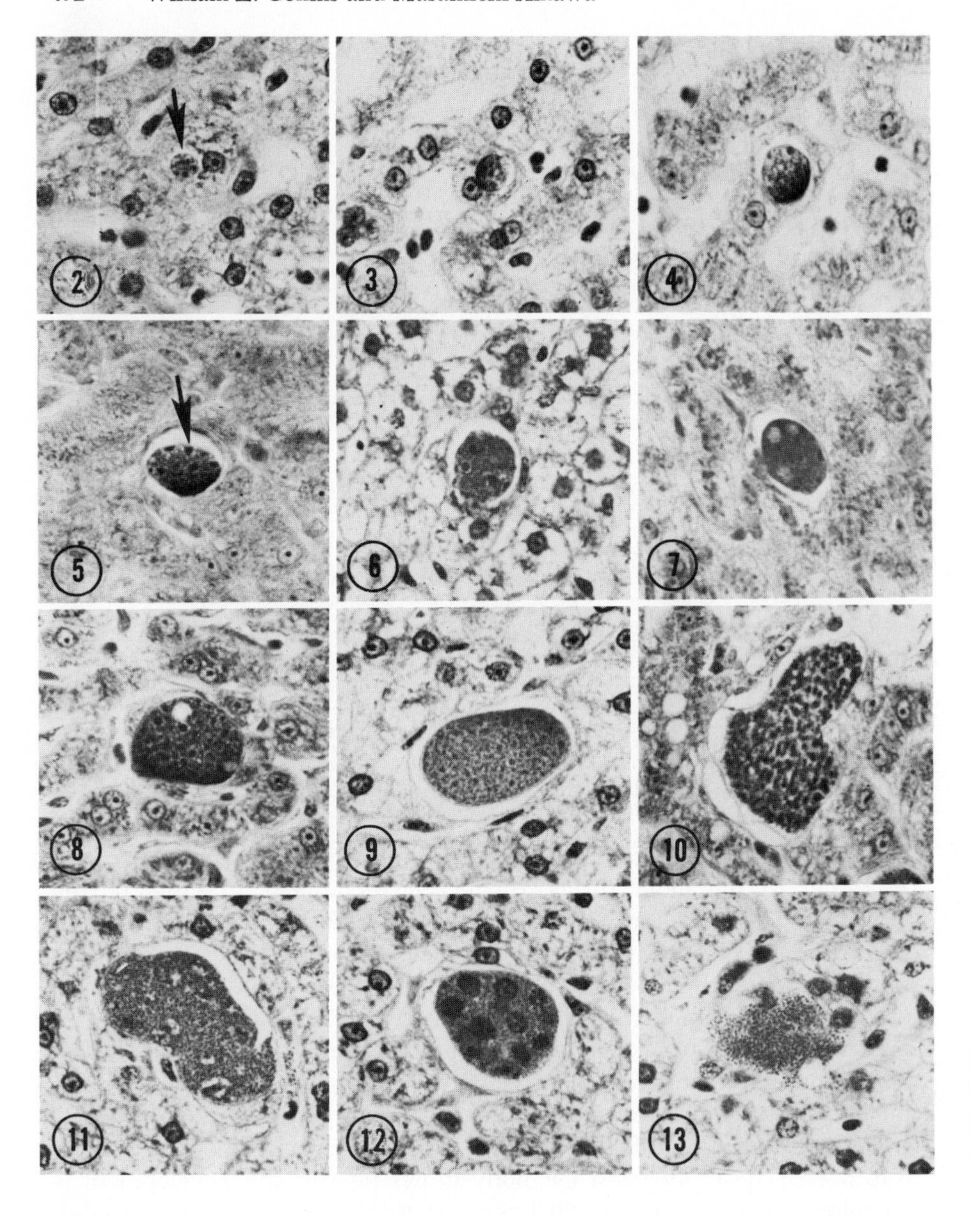

Figs. 2–13. Exoerythrocytic (EE) bodies of *Plasmodium knowlesi* in parenchymal cells of the liver of *Macaca mulatta* monkeys. × 475. **Fig. 2.** 48-Hour EE body slightly larger than liver cell nucleus. **Fig. 3.** 56-Hour EE body. **Fig. 4.** 64-Hour EE body. **Fig. 5.** 72-Hour EE body showing densely staining flocculi (FL). **Fig. 6.** 80-Hour EE body.

Table II
Diameters of Exoerythrocytic Bodies of *Plasmodium knowlesi* in Liver Cells of the *Macaca mulatta* Monkey

Hours after sporozoite inoculation	Mean diameter of midsections of 25 EE bodies (μm) [a]		
	Smallest	Largest	Mean
48	6	8	7.1
56	7	10	8.6
64	10	16.5	13.9
72	12.3	20.0	15.5
80	12.5	24.3	19.8
88	18.3	28.3	20.8
96	20.3	34.5	24.5
104	31.3	44.0	38.1
112	33.8	47.5	41.8
120	37.8	49.3	41.7

[a] Mean diameter = largest plus smallest diameter divided by 2.

that of the human malarias. However, the ultrastructure is much better known. Since Fulton and Flewett (1956) described the fine structure of the erythrocytic stages of the nonhuman primate malaria parasite, *P. knowlesi,* the ultrastructure of this group of organisms, particularly these stages, has been well established. The nonhuman primate plasmodia which have been studied by electron microscopy include the erythrocytic stages of *P. knowlesi* (Aikawa *et al.,* 1966), *P. cynomolgi* (Aikawa *et al.,* 1975), *P. simium* (Seed *et al.,* 1976), *P. brasilianum* (Sterling *et al.,* 1972), *P. simiovale* (Aikawa *et al.,* 1975), *P. fieldi* (Aikawa *et al.,* 1975), *P. inui* (H. N. Fremount, personal communication, 1975), *P. coatneyi* (Rudzinska and Trager, 1968), and *P. fragile* (Fremount and Miller, 1975), the exoerythrocytic stages of *P. cynomolgi* (Sodeman *et al.,* 1970), and the mosquito stages of *P. brasilianum* (Garnham *et al.,* 1963) and *P. cynomolgi* (Terzakis, 1971). With the exception of size and shape, the ultrastructure of the exoerythrocytic and mosquito stages of nonhuman primate parasites is essentially similar to that of the erythrocytic

Fig. 7. 88-Hour EE body showing vacuoles. **Fig. 8.** 96-Hour EE body with prominent vacuole and numerous flocculi. **Fig. 9.** 104-Hour EE body. **Fig. 10.** 112-Hour EE body. **Fig. 11.** 120-Hour EE body. **Fig. 12.** 120-Hour EE body. **Fig. 13.** 120-Hour EE body releasing merozoites.

stages. Therefore, only the fine structure of the erythrocytic stages will be discussed in this section.

In general, the ultrastructure of these different species of malaria parasites in red blood cells is basically the same. *Plasmodium brasilianum*, a parasite of New World monkeys, possesses a nucleolus and distinct mitochondria with cristae (Sterling *et al.*, 1972), while the parasites of the Old World monkeys do not possess these structures (Aikawa and Sterling, 1974).

The malaria parasite enters into the host cell by invaginating the host cell membrane by pressing the anterior end against the host cell (Fig. 14). After entering into the host cell, the parasite is located within a parasitophorous vacuole which separates the parasite from the host cytoplasm (Fig. 15). The uninucleate trophozoite of the erythrocytic stage is generally oval in shape, but often shows extensive protrusions and invaginations of the cytoplasm, assuming elongated, ring- to semiring shape. The irregular shape of the young trophozoite may account for the ring-shaped form seen in the blood smear by light microscopy.

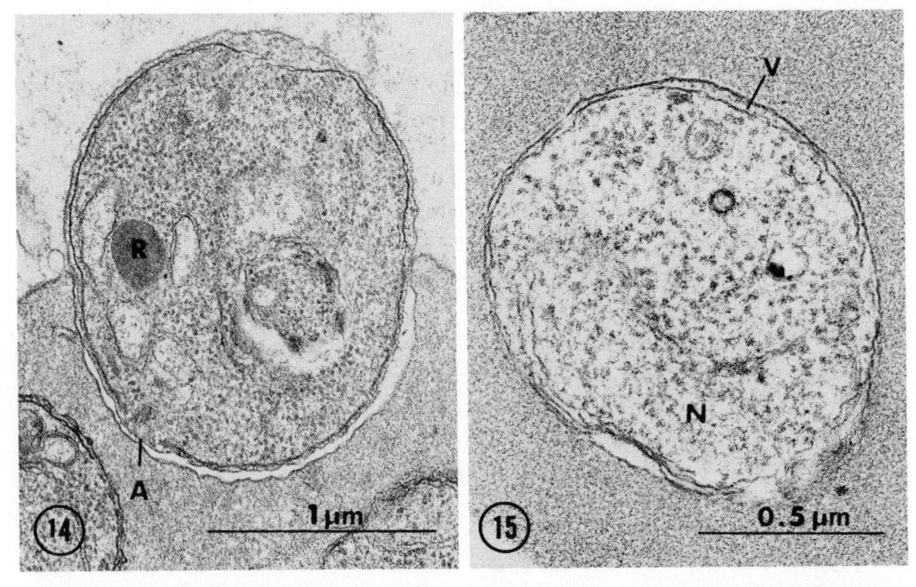

Fig. 14. A merozoite of *P. knowlesi* in the process of invading an erythrocyte. The anterior end (A) leads in the invagination of the erythrocyte plasmalemma. R, rhoptries. **Fig. 15.** A uninucleate trophozoite of *P. cynomolgi* in a parasitophorous vacuole (V). N, nucleus.

The uninucleate trophozoite possesses a nucleus, endoplasmic reticulum, ribosomes, food vacuoles, a cytostome, and mitochondria or mitochondria-equivalent bodies. Nonhuman primate malaria parasites infecting the Old World monkeys possess double membrane-bounded bodies instead of typical mitochondria (Figs. 16 and 17). The cytostome, which ingests host cell cytoplasm, is present in all of these nonhuman primate parasites (Fig. 19). The digestion of host cell cytoplasm occurs within food vacuoles derived from the cytostome. As the digestion process takes place, malaria pigment particles are formed within the food vacuoles.

During schizogony, nuclear division and differentiation of the cytoplasmic organelles are the two major events. Spindle fibers appear in the nucleus, radiating from poorly delineated electron-dense centriolar plaques located on the nuclear membrane (Fig. 18). As nuclear division progresses, the nucleus becomes more elongated, causing the centriolar plaques to become the long axis. The cytoplasm, meanwhile, begins to differentiate. First, randomly distributed segments of the thick membrane appear beneath the plasma membrane (Figs. 20 and 21). Also, pellicular microtubules of the merozoites and rhoptries appear. The area which is covered by the thick membrane begins to protrude outward to form new merozoites (Fig. 22). With the progression of merozoite budding, various organelles including nucleus, mitochondria, or mitochondria-equivalent together with endoplasmic reticulum and ribosomes migrate into the developing merozoites from the schizont.

The merozoite is ovoid to elongated in shape and is surrounded by a characteristic pellicular complex (Fig. 23). The anterior end is shaped like a truncated cone with its base demarcated by three electron-dense polar rings. Near the anterior end are two electron-dense rhoptries and many micronemes. A short duct from each of the rhoptries extends to the anterior (Fig. 23). These structures have been suggested to contain enzymes that dissolve the host cell membrane upon contact (Aikawa and Sterling, 1974), but this has not been proven. A nucleus is located in the center and there is a mitochondrion or mitochondrion-equivalent posterior to the nucleus (Fig. 23). Closely associated with the mitochondrion is a spherical body which is suggested to be an energy reservoir for the mitochondrion (Fig. 23).

The gametocyte is a uninucleate parasite with a three-layered pellicle (Fig. 24). It possesses a nucleus, ribosomes, endoplasmic reticulum, a cytostome, mitochondria, food vacuoles, and osmiophilic bodies (Fig. 25). The microgametocyte can be differentiated from the macrogametocyte by the number of ribosomes, the cytoplasm of the macrogametocyte containing more ribosomes than that of the microgametocyte (Figs. 24 and 25). This difference accounts for the more intense basophilia of the

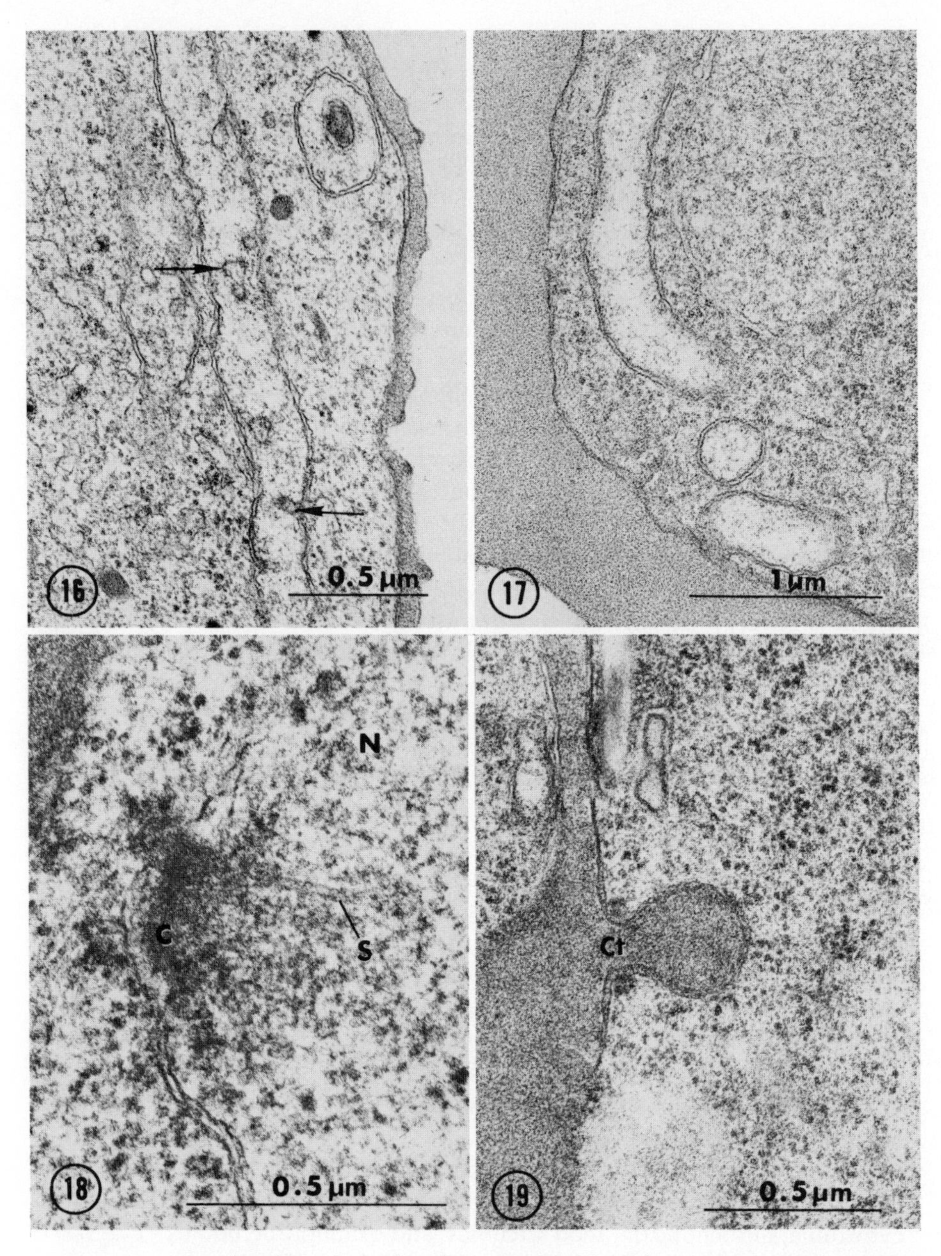

Fig. 16. Mitochondria of *P. brasilianum* with typical protozoan type cristae (arrow). **Fig. 17.** Mitochondria of *P. knowlesi* without prominent cristae. **Fig. 18.** Nuclear division of *P. brasilianum*. The nucleus (N) shows centricular plaques (C) and spindle fibers (S). **Fig. 19.** *Plasmodium knowlesi* ingesting host cell cytoplasm through a cytostome (Ct).

macrogametocyte cytoplasm than the microgametocyte cytoplasm when stained for light microscopy. Also, the macrogametocyte possesses more osmiophilic bodies than the microgametocyte.

Erythrocytes infected by human or nonhuman primate parasites show characteristic morphologic alterations (Table III). *Plasmodium coatneyi, P. fragile,* and *P. brasilianum* produce small excrescences on the erythrocyte plasma membrane (Fig. 27). These excrescences form focal junctions with the membrane of endothelial cells or with excrescences on other erythrocytes, suggesting that they are responsible for infected erythrocyte sequestration in deep organs (Aikawa *et al.,* 1972). Schüffner's dots seen by light microscopy in the erythrocytes infected by *P. vivax*-type primate malaria parasites are demonstrated by electron microscopy to be caveola-vesicle complexes along the erythrocyte plasma membrane (Fig. 26), while Maurer's clefts seen by light microscopy appear to correspond to narrow slitlike structures in the cytoplasm. The functional significance of these host cell changes produced by malaria parasites is still unknown and must await future investigation.

The sporogonic cycle is very similar to that seen in the human malarias. The length of time required for its completion varies from one species to another. Thus, it is possible to specifically separate the species based on the length of the developmental cycle and also on the size of the oocysts at certain periods, provided the temperature is held constant. The development of the oocysts is strongly temperature-oriented and lower temperatures can extend the developmental time into a number of weeks. Bennett *et al.* (1966) were able to differentiate several different subspecies of *P. cynomolgi* on the basis of oöcyst size and developmental time.

A comparison of the developmental time and the mean oocyst diameters for thirteen species of nonhuman primate malaria parasites is presented in Table IV. Some mosquitoes are refractory to infection and others will only partially support oocyst development. However, all of the thirteen species presented in the table develop readily in the mosquito species indicated. *Anopheles balabacensis balabacensis* appears to be the most universal experimental vector and has been studied with all except *P. brasilianum* and *P. simium*; results of studies using *A. freeborni* are presented for these species. It can be seen that the length of time required for completion of the sporogonic cycle varies from 10 days for *P. knowlesi* and *P. cynomolgi* to 17 days for *P. brasilianum.* It is also apparent that after different time intervals, there is a marked difference in the mean oocyst diameters for the different species. For example, at day 8, the mean diameter of *P. cynomolgi* oocysts is 49 μm whereas *P. fragile* and *P. jefferyi* have mean diameters of only 16 μm. *Plasmodium*

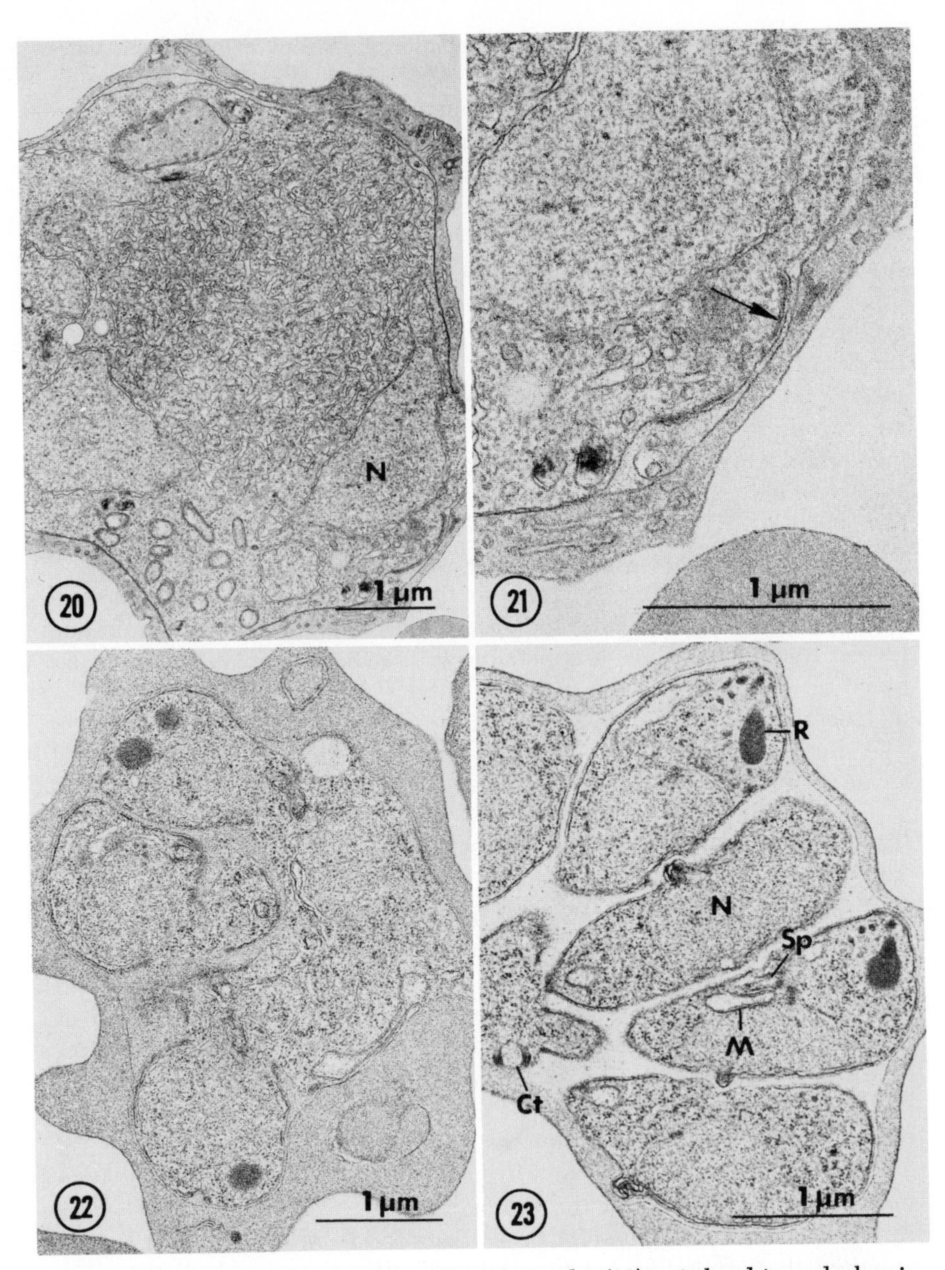

Fig. 20. A schizont of *P. simium* with four nuclei (N), mitchondria, endoplasmic reticulum, and ribosomes. **Fig. 21.** High magnification of Fig. 20. The areas covered by segments of thick inner membrane (arrow) protrude outward forming new merozoites. **Fig. 22.** Advanced schizogony of *P. knowlesi* with merozoite budding. **Fig. 23.** A cluster of merozoites of *P. knowlesi*. The merozoite is surrounded by a pellicle and possesses various organelles such as a nucleus (N), rhoptries (R), mitochondria (M), spherical body (Sp), and a cytostome (Ct).

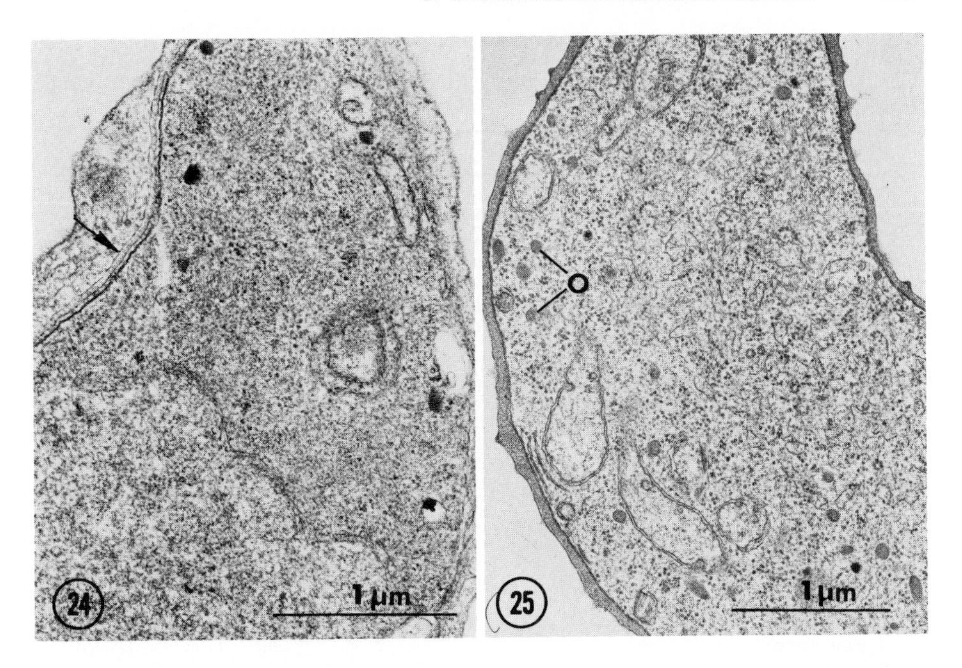

Fig. 24. A microgametocyte of *P. cynomolgi* surrounded by a three-layered membrane (arrow). The cytoplasm contains fewer ribosomes than the macrogametocyte (see Fig. 12). **Fig. 25.** A macrogametocyte of *P. brasilianum*. The cytoplasm contains abundant ribosomes and osmiophilic bodies (O).

schwetzi has much larger oocysts than any of the other species with a mean diameter of 82 μm at day 14; some individual oocysts of this species reach diameters of greater than 100 μm.

In addition to the length of the sporogonic cycle and oocyst size, there are morphologic features of the different species which serve to differentiate them. These include the size and arrangement of pigment, a residue from the macrogametocyte. The appearance of oocysts and sporozoites for several species of nonhuman primate malaria are presented in Figs. 28–35. Different characteristics of the oocysts offer a great deal of promise for specific and subspecific differentiation of the primate malarias.

A. Periodicity

Based on the length of time required for the completion of one schizogonic cycle, these parasites have been classified as having either quotidian (24-hour cycle), tertian (48-hour cycle), or quartan (72-hour cycle) periodicity. Only one species, *P. knowlesi*, has a quotidian cycle. Two, *P. brasilianum* and *P. inui*, are known to have quartan cycles. Garnham (1966) also thought that *P. girardi* would have a quartan

Table III

Changes in Erythrocytes Infected by Some Primate Malarial Parasites by Light and Electron Microscopy [a]

Parasites	Light microscopy	Electron microscopy: Cytoplasm: Clefts	Electron microscopy: Plasma membrane: Excrescences	Electron microscopy: Plasma membrane: Caveolae	Electron microscopy: Plasma membrane: C-V compl.
Vivax type					
P. vivax	Schüffner's dots	+	–	+	+
P. simium	Schüffner's dots	+	–	+	+
P. cynomolgi	Schüffner's dots	+	–	+	+
Ovale type					
P. simiovale	Schüffner's dots	+	±	+	+
P. fieldi	Schüffner's dots	+	±	+	+
Falciparum type					
P. falciparum	Maurer's clefts	+	+*	–	–
P. fragile	Faint stippling	+	+	+	–
P. coatneyi	Maurer's clefts	+	+*	+	–
Malariae type					
P. malariae	Ziemann's stippling	+	+*	–	–
P. brasilianum	Ziemann's stippling	+	+*	–	–
P. inui	Schüffner's-like stippling	+	–	+	±
Others					
P. knowlesi	Sinton and Mulligan's dots	+	–	+	–

[a] Key to symbols in table: +, structure present; ±, structure present occasionally; –, structure absent. C-V Compl., caveola-vesicle complexes. *Excresences are on both erythrocytes infected by asexual forms and gametocytes of *P. malariae* and *P. brasilianum*, but they are only on erythrocytes with asexual forms of *P. falciparum* and *P. coatneyi* (Aikawa et al., 1975).

periodicity. The remainder are all thought to have a 48-hour or tertian periodicity although the information on *P. lemuris* is meager.

B. Relapse

Infections in monkeys are characterized by periods of high parasitemias interspersed with periods of latency, often of considerable duration. This pattern is found in both blood-induced and sporozoite-induced infections. With blood-induced infections, this is caused by the reappearance of parasites derived from the erythrocytic cycle (recrudescence of the infection). With sporozoite-induced infections, this can be a recrudescence or a true "relapse" caused by the invasion of erythrocytes

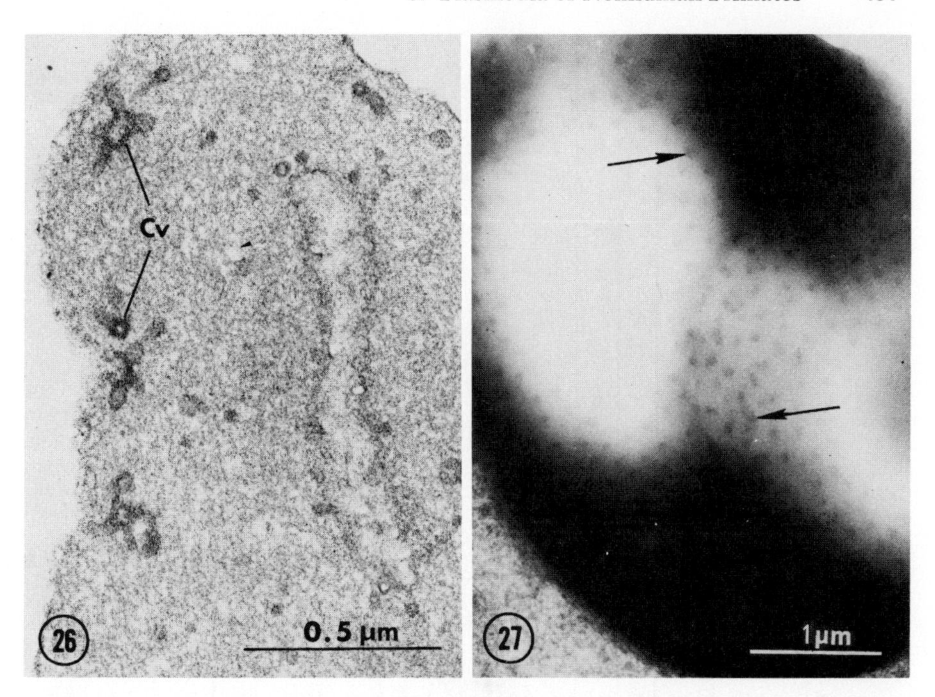

Fig. 26. An erythrocyte infected by *P. cynomolgi* showing caveola-vesicle complexes (Cv). **Fig. 27.** High voltage electron microg aph of an erythrocyte infected by *P. brasilianum* showing numerous excrescences (arrow) on the surface.

with merozoites from persisting schizonts in the liver. Only five primate malarias are known to have this relapse mechanism as part of their biology. These are *P. cynomolgi, P. fieldi,* and *P. simiovale* of monkeys and *P. vivax* and *P. ovale* of man. This relapse sometimes occurs after many months of latency following treatment with a schizonticidal drug. It has been postulated that these infections are due to latent forms derived from the initial sporozoite inoculation since there is no evidence of a cyclic stage in the exoerythrocytic schizogony (Garnham, 1967; Coatney *et al.*, 1971). Other species, *P. coatneyi, P. fragile, P. knowlesi, P. inui, P. gonderi,* and *P. hylobati* apparently do not possess this characteristic. Whether the other species have this relapse mechanism is not known.

III. Taxonomy

A discussion of the taxonomy of the primate malaria parasites was presented by Garnham (1973). This report excluded any discussion on *P. girardi* or *P. lemuris,* the parasites of lemurs. Garnham (1966) divided

Table IV

Comparison of the Mean Oocyst Diameters and the Length of Time Required for the Completion of the Sporogonic Cycle of Thirteen Species of Primate Malaria [a]

Plasmodium species	*Anopheles* species	Mean oocyst diameter (μm)						Sporozoites (day)
		Day 6	Day 8	Day 10	Day 12	Day 14	Day 16	
knowlesi	*b. balabacensis*	20	46	64	—	—	—	10
cynomolgi	*b. balabacensis*	22	49	67	—	—	—	10
coatneyi	*b. balabacensis*	19	35	54	—	—	—	11
simium	*freeborni*	19	26	42	46	—	—	12
hylobati	*b. balabacensis*	15	26	41	53	—	—	12
jefferyi	*b. balabacensis*	12	16	23	40	—	—	13
simiovale	*b. balabacensis*	16	27	44	67	—	—	13
gonderi	*b. balabacensis*	23	39	50	63	—	—	13
fieldi	*b. balabacensis*	15	26	43	62	63	—	14
schwetzi	*b. balabacensis*	31	46	62	77	82	—	14
inui	*b. balabacensis*	13	21	32	36	50	—	15
fragile	*b. balabacensis*	11	16	25	39	56	59	16
brasilianum	*freeborni*	11	15	21	30	39	48	17

[a] Extrinsic incubation period at 25°C.

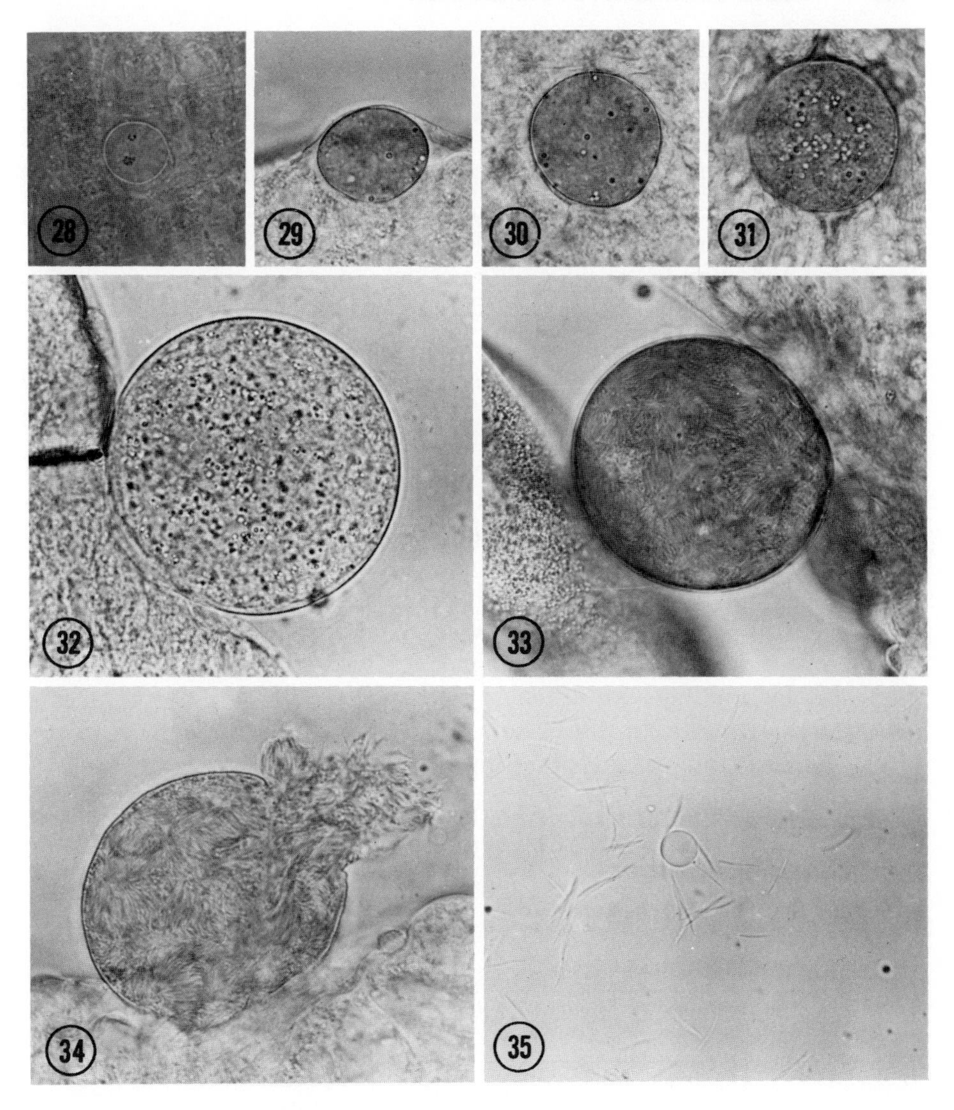

Figs. 28–35. Sporogonic stages of nonhuman primate malarias. × 475. **Fig. 28.** 7-Day oocyst of *P. coatneyi* showing two clumps of pigment. **Fig. 29.** 9-Day oocyst of *P. simium* under basement membrane of mosquito gut. **Fig. 30.** 8-Day oocyst of *P. simium*. **Fig. 31.** 10-Day oocyst of *P. simium*. **Fig. 32.** 12.5-Day oocyst of *P. schwetzi*. **Fig. 33.** 13-Day fully differentiated oocyst of *P. fieldi* showing nearly mature sporoizoites. **Fig. 34.** 12-Day rupturing oocyst of *P. simiovale* releasing sporozoites. **Fig. 35.** Sporozoites of *P. inui* released from salivary glands.

these parasites into three subgenera, *Laverania, Plasmodium,* and *Vinckeia.* However, separation of *Laverania* is based on only one characteristic of one stage of the parasite (crescentic gametocyte) and *Vinckeia* is based on small erythrocytic schizonts. Yet, this latter subgenus is actually a collection of many parasites about which very little is known and which possess many more distinct differences. Thus, at present, it appears suitable to consider the subgeneric classification as unsettled and open to considerable revision.

The classification of the species of *Plasmodium* found in the primates is based on the morphology of the erythrocytic stages (asexual cycle) in the primate host. Subspecific classification has been based on characteristics of the EE stage (*P. cynomolgi cynomolgi* vs. *P. cynomolgi bastianellii*) and geographic distribution (*P. inui inui* vs. *P. inui shortti*). There are, however, many differences between strains of monkey malarias, particularly *P. cynomolgi* and *P. inui,* due to their wide distribution and evolution in different primate hosts. Which of these differences are sufficient to warrant subspecific classification and which take precedence over others has not yet been determined. However, it appears that the subspecific classification will require specialized studies in laboratory animals and vector anophelines which, at present, are not available for routine study by most workers.

IV. Distribution and Host Range

The known natural hosts, natural vectors, and geographic range of the nonhuman primate malarias are presented in Table V. Two are found in Africa, two from the Malagasy Republic, two from South Amer-

Table V
Geographical Range and Natural Hosts for the Nonhuman Primate Malaria Parasites

Species	Periodicity	Geographical range	Natural hosts	Natural vectors
P. knowlesi	Quotidian	Malaysia	*Macaca fascicularis*	*A. hackeri*
			M. nemestrina	
			Presbytis melalophus	
P. coatneyi	Tertian	Malaysia	*Macaca fascicularis*	*A. hackeri*
		Philippines		
P. cynomolgi	Tertian	India	*Macaca fascicularis*	*A. balabacensis balabacensis*
		Sri Lanka	*M. nemestrina*	
		Malaysia	*M. radiata*	*A. hackeri*
		Taiwan	*M. cyclopis*	*A. b. introlatus*
		Indonesia (Java and the Celebes)	*M. mulatta*	*A. elegans*
			Presbytis cristatus	

		Assam	*P. entellus*	
P. eylesi	Tertian	Malaysia	*Hylobates lar*	Unknown
P. fieldi	Tertian	Malaysia	*Macaca nemestrina* *M. fascicularis*	*A. hackeri* *A. balabacensis introlatus*
P. fragile	Tertian	Sri Lanka	*Macaca sinica* *M. radiata*	*A. elegans*
P. gonderi	Tertian	Cameroons Zaire	*Cercocebus atys* *C. galeritus agilus* *Mandrillus leucophaeus*	Unknown
P. hylobati	Tertian	East Malaysia	*Hylobates moloch*	Unknown
P. jefferyi	Tertian	Malaysia	*Hylobates lar*	Unknown
P. pitheci	Tertian	East Malaysia	*Pongo pygmaeus*	Unknown
P. reichenowi	Tertian	Cameroons Sierra Leone Zaire Dem. Rep. of the Congo	*Pan troglodytes* *Gorilla gorilla*	Unknown
P. schwetzi	Tertian	Cameroons Sierra Leone Zaire Liberia Dem. Rep. of the Congo	*Pan troglodytes* *Gorilla gorilla*	Unknown
P. simiovale	Tertian	Sri Lanka	*Macaca sinica*	Unknown
P. simium	Tertian	Brazil	*Alouatta fusca* *Brachyteles arachnoides*	*A. cruzi*
P. sylvaticum	Tertian	East Malaysia	*Pongo pygmaeus*	Unknown
P. youngi	Tertian	Malaysia	*Hylobates lar*	Unknown
P. brasilianum	Quartan	Brazil Venezuela Colombia Peru Panama	Many species of the following genera: *Alouatta* *Ateles* *Brachyteles* *Callicebus* *Cebus* *Chiropotes* *Lagothrix* *Saimiri*	*A. cruzi*
P. inui	Quartan	Sri Lanka India Malaysia Philippines Indonesia Taiwan	*Macaca fascicularis* *M. nemestrina* *M. mulatta* *M. radiata* *M. cyclopis* *M. sinica* *Presbytis cristatus* *P. obscurus* *Cynopithecus niger*	*A. elegans* *A. leucosphyrus* *A. balabacensis introlatus*

ica, and the remaining fourteen in tropical Asia, extending from Sri Lanka to the Philippines. Since no nonhuman primates are found south of Wallace's line, the southern limit to their distribution is that part of Indonesia north of New Guinea. Seven species have been isolated from macaques. The natural mosquito hosts are less well known, the first being determined by Wharton and Eyles in 1961, who demonstrated that *A. hackeri* was a vector of *P. knowlesi* in Malaysia. Most of the efforts to determine the natural vectors have been in Malaysia, Sri Lanka, and Brazil. *Plasmodium coatneyi* holds the distinction of being initially isolated from its vector, *A. hackeri,* prior to its being found in a naturally infected primate.

Since field studies are often difficult, most efforts to study these malaria parasites have been made in laboratory animals. The rhesus monkey, *M. mulatta,* is the most widely used since it is susceptible to infection with all the Asian monkey malaria parasites as well as the African parasite, *P. gonderi.* In this host, the parasitemias often reach very high levels and, frequently, monkeys die of overwhelming infections. *Plasmodium knowlesi* is almost always fatal when the infection is induced via the inoculation of parasitized blood. A lower percentage (approximately 70%) of the animals die if the inoculation is with sporozoites. *Plasmodium coatneyi* and *P. fragile* are also lethal to this host, with mortalities of approximately 30%. The other species of malaria parasites rarely result in fatality even though the parasitemias may be very high and persist for many months. Splenectomy of the animal increases the parasitemia dramatically with concurrent increases in the mortality rate.

Many of the primate malarias are infective to and transmitted by a wide variety of coindigenous and exotic anophelines. The most prominent of these is *P. cynomolgi,* which has been experimentally transmitted by over twenty different species of mosquitoes. In our own laboratory, six different species of anophelines have successfully transmitted one or more species of the parasites. As indicated in Table VI, *A. balabacensis balabacensis* has transmitted eight different parasites to either *M. mulatta* or *Aotus trivirgatus* monkeys.

The prepatent periods have a considerable range for a particular host–parasite inoculation, probably due in large measure to the size of the sporozoite inoculum. Of 96 transmissions of *P. knowlesi* to *M. mulatta* monkeys, the prepatent periods ranged from 5 to 9 days with a mean of 7.1 days. With *P. inui,* the prepatent periods ranged from 10 to 28 days with a mean of 16.1 days. These differences in the prepatent period between the different species is directly related to (1) the length of time required for the completion of the exoerythrocytic cycle; (2) the length

Table VI

Summary of 354 Transmissions of Simian Malarias to Laboratory Monkeys via the Bites of Infected Mosquitoes

Species of *Plasmodium*	Primate host	No. trans.	Vector anopheline	Extrinsic incubation (days)	Prepatent period (days) Range	Mean
coatneyi	*Macaca mulatta*	13	*A. b. balabacensis*	12–18	10–15	12.8
cynomolgi	*M. mulatta*	158	*A. b. balabacensis* *A. atroparvus* *A. freeborni* *A. maculatus* *A. quadrimaculatus* *A. stephensi*	10–21	7–16	10.4
	Aotus trivirgatus	13	*A. b. balabacensis* *A. freeborni* *A. maculatus*	14–18	12–30	18.2
fieldi	*M. mulatta*	13	*A. b. balabacensis* *A. maculatus* *A. stephensi*	15–20	10–15	12.7
fragile	*M. mulatta*	2	*A. b. balabacensis*	18	17	17.0
	A. trivirgatus	4	*A. b. balabacensis*	17–22	20–27	24.5
gonderi	*M. mulatta*	9	*A. b. balabacensis* *A. freeborni* *A. maculatus* *A. stephensi*	15–19	9–17	12.7
inui	*M. mulatta*	37	*A. b. balabacensis* *A. maculatus* *A. stephensi*	13–22	10–28	16.1
knowlesi	*M. mulatta*	96	*A. b. balabacensis*	12–17	5–9	7.1
	A. trivirgatus	4	*A. b. balabacensis*	15	12–17	14.5
simiovale	*M. mulatta*	5	*A. b. balabacensis* *A. maculatus*	16–21	12–17	13.6

of the erythrocytic cycle (quotidian, tertian, or quartan); and (3) to the number of merozoites produced per cycle. In another host, *Aotus trivirgatus*, the prepatent periods were noticeably longer. Once established, however, the parasitemias in these animals often reached very high levels; with *P. knowlesi* and *P. fragile*, the infections had a high rate of mortality.

V. Infectivity to Man

Efforts to infect man with the malaria parasites of nonhuman primates have been of special interest. *Plasmodium knowlesi* was the first to be shown to be infectious to man. An infection was induced by the inoculation of parasitized blood (Knowles and Das Gupta, 1932). Subsequently, infections with this parasite were used by other workers for the treatment of general paresis (Van Rooyen and Pile, 1935; Chopra and Das Gupta, 1936). Ciucă *et al.* (1937a,b) reported the infection of 321 patients with the parasite, all by the inoculation of parasitized blood. After 170 passages, however, the infection became so virulent that its use had to be terminated (Ciucă *et al.*, 1955). One confirmed natural infection (Chin *et al.*, 1965) and another suspected infection (Fong *et al.*, 1971) with this parasite have been reported from Malaysia. Chin *et al.* (1968) showed that not only could the infection be transmitted to man via the bites of infected mosquitoes, but demonstrated man-to man and man-to-monkey transmission as well using *A. b. balabacensis* mosquitoes.

A considerable amount of study has been made on the transmission of *P. cynomolgi* to man following the accidental infection of several laboratory workers in 1960 (Eyles *et al.*, 1960). The results, summarized by Coatney *et al.* (1971) indicated that monkey-to-man, man-to-man, and man-to-monkey transmission via the bites of infected mosquitoes could be obtained. Although the parasitemias were low, the clinical symptoms were mildly severe.

Studies with volunteers indicated that *P. brasilianum, P. inui,* and *P. schwetzi* are also infectious to man although the parasitemias are low and the clinical symptoms mild (Contacos *et al.*, 1963, 1970; Coatney *et al.*, 1966). Both Caucasians and Negroes were susceptible to *P. knowlesi* and *P. brasilianum,* but Negroes were apparently refractory to *P. inui, P. schwetzi,* and *P. cynomolgi.* A natural infection with *P. simium* has been reported (Deane *et al.*, 1966a,b). In addition, Coatney *et al.* (1971) discussed the possible transmission of the gibbon parasite, *P. eylesi,* to man although it could not be confirmed by subpassage into gibbons.

It is apparent that man can be involved in the transmission cycle of these parasites. The morphology of some species is so similar to that of the human malarias that identification microscopically is unlikely. Only subpassage into susceptible laboratory animals can confirm the true identity of the causal agent. Serologic studies can indicate the probable species involved. However, serologic cross-reactions between the human and nonhuman species are common and identification on the basis of serologic results alone is questionable.

VI. Diagnosis

Diagnosis of the infection in a primate is based primarily on the appearance of the erythrocytic forms in a Giemsa-stained blood film. Confirmation of the infection can be made by serologic examination using fluorescent antibody techniques employing blood films of homologous and heterologous species. Chronic low-grade infections commonly have parasitemias of undetectable levels. Daily blood films may indicate the presence and species of malaria parasites. In the nonhuman primate, if identification is important, splenectomy can be performed. This will result in a marked increase in the parasitemia within a few weeks to levels which make species identification possible. In addition, treatment with immunosuppressant drugs, such as cortisone, may also result in increased parasitemias to detectable levels.

VII. Treatment

The antimalarial drugs which are used for the treatment of human malaria are also effective in the treatment of malaria infections in nonhuman primates. Dosage of treatment in macaques is normally three times the mg/kg rate as used in man. In smaller primates, such as the *Aotus* monkey, treatment is normally four times the mg/kg rate used for man.

Blood schizonticidal drugs such as the 4-aminoquinolines are effective in the treatment of all of the nonrelapsing malarias. For *P. cynomolgi*, *P. fieldi*, and *P. simiovale*, the known relapsing malarias, radical cure can be effected by the use of an 8-aminoquinoline such as primaquine in combination or following treatment with the blood schizonticidal drug.

ACKNOWLEDGMENTS

This study was supported in part by research grants (AI-10645 and AI-08970) from the U.S. Public Health Service, and by the U.S. Army R&D Command Control (DADA 17-70-C-0006). M.A. is a Research Career Development Awardee (AI-46237) from the U.S. Public Health Service.

REFERENCES

Aikawa, M., and Sterling, C. R. (1974). "Intracellular Parasitic Protozoa." Academic Press, New York.

Aikawa, M., Huff, C. G., and Sprintz, H. (1966). Comparative feeding mechanisms of avian and primate malarial parasites. *Mil. Med.* **131**, 969–983.

Aikawa, M., Rabbege, J. R., and Wellde, B. T. (1972). Junctional apparatus in erythrocytes infected with malarial parasites. *Z. Zellforsch. Mikrosk. Anat.* **124**, 72–75.

Aikawa, M., Miller, L. H., and Rabbege, J. (1975). Caveola-vesicle complexes in the plasmalemma of erythrocytes infected by *Plasmodium vivax* and *Plasmodium cynomolgi:* Unique structures related to Schüffner's dots. *Am. J. Pathol.* **79**, 285–300.

Bennett, G. F., Warren, McW., and Cheong, W. H. (1966). Biology of the simian malarias in Southeast Asia. IV. Sporogony of four strains of *Plasmodium cynomolgi*. *J. Parasitol.* **52**, 639–646.

Brumpt, E. (1939). Les parasites due paludisme des chimpanzés. *C. R. Seances Soc. Biol. Ses Fil.* **130**, 837–840.

Bück, G., Coudurier, J., and Quesnel, J. J. (1952). Sur deux nouveaux plasmodium observés chez un lémurien de Madagascar splénectomisé. *Arch. Inst. Pasteur Alger.* **30**, 240–243.

Chin, W., Contacos, P. G., Coatney, G. R., and Kimball, H. R. (1965). A naturally acquired quotidian-type malaria in man transferable to monkeys. *Science* **149**, 865.

Chin, W., Contacos, P. G., Collins, W. E., Jeter, M. H., and Alpert, E. (1968). Experimental mosquito-transmission of *Plasmodium knowlesi* to man and monkey. *Am. J. Trop. Med. Hyg.* **17**, 355–358.

Chopra, R. N., and Das Gupta, B. M. (1936). A preliminary note on the treatment of neuro-syphilis with monkey malaria. *Indian Med. Gaz.* **71**, 181–188.

Ciucă, M., Tomescu, P., and Badenski, G., with the collaboration of Badenski, A., Ionescu, P., and Teriteanu, M. (1937a). Contribution à l'étude de la virulence du *Pl. knowlesi* chez l'homme. Caractères de la maladie et biologie du parasite. *Arch. Roum. Pathol. Exp. Microbiol.* **10**, 5–28.

Ciucă, M., Ballif, L., Chelarescu, M., Lavrinenko, M., and Zotta, E. (1937b). Contributions à l'étude de l'action pathogène de *Pl. knowlesi* pour l'homme (considérations sur l'immunité naturelle et l'immunité acquise contre cette espèce de parasite). *Bull. Soc. Pathol. Exot.* **30**, 305–315.

Ciucă, M., Chelarescu, M., Sofletea, A., Constantinescu, P., Teriteanu, E., Cortez, P., Balanovschi, G. and Ilies M. (1955). "Contribution expérimentale à l'étude de l'immunité dans le paludisme." Ed. Acad. Repub. Pop. Roum.

Coatney, G. R., Chin, W., Contacos, P. G., and King, H. K. (1966). *Plasmodium inui*, a quartan-type malaria parasite of Old World monkeys transmissible to man. *J. Parasitol.* **52**, 660–663.

Coatney, G. R., Collins, W. E., Warren, McW., and Contacos, P. G. (1971). "The Primate Malarias." US Govt. Printing Office, Washington, D.C.

Contacos, P. G., Lunn, J. S., Coatney, G. R., Kilpatrick, J. W., and Jones, F. E. (1963). Quartan-type malaria parasite of New World monkeys transmissible to man. *Science* **142**, 676.

Contacos, P. G., Coatney, G. R., Orihel, T. C., Collins, W. E., Chin, W., and Jeter, M. H. (1970). Transmission of *Plasmodium schwetzi* from the chimpanzee to man by mosquito bite. *Am. J. Trop. Med. Hyg.* **19**, 190–196.

da Fonesca, F. (1951). Plasmodio de primata do Brasil. *Mem. Inst. Oswaldo Cruz* **49**, 543–551.

Deane, L. M., Deane, M. P., and Ferreira Neto, J. (1966a). A naturally acquired human infection by *Plasmodium simium* of howler monkeys. *Trans. R. Soc. Trop. Med. Hyg.* **60**, 563–564.

Deane, L. M., Deane, M. P., and Ferreira Neto, J. (1966b). Studies on transmission of simian malaria and on a natural infection in man with *Plasmodium simium* in Brazil. *Bull. W.H.O.* **35**, 805–808.

Dissanaike, A. S., Nelson, P., and Garnham, P. C. C. (1965a). Two new malaria parasites, *Plasmodium cynomolgi ceylonensis* subsp. nov. and *Plasmodium fragile* sp. nov. from monkeys in Ceylon. *Ceylon J. Med. Sci.* **14**, 1–9.

Dissanaike, A. S., Nelson, P., and Garnham, P. C. C. (1965b). *Plasmodium simiovale* sp. nov., a new simian malaria parasite from Ceylon. *Ceylon J. Med. Sci.* **14**, 27–32.

Eyles, D. E., Coatney, G. R., and Getz, M. E. (1960). Vivax-type malaria parasite of macaques transmissible to man. *Science* **132**, 1812–1813.

Eyles, D. E., Laing, A. B. G., and Fong, Y. L. (1962a). *Plasmodium fieldi* sp. nov., a new species of malaria parasite from the pig-tailed macaque in Malaya. *Ann. Trop. Med. Parasitol.* **56**, 242–247.

Eyles, D. E., Fong, Y. L., Warren, McW., Guinn, E., Sandosham, A. A., and Wharton, R. H. (1962b). *Plasmodium coatneyi*, a new species of primate malaria from Malaya. *Am. J. Trop. Med. Hyg.* **11**, 597–604.

Eyles, D. E., Fong, Y. L., Dunn, F. L., Guinn, E., Warren, McW., and Sandosham, A. A. (1964). *Plasmodium youngi*, n. sp., a malaria parasite of the Malayan gibbon, *Hylobates lar lar. Am. J. Trop. Med. Hyg.* **13**, 248–255.

Fong, Y. L., Cadigan, F. C., and Coatney, G. R. (1971). A presumptive case of naturally occurring *Plasmodium knowlesi* malaria in man in Malaysia. *Trans. R. Soc. Trop. Med. Hyg.* **65**, 839–840.

Fremount, H. N. and Miller, L. H. (1975). Devascular schizogony in *Plasmodium fragile:* Organ distribution and ultrastructure of red cells adherent to vascular endothelium. *Am. J. Trop. Med. Hyg.* **24**, 1–8.

Fulton, J. D., and Flewett, T. H. (1956). The relation of *Plasmodium berghei* and *P. knowlesi* to their respective red cell host. *Trans. R. Soc. Trop. Med. Hyg.* **50**, 150–156.

Garnham, P. C. C. (1966). "Malaria Parasites and Other Haemosporidia." Blackwell, Oxford.

Garnham, P. C. C. (1967). Relapses and latency in malaria. *Protozoology* **2**, 55–64. Festschrift in honor of H. E. Shortt on the occasion of his 80th birthday 1967. Supplement to *J. Helminthol.* Dec. 1967.

Garnham, P. C. C. (1973). Second roundtable discussion on taxonomic problems relating to malaria parasites. *J. Protozool.* **20**, 37–42.

Garnham, P. C. C., Bird, R. G., and Baker, J. R. (1963). Electron microscope studies of motile stages of malaria parasites. IV. The fine structure of the sporozoite of four species of *Plasmodium. Trans. R. Soc. Trop. Med. Hyg.* **57**, 27–31.

Garnham, P. C. C., Rajapaksa, N., Peters, W., and Killick-Kendrick, R. (1972). Malaria parasites of the orang-utan (*Pongo pygmaeus*). *Ann. Trop. Med. Parasitol.* **66**, 287–294.

Gonder, R., and von Berenberg-Gossler, H. V. (1908). Untersuchungen über malariaplasmodien der Affen, *Malaria-Intern. Arch. Leipzig* **1**, 47–56.

Gonder, R., and Rodenwaldt, E. (1910). Experimentelle untersuchungen über Affenmalaria. *Zentralbl. Bakteriol. Parasitenkd. Infektionskr. Hyg., Abt. 1: Orig.* **54**, 236–240.

Halberstaedter, L., and von Prowazek, S. (1907). Untersuchungen über die Malariaparasiten der Affen. *Arb. Gesundheitsamte* (*Berlin*) **26**, 37–43.

Huff, C. G., and Hoogstraal, H. (1963). *Plasmodium lemuris* n. sp. from *Lemur collaris. J. Infect. Dis.* **112**, 233–236.

Knowles, R., and Das Gupta, B. M. (1932). A study of monkey-malaria, and its experimental transmission to man. *Indian Med. Gaz.* **67**, 301–320.

Laveran, M. A. (1905). Haemocytozoa. Essai de classification. *Bull. Inst. Pasteur, Paris* **3**, 809–817.

Mayer, M. (1907). Ueber malaria beim Affen. *Med. Klin.* **3**, 579–580.

Reichenow, E. (1920). Ueber das Vorkommen der Malariaparasiten des Menschen bei den afrikanischen Menschenaffen. *Centralbl. Bakteriol., Parasitenkd., Infektionskr. Hyg., Abt. 1: Orig.* **85**, 207–216.

Rodhain, J. (1941). Sur un *Plasmodium* du gibbon *Hylobates lensciscus Geoff. Acta Biol. Belg.* **1,** 118–123.

Rodhain, J., and van den Berghe, L. (1936). Contribution à l'étude des plasmodiums des singes africains. *Ann. Soc. Belg. Med. Trop.* **16,** 521–531.

Rudzinska, M. A., and Trager, W. (1968). The fine structure of trophozoites and gametocytes in *Plasmodium coatneyi. J. Protozool.* **15,** 73–88.

Seed, T. M., Sterling, C. R., Aikawa, M., and Rabbege, J. (1976). *Plasmodium simium:* ultrastructure of erythrocytic phase. *Exp. Parasitol.* **39,** 262–276.

Shortt, H. E., and Garnham, P. C. C. (1948). Pre-erythrocytic stage in mammalian malaria parasites. *Nature (London)* **161,** 126.

Shortt, H. E., Garnham, P. C. C., and Malamos, B. (1948). The preerythrocytic stage of mammalian malaria. *Br. Med. J.* **1,** 192–194.

Sinton, J. A., and Mulligan, H. W. (1932–1933). A critical review of the literature relating to the identification of the malarial parasites recorded from monkeys of the families Cercopithecidae and Colobidae. *Rec. Malar. Surv. India* **3,** 357–380; 381–444.

Sluiter, C., Swellengrebel, N., and Ihle, J. (1922). "De Dierlijke Parasiten van den mensch en van onze huisdieren." Scheltema & Holkema's Boekhandel, Amsterdam.

Sodeman, T. M., Held, J. R., Contacos, P. G., Jumper, J. R., and Smith, C. S. (1969). Studies of the exoerythrocytic stages of simian malaria. IV. *Plasmodium brasilianum, J. Parasitol.* **55,** 963–970.

Sodeman, T., Schnitzer, B., Durkee, T., and Contacos, P. (1970). Fine structure of the exoerythrocytic stages of *Plasmodium cynomolgi. Science* **170,** 340–341.

Sterling, C. R., Aikawa, M., and Nussenzweig, R. S. (1972). Morphological divergence in mammalian malaria parasites: The fine structure of *Plasmodium brasilianum. Proc. Helminthol. Soc. Wash.* **39,** 109–129.

Terzakis, J. A. (1971). Transformation of the *Plasmodium cynomolgi* oocyst. *J. Protozool.* **18,** 62–73.

Van Rooyen, C. E., and Pile, G. R. (1935). Observations on infection by *Plasmodium knowlesi* (ape malaria) in the treatment of general paralysis of the insane. *Br. Med. J.* **2,** 662–666.

Warren, McW., Bennett, G. F., Sandosham, A. A., and Coatney, G. R. (1965). *Plasmodium eylesi* sp. nov., a tertian malaria parasite from the white-handed gibbon, *Hylobates lar. Ann. Trop. Med. Parasitol.* **59,** 500–508.

Warren, McW., Coatney, G. R., and Skinner, J. C. (1966). *Plasmodium jefferyi* sp. n. from *Hylobates lar* in Malaya. *J. Parasitol.* **52,** 9–13.

Wharton, R. H., and Eyles, D. E. (1961). *Anopheles hackeri,* a vector of *Plasmodium knowlesi* in Malaya. *Science* **134,** 279–280.

10

Plasmodia of Man

Karl H. Rieckmann and Paul H. Silverman

I. Introduction

Human malaria has been recognized in the earliest written records of man. It has been credited as the cause of the fall of civilizations and the reason for failure of military campaigns. Gorgas, in a letter to Sir Ronald

Ross in 1914, stated that the discovery that anopheline mosquitoes were responsible for transmission of the disease from man to man which led to control of the vector ". . . enabled us to build the canal in the Isthmus of Panama" (Manson-Bahr, 1963). The intriguing history of malaria has attracted numerous chroniclers and excellent summaries of these fascinating events include those by Manson-Bahr (1963), Garnham (1966), and Coatney *et al.* (1971).

Malaria is present in a wide belt around the earth which includes the tropical, subtropical, and temperate zones, i.e., wherever conditions are suitable for maintenance of the anopheline mosquito vectors. Antimalarial measures have eliminated the disease from some areas in which it was once endemic and have lowered its prevalence in many other areas. During the past several decades the methods utilized for this effort consisted primarily of vector control through the use of residual insecticides applied to indoor resting surfaces for killing of adult mosquitoes. Vector control was combined with programs of mass administration of antimalarial agents to the human populations when feasible and appropriate. The satisfactory results obtained during earlier years have become less apparent recently because of the emergence of a number of problems, including the resistance of mosquito vectors to insecticides and the resistance of parasites to chloroquine and related drugs.

The situation in May, 1975 was summarized by a World Health Organization Scientific Group:

> The present state of malaria control in many areas of the world gives no cause for complacency. In some countries setbacks to national control or eradication operations have led to a serious resurgence of infection while in others malaria remains virtually as prevalent as it has been for centuries. Estimates based on data from 30 countries indicate that, in the continent of Africa alone, 96 million cases of malaria occur annually; one million, usually those involving young children, end in death. These findings amply illustrate the magnitude of morbidity and mortality caused directly by malaria; unfortunately they give no indication of its less obvious effects.

The current views are well-stated by Coatney *et al.* (1971):

> The 1970's find man's struggle against the malaria parasite still unfinished. The dream of eradication is no longer considered attainable within the foreseeable future in parts of Central America, South America, Africa, and Asia. The XXIInd World Health Assembly meeting in Boston in 1969, called for a re-evaluation of the concept of eradication with a reversion to classical approaches of control where necessary. Giant strides have been made against the disease, especially through the eradication programs, but we have probably reached the limit imposed by the techniques available and the last quarter

of this century may need to be as innovative as was the first quarter, if we are to make real progress against the age-old-scourge—malaria.

II. Taxonomy

The current taxonomic status of human malaria parasites has been summarized by Garnham (1973) who reported the results of an international round table discussion held in 1970 and 1971.

Plasmodium vivax strains can be divided into two groups, depending upon the nature of their relapse characteristics. The St. Elizabeth-like strains are characterized by relapses which follow a primary attack by a long interval. Infections with the Chesson-like strains are characterized by short-term relapses. These two types had been previously identified as temperate zone (St. Elizabeth) or tropical zone (Chesson) types but it was agreed that these designations are inappropriate. Some isolates from Russia have both a long prepatent period and delayed relapse activity, and such strains have been tentatively given the subspecies rank of *P. vivax hibernans*.

Plasmodium ovale is regarded to be of a homogeneous nature irrespective of its geographical source. No appreciable differences have been observed among those isolates which have been studied in sufficient detail.

Plasmodium malariae was once considered to be synonymous with *P. rodhaini* which occurs in the chimpanzee. Although the synonym was previously abolished, recent data have raised the issue again.

Plasmodium falciparum belongs to the subgenus *Laverania*. (The other human malaria species belong to the subgenus *Plasmodium*.) It is apparent that *P. falciparum* is a heterogeneous species. Numerous differences in isolates from different geographical areas have been reported. The primary distinguishing characteristics have included drug susceptibility and the degree of infectivity for different species of *Anopheles*. However, the observed distinguishing biological and drug susceptibility characteristics have been considered inadequate to alter the current taxonomic status of various strains of *P. falciparum*.

III. Metabolic and Biochemical Characteristics

The development of resistance to antimalarial drugs and the increased interest in malaria immunology and in the *in vitro* cultivation of plasmodia for the production of antigens has highlighted the paucity of information available about the biochemistry and physiology of human malaria parasites. The current situation has been aptly described by a World Health Organization Scientific Group (1969).

Precise knowledge of the way in which the parasite respires, feeds, grows, and reproduces would be of inestimable value in the search for new antiparasitic drugs, in reaching an understanding of the antigens and antibodies that are formed in the course of erythrocytic infections, and in explaining the mechanism of pathological sequels to malaria.

Reviews by Thompson and Werbel (1972), Fletcher and Maegraith (1972), and Trager (1974) identify the numerous problems affecting studies on metabolic and biochemical characteristics of human malaria. (1) As a unicellular organism, the plasmodial parasite has biochemical processes which are more complicated than the specialized activities of metazoan cells. (2) As an obligate intracellular parasite the metabolic activities are intimately intertwined with those of the host cell which it inhabits during its development through a complex life cycle. (3) The complex life cycle involves two hosts, a poikilothermic invertebrate and homeothermic mammal, and various specialized tissues within each host. (4) Most of the metabolic and biochemical studies have been carried out with avian or rodent species and to a lesser extent with simian plasmodia. It is apparent that great caution must be exercised in applying even the general trends of the results obtained from nonhuman species to the four species that infect man. (5) Most of the metabolic studies have been performed with red blood cell stages and usually with either infected erythrocytes or on parasites freed by various techniques from host erythrocytes. The various methods have yielded different and often conflicting data about metabolic and biochemical characteristics.

It may well be that further elucidation of the basic metabolic mechanisms of human malaria parasites will have to await the development of long-term cultivation of these parasites *in vitro*. Only under highly controlled experimental conditions will it be feasible to carry out critical studies of required host factors, as well as biosynthetic pathways. Similarly, better isolation and characterization of parasite components (protein, nucleic acids, etc.) will be possible when adequate preparations of host-free parasites can be obtained.

IV. Cultivation of Malaria Parasites

The first attempts to grow human malaria parasites *in vitro* were made by Bass and Johns (1912). They added glucose to defibrinated whole blood infected with *P. falciparum* and observed maturation of trophozoites to schizonts during incubation; after 24 hours the erythrocytes and parasites showed obvious signs of degeneration and no evidence of reinvasion of red cells was observed. Short-term *in vitro* cultivation of *P. falciparum* through one asexual cycle has been used extensively for the *in vitro* evaluation of antimalarial drugs (Rieckmann *et al.*, 1968a), but serial propagation of parasites is needed for *in vitro* production of malaria antigen.

Long-term cultivation of parasitized red cells in buffered culture media and special gas mixtures has not been very successful (Bertagna *et al.*, 1972) until recently (Trager and Jensen, 1976). Erythrocytes deteriorate markedly within 24 hours during incubation and show a sharp drop in ATP, coenzyme A, and G6PD levels (Gomperts, 1969), all of which appear to be required by the malaria parasite for growth and replication. Intact and healthy red blood cells, free of biochemical lesions, appear to be necessary for the *in vitro* cultivation of the asexual stages of human plasmodia. In cultures to which fresh red cells were added, *P. falciparum* reinvaded uninfected cells and *in vitro* growth was supported for two or three cycles (Phillips *et al.*, 1972). Continuous *in vitro* cultivation of *P. falciparum* has now been reported by Trager and Jensen (1976). This important advance in the cultivation of human malaria parasites has been achieved with several strains of *P. falciparum* using outdated human blood from transfusion blood banks. Critical factors in successful cultivation appear to be low oxygen tension and static culture conditions.

Consideration has also been given to the possibility of growing the erythrocytic stages of the malaria parasite in cell lines which can be propagated continuously (Speer and Silverman, 1974). Garnham (1966) remarked on the ability of plasmodia to grow and develop in various exoerythrocytic sites, particularly the hemopoietic system. "In the primitive cells of the exoerythrocytic series, there is no haemoglobin, yet the schizont grows as well or better without it, for instance, more merozoites are produced in the erythroblastic schizont than in the erythrocytic." It was reported recently that the erythrocytic stages of *P. berghei* can grow in bone marrow cell lines (Speer *et al.*, 1976). If these findings are confirmed by further studies, similar approaches may be adopted in the cultivation of human malaria parasites.

Cultivation of exoerythrocytic stages of human malaria parasites has not yet been reported and may be difficult to achieve. However, the availability of various mammalian cell lines and the demonstration that exoerythrocytic forms of avian malaria parasites can mature in mouse liver cultures (Beaudoin *et al.*, 1974) have increased the possibility that this may be accomplished in the future.

V. Life Cycle

The four species of *Plasmodium* that infect man are *P. falciparum, P. vivax, P. malariae,* and *P. ovale.* On a global scale, *P. falciparum* and *P. vivax* are the most prevalent species. Some nonhuman primate malarias can be transmitted to man under experimental conditions, but this is not an important phenomenon in nature. Detailed knowledge of the life cycle of the different plasmodial species is important in understanding the clinical features, epidemiology, prevention, and treatment of malaria.

A. In Man (See Fig. 1)

Malaria infections are transmitted from person to person by various species of anopheline mosquitoes. An infected female mosquito bites the human host and injects sporozoites from its salivary glands. The sporozoites enter the blood stream and quickly reach tissue cells where they initiate the cycle of exoerythrocytic (EE) schizogony.

The development of tissue schizonts in liver parenchymal cells has been documented in either man or monkey for all four species of human malaria. The parasite grows within the cell, undergoes division and, eventually, the EE body contains several thousand merozoites. When the schizont is mature, the merozoites are released into the blood stream where they initiate the erythrocytic phase of the life cycle. The erythrocytic parasites may appear as early as 5½ days after the sporozoites are introduced in *P. falciparum* (Fairley, 1947), after 8 days in *P. vivax* (Fairley, 1947), after 15 days in *P. malariae* (Lupaşcu *et al.*, 1967), and after 9 days in *P. ovale* (Bray, 1957b). In *P. falciparum,* tissue schizonts have not been observed after the development of patent infections (Coatney *et al.*, 1971). On the other hand, tissue schizonts of *P. vivax* and *P. ovale* have been found in the liver of infected chimpanzees after the onset of patency (Rodhain, 1956, Bray, 1957a; Bray *et al.*, 1963) and it is presumed that the intermittent release of hepatic merozoites into the blood stream is responsible for the relapses observed during the course of these infections. Persistent cycling with the formation of secondary tissue schizonts is uncertain and it now seems more likely that relapses occur after the reactivation of dormant tissue forms (Contacos and Collins, 1973). Latent tissue schizonts or relapsing infections have not been observed in *P. malariae* (Ciucă *et al.*, 1964; Coatney *et al.*, 1971) and it is uncertain whether delayed parasitemias seen in these infections are caused by dormant tissue or blood forms.

The erythrocytic phase commences with the invasion of red cells by tissue merozoites. *Plasmodium vivax* prefers to invade young red blood cells (reticulocytes), *P. malariae* selects the more mature cells, and *P. falciparum* attacks red blood cells at all stages of maturity. The young intracellular trophozoite forms a central vacuole and, on microscopic examination, it usually has a ring-shaped appearance. As the parasite grows, both the nuclear chromatin dot and cytoplasm become more prominent, and pigment (hemozoin) accumulates as a result of incom-

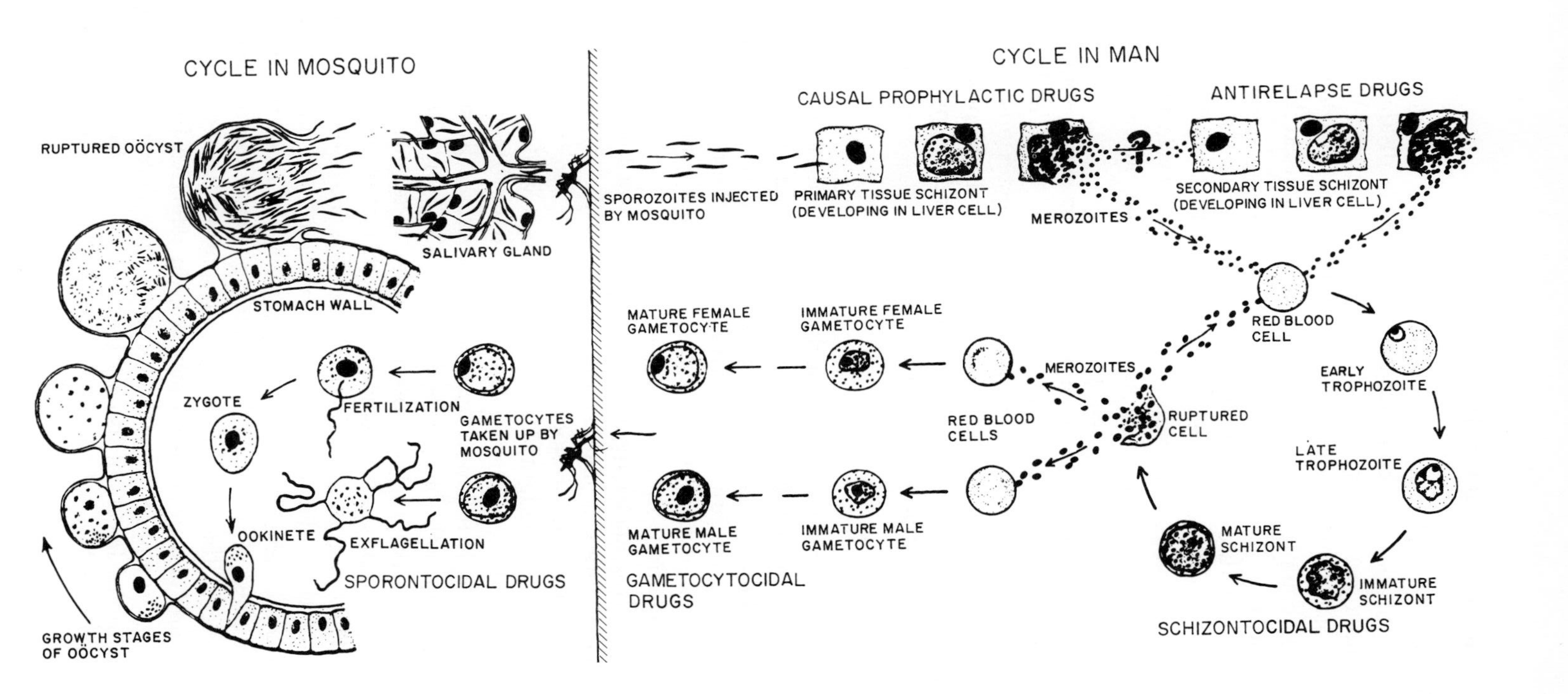

Fig. 1. Diagrammatic presentation of the life cycle of human malaria parasites and the nomenclature of drug groups acting against different stages of the life cycle. The formation of a second generation of tissue schizonts in the relapsing malaria species is now uncertain and it seems more likely that relapses occur after reactivation of dormant forms of the original generation of tissue schizonts. (Adapted from original World Health Organization publication.)

plete digestion of the host red cells. Before the start of nuclear division, the cytoplasm appears more compact and the vacuole disappears. The nucleus then undergoes division by mitosis. With successive nuclear division, the mass of cytoplasm also breaks up into various portions. When mitosis has been completed, each nuclear dot surrounded by a small amount of cytoplasm is known as a merozoite. The infected red cell containing such a mature schizont then ruptures, the merozoites are released into the blood stream, and they invade other erythrocytes to initiate another asexual erythrocytic cycle. The entire cycle takes about 48 hours in *P. falciparum, P. vivax,* and *P. ovale* (tertian malaria) and about 72 hours in *P. malariae* (quartan malaria).

After one or more generations of erythrocytic schizogony, some parasites do not develop into segmenters but become male and female gametocytes, sometimes referred to as microgametocytes and macrogametocytes, respectively. Mature gametocytes do not develop further in the human host but they are the only forms of the malaria parasite that are infective to susceptible anopheline mosquitoes.

B. In Mosquito (See Fig. 1)

Female mosquitoes require blood meals for maturation of their eggs. After feeding on an infected person whose blood contains mature gametocytes, the anopheline proceeds to digest the blood in its alimentary tract. In the midgut, mature gametocytes, freed of their red cell envelopes, are transformed to microgametes and macrogametes. By the process of exflagellation, the male gametocyte forms eight threadlike structures. The microgametes quickly break away from the parent body and one of them enters a macrogamete to form the zygote. It soon elongates and matures into a motile ookinete which forces its way into an epithelial cell of the midgut. Here it forms a cyst wall and, about 48 hours after the blood meal, the nucleus of the young oocyst undergoes meiotic division. This is followed by successive mitotic divisions, resulting in enormous growth of the oocyst and its protrusion into the hemocoel (body cavity) of the mosquito. After it matures, the oocyst bursts and releases thousands of spindle-shaped, motile sporozoites into the hemocoel. They migrate to all parts of the mosquito's body and some enter the cells of the salivary glands. They remain here until the mosquito takes a blood meal. Just prior to ingestion of blood, the infected mosquito inoculates salivary fluid containing some sporozoites; this initiates the cycle in the human host. Most sporozoites are retained in the salivary glands and mosquitoes may remain infective for as long as 12 weeks. The sporogonic cycle in the mosquito is dependent on environmental conditions. It cannot usually be completed if the temperature is below 16°C (60°F) or above 33°C (91°F). Between these limits, development proceeds faster at the higher temperatures. Thus, the entire cycle in the mosquito takes about 2 weeks in tropical lowlands but 3 weeks or longer in the cooler tropical highlands.

VI. Course of Infection

A. Clinical Attack

The typical clinical attack of malaria is associated with the rupture of schizont-infected red cells. No overt disease is associated with the exoerythrocytic or sexual stages of the infection. Peripheral blood films taken during a malarial paroxysm show a predominance of young ring forms. In nonimmune patients with falciparum malaria, only 10–100 trophozoites per cubic millimeter (mm^3) of blood are usually observed during the initial febrile episode. The fever threshold is higher in more immune persons and several hundred to several thousand parasites per cubic millimeter can be tolerated before any clinical symptoms or signs of the disease become apparent. The primary attack may occur before the first appearance of parasites in the peripheral blood. This, however, is rare and the onset of patent parasitemia, as determined by examination of approximately one-tenth of a mm^3 of blood, usually precedes the clinical attack by about one erythrocytic cycle.

The time interval between inoculation of sporozoites and the onset of a patent infection is referred to as the prepatent period and it is usually about 2 days shorter than the incubation period (Fig. 2). It should be noted that the release of merozoites from the liver into the peripheral blood circulation does not coincide with the onset of patency; one or more subpatent erythrocytic cycles are usually needed before parasites become detectable in the peripheral circulation. The mean prepatent period of the tertian malarias (*P. falciparum, P. vivax,* and *P. ovale*) is about 1½–2 weeks. These time intervals can vary considerably, depending on the sporozoite inoculum, parasite strain, and the person's immune status and racial background. For example, a prepatent period exceeding 2 weeks is often observed in falciparum infections induced by a small number of sporozoites (Powell and McNamara, 1970).

The onset of a primary attack is usually heralded by headache, general malaise, chills or fever. These symptoms are almost invariably present during the clinical episode and they may be accompanied by generalized myalgia, backache, nausea, vomiting, anorexia or diarrhea. A typical febrile paroxysm often starts in the afternoon or evening hours. It consists, successively, of a cold stage, associated with a shaking chill (usually less than 1 hour), a hot stage associated with a temperature of 40°–41°C (104°–106°F), and a flushed, dry skin (1–4 hours), and a crisis or sweating stage during which the temperature returns to normal

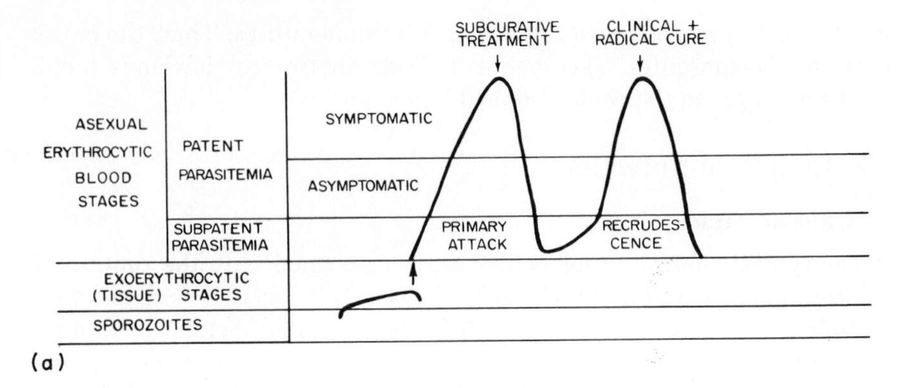

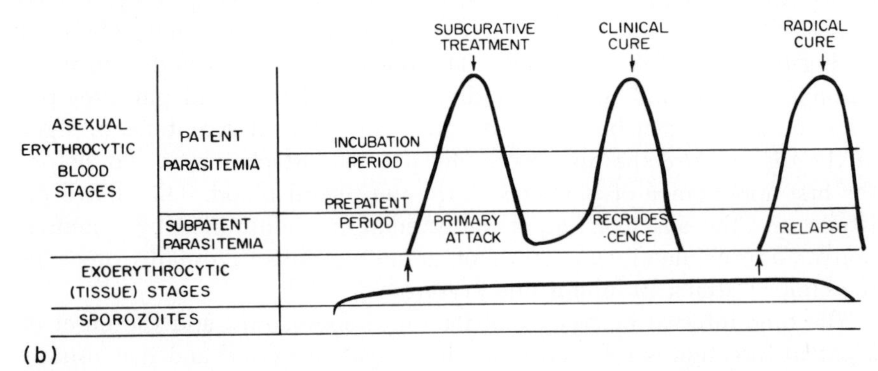

Fig. 2. Diagrammatic presentation of different courses of infection in (a) non-relapsing (e.g., *Plasmodium falciparum*) and (b) relapsing malaria (e.g., *Plasmodium vivax*).

(2–3 hours). Chills are often less pronounced in falciparum infections and severe "bed-shaking" chills are characteristic of vivax infections.

In a synchronous cycle, malaria paroxysms display a classical tertian or quartan periodicity and the intervals between paroxysms are relatively free of symptoms. Periodicity, however, is often not present during the initial stages of an infection and is frequently not apparent in falciparum malaria.

With each succeeding paroxysm, more and more red cells become parasitized. This may be limited by the preferential invasion of certain host red cells (e.g., the preference of *P. vivax* for reticulocytes). Parasite counts in vivax, malariae, and ovale infections rarely exceed 50,000/mm^3 of blood and the clinical course is relatively "benign." Parasites of *P. falciparum*, on the other hand, invade circulating red blood cells rather indiscriminately and about 50% of patients succumb to their infection after the count exceeds 500,000 parasites/mm^3. The progressive increase in the level of parasitemia may be interrupted by appropriate treatment and the development of host immunity to asexual blood forms.

B. Recrudescence of Infection

Patent parasitemia may recur after a variable period of subpatency due to persistence of the original infection (Fig. 2). Suboptimal prophylaxis or treatment of falciparum malaria with blood schizontocidal agents often results in the subsequent recrudescence of parasitemia. Recrudescences may be symptomatic or asymptomatic, depending on the degree of partial immunity acquired by patients during their previous exposure to malaria. They are usually not observed in falciparum malaria beyond a period of 18 months after original infection. In endemic areas, it is often difficult to differentiate between a new infection and recrudescence of an old infection.

C. Relapse of Infection

Relapses are commonly observed in vivax malaria after the original clinical attack. They are caused by intermittent reinvasion of the blood stream by exoerythrocytic parasites (Figs. 1 and 2). Different strains of *P. vivax* vary in their relapse pattern (Coatney *et al.*, 1971). Some strains, such as the Chesson strain (New Guinea–South Pacific), are characterized by early activity. The first relapse of a Chesson vivax infection occurs characteristically about 6 weeks after the primary attack and is followed by further relapses, gradually becoming more widely spaced, for a period of 18 months or more. Other strains may remain quiescent for many months after the initial attack. In infections with the St. Elizabeth strain (United States), the first relapse usually occurs about 9 months after the primary attack and this may be followed by further relapses for 2 years or longer.

D. Pathology and Clinical Complications

The pathogenic processes in malaria have been reviewed by Maegraith (1966), Neva *et al.* (1970), and Voller (1974) and some of them are referred to in this section.

Phagocytosis of parasites is a well known feature of malaria infections. Monocytes and polymorphonuclear leukocytes often contain malaria parasites and pigment, but fixed phagocytic cells of the reticuloendothelial system are probably more important in removing parasites from the bloodstream. Marked proliferation and hyperplasia of macrophages of the spleen and liver is accompanied by phagocytosis of free parasites and pigment as well as parasitized and nonparasitized red cells. This enhanced phagocytic activity is reflected, clinically, by enlargement of the spleen and liver in many malaria patients. These organs usually return to normal size after treatment of the infection. In persons with latent, subpatent infections of malaria, removal of the spleen often leads to recrudescence of parasitemia.

Massive enlargement of the spleen, associated with elevated serum levels of immunoglobin M and malaria antibodies, has been observed in some highly endemic malarious areas of Africa and New Guinea (Pitney, 1968). The term "tropical splenomegaly syndrome" or "big spleen disease" has been applied to this condition and, although its etiology is not known, it is felt that repeated exposure to malaria parasites is a prerequisite for the development of this syndrome.

Anemia and thrombocytopenia are often observed during the course of malaria infections. The degree of anemia is frequently much greater than what might be expected from hemolysis of infected red blood cells. Furthermore, the anemia is often more pronounced when the level of parasitemia has started to decline. It has been postulated that immunological mechanisms, such as red cell autoantibodies (Zuckerman, 1964) or binding of circulating malaria antigen–antibody complexes to uninfected red cells (Woodruff, 1973), might be involved in processes which enhance nonspecific erythrophagocytosis.

Most "malignant" clinical manifestations occur only during the course of severe falciparum infections and they are related primarily to pathological changes in the brain, kidneys, and lungs.

Cerebral malaria can cause a variety of clinical syndromes. The neurological abnormality may take the form of disturbances of consciousness, seizures, behavioral changes, focal manifestations, or movement disorders. One or more of the basic pathological changes observed in the brain may be responsible for the manifestations of cerebral malaria: (1) thrombosis of intracerebral vessels by infected, clumped erythrocytes. The "stickiness" of schizont-infected cells was thought to be primarily responsible for plugging of the small vessels. More recently, it has been postulated that the inability of parasitized red cells to change their shape as easily as normal cells prevents them from squeezing their way through some capillary blood vessels and, consequently, the flow of blood is interrupted (Miller *et al.*, 1972); (2) anoxia caused by anemia by stagnation of the blood flow and, particularly, by decreased oxygen-carrying capacity of parasitized erythrocytes; and, (3) edema associated with perivascular hemorrhage.

Nephritis and acute renal failure can occur during the course of acute malaria infections. Patients usually respond well to adequate therapy and the kidney damage is usually completely reversible. On the other hand, *P. malariae* infections are sometimes associated with progressive renal damage. Immunofluorescent studies strongly suggest that this syndrome is due to the deposition of circulating antigen–antibody complexes in the glomerular capillary walls (Hendrickse and Gilles, 1963). Patients with this syndrome usually respond poorly to treatment.

"Blackwater fever" is a urinary manifestation which has been associated with falciparum malaria and quinine administration. Severe hemolysis, possibly caused by a drug-dependent hypersensitivity reaction, results in the massive release of hemoglobin. If the concentration of hemoglobin exceeds the binding capacity of plasma haptoglobin, hemoglobin or its breakdown products are excreted by the kidney.

Pulmonary edema and intraalveolar hemorrhage may be observed in a few patients with acute falciparum malaria. These changes usually disappear spontaneously when response to treatment of the infection is satisfactory.

VII. Diagnosis

Identification of the plasmodial species causing the infection is mandatory if the clinician is to select appropriate chemotherapy and prevent death from falciparum malaria. The possibility of a plasmodial infection has to be considered in any febrile patient with a history of possible exposure to the malaria parasite. Before excluding such a possibility, about one-tenth of a cubic millimeter of blood must be examined carefully on two or three different occasions and found to be free of parasites. Examination of this quantity of blood is easily accomplished by examination of dehemoglobinized thick smears. Nevertheless, thin blood smears should also be collected in the event that difficulty is experienced in identifying the plasmodial species after examination of thick smears. The preparation, staining, and examination of thick and thin blood smears has been described and illustrated in considerable detail by Wilcox (1960). The morphological features of *P. vivax, P. falciparum,* and *P. malariae* in stained thin blood smears are shown in Fig. 3.

Ring forms or early trophozoites of the different species may be difficult to distinguish from one another. When only a few rings are present in the blood smear, species identification may be impossible. However, blood smears collected a few hours later should clarify the diagnosis.

Morphological differences between the species become more apparent as the parasites develop to mature trophozoites. In *P. falciparum,* host red cells are not larger than uninfected cells and older rings often resemble young ring forms of the other three species. The more mature

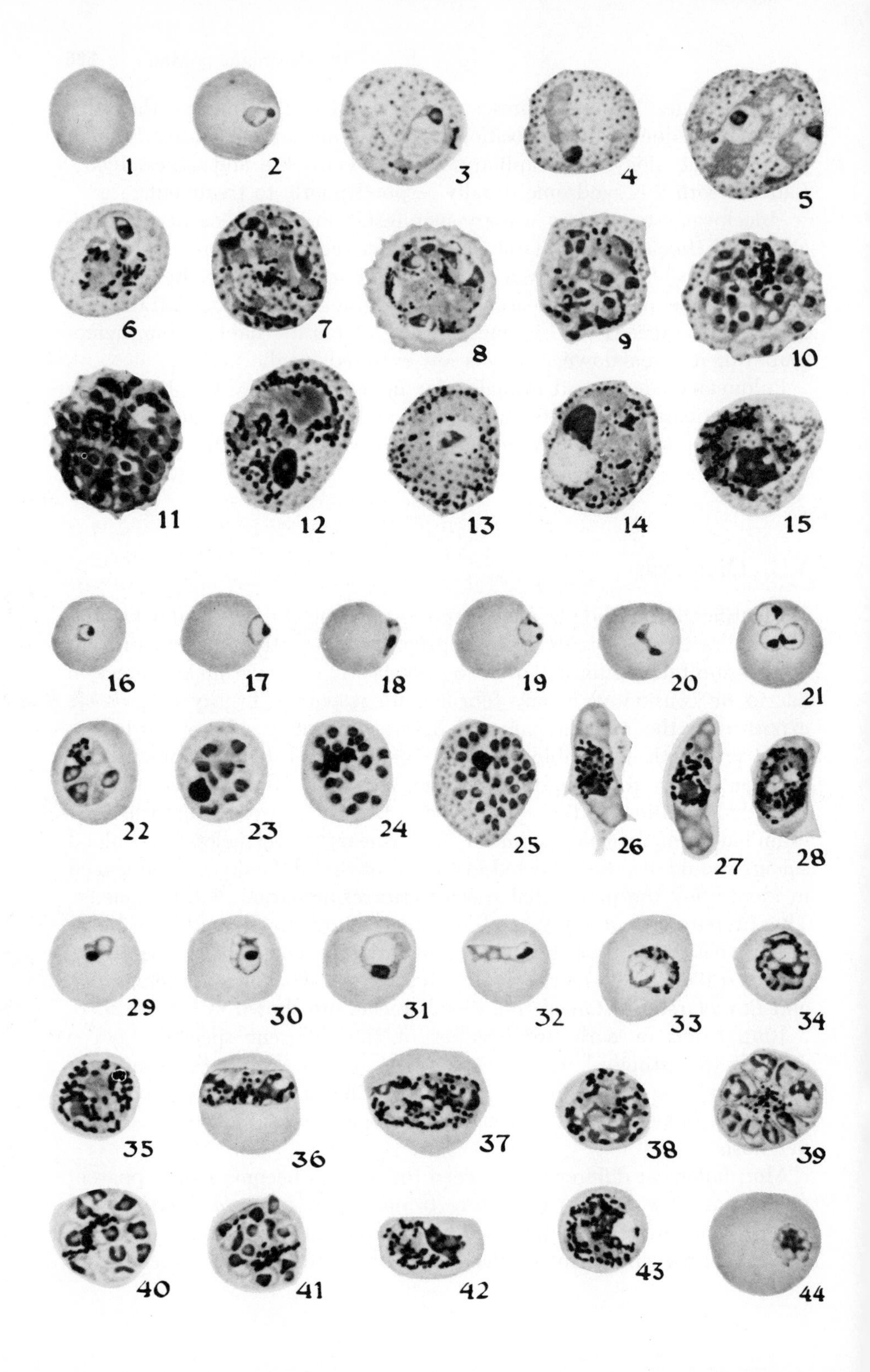
1
2
3
4
5
6
7
8
9
10
11
12
13
14
15
16
17
18
19
20
21
22
23
24
25
26
27
28
29
30
31
32
33
34
35
36
37
38
39
40
41
42
43
44

trophozoites of *P. falciparum* disappear from the peripheral blood circulation and undergo schizogony in the blood vessels of internal organs. In *P. vivax* and *P. ovale,* infected red cells are usually enlarged and polychromatophilic and exhibit eosinophilic stippling (Schüffner's dots); ovale-infected erythrocytes may also have an oval shape or a fimbriated margin. Parasites are larger, often contain some pigment, and the cytoplasm of vivax trophozoites has an amoeboid appearance. In *P. malariae,* infected red cells are normal in size and do not have Schüffner's dots. Although "band" forms are diagnostic, they are not as common as spherical trophozoites. These compact forms contain many pigment granules imparting a yellowish-brown (sometimes greenish) hue to the parasites.

Schizonts of *P. vivax, P. ovale,* and *P. malariae* are common in peripheral blood smears and this contrasts sharply with the "monotonous array" of ring forms usually seen in falciparum infections. Mature schizonts of *P. vivax* usually contain 16–24 merozoites and the host red cells are enlarged. Mature schizonts of *P. ovale* and *P. malariae* usually contain 8–10 merozoites arranged concentrically around the pigment. The enlarged host red cell and Schüffner's dots distinguish schizonts of *P. ovale* from those of *P. malariae.*

Gametocytes of *P. falciparum* are crescent-shaped, appear 7–10 days after onset of patent parasitemia, and remain in the peripheral blood, infective to mosquitoes, for several weeks. Gametocytes of the other three species are round or spherical in shape and are short-lived. Host red cells of *P. vivax* and *P. ovale* are enlarged and contain Schüffner's dots, whereas those of *P. malariae* are normal in size and show no stippling.

Fig. 3. Blood stages of *Plasmodium vivax* (1–15), *P. falciparum* (16–28), and *P. malariae* (29–44). (1) Uninfected red cell; (2) early trophozoite; (3 and 4) maturing trophozoites; (5) double infection with maturing trophozoites; (6) mature trophozoite; (7) early schizont; (8 and 9) maturing schizonts; (10 and 11) mature schizonts: (12) maturing gametocyte; (13) maturing microgametocyte; (14) mature macrogametocyte; (15) mature microgametocyte; (16 and 17) early trophozoites; (18–21) maturing trophozoites; (22) early schizont; (23) maturing schizont; (24 and 25) mature schizonts; (26 and 27) macrogametocytes (crescents); (28) microgametocyte (crescent); (29–31) early trophozoites; (32) "band" form of trophozoite; (33) maturing trophozoite; (35) early schizont; (36 and 37) "band" forms of early schizonts; (38) maturing schizont; (39–41) mature schizonts; (42) macrogametocyte; (43) microgametocyte; (44) uninfected red cell with superimposed blood platelet. All stages from peripheral blood smears except 22–25, which are from placental smears. (Reproduced from original color plates drawn by E. Bohlman, *in* "Manual of Medical Parasitology" by Clay G. Huff, University of Chicago Press, Chicago, Illinois.)

VIII. Immunity

A. Natural (Innate) Immunity

Host–parasite compatibility is not dependent on serum factors, but correlates with the presence of surface receptors for the specific parasite on the host's red cells (Butcher *et al.*, 1973). The well-recognized resistance of persons from West Africa or of West African ancestry to vivax malaria (Boyd and Stratman-Thomas, 1933) may be an example of this phenomenon.

In falciparum malaria, intraerythrocytic factors may be responsible for the slower development of partial immunity in white Americans than in black Americans (Powell *et al.*, 1972) and for the protection which sickle-cell trait carriers have against lethal infections with this species (Allison, 1963).

B. Development of Clinical Immunity

The acquisition of immunity to malaria has been well-documented in persons infected under experimental or natural conditions. The immune response seems to be directed mainly against the asexual erythrocytic phase of the malaria parasite and is expressed by suppression of symptoms or fever, reduction in the duration or intensity of parasitemia, or even complete clearance of parasites.

In adult nonimmune persons given intermittent subcurative antimalarial medication during the course of their falciparum infections, asymptomatic low-grade parasitemia usually develops within 6 to 8 weeks after onset of their infection. The actual duration of parasitemia seems to be a more important factor in stimulating the development of immunity than the density of parasitemia which develops during the course of the infection. The immune response during subsequent infections becomes progressively more pronounced, especially when patients are reinfected with homologous strains. Although less marked, the course of reinfection with heterologous strains is also modified. However, reinfections with homologous or heterologous strains rarely prevent the development of patent parasitemia and a "sterile" immunity is seldom observed. Nonimmune persons may also become partially immune without ever developing any significant fever or symptoms if they receive intermittent administration of antimalarial drugs that suppress but do not eliminate asexual parasites (Rieckmann, 1970).

The development of immunity to malaria in persons living in endemic areas depends on the level of transmission of the disease. When transmission is low, the community does not develop a high degree of immunity to the disease and persons of all ages can develop acute clinical infections. On the other hand, in areas of continuous and intense malaria transmission, young children develop the most severe clinical episodes and many succumb to fatal infections of falciparum malaria. As the survivors grow older, they tolerate the infections much better and, as adults, their infections are usually limited to asymptomatic, low-grade parasitemia.

C. Mechanisms of Acquired Immunity

Although the development of clinical immunity to malaria has been well-documented, comparatively little is known about the mechanisms of protective immunity in human malaria (McGregor, 1974). Our current knowledge concerning possible mechanisms in human malaria has been derived mainly from animal malaria studies (Brown, 1974; Cohen *et al.*, 1974).

Acquired immunity in human malaria seems to depend heavily on circulating humoral antibody, but the precise roles of specific antiplasmodial antibody are not well understood. In chronic malaria infections, up to one-third of the circulating immunoglobulin may consist of malaria antibody (Cohen *et al.*, 1961), but probably only a small portion of the specific antibody has any protective function (Cohen and Butcher, 1969). Protective antibodies appear to be mainly of the IgG class, although some activity is present in IgM fractions (Cohen and Butcher, 1971). The bivalent portion of the molecule [$F(ab)_2$], but not complement, is required for this immune response. In *P. knowlesi* infections in Rhesus monkeys, protective antibodies can be demonstrated to inhibit the reinvasion of erythrocytes by free merozoites (Cohen and Butcher, 1971) and to promote the phagocytosis of erythrocytes infected with mature parasites (Brown, 1973), but no effect on the maturation of parasites within the red cell has been observed in simian or human malaria (Rieckmann *et al.*, 1968a; Mitchell *et al.*, 1976).

Plasmodium falciparum antigens have been classified into three main groups—La or Lb (labile), R (resistant), and S (stable)—on the basis of their ability to withstand heating (Wilson *et al.*, 1969). La antigens appear to be the most immunogenic and, although antibodies are found predominantly in the IgG fraction of immune human sera, it is not known whether they contribute to protective immunity.

Relatively few studies have been conducted to investigate the possible role of cell-mediated immunity in human malaria. The participation of thymus-dependent T cells in the immune response to human malaria was demonstrated a few years ago when blood lymphocytes from partially immune persons were observed to undergo blastoid transformation in the presence of *P. falciparum* antigens (Kass *et al.*, 1971). This was confirmed in transformation studies involving incorporation of tritiated thymidine by whole lymphocyte preparations (Wyler and Oppenheim, 1974). Delayed dermal hypersensitivity has been observed in human (Makari, 1946) and nonhuman (Cabrera *et al.,* 1976; Taliaferro and Bloom, 1945) primates exposed to malaria antigens. The development of procedures to measure cell-mediated immune responses may, possibly, provide another approach for determining a person's previous exposure or susceptibility to malaria.

D. Vaccination Studies

The availability of an effective vaccine would undoubtedly facilitate the control of malaria. It was not until relatively recently, however, that the feasibility of developing such a vaccine was seriously considered. After the well known studies of Freund *et al.* (1945), more recent investigations have shown that active immunization by different forms of attenuated or killed malaria parasites can be effective in protecting rodent and simian hosts against malaria (Brown *et al.,* 1974; Corradetti *et al.,* 1969; Mitchell *et al.,* 1974; Nussenzweig *et al.,* 1967; Schenkel *et al.,* 1973; Simpson *et al.,* 1974). Furthermore, protective immunity against *P. falciparum* can be induced in owl monkeys and in man after exposure to x-irradiated parasitized red blood cells (Sadun *et al.,* 1969) or sporozoites (Clyde *et al.,* 1973; Rieckmann *et al.,* 1974a). These findings have generated a more optimistic view about the eventual production of an antimalarial vaccine (World Health Organization Scientific Group, 1975).

E. Serology

The antigenic composition of malaria parasites is complicated and, during the course of a chronic infection, infected persons develop various antibodies in response to these antigens. Although most of these antibody responses probably do not have any protective function, serological tests shed light on an individual's previous exposure to malaria and complement epidemiological investigations (World Health Organization Scientific Group, 1975). The two most widely used methods in current use are the indirect fluorescent antibody test and the indirect hemagglutination test (Ambroise-Thomas, 1974; Meuwissen, 1974). Potentially more sensitive serological tests include radioimmunoassay (T. Stewart, personal communication) and enzyme-linked immunosorbent assay (Voller *et al.,* 1975). Practical applications of serological malaria tests include: (1) determination of previous exposure to malaria in persons who were treated before a parasitological diagnosis was established or who are potential blood donors, and (2) epidemiological assessment and surveillance in areas where malaria transmission has been reduced by the use of drugs, insecticides, or other control measures.

IX. Chemotherapy

A. General Considerations

Although new drug regimens have been introduced recently to treat chloroquine-resistant infections of *P. falciparum,* the chemotherapy of the drug-sensitive malarias has not changed appreciably over the past decade (Powell, 1966; World Health Organization Scientific Group, 1967, 1973; Peters, 1970; Barrett-Connor, 1974). Different plasmodial species and different stages in the life cycle of a species show a differential susceptibility to antimalarial drugs. Consequently, the selection of the appropriate drug(s) is important in the prevention and treatment of malaria and in the reduction of further transmission of the disease.

Causal prophylactic drugs, acting against preerythrocytic tissue schizonts, prevent malaria among persons traveling in endemic areas. The 8-aminoquinoline drug, primaquine, is active against all four species of human malaria. However, its rapid degradation after being administered has limited the value of this drug as a causal prophylactic agent (Alving *et al.,* 1962). The dihydrofolate reductase inhibitors, chloroguanide and pyrimethamine, are effective against susceptible strains of *P. falciparum,* but they show a less pronounced activity against exoerythrocytic forms of *P. vivax.*

Antirelapse drugs, effective against latent tissue schizonts, prevent the recurrence of parasitemia in vivax and other relapsing malarias. The 8-aminoquinolines are still the only group of compounds which are of practical value in preventing relapses. However, they are not always effective in achieving radical cure of persons infected with some strains of *P. vivax,* e.g., the Chesson strain. In such cases, administration of large weekly doses of primaquine for 8 weeks may be more effective than the standard administration of smaller daily doses for 2 weeks (Alving *et al.,* 1960).

Blood schizontocidal drugs, effective against asexual erythrocytic parasites, are used either as suppressive prophylactics to prevent the development of clinical attacks or as curative agents to achieve clinical cure by eliminating parasites from the bloodstream. Suppression and cure of clinical attacks of malaria caused by *P. vivax, P. malariae, P. ovale,* and chloroquine-senstive strains of *P. falciparum* are readily achieved by administration of 4-aminoquinoline drugs, such as chloroquine or amodiaquine. These antimalarials are, however, not effective in the prevention or treatment of drug-resistant falciparum malaria and the use of alternative drug regimens is discussed in Section IX,B,2.

Gametocytocidal drugs, by their direct destruction of sexual forms, and sporontocidal drugs, by their interruption of sporogony in mosquitoes that feed on a treated person, prevent the transmission of malaria to other persons. They are especially important in lowering the transmission of falciparum malaria because gametocytes of *P. falciparum* can remain infective to mosquitoes for several weeks—gametocytes of the other species are short-lived and disappear within a few days after administration of blood schizontocidal drugs. Pyrimethamine and primaquine are the principal drugs that have been used extensively for this purpose. Pyrimethamine is only effective as a sporontocide if the asexual blood forms have not developed resistance to it. With the widespread occurrence of pyrimethamine-resistant strains of *P. falciparum,* the value of this drug as a sporontocidal agent has diminished. Primaquine, on the other hand, has pronounced gametocytocidal and sporontocidal activity against both drug-sensitive and drug-resistant strains of *P. falciparum* (Jeffery *et al.*, 1963; Rieckmann *et al.*, 1968b).

B. Drug Resistance

Resistance of malaria parasites has been defined as the "ability of a parasite strain to multiply or to survive in the presence of concentrations of a drug that normally destroy parasites of the same species or prevent their multiplication. Such a resistance may be relative (yielding to increased doses of the drug tolerated by the host) or complete (withstanding maximum doses tolerated by the host)" (World Health Organization, 1963). Although the drug sensitivity of all stages of the plasmodium should be considered in this context, the term "resistance" is applied primarily to the asexual blood forms, presumably because this stage in the life cycle of the parasite produces the acute clinical episodes during a malaria infection. Resistance of asexual erythrocytic parasites to drugs has been reported in all species of human plasmodia. However, the response of *P. falciparum* to antimalarials varies more widely than is observed with other species, and the appearance of chloroquine-resistant strains about 15 years ago introduced an era of renewed interest in the development of new antimalarial agents.

Drug resistance in human plasmodia has been a problem for some time. It was observed soon after chloroguanide and pyrimethamine were introduced for the prophylaxis and treatment of malaria about 25 years ago. Although these drugs were useful for the chemoprophylaxis of malaria and the prevention of its transmission, they exerted a slow effect against asexual blood forms and were not generally recommended for the treatment of febrile episodes of malaria. On the other hand, chloroquine and amodiaquine were rapidly effective against asexual erythrocytic parasites of all four species that infect man. In addition, the toxicity of these two 4-aminoquinolines is low at doses ordinarily employed for therapy of malaria and they have been generally regarded as the drugs of choice for treatment of the acute attack of malaria. Consequently, with the emergence of chloroquine-resistant strains of *P. falciparum* in South America and Southeast Asia the potential gravity of the situation quickly became apparent because of the known limitations of other available antimalarial drugs. Chloroquine-resistant falciparum malaria is now widespread in many areas of Southeast Asia and of South America. Fortunately, there is so far no evidence that it is present in Africa (World Health Organization Scientific Group, 1973).

1. Determination of the Drug Sensitivity of Parasites

A. IN VIVO. The sensitivity of human malaria parasites to drugs is usually determined by observing changes in the levels of parasitemia after administration of drugs to man, and, if clearance of parasitemia occurs, by noting whether there is a subsequent recrudescence of asexual parasites in the peripheral blood. The development of chloroquine resistance led to the adoption of a uniform procedure for determining the sensitivity of *P. falciparum* to chloroquine (World Health Organization Scientfic Group, 1967). According to this procedure, infected subjects receive 25 mg of chloroquine base per kilogram of body weight over a period of 3 days, or a total dose of 1.5 gm of chloroquine base for a 60-kg adult. The response of their asexual parasitemia is graded as follows: Sensitivity (S), clearance of asexual parasitemia within 7 days of the first day of treatment, without recrudescence; Resistance (RI), clearance of asexual parasitemia, as in sensitivity, followed by recrudescence of asexual parasites within 28 days of the first day of treatment; Resistance (RII), marked reduction of asexual parasitemia, but no clearance; Resistance (RIII), no marked reduction of asexual parasitemia.

The response of *P. falciparum* infections to antimalarial drugs in non-immune subjects differs from that observed in partially immunes; a non-immune person frequently shows a slower and less complete response of fever and parasitemia to the same drug dosage. Furthermore, certain other considerations frequently prevent an adequate assessment of the drug sensitivity of parasites by administration of drugs to man (World Health Organization Scientific Group, 1973).

A few years ago, infections with human malaria parasites were successfully established and maintained in the owl monkey or *Aotus trivirgatus* (Young *et al.*, 1966; Geiman and Meagher, 1967). This important discovery led to the evaluation of the activity of potential drugs against human malaria parasites in this primate host (Schmidt, 1969). The availability of these animals is restricted and, for the present, these hosts can only be used to a limited extent for this purpose.

B. IN VITRO. The problems inherent in conducting *in vivo* tests led to the development of an *in vitro* test for studying the response of *P. falciparum* to various schizontocidal drugs (Rieckmann *et al.*, 1968a). The test can be used under field conditions and consists of the addition of various concentrations of a drug to small quantities of parasitized blood and noting the extent to which maturation of trophozoites to schizonts is inhibited during a 24-hour period. The test has been used to determine the presence and prevalence of drug-resistant strains of *P. falciparum* in malarious areas, to obtain preliminary information concerning the blood schizontocidal activity of prospective antimalarial agents, and to determine the susceptibility of an infection to various antimalarial drugs (World Health Organization Scientific Group, 1973).

2. Chemotherapy of Chloroquine-Resistant Infections

The standard treatment for acute infections of chloroquine-resistant falciparum malaria used to be the administration of quinine for 10 to 14 days. In severe infections, intravenous quinine remains the treatment of choice at the start of therapy. Although acute clinical symptoms in patients with uncomplicated malaria subside after administration of quinine, some recrudescences of parasitemia occur within 1 month after the end of treatment (Tigertt, 1966). In an effort to achieve radical cure of such infections, pyrimethamine and one of the sulfonamides or sulfones were added to the 14-day quinine treatment schedule. Results obtained from various studies indicate that the recrudescence rate can be lowered by including one or more of these drugs in the treatment schedule. As quinine suffers from several disadvantages, a search was made for drugs which would be less toxic than quinine and which would be effective when administered over a shorter period of time.

Different combinations of dihydrofolate reductase inhibitors and sulfonamides have been investigated and some of them are being used in the treatment of chloroquine-resistant infections. The rationale for the use of these combinations is based on the sequential blockade in the malaria parasite of consecutive steps in the synthesis of folinic acid. The malaria parasite, unlike man, requires *p*-aminobenzoic acid for synthesis of dihydrofolic acid. This is then reduced by the enzyme dihydrofolate reductase to tetrahydrofolic acid (folinic acid). Sulfonamides interfere with the use of *p*-aminobenzoic acid by the parasite (Maier and Riley, 1942) and pyrimethamine irreversibly combines with dihydrofolate reductase (Rollo, 1955). As suggested by their mechanism of action, the combination of these two compounds exerts a synergistic effect against human malaria parasites (Hurly, 1959). At present, single doses of sulfadoxine–pyrimethamine or sulfalene–pyrimethamine are used very extensively in the chemothrapy of chloroquine-resistant malaria. The combination of sulfalene and trimethoprim has been administered as a single dose (Martin and Arnold, 1968) or as a 3-day course (Clyde *et al.*, 1971), but is generally considered to be less satisfactory than the sulfonamide–pyrimethamine combinations (Donno, 1974). Treatment of falciparum infections with these drug combinations has not always been successful (Chin *et al.*, 1966; Verdrager *et al.*, 1967) and such failures have been attributed to a host factor (or factors) which interferes in an undetermined way with the antimalarial activity of the drug combination (Clyde, 1972; Trenholme *et al.*, 1975; Williams *et al.*, 1975).

The antimalarial activity of tetracyclines against human malaria was recognized about 20 years ago (Ruiz-Sánchez *et al.*, 1956). Clearance of fever and parasitemia was generally slower than that noted after treatment with 4-aminoquinolines and, consequently, treatment with antibiotics gave no advantage over other therapeutic agents. With the apperance of chloroquine-resistant strains of *P. falciparum*, a reappraisal was made of the value of tetracycline in the treatment of drug-resistant malaria (Rieckmann *et al.*, 1971). Clinical and field studies have indicated that tetracycline is effective against chloroquine-resistant infections. Because of the slow remission of symptoms and parasitemia after treatment with tetracycline, rapidly acting blood schizontocidal drugs are added to the treatment schedule in order to adequately control acute clinical episodes (Colwell *et al.*, 1972; Rieckmann *et al.*, 1972). Administration of 3-day courses of quinine or amodiaquine at the onset of treatment, although noncurative, prevents the development of dangerously high levels of parasitemia.

The problems presented by chloroquine-resistant malaria led to an intensive search for new antimalarial drugs. This effort was spearheaded by the Walter Reed Army Institute for Research and has resulted, to date, in the screening of over 200,000 compounds. Some of these compounds were selected for further investigations in the *in vitro* system (Rieckmann *et al.*, 1968a) and the owl monkey model (Schmidt, 1973). The most promising of these agents were then evaluated in clinical studies with volunteers (Canfield and Rozman, 1974). Mefloquine hydrochloride was one of the drugs studied during these investigations. It has produced no serious toxic side effects and is remarkably active against drug-resistant falciparum malaria. A single dose of the drug exerted prolonged suppressive activity against a highly chloroquine-resistant strain of *P. falciparum* (Rieckmann *et al.*, 1974b) and cured volunteers infected with such a strain (Trenholme *et al.*, 1975b). Although preliminary findings are most encouraging, the value of this 4-quinolinemethanol compound in the widespread prophylaxis and treatment of drug-resistant malaria remains to be determined.

X. Epidemiology

A. Measurement of Transmission

Since the beginning of this century, much consideration has been given to understanding the factors which affect the transmission of malaria and to measuring the prevalence of the disease in the community (Russell *et al.*, 1963; Black, 1968). The interrelationship between human, mosquito, and climatic elements is complex and it may vary considerably from time to time and from place to place. Increase in the prevalence of malaria has been associated with migrations, wars, poverty, and poor agricultural practices. The activities of the community and its domestic animals, the location of housing in relation to mosquito breeding sites, and various environmental factors may profoundly influence the course of the disease. In many instances "man-made" malaria results from inadequate drainage of surface water after engineering projects, land reclamation or agricultural development. The effectiveness of an anopheline species as a vector of human malaria depends on its biting habits (preference for or availability of animals surrounding man, outside or inside dwellings, time of night, etc.), its preferred breeding site, its susceptibility to infection, and its longevity. Important climatic factors include temperature, humidity, rainfall, and wind. Mosquitoes feed once every 2 days under lowland tropical conditions, but only once every 3 days at cooler highland temperatures. The extrinsic sporogonic cycle in the mosquito cannot usually be completed if the environmental temperature is below 16°C or above 33°C, but between these limits, the developmental cycle is shorter at the higher environmental temperatures.

Traditionally, measurement of the incidence or prevalence of malaria in a community has been carried out by determining the proportion of persons with splenomegaly or with malaria parasites in their blood.

Determination of the spleen rate and the degree of splenic enlargement has probably been the most widely used method for measuring the endemicity of malaria. It has been carried out mainly in "stable" endemic areas where the amount of malaria transmission is generally high and not subject to marked fluctuations over a number of years. Under these conditions, different age groups in the population show more or less fixed degrees of splenomegaly and this has often been used to compare the malaria exposures of different population groups or to assess changing malaria conditions in the community over a period of years (World Health Organization, 1963).

Examination of thick blood smears for malaria parasites has been the other principal method of evaluating the epidemiological situation in a given area. Unfortunately, single mass blood surveys reflect the malaria situation only at the time of the examination; more frequent "longitudinal" surveys of samples of the population provide more information. Although samples from all age groups should be included, infants and younger children provide the most sensitive indicator of the degree of malaria transmission or of any changes produced by seasonal factors or control measures.

During the past two decades, increasing attention has been paid to developing new approaches to the quantitative epidemiology of malaria (Macdonald, 1957). Studies have centered around the measurement of the basic reproduction rate, i.e., the number of secondary infections originating from a primary infection. Factors which influence the reproduction rate include the density of mosquito vectors, their susceptibility to infection, their longevity, the frequency with which they obtain a human blood meal, the period of sporogony in the mosquito, and the duration and infectivity of untreated nonimmune individuals in the community. The net reproduction rate is lowered by the immune response of individuals in the locality.

The mathematical approach to malaria epidemiology is particularly useful in predicting and assessing the efficacy of measures to lower the transmission of the disease. In recent years, computer techniques have been developed to test the applicability of this quantitative approach to the field situation (Macdonald *et al.*, 1968). Although the value of such an approach is unquestioned (Bruce-Chwatt, 1969), further research is needed to formulate more realistic and useful models (Nájera, 1974).

The epidemiological pattern of malaria in different parts of the world may range from a very stable situation where no variation in transmission is observed over a number of years to a very unstable situation where the incidence of malaria shows marked fluctuations. In areas of stable malaria with a high level of transmission, young children have high parasite morbidity and mortality rates but, with increasing age, these rates diminish markedly and adults usually show no evidence of infection or have only low-grade asymptomatic parasitemia. At the other extreme, in areas with unstable malaria, the collective immunity of the population may be nonexistent or low and an epidemic exacerbation of the disease often affects all ages to a similar degree. Depending on the local conditions—

human, vector, and climatic–epidemiological patterns of the disease can vary markedly between these two extremes.

B. Epidemic Malaria

"Epidemic malaria" denotes, rather loosely, a periodic or occasional sharp increase in the malaria morbidity or mortality of a community in which malaria was previously unknown or in which unstable malaria is prevalent. Epidemic malaria may be caused by an increase in the density or longevity of the vector species, by a change in its behavior pattern, or the introduction of new gametocyte carriers into the area. Epidemics of malaria have been noted throughout the course of human history and, recently, serious outbreaks have occurred in Ethiopia, Haiti, and Sri Lanka. Localized outbreaks of the disease are observed periodically in many different parts of the world.

In 1960, the course of an epidemic exacerbation of malaria was studied in the Minj area of the highlands of Papua New Guinea. These observations were made by Mr. Stanley Christian and one of the authors (KHR) while they were working at the Malaria Center and Government Hospital in Minj. The area is located in a valley floor at an altitude of about 1650 m and has an equable climate throughout the year. The mean temperatures of 20°C (range 14°–27°C) shows little variation from month to month and the rainfall is distributed between a wet season (January–May) and a dry season (June–December). The principal vector of malaria, *Anopheles punctulatus farauti,* breeds in small, transient pools of water and the adult mosquitoes enter houses at night for their blood meal. Variations in the density of mosquitoes had been monitored for a number of years by window outlet traps located in 26 of these homes. Malaria in this area is unstable and severe epidemics affecting all age groups have been reported from time to time at the end of the wet season (T. E. T. Spencer, personal communication, 1956). During the dry season, the overall parasite rate is about 5% and the gametocyte rate is about 1%. *Plasmodium vivax* is the most prevalent species, but *P. falciparum* becomes relatively more common during epidemic exacerbations. Although the prevalence of malaria in the community had been estimated at different times, the only regular index of seasonal variations in the incidence of malaria was the number of malaria admissions into the Government Hospital. The age grouping of patients and the malaria species responsible for their infections were similar to the distribution observed in the general community (Peters and Christian, 1960).

The evolution of the epidemic is shown in Fig. 4. The rainfall during the early months of 1960 was appreciably heavier than for the same

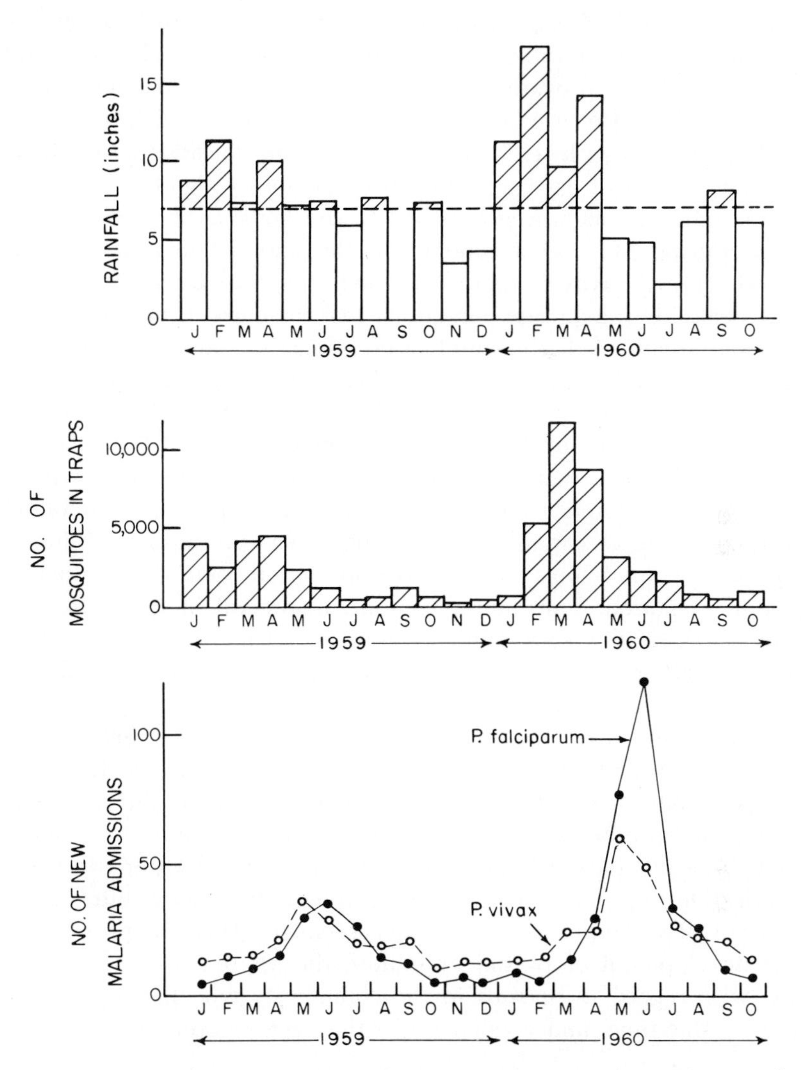

Fig. 4. Diagrammatic presentation of sequence of events in a malaria epidemic observed in the Minj highlands of Papua New Guinea during 1960.

period of the previous year and resulted in increased breeding of mosquitoes. The time needed for eggs to develop into mature adults probably explains the time lag observed between peak rainfall and peak mosquito populations caught in the outlet traps. During the first 3 months of 1960, mosquito densities seemed to reflect the intensity of rainfall a few weeks earlier. The severe intermittent rains during April were completely different from the earlier "steady" rains. The severe intermittent rains had caused the aquatic stages to be flushed away and this resulted in considerably fewer mosquitoes being caught in the traps during May. Under prevailing local conditions, mosquito densities were usually very low after a monthly rainfall of less than 7 inches. By drawing a baseline at this level, changes in mosquito densities correlate well with rainfall when it exceeds 7 inches per month (Fig. 4).

The number of malaria admissions to the hospital showed a marked increase during May, at a time when the anopheline population had declined sharply. Vivax infections reached their peak during May whereas the peak for falciparum infections was reached 1 month later. As the epidemic subsided, the number of falciparum and vivax infections were approximately the same and, subsequently, the original predominance of vivax infections was reestablished. The incubation interval in this epidemic was about 4 weeks for *P. vivax* and about 6 weeks for *P. falciparum*. This interval spans the period between the time when an infective mosquito feeds and the first day on which a person infected by this mosquito is capable of infecting other mosquitoes; it includes the period of sporogony, the incubation period, and the "gametocyte appearance" period. The longer incubation interval in *P. falciparum* is due to: (1) a 21-day sporogonic cycle in *P. falciparum* versus a 16-day cycle in *P. vivax* at the prevailing temperatures (Peters and Christian, 1960); and, (2) a 10-day delay in the appearance of mature gametocytes of *P. falciparum* after onset of clinical symptoms (see Section VII). With a small initial gametocyte reservoir, the increased vector population caused only a slight increase in malaria infections during the first two incubation intervals. After gaining momentum, the greatest number of vivax and falciparum infections occurred during the third incubation intervals. During the fourth and subsequent incubation intervals, the number of new symptomatic infections declined. This was probably due to the development of partial immunity, the approach of a "saturation point" with regard to persons in the community who had not been infected by this time, and a decrease in the vector population.

XI. Control of Malaria

Malaria has been eradicated from many areas of the world in which it was once endemic. By the end of 1974, malaria eradication had been achieved in countries with a population of about 790 million people. In the remaining malarious regions of the world, about 773 million people were living in areas where organized antimalarial measures were being carried out and 363 million people, living mostly in Africa, did not enjoy the benefit of such activities (World Health Organization, 1975).

Although past achievements are noteworthy, various problems have slowed down the progress of malaria eradication programs (Lepes, 1974; Palacios Fraire, 1975). In recent years increasing emphasis has been given to the concept of supporting control of malaria in communities where eradication of the disease is not yet feasible (World Health Organization Interregional Conference, 1974). This was prompted to a considerable degree by the lack of progress in some malaria eradication programs and the realization that the basic health services of many countries had to be strengthened before eradication programs could be started. The main objective of control programs is to reduce the malaria morbidity and mortality by the most efficient and economic means available to a particular community, area, or country. The control of malaria and other debilitating diseases in rural areas should lead to increased human productivity, contribute to improved living conditions, and stem the unfortunate migration to already overpopulated cities. In a control program, standard malaria eradication measures, such as the spraying of domiciles with residual insecticides, mass drug administration, and case detection, would have to be considered. In addition, other methods of control could be applied. These include reduction of mosquito breeding sources, the application of larvicides, the use of larvivorous fish, personal protection against malaria, and the administration of drugs to vulnerable groups such as children, pregnant women, or persons recruited to work on engineering or agricultural development schemes. Recent developments suggest that the use of ultralow-volume insecticide sprays, the release of sterile male mosquitoes, and biological control of mosquitoes by pathogens may also be useful under certain circumstances (Lofgren, 1974).

Further research is obviously needed to solve many of the technical problems and to provide new tools for achieving improved control or eradication of malaria. Nevertheless, despite technical and socioeconomic limitations, greater progress could be made in controlling malaria and other diseases with the relatively limited resources currently available in many tropical countries. New approaches are needed for a more effective and economic application of our current knowledge, particularly among population groups living in remote areas.

REFERENCES

Allison, A. C. (1963). Inherited factors in blood conferring resistance to protozoa. *In* "Immunity to Protozoa" (P. C. C. Garnham, A. E. Pierce, and I. Roitt, eds.), pp. 109–122. Blackwell, Oxford.

Alving, A. S., Johnson, C. F., Tarlov, A. R., Brewer, G. J., Kellermeyer, R. W., and

Carson, P. E. (1960). Mitigation of the haemolytic effect of primaquine and enhancement of its action against exoerythrocytic forms of the Chesson strain of *Plasmodium vivax* by intermittent regimens of drug administration. *Bull. W.H.O.* **22**, 621–631.

Alving, A. S., Powell, R. D., Brewer, G. J., and Arnold, J. D. (1962). Malaria, 8-aminoquinolines and haemolysis. *In* "Drugs, Parasites and Host" (L. G. Goodwin and R. H. Nimmo-Smith, eds.), pp. 83–97. Churchill, London.

Ambroise-Thomas, P. (1974). La réaction d'immunofluorescence dans l'étude séro-immunologique du paludisme. *Bull. W.H.O.* **50**, 267–276.

Barrett-Connor, E. (1974). Chemoprophylaxis of malaria for travelers. *Ann. Intern. Med.* **81**, 219–224.

Bass, C. C., and Johns, F. M. (1912). The cultivation of malaria plasmodia (*Plasmodium vivax* and *Plasmodium falciparum*) *in vitro*. *J. Exp. Med.* **16**, 567–599.

Beaudoin, R. L., Strome, C. P. A., and Clutter, W. G. (1974). Cultivation of avian malaria parasites in mammalian liver cells. *Exp. Parasitol.* **36**, 355–359.

Bertagna, P., Cohen, S., Geiman, Q. M., Haworth, J., Königk, E., Richards, W. H. G., and Trigg, P. I. (1972). Cultivation techniques for the erythrocytic stages of malaria parasites. *Bull. W.H.O.* **47**, 357–373.

Black, R. H. (1968). "Manual of Epidemiology and Epidemiological Services in Malaria Programmes." World Health Organ., Geneva.

Boyd, M. F., and Stratman-Thomas, W. K. (1933). Studies on benign tertian malaria. 4. On the refractoriness of Negros to inoculation with *Plasmodium vivax*. *Am. J. Hyg.* **18**, 485–489.

Bray, R. S. (1957a). Studies on malaria in chimpanzees. II. *Plasmodium vivax*. *Am. J. Trop. Med. Hyg.* **6**, 514–520.

Bray, R. S. (1957b). Studies on malaria in chimpanzees. IV. *Plasmodium ovale*. *Am. J. Trop. Med. Hyg.* **6**, 638–645.

Bray, R. S., Burgess, R. W., and Baker, J. R. (1963). Studies on malaria in chimpanzees. X. The presumed second generation of the tissue phase of *Plasmodium ovale*. *Am. J. Trop. Med. Hyg.* **12**, 1–12.

Brown, K. N. (1973). Antibody induced variation in malaria parasites. *Nature (London)* **242**, 49–50.

Brown, K. N. (1974). Antigenic variation and immunity to malaria. *Parasites Immunized Host: Mech. Surv., Ciba Found. Symp., 1974* pp. 35–51.

Brown, K. N., Brown, I. N., and Hills, L. A. (1970). Immunity to malaria. I. Protection against *Plasmodium knowlesi* shown by monkeys sensitized with drug-suppressed infections or by dead parasites in Freund's adjuvant. *Exp. Parasitol.* **28**, 304–317.

Bruce-Chwatt, L. J. (1969). George Mcdonald Memorial Lecture—Quantitative epidemiology of tropical diseases. *Trans. R. Soc. Trop. Med. Hyg.* **63**, 131–143.

Butcher, G. A., Mitchell, G. H., and Cohen, S. (1973). Mechanism of host specificity in malaria infection. *Nature (London)* **244**, 40–42.

Cabrera, E. J., Speer, C. A., Schenkel, R. H., Barr, M. L., and Silverman, P. H. (1976). Delayed dermal hypersensitivity in Rhesus monkeys (*Macaca mulatta*) immunized against *Plasmodium knowlesi*. *Z. Parasitenkd.* **50**, 31–42.

Canfield, C. J., and Rozman, R. S. (1974). Clinical testing of new antimalarial compounds. *Bull. W.H.O.* **50**, 203–212.

Chin, W., Contacos, P. G., Coatney, G. R., and King, H. K. (1966). The evaluation of sulfonamides, alone or in combination with pyrimethamine, in the treatment of multiresistant falciparum malaria. *Am. J. Trop. Med. Hyg.* **15**, 823–829.

Ciucă, M., Lupaşçu, G. Negulici, E., and Constantinescu, P. (1964). Recherches sur la transmission expérimentale de *P. malariae* à l'homme. *Arch. Roum. Pathol. Exp. Microbiol.* **23**, 763–776.

Clyde, D. F. (1972). Responsibility for failure of sulfonamides in falciparum malaria; host or parasite? *Trans. R. Soc. Trop. Med. Hyg.* **66**, 806.

Clyde, D. F., Miller, R. M., Schwartz, A. R., and Levine, M. M. (1971). Treatment of falciparum malaria with sulfalene and trimethoprim. *Am. J. Trop. Med. Hyg.* **20**, 804–810.

Clyde, D. F., Most, H., McCarthy, V. C., and Vanderberg, J. P. (1973). Immunization of man against sporozoite-induced falciparum malaria. *Am. J. Med. Sci.* **266**, 169–177.

Coatney, G. R., Collins, W. E., Warren, Mc W., and Contacos, P. G. (1971). "The Primate Malarias." U.S. Govt. Printing Office, Washington, D.C.

Cohen, S., and Butcher, G. A. (1969). Comments on immunization. *Mil. Med.* **134**, Spec. Issue, 1191–1197.

Cohen, S., and Butcher, G. A. (1971). Serum antibody in acquired malarial immunity. *Trans. R. Soc. Trop. Med. Hyg.* **65**, 125–135.

Cohen, S., McGregor, I. A., and Carrington, S. (1961). Gamma-globulin and acquired immunity to human malaria. *Nature (London)* **192**, 733–737.

Cohen, S., Butcher, G. A., and Mitchell, G. H. (1974). Mechanisms of immunity to malaria. *Bull. W.H.O.* **50**, 251–257.

Colwell, E. J., Hickman, R. L., Intraprasert, R., and Tirabutana, C. (1972). Minocycline and tetracycline treatment of acute falciparum malaria in Thailand. *Am. J. Trop. Med. Hyg.* **21**, 144–149.

Contacos, P. G., and Collins, W. E. (1973). Malarial relapse mechanism. *Trans. R. Soc. Trop. Med. Hyg.* **67**, 617–618.

Corradetti, A., Verolini, F., Bucci, A., and Pennacchio, A. E. (1969). Azione immunizzante protettiva dalla frazione non idrosolubile di *Plasmodium berghei* (Ceppo Istisan). *Parasitologia* **11**, 151–159.

Donno, L. (1974). Antifolic combination in the treatment of malaria. *Bull. W.H.O.* **50**, 223–230.

Fairley, N. H. (1947). Sidelights on malaria in man obtained by subinoculation experiments. *Trans. R. Soc. Trop. Med. Hyg.* **40**, 621–676.

Fletcher, A., and Maegraith, B. (1972). The metabolism of the malaria parasite and its host. *Adv. Parasitol.* **10**, 31–48.

Freund, J., Thomson, K. J., Sommer, H. E., Walter, A. W., and Schenkein, E. L. (1945). Immunization of Rhesus monkeys against malaria infection (*Plasmodium knowlesi*) with killed parasites and adjuvant. *Science* **102**, 202–204.

Garnham, P. C. C. (1966). "Malaria Parasites and other Haemosporidia." Blackwell, Oxford.

Garnham, P. C. C. (1973). Second roundtable discussion on taxonomic problems relating to malaria parasites. *J. Protozool.* **20**, 37–42.

Geiman, Q. M., and Meagher, M. J. (1967). Susceptibility of a new world monkey to *Plasmodium falciparum* from man. *Nature (London)* **215**, 437–439.

Gomperts, B. D. (1969). The biochemistry of red blood cells in health and disease. *Biol. Basis Med.* **3**, 81–127.

Hendrickse, R. G., and Gilles, H. M. (1963). The nephrotic syndrome and other renal diseases in children in Western Nigeria. *East Afr. Med. J.* **40**, 186–201.

Hurly, M. G. D. (1959). Potentiation of pyrimethamine by sulphadiazine in human malaria. *Trans. R. Soc. Trop. Med. Hyg.* **53**, 412–413.

Jeffery, G. M., Collins, W. E., and Skinner, J. C. (1963). Antimalarial drug trials on a multiresistant strain of *Plasmodium falciparum*. *Am. J. Trop. Med. Hyg.* **12,** 844–850.

Kass, L., Willerson, W. D., Jr., Rieckmann, K. H., and Carson, P. E. (1971). Blastoid transformation of lymphocytes in falciparum malaria. *Am. J. Trop. Med. Hyg.* **20,** 195–198.

Lepes, T. (1974). Review of research on malaria. *Bull. W.H.O.* **50,** 151–157.

Lofgren, C. S. (1974). Recent developments in methods of mosquito control. *Bull. W.H.O.* **50,** 323–328.

Lupasçu, G., Constantinescu, P., Negulici, E., Garnham, P. C. C. Bray, R. S., Killick-Kendrick, R., Shute, P. G., and Maryon, M. (1967). The late primary exo-erythrocytic stages of *Plasmodium malariae*. *Trans. R. Soc. Trop. Med. Hyg.* **61,** 482–489.

Macdonald, G. (1957). "The Epidemiology and Control of Malaria." Oxford Univ. Press, London and New York.

Macdonald, G., Cuellar, C. B., and Foll, C. V. (1968). The dynamics of malaria. *Bull. W.H.O.* **38,** 743–755.

McGregor, I. A. (1974). Mechanisms of acquired immunity and epidemiological patterns of antibody responses in malaria in man. *Bull. W.H.O.* **50,** 259–266.

Maegraith, B. G. (1966). Pathogenic processes in malaria. *In* "The Pathology of Parasitic Diseases" (A. E. R. Taylor, ed.), pp. 15–32. Blackwell, Oxford.

Maier, J., and Riley, E. (1942). Inhibition of antimalarial action of sulfonamides by p-aminobenzoic acid. *Proc. Soc. Exp. Biol. Med.* **50,** 152–154.

Makari, J. G. (1946). Intradermal test in malaria. *Am. J. Trop. Med.* **49,** 23–29.

Manson-Bahr, P. (1963). The Story of Malaria: The drama and actors. *Int. Rev. Trop. Med.* **2,** 329–290.

Martin, D. C., and Arnold, J. D. (1968). Treatment of acute falciparum malaria with sulfalene and trimethoprim. *J. Am. Med. Assoc.* **203,** 476–480.

Meuwissen, J. H. E. T. (1974). The direct haemagglutination test for malaria and its application to epidemiological surveillance. *Bull. W.H.O.* **50,** 277–286.

Miller, L. H., Chien, S., and Usami, S. (1972). Decreased deformability of *Plasmodium coatneyi* infected red cells and its possible relation to cerebral malaria. *Am. J. Trop. Med. Hyg.* **21,** 133–137.

Mitchell, G. H., Butcher, G. A., and Cohen, S. (1974). A merozoite vaccine effective against *Plasmodium knowlesi* malaria. *Nature* (*London*) **252,** 311.

Mitchell, G. H., Butcher, G. A., Voller, A., and Cohen, S. (1976). The effect of human immune IgG on the *in vitro* development of *Plasmodium falciparum*. *Parasitology* **72,** 149–162.

Nájera, A. J. (1974). A critical review of the field application of a mathematical model of malaria eradication. *Bull. W.H.O.* **50,** 449–457.

Neva, F. A., Sheagren, J. N., Shulman, N. R., and Canfield, C. J. (1970). Malaria: Host-defense mechanisms and complications. *Ann. Intern. Med.* **73,** 295–306.

Nussenzweig, R. S., Vanderberg, J., Most, H., and Orton, C. (1967). Protective immunity produced by injection of X-irradiated sporozoites of *Plasmodium berghei*. *Nature* (*London*) **216,** 160–162.

Palacios Fraire, S. (1975). Analysis of the principal problems impeding normal development of malaria eradication programs. *Bull., Pan Am. Health Org.* **9,** 283–294.

Peters, W. (1970). "Chemotherapy and Drug Resistance in Malaria." Academic Press, New York.

Peters, W., and Christian, S. H. (1960). Studies on the epidemiology of malaria in New Guinea. *Trans. R. Soc. Trop. Med. Hyg.* **54**, 529–548.

Phillips, R. S., Trigg, P. I., Scott-Finnigan, T. J., and Bartholomew, R. K. (1972). Culture of *Plasmodium falciparum in vitro:* A subculture technique used for demonstrating anti-plasmodial activity in serum from some Gambians, resident in an endemic malarious area. *Parasitology* **65**, 525–535.

Pitney, W. R. (1968). The tropical splenomegaly syndrome. *Trans. R. Soc. Trop. Med. Hyg.* **62**, 717–728.

Powell, R. D. (1966). The chemotherapy of malaria. *Clin. Pharmacol. Ther.* **7**, 48–76.

Powell, R. D., and McNamara, J. V. (1970). Infections with chloroquine-resistant *Plasmodium falciparum* in man: Prepatent periods, incubation periods, and relationship between parasitemia and the onset of fever in nonimmune persons. *Ann. N.Y. Acad. Sci.* **174**, 1027–1041.

Powell, R. D., McNamara, J. V., and Rieckmann, K. H. (1972). Clinical aspects of acquisition of immunity to falciparum malaria. *Proc. Helminthol. Soc. Wash.* **39**, Spec. Issue, 51–66.

Rieckmann, K. H. (1970). Asymptomatic malaria. *Lancet* **1**, 82–83.

Rieckmann, K. H., McNamara, J. V., Frischer, H., Stockert, T. A., Carson, P. E., and Powell, R. D. (1968a). Effects of chloroquine, quinine, and cycloguanil upon the maturation of asexual erythrocytic forms of two strains of *Plasmodium falciparum in vitro*. *Am. J. Trop. Med. Hyg.* **17**, 661–671.

Rieckmann, K. H., McNamara, J. V., Frischer, H., Stockert, T. A., Carson, P. E., and Powell, R. D. (1968b). Gametocytocidal and sporontocidal effects of primaquine and of sulfadiazine with pyrimethamine in a chloroquine-resistant strain of *Plasmodium falciparum*. *Bull. W.H.O.* **38**, 625–632.

Rieckmann, K. H., Powell, R. D., McNamara, J. V., Willerson, D., Jr., Kass, L., Frischer, H., and Carson P. E. (1971). Effects of tetracycline against chloroquine-resistant and chloroquine-sensitive strains of *Plasmodium falciparum*. *Am. J. Trop. Med. Hyg.* **20**, 811–815.

Rieckmann K. H., Willerson, W. D., Jr., Carson, P. E., and Frischer, H. (1972). Effects of tetracycline against drug-resistant falciparum malaria. *Proc. Helminthol. Soc. Wash.* **39**, Spec. Issue, 339–347.

Rieckmann, K. H., Carson, P. E., Beaudoin, R. L., Cassells, J. S., and Sell, K. W. (1974a). Sporozoite induced immunity in man against an Ethiopian strain of *Plasmodium falciparum*. *Trans. R. Soc. Trop. Med. Hyg.* **68**, 258–259.

Rieckmann, K. H., Trenholme, G. M., Williams, R. L., Carson, P. E., Frischer, H., and Desjardins, R. E. (1974b). Prophylactic activity of mefloquine hydrochloride (WR142490) in drug-resistant malaria. *Bull. W.H.O.* **51**, 375–377.

Rodhain, J. (1956). Paradoxical behavior of *Plasmodium vivax* in the chimpanzee. *Trans. R. Soc. Trop. Med. Hyg.* **50**, 287–293.

Rollo, I. M. (1955). The mode of action of sulphonamides, proguanil and pyrimethamine of *Plasmodium gallinaceum*. *Br. J. Pharmacol. Chemother.* **10**, 208–214.

Ruiz Sánchez, F., Ruiz Sánchez, A., and Naranjo Grande, E. (1956). The treatment of malaria with tetracycline. *Antibiot. Med. Clin. Ther. (N.Y.)* **3**, 193–196.

Russell, P. F., West, L. S., Maxwell, R. D., and Macdonald, G. (1963). "Practical Malariology." Oxford Univ. Press, London and New York.

Sadun, E. H., Wellde, B. T., and Hickmann, R. L. (1969). Resistance produced in owl monkeys (*Aotus trivirgatus*) by inoculation with irradiated *Plasmodium falciparum*. *Mil. Med.* **134,** Spec. Issue, 1165–1175.

Schenkel, R. H., Simpson, G. L., and Silverman, P. H. (1973). Vaccination of Rhesus monkeys (*Macaca mulatta*) against *Plasmodium knowlesi* by the use of non-viable antigen. *Bull. W.H.O.* **48,** 597–604.

Schmidt, L. H. (1969). Chemotherapy of the drug-resistant malarias. *Annu. Rev. Microbiol.* **23,** 427–454.

Schmidt, L. H. (1973). Infections with *Plasmodium falciparum* and *Plasmodium vivax* in the owl monkey—model systems for basic biological and chemotherapeutic studies. *Trans. R. Soc. Trop. Med. Hyg.* **67,** 446–474.

Simpson, G. L., Schenkel, R. H., and Silverman, P. H. (1974). Vaccination of Rhesus monkeys against malaria by use of sucrose density gradient fraction of *Plasmodium knowlesi* antigen. *Nature (London)* **247,** 304–306.

Speer, C. A., and Silverman, P. H. (1974). Cultivation of the malaria parasite, *Plasmodium berghei,* in Leydig cell tumor cultures. *In Vitro* **10,** 370.

Speer, C. A., Silverman, P. H., and Schiewe, S. G. (1976). Cultivation of the erythrocytic stages of *Plasmodium berghei* in primary bone marrow cells. *J. Parasitol.* **62,** 657–663.

Taliaferro, W. H., and Bloom, W. (1945). Inflammatory reactions in the skin of normal and immune canaries and monkeys after the local injection of malarial blood. *J. Infect. Dis.* **77,** 109–138.

Thompson, P. E., and Werbel, L. M. (1972). "Antimalarial Agents: Chemistry and Pharmacology." Academic Press, New York.

Tigertt, W. D. (1966). Present and potential malaria problems. *Mil. Med.* **131,** 853–856.

Trager, W. (1974). Some aspects of intracellular parasitism. *Science* **183,** 269–272.

Trager, W., and Jensen, J. B. (1976). Human malaria parasites in continuous culture. Science **191,** 673–675.

Trenholme, G. M., Williams, R. L., Frischer, H., Carson, P. E., and Rieckmann, K. H. (1975a). Host failure in treatment of malaria with sulfalene and pyrimethamine. *Ann. Intern. Med.* **82,** 219–223.

Trenholme, G. M., Williams, R. L., Desjardins, R. E., Frischer, H., Carson, P. E., and Rieckmann, K. H. (1975b). Mefloquine (WR142490) in the treatment of human malaria. *Science* **190,** 792–794.

Verdrager, J., Riche, A., and Chheang, C. M. (1967). "Traitement du paludisme a *Plasmodium falciparum par* l'association sulformethoxine-pyrimethamine," WHO/Mal/67, 629 (WHO cyclostyled report). World Health Organ., Geneva.

Voller, A. (1974). Immunopathology of malaria. *Bull. W.H.O.* **50,** 177–183.

Voller, A., Huldt, G., Thors, C., and Engvall, E. (1975). New serological tests for malaria antibodies. *Br. Med. J.* **1,** 659–661.

Wilcox, A. (1960). "Manual of the Microscopical Diagnosis of Malaria in Man." U.S. Govt. Printing Office, Washington, D.C.

Williams, R. L., Trenholme, G. M., Carson, P. E., Frischer, H., and Rieckmann, K. H. (1975). Acetylator phenotype and response of individuals infected with a chloroquine-resistant strain of *Plasmodium falciparum* to sulfalene and pyrimethamine. *Am. J. Trop. Med. Hyg.* **24,** 734–739.

Wilson, R. J. M., McGregor, I. A., Hall, P. J., Williams, K., and Bartholomew, R. (1969). Antigens associated with *Plasmodium falciparum* infections in man. *Lancet* **2,** 201–205.

Woodruff, A. W. (1973). Mechanisms involved in anaemia associated with infection and splenomegaly in the tropics. *Trans. R. Soc. Trop. Med. Hyg.* **67**, 313–325.

World Health Organization (1963). "Terminology of Malaria and Malaria Eradication," Report of a Drafting Committee. World Health Organ., Geneva.

World Health Organization. (1975). The malaria situation in 1974. *WHO Chron.* **29**, 474–481.

World Health Organization Interregional Conference. (1974). Malaria control in countries where time-limited eradication is impracticable at present. *WHO, Tech. Rep. Ser.* **537.**

World Health Organization Scientific Group. (1967). Chemotherapy of malaria. *WHO, Tech. Rep. Ser.* **375.**

World Health Organization Scientific Group. (1969). Parasitology of malaria. *WHO, Tech. Rep. Ser.* **433.**

World Health Organization Scientific Group. (1973). Chemotherapy of malaria and resistance to antimalarials. *WHO, Tech. Rep. Ser.* **529.**

World Health Organization Scientific Group. (1975). Developments in malaria immunology. *WHO, Tech. Rep. Ser.* **579.**

Wyler, D. J., and Oppenheim, J. J. (1974). Lymphocyte transformation in human *Plasmodium falciparum* malaria. *J. Immunol.* **113**, 449–454.

Young, M. D., Porter, J. A., Jr., and Johnson, C. M. (1966). *Plasmodium vivax* transmitted from man to monkey to man. *Science* **153**, 1006–1007.

Zuckerman, A. (1964). Autoimmunization and other types of indirect damage to host cells as factors in certain protozoan diseases. *Exp. Parasitol.* **15**, 138–183.

Index

E

F

H

I

N

S

T

U

A 7
B 8
C 9
D 0
E 1
F 2
G 3
H 4
I 5
J 6